Lecture Notes in Computer Science 3383

Commenced Publication in 1973
Founding and Former Series Editors:
Gerhard Goos, Juris Hartmanis, and Jan van Leeuw

T0140250

János Pach (Ed.)

Graph Drawing

12th International Symposium, GD 2004
New York, NY, USA, September 29-October 2, 2004
Revised Selected Papers

 Springer

Volume Editor

János Pach
City University of New York
City College, NY, USA
and
Hungarian Academy of Sciences
Rényi Institute, Budapest, Hungary
E-mail: pach@cs.nyu.edu

Library of Congress Control Number: 2005920703

CR Subject Classification (1998): G.2, F.2, I.3, E.1

ISSN 0302-9743
ISBN 3-540-24528-6 Springer Berlin Heidelberg New York

Springer is a part of Springer Science+Business Media

springeronline.com

© Springer-Verlag Berlin Heidelberg 2004
Printed in Germany

Typesetting: Camera-ready by author, data conversion by Olgun Computergrafik
Printed on acid-free paper   SPIN: 11384861   06/3142   5 4 3 2 1 0

Preface

The 12th International Symposium on Graph Drawing (GD 2004) was held during September 29–October 2, 2004, at City College, CUNY, in the heart of Harlem, New York City. GD 2004 attracted 94 participants from 19 countries.

In response to the call for papers, the program committee received 86 regular submissions describing original research and/or system demonstrations. Each submission was reviewed by at least three program committee members and comments were returned to the authors. Following extensive e-mail discussions, the program committee accepted 39 long papers (11 pages each in the proceedings) and 12 short papers (6 pages each). In addition, 4 posters were displayed and discussed in the conference exhibition room (2 pages each in the proceedings).

The program committee of GD 2004 invited two distinguished lecturers. Professor Paul Seymour from Princeton University presented a new characterization of claw-free graphs (joint work with Maria Chudnovsky). Professor Erik Demaine from MIT reported on his joint work with Fedor Fomin, MohammadTaghi Hajiaghayi and Dimitrios Thilikos, concerning fast (often subexponential) fixed-parameter algorithms and polynomial approximation schemes for broad classes of NP-hard problems in topological graph theory. A survey of the subject by Professors Demaine and Hajiaghayi is included in this volume.

As usual, the annual graph drawing contest was held during the conference. This time the contest had two distinct tracks: the graph drawing challenge and the freestyle contest. A report is included in the proceedings.

Many people in the graph drawing community contributed to the success of GD 2004. First of all, special thanks are due to the authors of submitted papers, demos, and posters, and to the members of the program committee as well as to the external referees. Many thanks to organizing committee members Gary Bloom, Peter Brass, Stephen Kobourov, and Farhad Shahrokhi. My very special thanks go to Hanna Seifu who was in charge of all local arrangements, Robert Gatti who developed the software used for registration and paper submission, and John Weber and Eric Lim who designed the logo, the webpage, and the brochures of the conference. I am very much indebted to Dr. Joseph Barba and Dr. Mohammad Karim, present and former Deans of the School of Engineering, and to Dr. Gregory H. Williams, President of the City College of New York, for their continuing support.

Thanks are due to our "gold" sponsors, the City College of New York, the University of North Texas at Denton, and Tom Sawyer Software, and to our "silver" sponsors, ILOG, the DIMACS Center for Discrete Mathematics and Theoretical Computer Science, and the Computer Science Program at the CUNY Graduate Center. Springer and World Scientific Publishing contributed to the success of GD 2004 by sending selections of their recent publications in the subject.

The 13th International Symposium on Graph Drawing (GD 2005) will be held in Limerick, Ireland, 12–14 September, 2005, with Peter Eades and Patrick Healy as conference co-chairs.

December 2004   János Pach
                                                        New York and Budapest

Sponsoring Institutions

Tom Sawyer Software
City College of New York, CUNY
University of North Texas at Denton
DIMACS Center for Discrete Math. and Theoretical Computer Science
ILOG
The Graduate Center of the City University of New York

Organization

Program Committee

Franz-J. Brandenburg	Universität Passau
Stephen G. Eick	SSS Research, Inc.
Genghua Fan	Fuzhou University
Emden Gansner	AT&T Labs
Giuseppe Liotta	Università degli Studi di Perugia
Patrice de Mendez	Centre Nat. de la Recherche Scientifique (Paris)
Takao Nishizeki	Tohoku University
János Pach	City College and Courant Inst., NY (chair)
László Székely	University of South Carolina
Roberto Tamassia	Brown University
Géza Tóth	Alfréd Rényi Institute of Mathematics
Imrich Vrťo	Slovak Academy of Sciences

Contest Committee

Franz-J. Brandenburg	Universität Passau
Christian Duncan	University of Miami
Emden Gansner	AT&T Labs
Stephen Kobourov	University of Arizona (chair)

Steering Committee

Franz-J. Brandenburg	Universität Passau
Giuseppe Di Battista	Università degli Studi Roma
Peter Eades	National ICT Australia Ltd., Univ. of Sydney
Patrick Healy	University of Limerick
Giuseppe Liotta	Università degli Studi di Perugia
Takao Nishizeki	Tohoku University
János Pach	City College and Courant Inst., NY
Pierre Rosenstiehl	Centre Nat. de la Recherche Scientifique (Paris)
Roberto Tamassia	Brown University (chair)
Ioannis G. Tollis	Foundation for Research and Techn. (FORTH – Hellas)
Sue Whitesides	McGill University

Organizing Committee

Gary Bloom   City College, CUNY
Peter Braß   City College, CUNY (co-chair)
Stephen Kobourov   University of Arizona, Tucson
János Pach   City College, CUNY (chair)
Farhad Shahrokhi   University of North Texas (co-chair)

External Referees

Christian Bachmaier   Linyuan Lu   Ondrej Sýkora
Nicolas Bonichon   András Lukács   Konrad Swanepoel
Shen Bau   Endre Makai   Gábor Tardos
Walter Didimo   Stephen C. North   Ioannis Tollis
Daniel Dix   Maurizio Patrignani   Csaba Tóth
Adrian Dumitrescu   Rom Pinchasi   Pavel Valtr
Peter Eades   Maurizio Pizzonia   Hua Wang
Michael Forster   Richard Pollack   Colin Ware
Emilio Di Giacomo   Radoś Radoičić   Sue Whitesides
Gyula Károlyi   Marcus Raitner   Nick Wormald
Yehuda Koren   Falk Schreiber
Martin Loebl   Zixia Song

Table of Contents

Software Demonstrations

Posters

Graph Drawing Contest

Invited Talk

Reconfiguring Triangulations
with Edge Flips and Point Moves*

Greg Aloupis[1], Prosenjit Bose[2], and Pat Morin[2]

[1] School of Computer Science, McGill University
athens@cgm.cs.mcgill.ca
[2] School of Computer Science, Carleton University
{jit,morin}@scs.carleton.ca

Abstract. We examine reconfigurations between triangulations and near-triangulations of point sets, and give new bounds on the number of *point moves* and *edge flips* sufficient for any reconfiguration. We show that with $O(n \log n)$ edge flips and point moves, we can transform any geometric near-triangulation on n points to any other geometric near-triangulation on n possibly different points. This improves the previously known bound of $O(n^2)$ edge flips and point moves.

1 Introduction

An *edge flip* is a graph operation that is defined on (near)-triangulations[1]. An edge flip on a triangulation is simply the deletion of an edge, followed by the insertion of another edge such that the resulting graph remains a triangulation. The definition of an edge flip gives rise to several natural questions: Does there always exist a sequence of flips that reconfigures a given triangulation to any other triangulation? Are there bounds on the lengths of such sequences if they exist? Can these sequences be computed? These questions have been studied in the literature in many different settings. In particular, Wagner [19] proved that given any two n-vertex triangulations G_1 and G_2, there always exists a finite sequence of edge flips that reconfigures G_1 into a graph isomorphic to G_2. Subsequently, Komuro [10] showed that in fact $O(n)$ edge flips suffice. Recently, Bose et al. [2] showed that $O(\log n)$ simultaneous edge flips suffice and are sometimes necessary. This setting of the problem is referred to as the combinatorial setting since the triangulations are only embedded combinatorially, i.e. only the cyclic order of edges around each vertex is defined.

In the geometric setting, the graphs are embedded in the plane with edges represented by straight line segments. Pairs of edges can only intersect at their endpoints. Edge flips are still valid operations in this setting, except that now the edge that is added must be a line segment that cannot properly intersect any of the existing edges of the graph. This implies that there are valid edge flips

---

* Research supported in part by the Natural Science and Engineering Council of Canada.

[1] A triangulation is a plane graph where every face is a triangle. In a *near*-triangulation, the outer face may not be a triangle.

J. Pach (Ed.): GD 2004, LNCS 3383, pp. 1–11, 2004.

in the combinatorial setting that are no longer valid in the geometric setting. Lawson [12] showed that given any two geometric near-triangulations N_1 and N_2 embedded on the same n points in the plane, there always exists a finite sequence of edge flips that transforms the edge set of N_1 to the edge set of N_2. Hurtado, Noy and Urrutia [9] showed that $O(n^2)$ flips are always sufficient and that $\Omega(n^2)$ flips are sometimes necessary.

Note that in the geometric setting, only the near-triangulations that are defined on the *specified point set* can be attained via edge flips. For example, no planar K_4 can be drawn on a convex set of four points without introducing a crossing.

In order to resolve the discrepancy between the combinatorial and geometric settings, Abellanas et al. [1] introduced a geometric operation called a *point move*. A point move on a geometric triangulation is simply the modification of the coordinates of one vertex such that after the modification the graph remains a geometric triangulation. That is, the move is valid provided that after moving the vertex to a new position, no edge crossings are introduced. They also showed that with $O(n^2)$ edge flips and $O(n)$ point moves, any geometric triangulation on n points can be transformed to any other geometric triangulation on n possibly different points.

The question which initiated our investigation is whether or not $O(n^2)$ edge flips are necessary. In this paper, we show that with $O(n \log n)$ edge flips and point moves, we can transform any geometric near-triangulation on n points to any other geometric near-triangulation on n possibly different points. Next, we show that if we restrict our attention to geometric near-triangulations defined on a fixed point set of size n, the problem is just as difficult even with the use of point moves. Finally, we show that with a slightly more general point move, we can remove the extra log factor from our main result.

2 Results

In the remainder of the paper, all triangulations and near-triangulations are geometric. It is assumed that the outer face any given near-triangulation is convex, and that any two near-triangulations involved in a reconfiguration have the same number of points on the convex hull.

We assume that the n vertices of any given triangulation are in general position. It is not difficult to see that $O(n)$ point moves can reconfigure a triangulation to this form. We begin with some basic building blocks that will allow us to prove the main theorems.

Lemma 1. [2] *A reconfiguration between two triangulations of the same point set that is in convex position can be done with $O(n)$ edge flips.*

Lemma 2. [9] *Let v_1, v_2 and v_3 be three consecutive vertices on the outer face of a near-triangulation T_1. Let C be the path from v_1 to v_3 on the convex hull of all vertices but v_2. A near-triangulation T_2 containing all edges of C may be constructed from T_1 with t edge flips, where t is the number of edges initially intersecting C in T_1.*

Lemma 3. *Given a near-triangulation T, any vertex $p \in T$ with degree $d > 3$ that is inside the convex hull of the vertices of T can have its degree reduced to 3 with $d - 3$ edge flips.*

Proof. Let P be the polygon that is the union of all triangles incident to p. By Meister's *two-ears theorem* [13], if P has more than three vertices, then it has at least two disjoint ears[2]. At most one of them can contain p. Therefore p and one of the ears form a convex quadrilateral. We may flip the edge from p to the tip of the ear, effectively cutting the ear from P and reducing the number of vertices of P by one. This process may be continued until P is reduced to a triangle that contains p as desired.   □

Lemma 4. *Given a near-triangulation T, any vertex $p \in T$ with degree 3 that is inside the convex hull of the vertices of T can be moved to a new position in the triangulation along a straight path crossing t edges, using at most $2t$ edge flips and $2t + 1$ point moves, assuming the path does not cross through any vertices.*

Proof. Suppose that p is joined by edges to vertices v_1, v_2 and v_3. Without loss of generality, let edge v_2v_3 intersect the path that p must follow, and let this path continue into triangle $v_2v_3v_4$, as shown in Figure 1.

Clearly p can be moved anywhere within triangle $v_1v_2v_3$ without the need of any edge flips. Then it can be moved along its path, as close to edge v_2v_3 as necessary, so that the quadrilateral $pv_2v_3v_4$ becomes convex. This allows edge v_2v_3 to be flipped into edge pv_4. Now p may continue along its path. As soon as it enters $v_2v_3v_4$, edge pv_1 may be flipped into v_2v_3. Now, with two edge flips and two point moves, p has crossed through the first edge intersecting its path, and still has degree 3. By the same argument, p may traverse its entire path with two edge flips and two point moves for each intersecting edge. One additional point move is required in the last triangle. Note that only three edges in the original and final triangulations will be different.   □

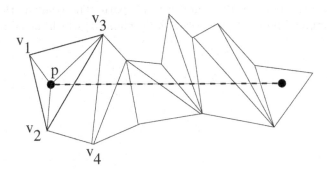

Fig. 1. A vertex p and a straight path that it must move along (dashed). p can pass through any edge with two edge flips.

---

[2] A triangle, defined by three consecutive vertices of a polygon, is an ear if it is empty and the vertices form a convex angle. The second vertex is the *tip* of the ear.

Lemmata 3 and 4 imply the following result:

Lemma 5. *Given a near-triangulation T, any vertex in the interior of the convex hull of the vertices of T with degree d can be moved to a new position in the triangulation along a path crossing t edges, using $O(d + t)$ edge flips and point moves.*

Lemma 6. *An edge can be constructed between a convex hull vertex and any other vertex in a triangulation using $O(n)$ edge flips, with the aid of one moving point that is moved $O(n)$ times.*

Proof. Let v_1 be the hull vertex. First suppose that the second vertex is an interior point. Then it will play the role of the moving point, and we will label it p. We can move p directly towards v_1, until it is located within a triangle that has v_1 as a vertex. Now v_1 and p must be joined with an edge. Next we move p back along the same line to its original position, always maintaining edge v_1p. To do this, we consider the set of triangles that intersect p's path, as in Lemma 4. The point p can always enter a triangle intersecting the path back to its original location. The difference is that once it has crossed an intersecting edge, we do not restore the edge. This means that p will accumulate edge degree. An issue that needs to be taken care of is that of maintaining a triangulation when p is about to lose visibility to another vertex. This occurs when one of its incident edges is about to overlap with another edge in the triangulation, as shown in Figure 2.

Suppose that edge pv_3 is about to overlap with edge v_3v_4. Vertices v_3 and v_4 cannot be on opposite sides of the remaining path that p must traverse, otherwise v_3v_4 may be flipped. The point p must share an edge with v_4 in this configuration. Points p and v_3 are also part of another triangle, along with some vertex v^* which may be anywhere on the path from v_1 to v_3. These two triangles must form a convex quadrilateral $pv^*v_3v_4$, otherwise p would have already lost visibility to v^*. Thus pv_3 may be flipped into v_4v^*, which means that v_3 is removed from the polygon that intersects p's path. The result is that when p reaches its original position, it leaves a *fan*[3] behind it, which includes edge v_1p.

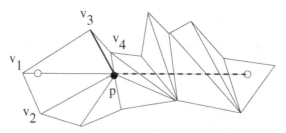

Fig. 2. Maintaining a triangulation while extending edge v_1p: p has moved from a position close to v_1 (shown white), and still has to traverse the dashed segment to its original position. Edge pv_3 causes a problem if p is to continue.

---

[3] A fan is a star-shaped polygon with a vertex as its kernel.

Overall one edge flip is used when p enters a new triangle, and at most one flip is used for every edge that attaches to p.

If both vertices of the edge that we wish to construct are on the hull, then we can take any point p within the hull and move it close to v_1 and onto the segment between the two hull vertices. p can then move along this segment to the second hull vertex until it is connected to both. At this moment, p may be perturbed so that the three vertices form a triangle. This triangle might contain other edges incident to p. Lemma 2 implies that these edges may be removed so that the desired edge can be constructed with $O(n)$ edge flips.   □

2.1   Triangulations

With the basic building blocks in place, we now prove one of our main results.

Theorem 1. *With $O(n \log n)$ edge flips and point moves, we can transform any geometric triangulation on n points to any other geometric triangulation on n possibly different points.*

Proof. We transform one triangulation to another via a canonical configuration. As shown in Figure 3, the interior vertices form a *backbone* (i.e. their induced subgraph is a path). The top of the backbone is joined to the topmost hull vertex v_1, and all interior vertices are joined to the other two hull vertices, v_L and v_R.

The canonical configuration is constructed in a divide-and-conquer manner. We perform a radial sweep from v_1, to find the median vertex interior to the convex hull, v_M. After constructing edge $v_1 v_M$ we move v_M directly away from v_1 towards the base $v_L v_R$, maintaining $v_1 v_M$ until triangle $v_M v_L v_R$ contains no interior points. By Lemma 6, we use $O(n)$ operations to accomplish this. Now, we transform $v_1 v_M v_L$ and $v_1 v_M v_R$ into backbone configurations by induction since they are smaller instances of the same problem. The resulting configuration is shown in Figure 4.

We now show that the two sides may be merged using $O(n)$ operations. As shown in Figure 5a, we first move the lowest vertex of a backbone into a position that is close to the base and is along the extension of edge $v_1 v_M$. This requires one edge flip. The vertices on the left/right backbones are processed in

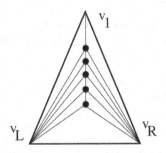

Fig. 3. The canonical configuration used for triangulations.

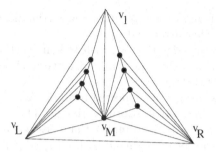

Fig. 4. The configuration of a triangulation prior to merging the backbones on each side of the median vertex v_M.

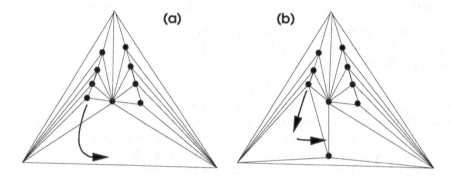

Fig. 5. Merging two backbones into one.

ascending order, and are always moved just above the previous processed vertex, as shown in Figure 5b. Each vertex will require two point moves and one edge flip. Thus $v_1 v_L v_R$ is reconfigured into canonical form, and by a simple recurrence the number of edge flips and point moves used is $O(n \log n)$. It is trivial to move a canonical triangulation to specific coordinates using n point moves. Thus the transformation between any two triangulations may be completed.   □

2.2   Near-Triangulations

If the initial graph is a near-triangulation, Theorem 1 does not directly apply. Some care must be taken to handle a non-triangular outer face. Details are given in the proof of the following theorem:

Theorem 2. *With $O(n \log n)$ edge flips and point moves, we can transform any geometric near-triangulation on n points to any other geometric near-triangulation on n possibly different points.*

Proof. As in the case with triangulations, we transform one near-triangulation to another via a canonical configuration. In the primary canonical configuration, shown in Figure 6, one chosen hull vertex (v_1) is joined by chords to all other

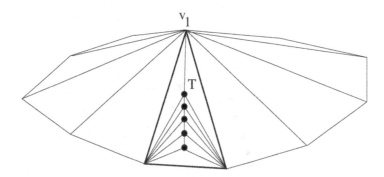

Fig. 6. The primary canonical configuration used for near-triangulations.

hull vertices. Thus v_1 is in the kernel of a convex fan. Every triangle of the fan, except for one, is empty. All interior vertices, located in the non-empty triangle T, are in the canonical configuration of a triangulation.

We first construct all edges of the top-level fan configuration, leaving interior vertices in their original positions. Then within each triangle of the fan, we rearrange the interior vertices into a canonical triangulation. Finally, we merge all triangles of the fan, so that all interior points move to a single triangle and are in canonical form.

To construct the fan chords, we always divide the problem into two roughly equal parts. We begin by constructing two chords as follows: perform a radial sweep from v_1 to successive hull vertices v_i $\{2 \le i \le n - 1\}$, always keeping fewer than $\frac{n}{3}$ vertices in the swept region. Let v_j be the last hull vertex for which this holds. Construct chords v_1v_j and v_1v_{j+1}. The unswept region not including triangle $v_1v_jv_{j+1}$ contains fewer than $\frac{2n}{3}$ vertices. The swept region contains fewer than $\frac{n}{3}$ vertices. Triangle $v_1v_jv_{j+1}$ may contain an arbitrary number of vertices, but this is not a sub-problem (we will not look at this region again during the construction of the fan). Now we can continue a new sweep on each side of $v_1v_jv_{j+1}$. Construction of the two chords could take $O(n)$ edge flips and point moves, as described in Lemma 6. However the even split of the sub-problems ensures that the total number of operations is $O(n \log n)$.

Each fan triangle $v_1v_iv_{i+1}$, containing k_i interior points, can be reconfigured into a backbone structure with $O(k_i \log k_i)$ operations, by Theorem 1. Thus the total number of edge flips and point moves used to reconfigure all triangles of the fan into backbone structures is $O(n \log n)$.

Now we are left only with the task of merging the fan triangles so that only one of them will contain all interior points. We can add k_i interior points of a canonical triangulation to an adjacent canonical triangulation using $O(k_i)$ edge flips and point moves. The k_i points are processed in descending order and are always added to the top of the adjacent triangulation, as shown in Figure 7.

Thus we obtain one triangle in canonical form next to an empty triangle. It is just as easy to merge two canonical triangles separated by an empty triangle. If we encounter two or more adjacent empty fan triangles, we may use Lemma 1

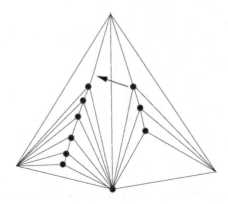

Fig. 7. Merging two adjacent fan triangles.

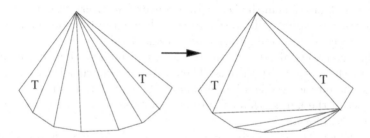

Fig. 8. Handling multiple adjacent empty fan triangles. Triangles marked (T) contain triangulations.

to reconfigure them so that they will not affect the fan-merging process (see Figure 8). By the above arguments, once we select the triangle that is to finally contain all of the interior points (the median triangle is a good choice), we can iteratively merge its neighboring triangles onto it using a total of $O(n)$ edge flips and point moves.

Finally we are left with a single triangle containing all interior points in canonical form. On either side, we may have an arbitrary triangulation (resulting from handling multiple adjacent empty fan triangles), but the vertices will be in convex position. By Lemma 1 they may be moved to our desired configuration using $O(n)$ edge flips.

We must still show that this primary canonical configuration can be moved to specific coordinates. This can be done with $O(n)$ point moves, though space restrictions prevent us from going into any detail.   □

2.3   Remarks

If two triangulations have the same point set, the problem is no easier than the general problem. Suppose that there exists an algorithm that can transform a triangulation T_1 on a given n-point set to a triangulation T_2 on the same point

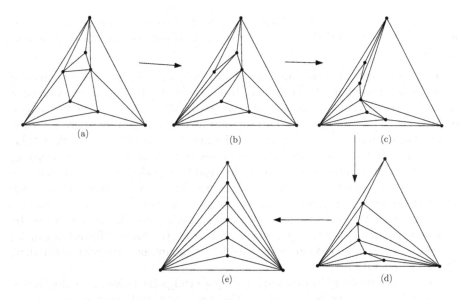

Fig. 9. Problem on fixed point set is not easier.

set using $F_n = o(n \log n)$ edge flips and point moves. Then this algorithm can be used to transform a triangulation on one point set to any other triangulation on a possibly different point set with $F_n + O(n)$ edge flips and point moves. This argument is summarized in Figure 9. Let Figure 9a be the input triangulation. With F_n flips and moves, move to the triangulation in Figure 9b where every vertex is adjacent to the lower left vertex v_ℓ of the outer face.

Now consider the triangulated polygon, P, that consists of edges not adjacent to v_ℓ. Notice that if we perform a radial sweep from v_ℓ, the boundary of P is monotonic. At least two of the triangles in P are disjoint ears, which means there must exist an ear tip that is an interior vertex and is also joined to v_ℓ by an edge in the original triangulation. We may move this point directly towards v_ℓ and cut the ear from P. This still leaves a monotone polygon P'. By continuously locating such ears, and moving them to a predefined convex position, we can obtain the configuration illustrated in Figure 9c. The monotonicity of P (and its descendants) and the convexity of the final configuration of interior points guarantee that no edge crossings will occur. This process requires a linear number of point moves.

Next, by Lemma 1, we can use $O(n)$ edge flips to obtain the triangulation where the lower right vertex of the outer face is adjacent to every vertex, as illustrated in Figure 9d. From here, it is trivial to move to the canonical configuration.

We conclude with the following:

Theorem 3. *If an algorithm exists that can reconfigure between any two geometric triangulations of the same point set with $o(n \log n)$ edge flips and point* *moves, then we can also transform any geometric triangulation on n points to any other geometric triangulation on n different points with o(n log n) flips and moves.*

It is tempting to try to find a fast algorithm that will construct a monotone path, as illustrated in the transition from Figure 9a to Figure 9b. Consider the polygon that is the union of all triangles incident to the lower left vertex of Figure 9b. By continuously cutting ears of this polygon, we may get to a triangulation that is *similar* to that of Figure 9a, using $O(n)$ edge flips. The similarity is that all neighbors of the lower left vertex will be in convex position. However, we have little control over the resulting positions of the remaining edges if we use only $O(n)$ operations. It is possible to create triangulations for which the reversal of this ear-cutting technique is not possible. In fact, Figure 9c serves as an example, if we add a few more vertices inside the large triangle. In this figure none of the edges directly visible from the lower left vertex can be flipped, so there is no obvious way to achieve a monotone path with fewer than $O(n \log n)$ operations.

We finally consider the following more powerful point move as an alternative to the point move studied so far. In this more powerful point move, we can delete an interior vertex of degree three (and all its incident edges), and create a new vertex of degree three inside another triangle of the triangulation. With this type of move we can reconfigure triangulations using $O(n)$ operations. We simply select a triangle incident to a hull edge and create a backbone inside. This is done by continuously selecting a vertex of constant degree from outside the triangle, reducing its degree to three, and moving it to the lower end of the backbone.

References

1. Abellanas, M., Bose, P., Garcia, A., Hurtado, F., Ramos, P., Rivera-Campo, E., Tejel, J.: On local transformations in plane geometric graphs embedded on small grids. In Proceedings of the International Workshop on Computational Geometry and Applications (CGA) **2** (2004) 22–31
2. Bose, P., Czyzowicz, J., Gao, Z., Morin, P., Wood, D.: Parallel diagonal flips in plane triangulations. Tech. Rep. TR-2003-05, School of Computer Science, Carleton University, Ottawa, Canada (2003)
3. Brunet, R., Nakamoto, A., Negami, S.: Diagonal flips of triangulations on closed surfaces preserving specified properties. J. Combin. Theory Ser. B **68(2)** (1996) 295–309
4. Cortés, C., Grima, C., Marquez, A., Nakamoto, A.: Diagonal flips in outer-triangulations on closed surfaces. Discrete Math. **254(1-3)** (2002) 63–74
5. Cortés, C., Nakamoto, A.: Diagonal flips in outer-torus triangulations. Discrete Math. **216(1-3)** (2000) 71–83
6. Galtier, J., Hurtado, F., Noy, M., Pérennes, S., Urrutia, J.: Simultaneous edge flipping in triangulations. Internat. J. Comput. Geom. Appl. **13(2)** (2003) 113–133
7. Gao, Z., Urrutia, J., Wang, J.: Diagonal flips in labelled planar triangulations. Graphs Combin. **17(4)** (2001) 647–657

8. Hurtado F., Noy, M.: Graph of triangulations of a convex polygon and tree of triangulations. Comput. Geom. **13(3)** (1999) 179–188
9. Hurtado, F., Noy, M., Urrutia, J. Flipping edges in triangulations. Discrete Comput. Geom. **22(3)** (1999) 333–346
10. Komuro, H.: The diagonal flips of triangulations on the sphere. Yokohama Math. J. **44(2)** (1997) 115–122
11. Komuro, H., Nakamoto, A., Negami, S.: Diagonal flips in triangulations on closed surfaces with minimum degree at least 4. J. Combin. Theory Ser. B **76(1)** (1999) 68–92
12. Lawson, C.: Software for c_1 surface interpolation. In J. Rice, ed., Mathematical Software III, pp. 161–194, Academic Press, New York (1977)
13. Meisters, G.: Polygons have ears. American Mathematical Monthly **82** (1975) 648–651
14. Nakamoto, A., Negami, S.: Diagonal flips in graphs on closed surfaces with specified face size distributions. Yokohama Math. J. **49(2)** (2002) 171–180
15. Negami, S.: Diagonal flips in triangulations of surfaces. Discrete Math.**135(1-3)** (1994) 225–232
16. Negami, S.: Diagonal flips in triangulations on closed surfaces, estimating upper bounds. Yokohama Math. J. **45(2)** (1998) 113–124
17. Negami, S.: Diagonal flips of triangulations on surfaces, a survey. Yokohama Math. J. **47** (1999) 1–40
18. Negami, S., Nakamoto, A.: Diagonal transformations of graphs on closed surfaces. Sci. Rep. Yokohama Nat. Univ. Sect. I Math. Phys. Chem. **40** (1993) 71–97
19. Wagner, K.: Bemerkung zum Vierfarbenproblem. Jber. Deutsch. Math.-Verein. **46** (1936) 26–32
20. Watanabe, T., Negami, S.: Diagonal flips in pseudo-triangulations on closed surfaces without loops. Yokohama Math. J. **47** (1999) 213–223

Drawing Power Law Graphs

Reid Andersen, Fan Chung*, and Lincoln Lu

University of California, San Diego

Abstract. We present methods for drawing graphs that arise in various information networks. It has been noted that many realistic graphs have a power law degree distribution and exhibit the small world phenomenon. Our methods are influenced by recent developments in the modeling of such graphs.

1 Introduction

Several research groups have observed that many networks, including Internet graphs, call graphs and social networks, have a *power law* degree distribution, where the fraction of nodes with degree k is proportional to $k^{-\beta}$ for some positive exponent β [8]. Many networks also exhibit a so-called "small world phenomenon" consisting of two distinct properties — small average distance between nodes, and a clustering effect where two nodes sharing a common neighbor are more likely to be adjacent. It was shown in [2] that a random power law graph has small average distance and small diameter. However, random power law graphs do not adequately capture the clustering effect.

To model the small world phenomenon, several researchers have introduced random graph models with additional geometric structure. Kleinberg [7] proposed a model where a grid graph G is augmented with random edges between nodes u, v with probability proportional to $[d_G(u,v)]^{-r}$ for some constant r. Fabrikant, Koutsoupias and Paradimitriou [4] proposed a model where vertices are points in the Euclidean plane and edges are added by optimizing a function involving both Euclidean distance and graph distance to a central node.

Chung and Lu [3] introduced a hybrid graph model where a random power law graph called the "global" graph is added to a "local graph" having a certain kind of local connectivity. In [1] an efficient algorithm was presented for extracting a highly connected local graph from an arbitrary graph. For a graph generated by the hybrid model, this algorithm recovers the original local graph up to a small error.

In this paper, we present a drawing method using the algorithm for extracting local graphs. This algorithm may be useful for drawing graphs similar to those produced by the hybrid model. A graph from the hybrid model contains a random power law graph which will not be amenable to most drawing methods, but also contains a local graph which can be more geometric in nature. The recovery theorem in [1] guarantees that when applied to a graph from the hybrid model, our algorithm produces a layout which depends largely on the local graph.

* Research supported in part by NSF Grants DMS 0100472 and ITR 0205061.

2   Preliminaries

2.1   Weighted Graphs and Quotient Graphs

Although our input graphs are unweighted, our algorithm will form weighted graphs by collapsing connected components into single vertices. A weighted graph is a simple graph G together with a vertex weight function $w_G(v)$ and an edge weight function $\phi_G(e)$. Suppose that $V(G)$ has a partition $V(G) = C_1 \cup C_2 \cup \cdots \cup C_k$. The quotient graph Q is defined as follows. The vertices of Q are communities $C_1, \ldots C_k$, and we set

$$w_Q(C_k) = \sum_{u \in C_i} w_G(u).$$

$$\phi_Q(C_i, C_j) = \sum_{u \in C_i, v \in C_j} \phi_G(u, v).$$

There is an edge between C_i and C_j if $\phi_Q(C_i, C_j) > 0$.

2.2   Local Flow and Local Graphs

Given a weighted graph with edge capacity function ϕ, we will define a notion of local connectivity between vertices. We will say a path is *short* if it has length less than or equal to ℓ. A short flow is a positive linear combination of short paths where no edge carries more than its capacity. The maximum short flow problem can be viewed as a linear program, and can be computed in polynomial time using nontrivial but relatively efficient algorithms for fractional packing (See 2.3).

Definition 1 (Short Flow). *A short flow is a feasible solution to the following linear program. The flow connectivity $f(u, v)$ between two vertices is the maximum value of any short flow, which is the optimum value of the following LP problem. Let P_ℓ be the collection of short u-v paths, and let P_e be the collection of short u-v paths which intersect the edge e.*

$$\textit{maximize} \quad \sum_{p \in P_\ell} f_p \tag{1}$$

$$\textit{subject to} \quad \sum_{p \in P_e} f_p \leq \phi(e) \qquad \textit{for each } e \in L$$

$$f_p \geq 0 \qquad \textit{for each } p \in P_\ell$$

We say two vertices u and v are (f, ℓ)-connected if there exists a short flow between them of size at least f. We a say a graph L is an (f, ℓ)-local graph if for each edge $e = (u, v)$ in L, the vertices u and v are (f, ℓ)-connected in L.

2.3   Computing the Maximum Short Flow

Finding the maximum short flow between u and v in a graph G with given edge capacities $\phi(e)$ can be viewed as a fractional packing problem, which has the form

$$\max\{\, \mathbf{c}^{\mathrm{T}}\mathbf{x} \mid A\mathbf{x} \leq \mathbf{b}, \mathbf{x} \succeq \mathbf{0} \,\}.$$

To view the maximum short flow as a fractional packing problem, first let $G(u, v)$ be a subgraph containing all short paths from u to v. For example, we may take $G(u, v) = N_{\ell/2}(u) \cup N_{\ell/2}(v)$. Let A be the incidence matrix where each row represents an edge in $G(u, v)$ and each column represents a short path from u to v. Let $\mathbf{b} = \phi$, and $\mathbf{c} = \mathbf{1}$.

Using the algorithm of Garg and Könemann in [5] for general fractional packing problems, one can obtain a $(1 - \epsilon)^{-2}$-approximation to the maximum short flow in time $O(M^2 \ell \lceil \frac{1}{\epsilon} log_{1+\epsilon} M \rceil)$, where M is the number of edges in $G(u, v)$.

3   Extracting the Local Graph

For a given graph, we wish to extract the largest (f, ℓ)-local subgraph. We define $L_{f,\ell}(G)$ to be the union of all (f, ℓ)-local subgraphs in G. By definition, the union of two (f, ℓ)-local graphs is an (f, ℓ)-local graph, and so $L_{f,\ell}(G)$ is in fact the unique largest (f, ℓ)-local subgraph in G. We remark that $L_{f,\ell}(G)$ is not necessarily connected. The simple greedy algorithm Extract computes $L_{f,\ell}(G)$ in any graph G using $O(m^2)$ max-short-flow computations, where m is the number of edges in G. The number of max-short-flow computations can be reduced by using a standard random sampling approach if we are willing to accept approximate local graphs. We say L is an α-approximate (f, ℓ)-local graph if $L(f, \ell) \subseteq L$, and at most an α-fraction of the edges in L are not (f, ℓ)-connected. The algorithm Approximate Extract computes a series of approximate local graphs.

Extract:
Input: G, f, ℓ
If there is an edge $e = (u, v) \in G$ where u, v are not (f, ℓ)-connected in G,
      remove e from G.
When no further edges can be removed, output G.

Approximate Extract:
Input: $G, \ell, \{f_1 \leq \cdots \leq f_k\}$
Let m be the number of edges in G.
For $i = 1 \ldots k$:
      Repeat until no edge is removed for $\frac{1}{\alpha} \log \frac{mk}{\delta}$ consecutive attempts:
            Pick an edge $e = (u, v)$ from G uniformly at random.
            If u, v are not (f_i, ℓ)-connected, remove (u, v) from G.
      Let $L_i = G$, reset m to be the number of edges in L_i,
            and proceed to compute L_{i+1}.
Stop when graphs $L_1 \supseteq \cdots \supseteq L_k$ have been output.

Since at most m edges are removed from G and there are at most $\frac{1}{\alpha} \log \frac{mk}{\delta}$ attempted removals for every edge removed, Approximate Extract performs at most $\frac{m}{\alpha} \log \frac{mk}{\delta}$ max-short-flow computations.

Theorem 1. *Given G, ℓ, and $\{f_1 \leq \cdots \leq f_k\}$, let $L_1 \supseteq \cdots \supseteq L_k$ be the output of* Approximate Extract. *With probability at least $1 - \delta$, each of the graphs L_i is an α-approximate (f_i, ℓ)-local graph.*

Proof: Given $i \in [1, k]$, let $e_1 \ldots e_J$ be the edges removed from L_{i-1} to obtain L_i. Let m_i be the number of edges in L_{i-1} and note that $J \leq m_i$. Let T_j be the number of attempts between the removal of the e_{j-1} and e_j. If L_i is not an α-approximate local graph, then some T_j must be at least $\frac{1}{\alpha} \log \frac{m_i k}{\delta}$ when at least an α-fraction of the edges remaining in L_i were not (f_i, ℓ)-connected. For a given j, this occurs with probability at most

$$(1 - \alpha)^{T_j} \leq e^{-\alpha T_j} \leq e^{-\log \frac{m_i k}{\delta}} \leq \delta m_i^{-1}/k.$$

Since $J \leq m_i$, the probability that this occurs for any T_j is at most δ/k. The probability that a bad T_j occurs for any L_i is at most δ, and the result follows.

4   An Algorithm for Drawing Power Law Graphs

In this section we describe a framework for producing drawings of power law graphs that reflect local connectivity. In the algorithm Local Draw below, a local subgraph is used to determine the layout of the vertices. Our algorithm uses as a subroutine a standard force-directed drawing method which we describe in section 4.2, but other methods can be used in its place. The algorithm is motivated by the structure of power law graphs, but can be applied to general graphs as well.

4.1   The Algorithm

Local Draw:
Given an input graph G, compute the local graph $L_{f,\ell}$ for some choice of f and ℓ using Extract or Approximate Extract. Let $\Pi_{f,\ell}$ be the partition induced by the connected components $C_1 \ldots C_k$ of $L_{f,\ell}$, and let Q be the quotient graph of G with respect to this partition. Use the force-based drawing algorithm to produce drawings of each component $C_1 \ldots C_k$ and Q separately. To combine into a single drawing, let $q_1 \ldots q_k$ be the coordinates of the vertices in Q corresponding to $C_1 \ldots C_k$, and let

$$r_i = \frac{1}{2} \min_j \|q_i - q_j\|.$$

Scale each drawing of C_i by r_i, and place at location q_i to create a new drawing which only contains edges in $L_{f,\ell}$. Apply the force-based algorithm to this drawing to determine the final layout of the vertices, and then add back the edges in $G \setminus L_{f,\ell}$.

4.2   A Force-Directed Drawing Method

Our algorithms use a standard force-based drawing method, modified for use on graphs with vertex weights $w(v)$ and edge weights $\phi(e)$. We define a repulsive force between every pair of vertices, where the force acting on vertex u due to vertex v is

$$R_{u,v} = \frac{1}{n^2} \frac{u - v}{\|u - v\|^2} w(u) w(v)$$

Each edge also acts as a spring, with the force on a vertex u from the edge $e = (u, v)$ defined to be

$$S_{u,v} = \frac{1}{n}(v - u)\phi(e)$$

To keep the drawing in a bounded area, we place all vertices within the unit circle and define a force between each vertex and the boundary of the circle.

$$B_u = -\frac{u}{\|u\|} \frac{1}{(1 - \|u\|)} w(u)$$

The standard force-based approach is to compute the sum of the forces acting on each vertex and move in the resulting direction at each time step.

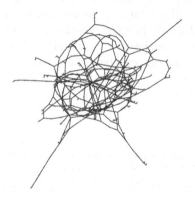

Fig. 1. Local Draw applied to the giant component of random graph $G(n, p)$ with $n = 500$ and $p = 0.004$.

Fig. 2. Local Draw applied to the induced subgraph of G: the collaboration graph on authors with Erdős number exactly 2.

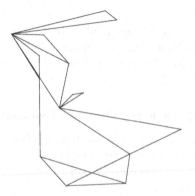

Fig. 3. The quotient graph of G.

Fig. 4. The largest connected component in the local graph of G, of size 15.

5   Implementation and Examples

We have implemented **Extract** and **Local Draw** and experimented on several examples. Figure 1 is a drawing of a sparse random graph, generated from the Erdős-Rényi model $G(n, p)$ with $n = 500$ and $p = .004$. Jerry Grossman [6] has graciously provided data from a collaboration graph of the second kind, where each vertex represents an author and each edge represents a joint paper with two authors. Our example graph G is the largest component of the induced subgraph on authors with Erdős number exactly 2. This graph contains 834 vertices. We applied **Local Draw** to G with parameters $(f = 2, \ell = 3)$, obtaining the drawing in Figure 2, and in the process obtaining the quotient graph shown in Figure 3 and the local graph. The largest connected component of the local graph is shown in Figure 4.

References

1. R. Andersen, F. Chung and L. Lu, Analyzing the small world phenomenon using a hybrid model with local network flow, *Proceedings of the Third Workshop on Algorithms and Models for the Web-Graph* (2004).
2. F. Chung and L. Lu, Average distances in random graphs with given expected degree sequences, *Proceedings of National Academy of Science*, **99** (2002).
3. F. Chung and L. Lu, The small world phenomenon in hybrid power law graphs *Complex Networks*, (Eds. E. Ben-Naim et. al.), Springer-Verlag, (2004).
4. A. Fabrikant, E. Koutsoupias and C. H. Papadimitriou, Heuristically optimized trade-offs: a new paradigm for power laws in the Internet, *STOC* 2002.
5. N. Garg, J. Könemann, Faster and simpler algorithms for multicommodity flow and other fractional packing problems. *Technical Report, Max-Planck-Institut fur Informatik, Saarbrucken, Germany* (1997).
6. Jerry Grossman, Patrick Ion, and Rodrigo De Castro, Facts about Erdős Numbers and the Collaboration Graph, http://www.oakland.edu/~grossman/trivia.html.
7. J. Kleinberg, The small-world phenomenon: An algorithmic perspective, *Proc. 32nd ACM Symposium on Theory of Computing*, 2000.
8. M. Mitzenmacher, A Brief History of Generative Models for Power Law and Lognormal Distributions, *Internet Math.* 1 (2003), no. 2.

Hexagonal Grid Drawings:
Algorithms and Lower Bounds⋆

Shabnam Aziza and Therese Biedl

School of Computer Science, University of Waterloo,
Waterloo, Ontario N2L 3G1, Canada
{saziza,biedl}@uwaterloo.ca

Abstract. We study drawings of graphs of maximum degree six on the hexagonal (triangular) grid, with the main focus of keeping the number of bends small. We give algorithms that achieve $3.5n + 3.5$ bends for all simple graphs. We also prove optimal lower bounds on the number of bends for K_7, and give asymptotic lower bounds for graph classes of varying connectivity.

1 Introduction

There are numerous algorithms to draw 4-graphs (graphs of maximum degree of at most four) on the 2D rectangular (orthogonal) grid [3, 8–11]. All 4-graphs can be drawn with at most $2n + 2$ bends [3], and there are arbitrarily large graphs that need $\frac{11}{6}n$ bends [1]. In 3D, orthogonal drawings exist for all 6-graphs [4, 13–15]. In this paper, we study *hexagonal drawings*, which are embeddings of 6-graphs in the 2D *hexagonal grid*. We consider the hexagonal grid to consist of horizontal gridlines (rows), vertical gridlines (columns) and diagonals; this is the same grid as the "standard" hexagonal grid (with 60° angles) after a shear in the x-direction. See also Fig. 1.

Only few results are known for hexagonal drawings. The algorithm by Tamassia [10] to obtain bend-minimum orthogonal drawings of planar graphs can be extended to the hexagonal grid as well. Kant [6] showed how to draw 3-connected

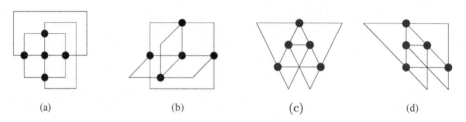

(a)	(b)	(c)	(d)

Fig. 1. Different types of grid drawings of K_5: (a) 2D orthogonal, (b) 3D orthogonal, (c) hexagonal with 60° angles, (d) hexagonal with diagonals.

⋆ Research supported by NSERC. These results appeared as part of the MMath thesis of the first author at University of Waterloo.

J. Pach (Ed.): GD 2004, LNCS 3383, pp. 18–24, 2004.

cubic planar graphs on the hexagonal grid. Tollis [12] uses a similar grid called tri-hexagonal grid for wiring VLSI layouts.

In this paper, we provide an algorithm that draws all graphs with maximum degree 6 on the hexagonal grid, and achieves $3.5n + 3.5$ bends. We also study lower bounds on the number of bends for K_7 and for larger graphs.

2 Algorithms

Our algorithms are inspired by the algorithm of Biedl and Kant [3] for drawing biconnected 4-graphs on the orthogonal grid. They use a vertex ordering known as *st-ordering*, which is an ordering $v_1, v_2, \ldots, v_n$ of the vertices such that each v_i, $2 \leq i \leq n-1$, has at least one predecessor, i.e., a neighbour v_h with $h < i$ and at least one successor, i.e., a neighbour v_j with $j > i$. The edges from v_i to its predecessors [successors] are called incoming [outgoing] edges of v_i. The number of incoming [outgoing] edges of vertex v_i are denoted by $indeg(v_i)$ [$outdeg(v_i)$]. For any biconnected graph, and any two vertices s, t, an st-ordering exists with $v_1 = s$ and $v_n = t$ [7] and can be computed in linear time [5].

Assume from now on that G is a biconnected 6-graph without loops, and $v_1, \ldots, v_n$ is an st-ordering of G. Let G_j be the graph induced by $v_1, \ldots, v_j$. An edge (v_i, v_k) with $i \leq j < k$ is called an *unfinished edge of G_j*. For $j = 1, \ldots, n$, we create a drawing of G_j such that every unfinished edge "ends in a free ray", i.e., every unfinished edge is drawn up to a point, and there exists a ray (along a grid line) from this point that does not contain any vertex or edge segment in it, and is the free ray for only one unfinished edge. These rays must go in direction north (N), north-west (NW) or west (W). Fig. 2 shows a suitable drawing of G_1 and illustrates the invariant.

Now assume that we have a suitable drawing of G_{j-1}. If (v_i, v_j) is an incoming edge of v_j, then there is an unfinished edge at v_i, and hence a ray associated with it. (We choose one arbitrarily if there is more than one ray.) We add (if needed) bends in these rays, and new line segments in new grid lines that are fully outside the drawing of G_{j-1} in such a way that all incoming edges meet in one grid point at which we place v_j. Then we assign rays to outgoing edges of v_j, adding more bends (if needed) to enforce that rays go into one of the three allowed directions. The specific drawing of v_j depends on the number and directions of rays of incoming edges of v_j. There are many cases here; Fig. 3 shows some of them.

Studying all cases yields that each vertex $v \neq v_1, v_n$ needs at most $indeg(v)+1$ bends, which leads to a bound of $m + n + O(1) \leq 4n + O(1)$. But this is not

Fig. 2. Embedding of the first vertex, and the maintained invariant.

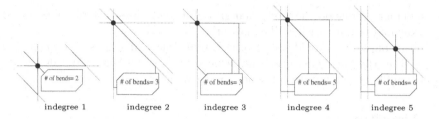

Fig. 3. Some of the cases of embedding vertices. Dotted lines indicate new grid lines.

tight. To obtain a better bound on the number of bends, we developed a potential function argument, which to our knowledge is a new idea in graph drawing. We assign potential p_d, $d \in \{N, NW, W\}$ to each unfinished edge whose free ray ends in direction d. Let $\phi(j)$ be the sum of the potentials of the drawing of G_j. Then the amortized cost of a vertex v_j, $1 \le j \le n$, is,

$$\hat{c}(v_j) = \text{number of bends added when placing vertex } v_j + \phi(j) - \phi(j-1).$$

Hence the total number of bends is $\sum_{v \in V} \hat{c}(v)$.

Theorem 1. *Any biconnected 6-graph without loops has a hexagonal grid drawing with at most $3.5n + 3.5$ bends.*

Proof. We choose as potentials $p_N = p_W = \frac{5}{8}$ and $p_{NW} = \frac{1}{8}$. The amortized cost at v_i is determined uniquely from the number of bends needed when placing v_i and the directions of the incoming and outgoing edges. Let $\hat{c}_i$ be the maximum amortized cost of a vertex with indegree i. Going through the cases, one obtains $\hat{c}_0 = 5.25, \hat{c}_1 = 3.5, \hat{c}_2 = 3.25, \hat{c}_3 = 3, \hat{c}_4 = 3.25, \hat{c}_5 = 3.5$ and $\hat{c}_6 = 5.25$. (The cases in Fig. 3 are some of those where these bounds are tight.) So the number of bends is

$$\sum_{v \in V} \hat{c}(v) \le \sum_{i=0}^{6} \hat{c}_i n_i = 5.25n_0 + 3.5n_1 + 3.25n_2 + 3n_3 + 3.25n_4 + 3.5n_5 + 5.25n_6.$$

Since $n_0 = 1$ and $n_6 \le 1$ for an st-ordering of a 6-graph, this yields the desired bound.

Our algorithm can be expanded with similar techniques as in [3] to handle graphs that are not biconnected or that have loops. The proofs of the following theorems are omitted.

Theorem 2. *Any connected 6-graph without loops with $n \ge 3$ can be drawn on the hexagonal grid with at most $4.2n$ bends.*

Theorem 3. *Every simple connected 6-graph can be drawn on the hexagonal grid with at most $3.5n + 3.5$ bends.*

Theorem 4. *Any biconnected 6-graph can be drawn on the hexagonal grid with at most $3.5n + 3.5 + \frac{1}{4}\ell$ bends, where ℓ is the number of loops in the graph.*

The area of our construction may be exponential, since in some cases we add $O(w)$ new rows to a drawing of width w, or $O(h)$ many columns to a drawing of height h. The area can be reduced to quadratic if we allow more bends.

Theorem 5. *Every biconnected 6-graph without loops can be drawn on the hexagonal grid with at most $6n + 2$ bends and area $O(n^2)$.*

3   Lower Bounds

We now turn to lower bounds on the number of bends of hexagonal grid drawings. We start with K_7, the complete graph on 7 vertices. This graph requires 20 bends in a 3D orthogonal layout [14]. In the hexagonal grid, it can be drawn with 18 bends (see Fig. 4), and as we show now, this bound is tight. So assume that an arbitrary drawing of K_7 is fixed. Let r be the number of rows that are *truly used*, i.e., they contain either a vertex or a segment of an edge. Similarly let c and d be the number of truly used rows, columns, and diagonals. We first show $r + c + d \geq 18$, for which by symmetry it suffices to show $r + d \geq 12$.

Lemma 1. *In any hexagonal drawing of K_7, $r + d \geq 12$.*

Proof. We use a cut-argument similar as in [1]. A *(vertical) cut* is a vertical line that does not coincide with a column. An (x, y)-*cut* is a cut with x vertices on one side and y vertices on the other. The edges between vertices on different sides are called *cut-edges*. Each cut-edge has at least one segment crossed by the cut, and hence truly uses a grid line that crosses the cut (i.e., a row or a diagonal). A $(3, 4)$-cut in K_7 has 12 cut-edges, and hence immediately implies that $r + d \geq 12$. However, such a cut need not always exist. We distinguish cases.

Assume there is a column c_6 with exactly six vertices in it. We consider two cuts; one cut c_l immediately to the left of c_6 and one cut c_r immediately to the right of c_6. We assume that the seventh vertex is to the left of c_6, so c_l has 6 cut-edges. Let (u, v) be an edge for which u, v are both in c_6, but not consecutive in c_6. Then (u, v) cannot be drawn as a straight line. We call (u, v) a *non-straight edge* and note that there are $\binom{6}{2} - 5 = 10$ non-straight edges. See also Fig. 4.

The drawing of each non-straight edge thus must leave column c_6 and then return to it. Say it leaves towards the left side, then it crosses the cut c_l when it leaves, and crosses the cut c_l again when it returns. If a non-straight edges go

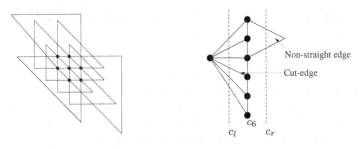

Fig. 4. K_7 drawn with 18 bends, and the case with 6 vertices in the same column.

left, then there are therefore at least $2a$ edge segments that cross cut c_l and truly use a row or a diagonal. Since c_l also has 6 cut-edges, we have $r + d \geq 2a + 6$.

The other $10 - a$ non-straight edges go right and cross c_r twice, so $r + d \geq 2(10 - a)$. Altogether therefore $r + d \geq \max\{2a + 6, 2(10 - a)\}$. Since the value of a is unknown, we take the minimum over all possible values, and get $r + d \geq \min_{0 \leq a \leq 10} \max\{2a + 6, 2(10 - a)\} = 14$.

All other cases are treated similarly: we count how many cut-edges are on c_l and c_r and how many non-straight edges there are, and apply the formula to compute the lower bound on $r + d$. In all cases, we get $r + d \geq 12$. For space reasons, we omit the details of these cases, but list the cases so the reader can verify that all cases have been covered.

- A column c contains 7 vertices.
- A column c contains 5 vertices, and there are 2 vertices on one side of c.
- A column c contains 5 vertices, and there is 1 vertex on each side of c.
- A column c contains 4 vertices, and there are 3 vertices on one side of c. (In this case c_l or c_r is a $(3, 4)$-cut.)
- A column c contains 4 vertices, and there are 1 or 2 vertices on one side of c.
- A column c contains 3 vertices, and there are 3 or 4 vertices on one side of c. (In this case c_l or c_r is a $(3, 4)$-cut.)
- A column c contains 3 vertices, and there are 2 vertices on each side of c.
- All columns contain at most 2 vertices. Applying a scan from left to right, one can show that then there must exist a $(3, 4)$-cut.   □

We can relate $r + c + d$ to the number of bends, similarly as in [2].

Lemma 2. *In a hexagonal drawing of a graph with n vertices and m edges, let b be the number of bends and let r, c, d be the number of truly used rows, columns and diagonals. Then $b \geq r + c + d - 3n + m$.*

Combining this with $r + c + d \geq 18$ give the lower bounds for K_7.

Theorem 6. *Any hexagonal drawing of K_7 has at least 18 bends.*

Using K_7 and other small graphs, we can build arbitrarily large graphs that also have a large lower bound on the number of bends, similarly as done in [1] for orthogonal 2D drawings and in [14] for orthogonal 3D drawings. We give the detailed construction for one (illustrative) case.

Theorem 7. *For any n, there is a 3-connected simple graph on $n' > n$ vertices that requires at least $1.87n'$ bends.*

Proof. The graph for this lower bound is illustrated in the bottom left entry of Table 1. We take an even number c of copies of K_7 and place half of them in the first row and half of them in the second row. In each copy of K_7, we subdivide one edge; then we identify the subdivision vertex in the first row with the one in the second row. Also, in each copy we delete an edge (not incident to the subdivision vertex). Then we add an edge between the ith and the $((i \mod \frac{c}{2}) + 1)$st copy of K_7 in each row.

Table 1. Lower bounds for arbitrarily large graphs.

	Simple	Multigraph	Graph with loops
1-connected	$\frac{34}{15}n = 2.27n$	$\frac{16}{5}n = 3.2n$	$4n$
2-connected	$\frac{15}{7}n = 2.14n$	$3n$	$4n$
3-connected	$\frac{28}{15}n = 1.87n$	$2n$	$2.5n$

Recall that K_7 needs 18 bends. Subdividing an edge lowers this to 17 bends, because the subdivision vertex could take the place of a bend. Deleting an edge lowers this to 14 bends since one can show that in any hexagonal drawing, we can add an edge while adding at most 3 bends. Thus each copy needs at least 14 bends, so in total we need at least $28c$ bends for $n = 15c$ vertices, and the total number of bends is $\frac{28n}{15} = 1.87n$.   □

Similar (and often easier) constructions can be used to obtain lower bounds for various types of graphs and connectivity; see Table 1.

References

1. T. Biedl. New lower bounds for orthogonal drawings. *J. Graph Algorithms Appl.*, 2(7):1–31, 1998.
2. T. Biedl. Relating bends and size in orthogonal graph drawings. *Inform. Process. Lett.*, 65(2):111–115, 1998.
3. T. Biedl and G. Kant. A better heuristic for orthogonal graph drawings. *Comput. Geom.*, 9(3):159–180, 1998.
4. P. Eades, A. Symvonis, and S. Whitesides. Three-dimensional orthogonal graph drawing algorithms. *Discrete Appl. Math.*, 103(1-3):55–87, 2000.
5. S. Even and R. E. Tarjan. Computing an *st*-numbering. *Th. Comput. Sci.*, 2:339–344, 1976.
6. G. Kant. Hexagonal grid drawings. In *Graph-theoretic Concepts in Computer Science*, vol. 657 of *Lect. Notes in Comput. Sci.*, pages 263–276. Springer, 1993.
7. A. Lempel, S. Even, and I. Cederbaum. An algorithm for planarity testing of graphs. In *Theory of Graphs (Internat. Sympos., Rome, 1966)*, pages 215–232. New York, 1967.
8. A. Papakostas and I. G. Tollis. Algorithms for area-efficient orthogonal drawings. *Comput. Geom.*, 9(1-2):83–110, 1998.
9. J. Storer. On minimal-node-cost planar embeddings. *Networks*, 14(2):181–212, 1984.

10. R. Tamassia. On embedding a graph in the grid with the minimum number of bends. *SIAM J. Comput.*, 16(3):421–444, 1987.
11. R. Tamassia and I. G. Tollis. Planar grid embedding in linear time. *IEEE Transactions on Circuits and Systems*, 36(9):1230–1234, 1989.
12. I. Tollis. Wiring layouts in the tri-hexagonal grid. *Constraints*, 3(1):87–120, 1998.
13. D. R. Wood. An algorithm for three-dimensional orthogonal graph drawing. In *Graph Drawing*, volume 1547 of *Lecture Notes in Comput. Sci.*, pages 332–346. Springer, 1998.
14. D. R. Wood. Lower bounds for the number of bends in three-dimensional orthogonal graph drawings. *J. Graph Algorithms Appl.*, 7(1):33–77, 2003.
15. D.R. Wood. Optimal three-dimensional orthogonal graph drawing in the general position model. *Th. Comput. Sci.*, 299(1-3):151–178, 2003.

Improved Bounds for the Number of ($\leq k$)-Sets, Convex Quadrilaterals, and the Rectilinear Crossing Number of K_n

József Balogh[1,*] and Gelasio Salazar[2,**]

[1] Department of Mathematics, The Ohio State University,
Columbus OH 43210 USA
jobal@math.ohio-state.edu
[2] IICO-UASLP, Av. Karakorum 1470, Lomas 4ta. Seccion,
San Luis Potosi, SLP, Mexico 78210
gsalazar@cactus.iico.uaslp.mx
http://cactus.iico.uaslp.mx/~gsalazar

Abstract. We use circular sequences to give an improved lower bound on the minimum number of ($\leq k$)-sets in a set of points in general position. We then use this to show that if S is a set of n points in general position, then the number $\square(S)$ of convex quadrilaterals determined by the points in S is at least $0.37553\binom{n}{4} + O(n^3)$. This in turn implies that the rectilinear crossing number $\overline{\mathrm{cr}}(K_n)$ of the complete graph K_n is at least $0.37553\binom{n}{4} + O(n^3)$. These improved bounds refine results recently obtained by Ábrego and Fernández-Merchant, and by Lovász, Vesztergombi, Wagner and Welzl.

1 Introduction

Our aim in this work is to present some selected results and sketches of proofs of our recent work [5] on the use of circular sequences in the problems described in the title. For the reader familiar with the application of circular sequences to these closely related problems, we give in Subsection 1.4 a brief account of what we perceive is the main achievement hereby reported.

It is well-known that the rectilinear crossing number $\overline{\mathrm{cr}}(K_n)$ of the complete graph K_n is closely related to the minimum number $\square(S)$ of convex quadrilaterals in a set S of n points in general position.

Observation 1 *For each positive integer n,*

$$\overline{\mathrm{cr}}(K_n) = \min_{|S|=n} \square(S),$$

with the minimum taken over all point sets S with n elements in general position.

* Supported by NSF Grant DMS-0302804. Partially supported by City of Morahalom, Hungary.
** Supported by FAI-UASLP.

Working independently, Ábrego and Fernández-Merchant [1], and Lovász, Vesztergombi, Wagner and Welzl [13] recently explored the close connection between $\Box(S)$ and the number $\eta_{\leq k}(S)$ of $(\leq k)$-sets of S. The following result is implicitly proved in [1], and the connection with $(\leq k)$-sets was particularly emphasized in [13].

Theorem 1 ([1] and [13]). *Let S be a set of n points in the plane in general position. Then*

$$\Box(S) = \sum_{1 \leq k < (n-2)/2} (n - 2k - 3)\eta_{\leq k+1}(S) + O(n^3),$$

where $\eta_{\leq j}(S)$ denotes the number of $(\leq j)$-sets of S.

We recall that the *rectilinear crossing number* $\overline{cr}(G)$ of a graph G is the minimum number of pairwise intersections of edges in a drawing of G in the plane in which every edge is drawn as a straight segment. We also recall that if S is a set of points in the plane in general position, then a *k-set* is a subset T of S with $|T| = k$, and such that T can be separated from its complement $T \setminus S$ by a line. An *i-set* with $1 \leq i \leq k$ is a $(\leq k)$-*set*. As we mentioned above, we use $\eta_{\leq k}(S)$ to denote the number of $(\leq k)$-sets of S.

In this paper we follow the approach, via circular sequences, used by Ábrego and Fernández-Merchant and (independently) by Lovász, Vesztergombi, Wagner and Welzl, to give improved lower bounds for $\eta_{\leq k}(S)$. In view of Observation 1 and Theorem 1, these refined bounds immediately imply improved bounds for $\Box(S)$ (for any set S) and for $\overline{cr}(K_n)$.

1.1   The Relationship Between $\Box(S)$ and Circular Sequences

Let S be a set of n points in general position in the plane. In [1] and [13], it is shown that $\Box(S)$ is closely related to $\eta_{\leq k}(S)$.

While the important problem of determining, for each k, the maximum number of k-sets remains tantalizingly open (the best current bounds are $O(nk^{1/3})$ and $ne^{\Omega(\log k)}$ (see [8] and [18], respectively), it is known that the maximum number of $(\leq k)$-sets of an n-point set S in the plane is nk (this is attained iff S is in convex position; see [3] and [21]).

In [13] and [21], it is shown that if S is a collection of points in general position, then $\Box(S)$ is a linear combination of $\{\eta_{\leq j}(S)\}$. Indeed, Theorem 1 above is a direct consequence of Lemma 9 in [13].

Theorem 1 is exploited in [13] by finding a nontrivial lower bound for $\eta_{\leq k}(S)$ for every $k < n/2$ and every set S of n points in general position (and using an even better bound for k close to $n/2$, which follows from the results in [20]). See Theorems 2 and 4 in [13]. To obtain the bound in their Theorem 2, they follow the approach of circular sequences.

A *circular sequence on n elements Π* is a sequence $(\pi_0, \pi_1, \ldots, \pi_{\binom{n}{2}})$ of permutation of the set $\{1, 2, \ldots, n\}$, where π_0 is the identity permutation $(1, 2, \ldots, n)$,

$\pi_{\binom{n}{2}}$ is the reverse permutation $(n, n-1, \ldots, 1)$, and any two consecutive permutations differ by exactly one transposition of two elements in adjacent positions. A transposition that occurs between elements in positions i and $i+1$, or between elements in positions $n - i$ and $n - i + 1$ is i-critical. A transposition is $(\leq k)$-critical if it is critical for some $i \leq k$. We denote the number of $(\leq k)$-critical transpositions in Π by $\chi_{\leq k}(\Pi))$, and use $\mathbf{X}_{\leq k}(n)$ to denote the minimum of $\chi_{\leq k}(\Pi)$ taken over all circular sequences Π on n elements.

Circular sequences can be used to encode any set S of points in general position as follows (see [12]). Let L be a (directed) line that is not orthogonal to any of the lines defined by pairs of points in S. We label the points in S as $p_1, p_2, \ldots, p_n$, according to the order in which their orthogonal projections appear along L. As we rotate L (say counterclockwise), the ordering of the projections changes precisely at the positions where L passes through a position orthogonal to the line defined by some pair of points r, s in S. At the time the projection change occurs, r and s are adjacent in the ordering. and the ordering changes by transposing r and s. By keeping track of all permutations of the projections as L is rotated by $180°$, we obtain a circular sequence Π_S.

The crucial observation is that $(\leq k)$-sets are in one-to-one correspondence with $(\leq k)$-critical transpositions of Π_S.

Observation 2 *Let S be a set of n points in the plane in general position, and let $k < n/2$. Then*

$$\eta_{\leq k}(S) = \chi_{\leq k}(\Pi_S).$$

Combining Theorem 1 and Observation 2 and recalling the definition of $\mathbf{X}_{\leq k}(n)$, one immediately obtains the following statement, obtained independently in [1] and [13].

Theorem 2 ([1] and [13]). *Let S be a set of n points in the plane in general position. Then*

$$\Box(S) = \sum_{1 \leq k < (n-2)/2} (n - 2k - 3)\chi_{\leq k+1}(\Pi_S) + O(n^3)$$

$$\geq \sum_{1 \leq k < (n-2)/2} (n - 2k - 3)\mathbf{X}_{\leq k+1}(n) + O(n^3).$$

Having reduced the problem of bounding $\Box(S)$ to the problem of bounding $\mathbf{X}_{\leq k}(n)$, Ábrego and Fernández-Merchant [1], and independently Lovász, Vesztergombi, Wagner and Welzl [13], then proceeded to the (combinatorial) problem of deriving good estimates for $\mathbf{X}_{\leq k}(n)$.

1.2   Previous Estimates for $\mathbf{X}_{\leq k}(n)$ and Their Consequences

In [1] and [13], the following was proved:

$$\mathbf{X}_{\leq k}(n) \geq 3\binom{k + 1}{2}, \text{ for every positive } n \text{ and every } k < n/2. \qquad (1)$$

In [1], this result is applied together with Theorem 2, to obtain the following.

Theorem 3 (Ábrego and Fernández-Merchant [1]). *If S is any set of n points in general position, then*

$$\Box(S) \geq \frac{1}{4}\left\lfloor\frac{n}{4}\right\rfloor\left\lfloor\frac{n-1}{4}\right\rfloor\left\lfloor\frac{n-2}{4}\right\rfloor\left\lfloor\frac{n-3}{4}\right\rfloor = 0.375\binom{n}{4} + O(n^3). \qquad (2)$$

As a corollary, they obtain $\overline{\mathrm{cr}}(K_n) \geq 0.375\binom{n}{4} + O(n^3)$.

We observe that the bound $\mathbf{X}_{\leq k}(n) \geq 3\binom{k+1}{2}$ is sharp for $k \leq n/3$ (see Example 3 in [13]). Therefore, any improvement on $\Box(S)$ based on the approach of circular sequences must necessarily rely on bounds for $\mathbf{X}_{\leq k}(n)$ that are strictly better than $3\binom{k+1}{2}$ for (some subset of) the interval $n/3 < k < (n-2)/2$. Prior to the present paper, the only such bound reported is the following, which is derived in [13] using a result from [20]:

$$\mathbf{X}_{\leq k}(n) \geq \frac{n^2}{2} - n\sqrt{n^2 - 4k^2} + O(n). \qquad (3)$$

Now (3) is strictly better than (1) for k sufficiently close to $n/2$, namely for $k > k_0(n) := \sqrt{(2\sqrt{13}-5)/9}n \approx 0.4956n + O(\sqrt{n})$. Combining (1) (which is also proved in [13] independently of [1]) and (3), and applying Theorem 2, the following was proved in [13].

Theorem 4 (Lovász, Vesztergombi, Wagner and Welzl [13]). *If S is any set of n points in general position, then*

$$\Box(S) > 0.37501\binom{n}{4} + O(n^3).$$

Again, in view of Observation 1 this immediately yields an improved bound for $\overline{\mathrm{cr}}(K_n)$.

Although numerically the improvement (of roughly $1.088 \cdot 10^{-5}$) given in Theorem 4 over 0.375 may seem marginal, conceptually it is most relevant, since it shows that the rectilinear and the ordinary crossing number of K_n (which considers drawings in which the edges are not necessarily straight segments) are different on the asymptotically relevant term n^4. This last observation follows since there are (non-rectilinear) drawings of K_n with exactly $(1/4)\lfloor n/4\rfloor\lfloor(n-1)/4\rfloor\lfloor(n-2)/4\rfloor\lfloor(n-3)/4\rfloor = 0.375\binom{n}{4} + O(n^3)$ crossings. No better (non-rectilinear) drawings of K_n are known, and consequently the (non-rectilinear) crossing number of K_n has been long conjectured to be exactly $(1/4)\lfloor n/4\rfloor\lfloor(n-1)/4\rfloor\lfloor(n-2)/4\rfloor\lfloor(n-3)/4\rfloor$ (see for instance [10]).

1.3 Our Results: Improved Bound for $\mathbf{X}_{\leq k}(n)$ and Its Consequences

The core of this paper is an improved bound on the minimum number $\mathbf{X}_{\leq k}(n)$ of $(\leq k)$-critical transpositions in any circular sequence on n elements. Our bound is given in terms of two functions $F(k, n)$ and $s(k, n)$ defined as follows.

For all positive integers k, n such that $k < n$, let

$$F(k,n) := \left(2 - \frac{1}{s(k,n)}\right) k^2 - \left(\frac{(s(k,n) - 1)^2}{s(k,n)}\right) k(n - 2k - 1)$$

$$+ \left(\frac{s(k,n)^4 - 7s(k,n)^2 + 12s(k,n) - 6}{12s(k,n)}\right)(n - 2k - 1)^2,$$

where

$$s(k,n) := \left\lfloor \frac{1}{2}\left(1 + \sqrt{\frac{1 + 6\left(\frac{k}{n}\right) - \left(\frac{9}{n}\right)}{1 - 2\left(\frac{k}{n}\right) - \left(\frac{1}{n}\right)}}\right) \right\rfloor.$$

Using this notation, our main result is the following.

Theorem 5 (Main result). *For every positive integer n and every $k < n/2$,*

$$\mathbf{X}_{\leq k}(n) \geq F(k,n) + O(n).$$

This bound is better than the bounds in (1) and (3) for $k > k_1(n) := (1/162)\left(-71 + 71n + \sqrt{19n^2 - 38n + 19}\right) \approx 0.465178n + O(\sqrt{n})$ (see [5]).

The full proof of Theorem 5 is given in [5]. We present a sketch of the general ideas in the proof in Section 2.

By Observation 2, the refined bound for $\mathbf{X}_{\leq k}(n)$ given in Theorem 5 immediately implies improved bounds for $\eta_{\leq k}(S)$, for $k \geq k_1(n)$.

Moreover, in view of Theorem 2, Theorem 5 also gives improved bounds for $\square(S)$, for any set S of n points in general position.

The corresponding calculations (which are somewhat tedious but by no means difficult) are sketched in Section 3, where the following is established.

Proposition 1. *For every positive integer n and every $k < n/2$,*

$$\sum_{1 \leq k < (n-2)/2} (n - 2k - 3) \cdot \max\left\{3\binom{k+2}{2}, F(k+1, n)\right\} \geq 0.37553\binom{n}{4} + O(n^3).$$

By applying Theorem 5 and Proposition 1 to Theorem 2, we obtain the following.

Corollary 1. *If S is a set of n points in the plane in general position, then*

$$\square(S) \geq 0.37553\binom{n}{4} + O(n^3).$$

In view of Observation 1, we also have the following.

Corollary 2. *For each positive integer n,*

$$\overline{\mathrm{cr}}(K_n) \geq 0.37553\binom{n}{4} + O(n^3).$$

To put this improved lower bound on $\overline{\mathrm{cr}}(K_n)$ into context, first we should point out that the lower bounds on $\overline{\mathrm{cr}}(K_n)$ proved in [1] and [13] represent a remarkable improvement over the previous best general lower bounds. Previous to the successful use of the approach of circular sequences (Edelsbrunner et al. [9] also claimed to have proved that $\mathbf{X}_{\leq k}(n) \geq 3\binom{k+1}{2}$, but their argument seems to have a gap), the best lower bound known was $\overline{\mathrm{cr}}(K_n) \geq 0.3288\binom{n}{4}$ [19].

The improved lower bounds on $\overline{\mathrm{cr}}(K_n)$ reported in [1] and [13] are particularly attractive since they are remarkably close to the best upper bound currently known, namely $\overline{\mathrm{cr}}(K_n) \leq 0.3807\binom{n}{4}$ [2]. This bound was obtained using a computer-generated base case. The best known upper bound derived "by hand" (quoting [13]), namely $\overline{\mathrm{cr}}(K_n) \leq 0.3838\binom{n}{4}$, was obtained by Brodsky, Durocher, and Gethner [6].

We also mention that the exact crossing number of K_n is known for $n \leq 16$. For all $n \leq 9$, the exact value of $\overline{\mathrm{cr}}(K_n)$ can be found for instance in [22]. For $n = 10$ it was determined by Brodsky, Durocher, and Gethner [7], for $n = 11$ and 12 it was calculated by Aichholzer, Aurenhammer, and Krasser [2], and quite recently Aichholzer and Krasser determined it for $n = 13, 14, 15, 16$ (private communication). The most current information on the rectilinear crossing number of K_n for specific values of n is given in Aichholzer's comprehensive web page http://www.igi.tugraz.at/oaich/triangulations/crossing.html.

From Corollary 2, the best bounds currently known for $\overline{\mathrm{cr}}(K_n)$ are as follows:

$$0.37553\binom{n}{4} + O(n^3) \leq \overline{\mathrm{cr}}(K_n) \leq 0.3807\binom{n}{4} + O(n^3).$$

1.4   A Brief Discussion on the Main New Results

From our own perspective, the most important contribution of this work is perhaps not the closing of the gap between the lower and upper bounds for $\square(S)$ and $\overline{\mathrm{cr}}(K_n)$, but the evidence that the technique of circular sequences can be further pushed to yield (substantial, we think) improved results. Indeed, by using exclusively circular sequences we could show that the number of $(\leq k)$-sets is strictly greater than $3\binom{k+1}{2}$ for $k \geq k_1 n \approx 0.465n$, thus closing the gap for roughly 20% of the interval for which this was previously unknown. This success gives us hope that even better results can be obtained by alternative approaches within the technique of circular sequences.

2   Bounding the Number of $(\leq k)$-Critical Transpositions: Sketch of Proof of Theorem 5

Our strategy to prove Theorem 5 is as follows. First we show that the number of $(\leq k)$-critical transpositions in *any* circular sequence $\varPi$ on n elements is bounded by below by a function that depends on the solution of a maximization problem over a certain family of digraphs. This is done in Section 2.1 (see Proposition 2). Then, in Section 2.2, we find an upper bound for the solution of the maximization problem over this set of digraphs (see Proposition 5).

We will conclude this section with the (by then obvious) observation that Theorem 5 follows from Propositions 2 and 5.

2.1   Bounding the Number of ($\leq k$)-Critical Transpositions in Terms of the Solution of a Digraph Optimization Problem

Our lower bound for the number of ($\leq k$)-critical transpositions in a circular sequence is given in terms of the maximum of an objective function taken over a certain set of digraphs which we now proceed to define. We use $\overrightarrow{uv}$ to denote the directed edge from vertex u to vertex v. The indegree and outdegree of vertex u in the digraph D are denoted $[u]_D^-$ and $[u]_D^+$, respectively.

Definition. Let k, m be integers such that $2 \leq m < k$. A digraph D with vertex set $\{v_1, v_2, \ldots, v_k\}$ is a (k, m)-*digraph* if it satisfies the following conditions:

   (i) There is some vertex v_i such that $[v_i]_D^- = 0$.
   (ii) For every $i \in \{1, \ldots, k\}$, $[v_i]_D^+ \leq [v_i]_D^- + (m - 1)$.
   (iii) There is a one-to-one *ordering map* $f_D : \{1, 2, \ldots, k\} \to \{1, 2, \ldots, k\}$, such that, for all $i, j \in \{1, 2, \ldots, k\}$, if $\overrightarrow{v_i v_j}$ is in D then $f_D(i) < f_D(j)$.

We let $\mathcal{D}_{k,m}$ denote the set of all (k, m)-digraphs.

The following is one of the core statements of this work. For the sake of brevity, we omit its proof (see [5]).

Proposition 2. *Let Π be any circular sequence on n elements and let $k < n/2$. Define $m := n - 2k$. Then*

$$\chi_{\leq k}(\Pi) \geq 2k^2 + km$$

$$- \max_{D \in \mathcal{D}_{k,m}} \left\{ 2 \sum_{1 \leq i \leq k} [v_i]_D^- + \sum_{1 \leq i \leq k} \min \left\{ [v_i]_D^- - [v_i]_D^+ + (m - 1), m \right\} \right\}.$$

2.2   Bounding the Solution of the Digraph Optimization Problem

The next step is to find a (good) upper bound for the maximization problem in Proposition 2. We achieve this in two steps. First we find a digraph $D_0(k, m)$ in which the maximum is attained, and then we estimate the value of the objective function at $D_0(k, m)$.

Given the nature of the maximization problem in Proposition 2, it is natural to expect that the objective function is maximized in the digraph $D_0(k, m)$ (with vertex set $\{v_1, v_2, \ldots, v_k\}$) in which $[v_i]_{D_0(k,m)}^+$ is maximum possible for each i (subject to the conditions that define $\mathcal{D}_{k,m}$), and in which the $[v_i]_{D_0(k,m)}^+$ directed edges leaving each v_i have endpoints $v_{i+1}, v_{i+2}, \ldots, v_{i+[v_i]_{D_0(k,m)}^+}$ (informally speaking, "there are no gaps"). It can be proved that this is indeed the case, but the proof is long and somewhat technical. For the sake of brevity, we omit the proof of the following statement, and refer the interested reader to [5].

Proposition 3. *The optimal value of the maximization problem in Proposition 2 is attained at the digraph $D_0(k, m)$ with vertex set $\{v_1, v_2, \ldots, v_k\}$ defined as follows:*

(1) $[v_1]^-_{D_0(k,m)} = 0;$

(2) $[v_i]^+_{D_0(k,m)} = \min\{[v_i]^-_{D_0(k,m)} + (m-1), k - i\}$, for every $i \geq 1$; and

(3) For all i, j such that $1 \leq i < j \leq k$, the directed edge $\overrightarrow{v_i v_j}$ is in $D_0(k, m)$ if and only if $i + 1 \leq j \leq i + [v_i]^+_{D_0(k,m)}$. ∎

For the rest of the section, we denote $D_0(k, m)$ simply by D_0.

In view of this and Proposition 2, our next goal is to estimate a bound for

$$2 \sum_{1 \leq i \leq k} [v_i]^-_{D_0(k,m)} + \sum_{1 \leq i \leq k} \min\left\{ [v_i]^-_{D_0(k,m)} - [v_i]^+_{D_0(k,m)} + (m-1), m \right\}.$$

We note that this expression is given in terms of $[v_i]^-_{D_0}$ and $[v_i]^+_{D_0}$. Moreover, in view of the properties of D_0, each $[v_i]^+_{D_0}$ is fully determined by $[v_i]^-_{D_0}$. Thus our first step is to determine (exactly) $[v_i]^-_{D_0}$ for each i. The value of $[v_i]^-_{D_0}$ is given in terms of functions S_m and T_m defined as follows.

For each real number $x \geq 1$, we let $S_m(x)$ denote the (unique) positive integer such that $1 + (S_m(x) - 1)S_m(x)(m-1)/2 \leq x < S_m(x)(S_m(x) + 1)(m-1)/2$. If $i \geq 1$ is an integer, then we let $T_m(i), U_m(i)$ denote the (unique) integers that satisfy $0 \leq T_m(i) \leq m - 2$, $0 \leq U_m(i) \leq S_m(i) - 1$, and such that $i = 1 + (S_m(i) - 1)S_m(i)(m-1)/2 + S_m(i)T_m(i) + U_m(i)$.

The following statement can be proved by induction on i (see [5]).

Proposition 4. For each integer i such that $1 \leq i \leq k$, we have $[v_i]^-_{D_0} = (S_m(i) - 1)(m-1) + T_m(i)$. ∎

Once we have the exact value of $[v_i]^-_{D_0}$ for every i, we then proceed to estimate an upper bound for the objective function in Proposition 2, evaluated at D_0. The arguments and calculations needed to prove this bound are not difficult, but somewhat technical and long. We omit the proof of this statement, and refer once again the interested reader to [5]. The upper bound obtained is the right hand side in the inequality in our next statement. Since the objective function is maximized at D_0, we finally conclude the following.

Proposition 5.

$$\max_{D \in \mathcal{D}_{k,m}} \left\{ 2 \sum_{1 \leq i \leq k} [v_i]^-_D + \sum_{1 \leq i \leq k} \min\left\{ [v_i]^-_D - [v_i]^+_D + (m-1), m \right\} \right\} \leq$$
$$\frac{k^2}{S_m(k)} + \frac{(S_m(k)^2 - S_m(k) + 1)}{S_m(k)}(m-1)k$$
$$- \left(\frac{S_m(k)^4 - 7S_m(k)^2 + 12S_m(k) - 6}{12S_m(k)} \right) (m-1)^2 + O(k),$$

where

$$S_m(k) = \left\lfloor \frac{1 + \sqrt{1 + \dfrac{8(k-1)}{m-1}}}{2} \right\rfloor.$$

∎

2.3 Proof of Theorem 5

We recall that $m = n - 2k$, and so $s(k,n) = S_m(k)$. Therefore Theorem 5 is an immediate consequence of Propositions 2 and 5 (note that we also used the obvious inequality $km \ge k(m-1)$). ∎

3 Proof of Proposition 1

Our first observation is that, for sufficiently large n, $F(k,n) > 3\binom{k+1}{2}$ for every $k > k_1(n)$ (see Appendix in [5]). We also note that if we define

$$\widetilde{s}(x) := \left\lfloor \frac{1}{2}\left(1 + \sqrt{\frac{1+6x}{1-2x}}\right)\right\rfloor,$$

then it is easy to check that $\widetilde{s}(k/n) = s(k,n)$ (and, moreover, $\widetilde{s}(k/n) = s(k+1,n)$) for all but at most $O(\sqrt{n})$ values of k.

These observations imply that

$$\sum_{k=1}^{(n-2)/2-1} (n-2k-3) \cdot \max\left\{3\binom{k+2}{2}, F(k+1,n)\right\}$$

$$\ge 3\sum_{k=1}^{\lfloor k_1(n)\rfloor} (n-2k-3)\binom{k+2}{2} + \sum_{k=\lfloor k_1(n)\rfloor+1}^{(n-2)/2-1} (n-2k-3)F(k+1,n)$$

$$\ge \frac{3}{2}n^3 \cdot \left(\sum_{k=1}^{\lfloor k_1(n)\rfloor}\left(1-2\left(\frac{k}{n}\right)\right)\left(\frac{k}{n}\right)^2\right) +$$

$$n^3 \cdot \left(\sum_{k=\lfloor k_1(n)\rfloor+1}^{(n-2)/2-1}\left(1-2\left(\frac{k}{n}\right)\right)\frac{F(k+1,n)}{n^2}\right) + O(n^3)$$

$$\frac{3}{2}n^4 \cdot \left(\int_0^{c_1}(1-2x)x^2\,dx\right) + n^4 \cdot \left(\int_{c_1}^{1/2}(1-2x)\widetilde{f}(x)\,dx\right) + O(n^3),$$

where $c_1 := 0.465178$ (recall that $k_1(n) \approx 0.465178n + O(\sqrt{n})$), and

$$\widetilde{f}(x) := \left(2 - \frac{1}{\widetilde{s}(x)}\right)x^2 - \left(\frac{(\widetilde{s}(x)-1)^2}{\widetilde{s}(x)}\right)x(1-2x)$$

$$+ \left(\frac{\widetilde{s}(x)^4 - 7\widetilde{s}(x)^2 + 12\widetilde{s}(x) - 6}{12\widetilde{s}(x)}\right)(1-2x)^2.$$

To complete the proof, we note that a numerical evaluation of the integrals in the previous inequality yields

$$\frac{3}{2} \int_0^{c_1} (1 - 2x)x^2 \, dx + \int_{c_1}^{1/2} (1 - 2x)\widetilde{f}(x) \, dx \approx \frac{0.37553}{24}. \qquad \blacksquare$$

References

1. B.M. Ábrego and S. Fernández-Merchant, A lower bound for the rectilinear crossing number, Manuscript (2003).
2. O. Aichholzer, F. Aurenhammer, and H. Krasser, On the crossing number of complete graphs, *Proc. 18th Ann. ACM Symp. Comp. Geom., Barcelona, Spain* (2002), 19–24.
3. N. Alon and E. Győri, The number of small semispaces of a finite set of points in the plane, *J. Combin. Theory Ser. A* **41** (1986), 154–157.
4. A. Andrzejak, B. Aronov, S. Har-Peled, R. Seidel, and E. Welzl, Results on k-sets and j-facets via continuous motion, *Proc. 14th Ann. ACM Sympos. Comput. Geom.* (1998), 192–198.
5. J. Balogh and G. Salazar, On k-sets, convex quadrilaterals, and the rectilinear crossing number of K_n. Manuscript (2004). Submitted.
6. A. Brodsky, S. Durocher, and E. Gethner, Toward the Rectilinear Crossing Number of K_n: New Drawings, Upper Bounds, and Asymptotics, *Discrete Math.* **262** (2003), 59–77.
7. A. Brodsky, S. Durocher, and E. Gethner, The rectilinear crossing number of K_{10} is 62. *Electron. J. Combin.* **8** (2001), Research Paper 23, 30 pp.
8. T. Dey, Improved bounds on planar k-sets and related problems, *Discr. Comput. Geom.* **19** (1998), 373–382.
9. H. Edelsbrunner, N. Hasan, R. Seidel, and X.J. Shen, Circles through two points that always enclose many points, *Geometriae Dedicata* **32** (1989), 1–12.
10. P. Erdős and R. K. Guy, Crossing number problems, *Amer. Math. Monthly* **80** (1973), 52–58.
11. P. Erdős, L. Lovász, A. Simmons, and E. G. Strauss, Dissection graphs of planar point sets, *A Survey of Combinatorial Theory*, North Holland, Amsterdam (1973), 139–149.
12. J. E. Goodman and R. Pollack, On the combinatorial classification of nondegenerate configurations in the plane, *J. Combin. Theory Ser. A* **29** (1980), 220–235.
13. L. Lovász, K. Vesztergombi, U. Wagner, and E. Welzl, Convex Quadrilaterals and k-Sets. Microsoft Research Technical Report MSR-TR-2003-06 (2003).
14. R. E. Pfiefer, The historical development of J. J. Sylvester's problem, *Math. Mag.* **62** (1989), 309–317.
15. J. Spencer and G. Tóth, Crossing numbers of random graphs, *Random Structures and Algorithms* **21** (2003), 347–358.
16. E. R. Scheinerman and H. S. Wilf, The rectilinear crossing number of a complete graph and Sylvester's "four point problem" of geometric probability, *Amer. Math. Monthly* **101** (1994), 939–943.
17. J. J. Sylvester, On a special class of questions on the theory of probabilities, *Birmingham British Assoc.*, Report 35 (1865) 8–9.
18. G. Tóth, Point sets with many k-sets, *Discr. Comput. Geom.* **26** (2001), 187–194.

19. U. Wagner, On the rectilinear crossing number of complete graphs, *Proc. 14th ACM-SIAM Sympos. Discr. Alg.* (2003), 583–588.
20. E. Welzl, More on k-sets of finite sets in the plane, *Discr. Comput. Geom.* **1** (1986), 95–100.
21. E. Welzl, Entering and leaving j-facets, *Discr. Comput. Geom.* **25** (2001), 351–364.
22. A. White and L.W. Beineke, Topological graph theory. In *Selected Topics in Graph Theory* (L.W. Beineke and R.J. Wilson, eds.), pp. 15–49. Academic Press (1978).

On the Realizable Weaving Patterns
of Polynomial Curves in $\mathbb{R}^3$

Saugata Basu[1,*], Raghavan Dhandapani[2], and Richard Pollack[2,**]

[1] School of Mathematics, Georgia Institute of Technology, Atlanta, GA 30332, USA
saugata@math.gatech.edu
[2] Courant Institute of Mathematical Sciences, NYU, New York, NY 10012, USA
raghavan@cs.nyu.edu, pollack@cims.nyu.edu

Abstract. We prove that the number of distinct weaving patterns produced by n semi-algebraic curves in $\mathbb{R}^3$ defined coordinate-wise by polynomials of degrees bounded by some constant d, is bounded by $2^{O(n \log n)}$, where the implied constant in the exponent depends on d. This generalizes a similar bound obtained by Pach, Pollack and Welzl [3] for the case when $d = 1$.

1 Introduction

In [3], Pach, Pollack and Welzl considered weaving patterns of n lines in $\mathbb{R}^3$ and showed that asymptotically only a negligible fraction of possible weaving patterns are realizable by straight lines in $\mathbb{R}^3$ (see Remark 2 below). In this paper, we consider weaving patterns produced by polynomial curves in $\mathbb{R}^3$. Since, such curves are much more flexible than lines, it is reasonable to expect a much bigger number of realizable weaving patterns. In this paper, we prove that the number of distinct weaving patterns, realized by polynomial curves with degrees bounded by some constant d, is still asymptotically negligible.

Crossing patterns of semi-algebraic sets of fixed description complexity were considered in [1], where Ramsey type results are proved for such arrangements. However, since semi-algebraic curves in $\mathbb{R}^3$ (unlike lines) need not satisfy simple above-below relationships and can intertwine in complicated ways, it is not immediately clear whether the framework in [1] is applicable in our setting.

The rest of the paper is organized as follows. In Section 2, we define weaving patterns for polynomial curves and state the main result of the paper (Theorem 1). Since, the projections to the plane of curves defined by polynomials in $\mathbb{R}^3$ can have complicated patterns of intersection, defining what is meant by a weaving pattern for such curves requires some care. In Section 3, we recall some basic facts from [2]. The main tools used in the proof of Theorem 1, are Cylindrical Algebraic Decomposition, and a bound on the number of connected components of the realizations of all realizable sign conditions on a family of polynomials

* Supported in part by NSF Career Award 0133597 and a Sloan Foundation Fellowship.
** Supported in part by NSF grant CCR-0098246.

(see Theorem 2). We give here the basic definitions, and state the results that we need, referring the reader to [2] for details. In Section 4 we prove Theorem 1. Finally, in Section 5 we compare the number of weaving patterns realizable by polynomial curves of fixed degrees with the total number of weaving patterns.

2   Weaving Patterns in $\mathbb{R}^3$

Let $\gamma_1, \ldots, \gamma_n : (-\infty, \infty) \to \mathbb{R}^3$ be n semi-algebraic curves given by

$$\gamma_i(s) = (x_i(s), y_i(s), z_i(s)), \ 1 \leq i \leq n,$$

where x_i, y_i, z_i are polynomials whose degrees are bounded by d. We will assume that the curves are not self-intersecting in $\mathbb{R}^3$ and the images of γ_i and γ_j do not intersect, unless $i = j$.

Let $\pi : \mathbb{R}^3 \to \mathbb{R}^2$ denote the projection sending $(x, y, z) \mapsto (x, y)$. For $1 \leq i < j \leq n$, let

$$M_{ij} = \{\pi(\gamma_i(s_{ij}^1)), \ldots, \pi(\gamma_i(s_{ij}^{\ell_{ij}}))\} \subset \mathbb{R}^2, \ s_{ij}^1 < \cdots < s_{ij}^{\ell_{ij}},$$

denote the finite set of ℓ_{ij} isolated points of intersections of $\pi(\text{image}(\gamma_i))$ and $\pi(\text{image}(\gamma_j))$. Also, let

$$M_{ii} = \{\pi(\gamma_i(s_{ii}^1)), \ldots, \pi(\gamma_i(s_{ii}^{\ell_{ii}}))\} \subset \mathbb{R}^2, \ s_{ii}^1 < \cdots < s_{ii}^{\ell_{ii}},$$

and such that $\pi(\gamma_i(s_{ii}^k)) = \pi(\gamma_i(s)), s \neq s_{ii}^k \Rightarrow s > s_{ii}^k$.

We assume that each of the intersection points correspond to a normal crossing. In particular, for $p \in M_{ij}$ (respectively, $p \in M_{ii}$) $\pi^{-1}(p) \cap (\text{image}(\gamma_i) \cup \text{image}(\gamma_j))$ (respectively, $\pi^{-1}(p) \cap \text{image}(\gamma_i)$) consists of exactly two points. This is not a very strong assumption, since for every finite family of smooth algebraic curves, almost all linear projections, π, satisfy these assumptions. The set of bad projections is a Zariski closed subset in the space of all linear projections.

For $1 \leq i < j \leq n$, and $1 \leq k \leq \ell_{ij}$, we define $V_{ij}^k \in \{+1, -1\}$ in the following way.

$V_{ij}^k = +1$ if $z_i(s_{ij}^k) > z_j(s)$ where $s \in \mathbb{R}$ is such that $\pi(\gamma_j(s)) = \pi(\gamma_i(s_{ij}^k))$,
$\quad\quad = -1$ else.

In other words, V_{ij}^k is $+1$ if image(γ_i) lies *above* image(γ_j) over $\pi(\gamma_i(s_{ij}^k))$, which is a point of intersection of the projections of the images of the two curves, γ_i, γ_j, to the XY-plane.

Similarly, we define for each $1 \leq i \leq n$, and $1 \leq k \leq \ell_{ii}$, $V_{ii}^k \in \{+1, -1\}$ as follows.

$V_{ii}^k = +1$ if $z_i(s_{ii}^k) > z_j(s)$ where $s \neq s_{ii}^k$ is such that $\pi(\gamma_i(s)) = \pi(\gamma_i(s_{ij}^k))$,
$\quad\quad = -1$ else.

Now consider the union of the projections of the images of the curves, namely

$$\pi(\text{image}(\gamma_1)), \ldots, \pi(\text{image}(\gamma_n)),$$

as a planar embedding of a planar graph (self loops allowed), whose vertices are at the points, $M_{ij}^k, 1 \leq i \leq j \leq n, 1 \leq k \leq \ell_{ij}$, and whose edges are the various curve segments joining the vertices. Two such graph embeddings are said to be equivalent, if one can be mapped to the other by a homeomorphism of the plane. Given an ordered set of curves, $\Gamma = \{\gamma_1, \ldots, \gamma_n\}$, satisfying the assumptions stated above, we denote by $G(\Gamma)$ the equivalence class of the corresponding embedded graph in the XY-plane. Finally, we call $G(\Gamma)$ along with the labeling of each of its vertex, M_{ij}^k by $V_{ij}^k \in \{+1, -1\}$, to be the weaving pattern produced by Γ.

In this paper we address the following question. How many distinct weaving patterns can be produced by n algebraic curves, $\gamma_1, \ldots, \gamma_n : (-\infty, \infty) \to \mathbb{R}^3$ where $\gamma_i(s) = (x_i(s), y_i(s), z_i(s))$, and x_i, y_i, z_i are polynomials whose degrees are bounded by some constant d ?

We prove the following theorem.

Theorem 1. *The number of distinct weaving patterns produced by Γ is bounded by $2^{O(n \log n)}$, where the constant in the exponent depends on d.*

This generalizes the bound proved in [3], which is the special case when $d = 1$. Also, note that $\pi(\text{image}(\gamma_1)), \ldots, \pi(\text{image}(\gamma_n))$, can have $\binom{n}{2}d^2$ crossing points and hence the number of possible weaving patterns could be potentially as large as $2^{\binom{n}{2}d^2}$. However, its clear from Theorem 1 only a negligible fraction of these are realizable by curves defined by polynomials with degrees bounded by d.

3   Preliminaries

In this section, we recall a few notions from semi-algebraic geometry that we will need in the proof of Theorem 1. More details, including proofs of the results stated below, can be found in [2].

3.1   Realizable Sign Conditions and Associated Bounds

A *sign condition* is an element of $\{0, 1, -1\}$. We denote for $x \in \mathbb{R}$

$$\begin{cases} \text{sign}(x) = 0 & \text{iff } x = 0, \\ \text{sign}(x) = 1 & \text{iff } x > 0, \\ \text{sign}(x) = -1 & \text{iff } x < 0. \end{cases}$$

Let $\mathcal{Q} \subset \mathbb{R}[X_1, \ldots, X_k]$, A *sign condition* on $\mathcal{Q}$ is an element of $\{0, 1, -1\}^{\mathcal{Q}}$. We say that $\mathcal{Q}$ *realizes* the sign condition σ at $x \in \mathbb{R}^k$ if

$$\bigwedge_{Q \in \mathcal{Q}} \text{sign}(Q(x)) = \sigma(Q).$$

The *realization of the sign condition* σ is

$$\mathcal{R}(\sigma) = \{x \in \mathbb{R}^k \mid \bigwedge_{Q \in \mathcal{Q}} \text{sign}(Q(x)) = \sigma(Q)\}.$$

The sign condition σ is *realizable* if $\mathcal{R}(\sigma)$ is non-empty. The set $\mathrm{Sign}(\mathcal{Q}) \subset \{0, 1, -1\}^{\mathcal{Q}}$ is the set of all realizable sign conditions for $\mathcal{Q}$ over $\mathbb{R}^k$.

For $\sigma \in \mathrm{Sign}(\mathcal{Q})$, let $b_0(\sigma)$ denote the number of connected components of

$$\mathcal{R}(\sigma) = \{x \in \mathbb{R}^k \mid \bigwedge_{Q \in \mathcal{Q}} \mathrm{sign}(Q(x)) = \sigma(Q)\}.$$

Let $b_0(\mathcal{Q}) = \sum_\sigma b_0(\sigma)$. We write $b_0(d, k, s)$ for the maximum of $b_0(\mathcal{Q})$ over all $\mathcal{Q}$, where $\mathcal{Q}$ is a finite subset of $\mathbb{R}[X_1, \ldots, X_k]$ whose elements have degree at most d, $\#(\mathcal{Q}) = s$.

The following theorem [2] gives an upper bound on $b_0(d, k, s)$ which we will use later in the paper.

Theorem 2.

$$b_0(d, k, s) \leq \sum_{1 \leq j \leq k} \binom{s}{j} 4^j d(2d - 1)^{k-1}.$$

3.2 Cylindrical Decomposition

Cylindrical Algebraic Decomposition is a classical tool used in the study of, as well as in algorithms for computing, topological properties of semi-algebraic sets. We give here the basic definitions and properties of Cylindrical Algebraic Decomposition referring the reader to [2] for greater details.

A *cylindrical decomposition* of $\mathbb{R}^k$ is a sequence $\mathcal{S}_1, \ldots, \mathcal{S}_k$ where, for each $1 \leq i \leq k$, $\mathcal{S}_i$ is a finite partition of $\mathbb{R}^i$ into semi-algebraic subsets, called the *cells of level i*, which satisfy the following properties:

Each cell $S \in \mathcal{S}_1$ is either a point or an open interval.
For every $1 \leq i < k$ and every $S \in \mathcal{S}_i$, there are finitely many continuous semi-algebraic functions

$$\xi_{S,1} < \ldots < \xi_{S,\ell_S} : S \longrightarrow \mathbb{R}$$

such that the cylinder $S \times \mathbb{R} \subset \mathbb{R}^{i+1}$ is the disjoint union of cells of $\mathcal{S}_{i+1}$ which are:
either the graph of one of the functions $\xi_{S,j}$, for $j = 1, \ldots, \ell_S$:

$$\{(x', x_{j+1}) \in S \times \mathbb{R} \mid x_{j+1} = \xi_{S,j}(x')\},$$

or a band of the cylinder bounded from below and from above by the graphs of the functions $\xi_{S,j}$ and $\xi_{S,j+1}$, for $j = 0, \ldots, \ell_S$, where we take $\xi_{S,0} = -\infty$ and $\xi_{i,\ell_S+1} = +\infty$:

$$\{(x', x_{j+1}) \in S \times \mathbb{R} \mid \xi_{S,j}(x') < x_{j+1} < \xi_{S,j+1}(x')\}.$$

A *cylindrical decomposition adapted to a finite family of semi-algebraic sets* $T_1, \ldots, T_\ell$ is a cylindrical decomposition of $\mathbb{R}^k$ such that every T_i is a union of cells.

Given a finite set $\mathcal{P}$ of polynomials in $\mathbb{R}[X_1, \ldots, X_k]$, a subset S of $\mathbb{R}^k$ is $\mathcal{P}$-*semi-algebraic* if S is the realization of a quantifier free formula with atoms $P = 0$, $P > 0$ or $P < 0$ with $P \in \mathcal{P}$. A subset S of $\mathbb{R}^k$ is $\mathcal{P}$-*invariant* if every polynomial $P \in \mathcal{P}$ has a constant sign (> 0, < 0, or $= 0$) on S. A *cylindrical decomposition of* $\mathbb{R}^k$ *adapted to* $\mathcal{P}$ is a cylindrical decomposition for which each cell $C \in \mathcal{S}_k$ is $\mathcal{P}$-invariant. It is clear that if S is $\mathcal{P}$-semi-algebraic, a cylindrical decomposition adapted to $\mathcal{P}$ is a cylindrical decomposition adapted to S.

Given a family of polynomials $\mathcal{P} \subset \mathbb{R}[X_1, \ldots, X_k]$, there exists another family of polynomials $\mathrm{Elim}_{X_k}(\mathcal{P})$ (see [2], page 145, for the precise definition of Elim) having the following property.

We denote, for $i = k - 1, \ldots, 1$,

$$C_i(\mathcal{P}) = \mathrm{Elim}_{X_{i+1}}(C_{i+1}(\mathcal{P})),$$

with $C_k(\mathcal{P}) = \mathcal{P}$, so that

$$C_i(\mathcal{P}) \subset \mathbb{R}[X_1, \ldots, X_i].$$

The semi-algebraically connected components of the sign conditions on the family,

$$C(\mathcal{P}) = \cup_{i \leq k} C_i(\mathcal{P})$$

are the cells of a cylindrical decomposition adapted to $\mathcal{P}$. We call $C(\mathcal{P})$ the *cylindrifying family of polynomials associated to* $\mathcal{P}$.

Moreover, if s is a bound on $\#(\mathcal{P})$, and d a bound on the degrees of the elements of $\mathcal{P}$, $\#(\mathrm{Elim}_{X_k}(\mathcal{P}))$ is bounded by $O(s^2 d^3)$. Moreover, the the degrees of the polynomials in $\mathrm{Elim}_{X_k}(\mathcal{P})$ with respect to $X_1, \ldots, X_{k-1}$ is bounded by $2d^2$.

Remark 1. The set $C_i(\mathcal{P})$ has the following additional property. For $\sigma \in \mathrm{sign}(C_i(\mathcal{P}))$ and C a connected component of $\mathcal{R}(\sigma, \mathbb{R}^i)$, for each $x = (x_1, \ldots, x_i) \in C$, the family $\bigcup_{i < j \leq k} C_j(\mathcal{P})(x_1, \ldots, x_i)$ is the cylindrifying family of polynomials associated to $\mathcal{P}(x_1, \ldots, x_i)$, and moreover the induced cylindrical decompositions *have the same structure*. More precisely, this means that there is a 1-1 correspondence between the cylindrical cells as x varies over C.

4   Proof of the Main Result

For $1 \leq i \leq n$, let

$$P_i = \sum_{j=0}^{d} A_{i,j} T_i^j \in \mathbb{R}[\bar{A}_i, T_i],$$

$$Q_i = \sum_{j=0}^{d} B_{i,j} T_i^j \in \mathbb{R}[\bar{B}_i, T_i],$$

$$R_i = \sum_{j=0}^{d} C_{i,j} T_i^j \in \mathbb{R}[\bar{C}_i, T_i],$$

where we denote by $\bar{A}_i$ (respectively, $\bar{B}_i, \bar{C}_i$) the vector of variables, $(A_{i,0}, ..., A_{i,d})$ (respectively, $(B_{i,0}, \ldots, B_{i,d}), (C_{i,0}, \ldots, C_{i,d})$).

Similarly, we denote by $\bar{A}$ (respectively, $\bar{B}, \bar{C}$) the vector of variables,

$$(A_{1,0}, \ldots, A_{1,d}, \ldots, A_{n,0}, \ldots, A_{n,d})$$

(respectively, $(B_{1,0}, ..., B_{1,d}, ..., B_{n,0}, ..., B_{n,d}), (C_{1,0}, ..., C_{1,d}, ..., C_{n,0}, ..., C_{n,d})$).
We denote by γ_i the triple (P_i, Q_i, R_i). For fixed values $(\bar{a}_i, \bar{b}_i, \bar{c}_i)$, the triples $\gamma_i(\bar{a}_i, \bar{b}_i, \bar{c}_i) = (P_i(\bar{a}_i, T_i), Q_i(\bar{b}_i, T_i), R_i(\bar{c}_i, T_i)), 1 \le i \le n$ gives rise to an ordered set of curves in $\mathbb{R}^3$, which we denote by $\Gamma(\bar{a}, \bar{b}, \bar{c})$. Let $WP(\bar{a}, \bar{b}, \bar{c})$ denote the weaving pattern produced by $\Gamma(\bar{a}, \bar{b}, \bar{c})$. We want to bound the cardinality of the set,

$$\{WP(\bar{a}, \bar{b}, \bar{c}) \mid (\bar{a}, \bar{b}, \bar{c}) \in \mathbb{R}^{3(d+1)n}\}.$$

Now, consider the following family of polynomials:

$$\mathcal{A}_i = \{X - P_i(\bar{A}_i, T_i), Y - Q_i(\bar{B}_i, T_i), Z - R_i(\bar{C}_i, T_i)\} \subset \mathbb{R}[\bar{A}_i, \bar{B}_i, \bar{C}_i, X, Y, Z, T_i].$$

Let $\mathcal{B}_i = \text{Elim}_{T_i}(\mathcal{A}_i) \subset \mathbb{R}[\bar{A}_i, \bar{B}_i, \bar{C}_i, X, Y, Z]$, and let

$$\mathcal{B} = \bigcup_{1 \le i \le n} \mathcal{B}_i.$$

Notice, if we specialize $(\bar{A}_i, \bar{B}_i, \bar{C}_i)$ to some $(\bar{a}_i, \bar{b}_i, \bar{c}_i) \in \mathbb{R}^{3(d+1)}$, the image of the curve $\gamma_i(\bar{a}_i, \bar{b}_i, \bar{c}_i) \in \mathbb{R}^3$ is a $\mathcal{B}_i(\bar{a}_i, \bar{b}_i, \bar{c}_i)$-semi-algebraic set.

The following proposition relates the weaving pattern, $WP(\bar{a}, \bar{b}, \bar{c})$ to a cylindrical decomposition of R^3 adapted to the family $\mathcal{B}(\bar{a}, \bar{b}, \bar{c})$.

Proposition 1. *Let $(\bar{a}, \bar{b}, \bar{c}) \in \mathbb{R}^{3(d+1)n}$. The weaving pattern, $WP(\bar{a}, \bar{b}, \bar{c})$ is determined by the cylindrical decomposition induced by the cylindrifying family of polynomials associated to $\mathcal{B}(\bar{a}, \bar{b}, \bar{c})$.*

In particular, if two points $(\bar{a}, \bar{b}, \bar{c}), (\bar{a}', \bar{b}', \bar{c}') \in \mathbb{R}^{3(d+1)n}$, are such that the cylindrical decompositions induced by the cylindrifying families of polynomials associated to $\mathcal{B}(\bar{a}, \bar{b}, \bar{c})$ and $\mathcal{B}(\bar{a}', \bar{b}', \bar{c}')$ have the same structure, then $WP(\bar{a}, \bar{b}, \bar{c}) = WP(\bar{a}', \bar{b}', \bar{c}')$.

Proof. The proposition is a consequence of the definition of weaving pattern, the definition of cylindrifying families of polynomials, and the fact that the images of the curves, $\gamma_i(\bar{a}_i, \bar{b}_i, \bar{c}_i)$, are all $\mathcal{B}(\bar{a}, \bar{b}, \bar{c})$-semi-algebraic sets.

We now prove Theorem 1.

Proof. Let $\mathcal{C}_1 = \text{Elim}_Z(\mathcal{B}), \mathcal{C}_2 = \text{Elim}_Y(\mathcal{C}_1)$, and $\mathcal{C}_3 = \text{Elim}_X(\mathcal{C}_2)$.

The set $\mathcal{C}_3$ has the following property which is a consequence of Remark 1 in Section 3. Let C be a connected component of the realization of a realizable sign condition of $\mathcal{C}_3$. Then, for each $(\bar{a}, \bar{b}, \bar{c}) \in C$, $\mathcal{B}(\bar{a}, \bar{b}, \bar{c}) \cup \mathcal{C}_1(\bar{a}, \bar{b}, \bar{c}) \cup \mathcal{C}_2(\bar{a}, \bar{b}, \bar{c})$ is the cylindrifying family of polynomials associated to $\mathcal{B}(\bar{a}, \bar{b}, \bar{c})$ and moreover the cylindrical decompositions induced have the same structure as $(\bar{a}, \bar{b}, \bar{c})$ varies over C.

Since, by Proposition 1, for fixed $(\bar{a}, \bar{b}, \bar{c}) \in \mathbb{R}^{3(d+1)n}$ the weaving pattern of $\Gamma(\bar{a}, \bar{b}, \bar{c})$ is determined by any Cylindrical Decomposition of $\mathbb{R}^3$ adapted to $\mathcal{B}(\bar{a}, \bar{b}, \bar{c})$, then by the previous observation, the number of distinct weaving patterns is clearly bounded by $b_0(\mathcal{C}_3)$, which we now proceed to bound from above.

From the bounds stated in Section 3, we have that for $1 \leq i \leq n$, $\#(\mathcal{B}_i) = O(d^3)$ and the degrees of the polynomials in $\mathcal{B}_i$ are bounded by $O(d^2)$. Hence, $\#(\mathcal{B}) = O(nd^3)$. Since, $\mathcal{C}_3$ is obtained from $\mathcal{B}$ after three successive Elim operations, we get that, $\#\mathcal{C}_3 = (nd)^{O(1)}$ and the degrees of the polynomials in $\mathcal{C}_3$ is bounded by $d^{O(1)}$). The number of variables in the polynomials in $\mathcal{C}_3$ is $3(d+1)n$. Using the bound in Theorem 2, we get that $b_0(\mathcal{C}_2)$ is bounded by

$$(nd)^{O(dn)} = 2^{O(n \log n)}.$$

5   Most Weaving Patterns Are Not Realizable

We have the following theorem which generalizes Theorem 3 in [3].

Theorem 3. *The number of weaving patterns realizable by polynomial curves of degrees bounded by a constant, divided by the total number of weaving patterns of n curves whose projections are allowed to intersect at most a constant number of times, tends to 0 exponentially fast, as $n \to \infty$.*

Proof. By Theorem 1, the number of distinct weaving patterns produced by such curves is bounded by $2^{O(n \log n)}$. On the other hand, considering n lines in the plane in general position, and counting all possible ways of labeling the $\binom{n}{2}$ crossings, we see that there are at least $2^{\binom{n}{2}}$ possible weaving patterns.

Remark 2. The proof of the upper bound in Theorem 3 in [3] does not seem to consider the fact that the projections of different sets of n lines in $\mathbb{R}^3$ to the plane, can produce arrangements which are combinatorially distinct, and these would produce distinct weaving patterns by definition. In fact, obtaining good control on this number complicates the proof of Theorem 1 in this paper. However, since the number of combinatorially distinct arrangements of n lines in $\mathbb{R}^2$ is still bounded by $2^{O(n \log n)}$, the proof of the theorem in [3] is still valid.

References

1. N. ALON, J. PACH, R. PINCHASI, R. RADOICIC, M. SHARIR, *Crossing Patterns of Semi-algebraic Sets*, Preprint.
2. S. BASU, R. POLLACK, M.-F. ROY, *Algorithms in Real Algebraic Geometry*, Springer-Verlag, 2003.
3. J. PACH, R. POLLACK, E. WELZL, *Weaving Patterns of Lines and Line Segments in Space*, Algorithmica, 9:561-571, 1993.

Drawing the AS Graph in 2.5 Dimensions*

Michael Baur[1], Ulrik Brandes[2], Marco Gaertler[1], and Dorothea Wagner[1]

[1] University of Karlsruhe, Department of Computer Science,
76128 Karlsruhe, Germany
{baur,gaertler,dwagner}@ilkd.uni-karlsruhe.de
[2] University of Konstanz, Department of Computer & Information Science,
78457 Konstanz, Germany
Ulrik.Brandes@uni-konstanz.de

Abstract. We propose a method for drawing AS graph data using 2.5D graph visualization. In order to bring out the pure graph structure of the AS graph we consider its core hierarchy. The k-cores are represented by 2D layouts whose interdependence for increasing k is displayed by the third dimension. For the core with maximum value a spectral layout is chosen thus emphasizing on the most important part of the AS graph. The lower cores are added iteratively by force-based methods. In contrast to alternative approaches to visualize AS graph data, our method illustrates the entire AS graph structure. Moreover, it is generic with regard to the hierarchy displayed by the third dimension.

1 Introduction

Current research activities in computer science and physics are aiming at understanding the dynamic evolution of large and complex networks like the physical internet, World Wide Web, peer-to-peer systems and the relation between autonomous systems (AS). The design of adequate visualization methods for such networks is an important step towards this aim. As these graphs are on one hand large or even huge, on the other hand evolving, customized visualizations concentrating on their intrinsic structural characteristics are required.

In this paper we propose a layout method that brings out the pure structure of an autonomous systems (AS) graph. More precisely, we focus on the core hierarchy of AS graphs. A 2D layout is obtained by first choosing a spectral layout to display the core with maximum value and then adding the lower cores iteratively by force-based methods. Using 2.5D graph visualization, we then represent the core hierarchy by stacking the induced 2D layouts of the k-cores for increasing k on top of each other in the third dimension. Visualizations in 2.5D have been proposed frequently for network data, for example to display other graph hierarchies [6, 9] or evolving graphs over time [4].

A few samples of visualizations of AS graphs are already available. However, they either focus on the geographic location of the AS [8], on the routing structure seen from a selected AS [2, 7] or on a high level view created by clustering

* The authors gratefully acknowledge financial support from DFG under grant WA 654/13-2 and BR 2158/1-2, and from the European Commission within FET Open Projects COSIN (IST-2001-33555) and DELIS (contract no. 001907).

the nodes [13]. In contrast, our method displays the entire AS graph structure without using external information. Previous attempts to analyze the structure of the AS graph propose the existence of meaningful central nodes that are highly connected to a large fraction of the graph [11]. It seems that this structural peculiarity is interpreted very well by the notion of k-cores [14, 1]. This concept is already rudimentary used for initial cleaning in [12]. Accordingly, our approach is based on the hierarchical core decomposition of the AS graph. Moreover, other kind of hierarchies can be used instead.

We consider AS graphs from different dates between 2001 and 2003 to demonstrate the usefulness of our method as means for analyzing the relation between ASes. Also graphs obtained by the Internet Topology Generator INET 3.0 [15] are consulted.

The new 2.5D visualization method for AS graphs is explained in Section 2. In Section 3 we present and discuss the results obtained for various AS graph data sets and Section 4 gives the conclusions.

2   Layout Method

Layout Paradigm. We assume a hierarchical decomposition based on the k-core concept. The k-core of a graph is defined as the unique subgraph obtained by recursively removing all nodes of degree less than k. A node has *coreness* ℓ, if it belongs to the ℓ-core but not to the $(\ell + 1)$-core. The ℓ-core layer is the collection of all nodes having coreness ℓ. The *core* of a graph is the k-core such that the $(k + 1)$-core is empty. In general, the core decomposition can result in disconnected parts. For the AS graph, all k-cores stay connected which is an advantage of the core hierarchy.

However, abstraction to the levels of hierarchy is normally accompanied by a loss of information that should be avoided. Therefore, we establish the following layout paradigm: First, all nodes and edges are displayed, second, the levels of hierarchy are emphasized, and third, the inter- and intra-level connections are made clear.

We propose an incremental algorithm to produce a 2D layout satisfying our layout paradigm. This layout is afterwards transformed into 2.5D in a canonical way using the core hierarchy. First a generic method to generate a 2D layout of a hierarchical decomposition of the graph is introduced, followed by the specification of parameters that can be chosen to fulfill certain requirements and requests induced by the structure of AS graphs.

Generic Algorithm. The first step of the algorithm constructs a spectral layout for the highest level of the hierarchy. Then, iteratively, the lower levels are added using a combination of barycentric and force-directed placement. Algorithm 1 gives a formal description of this procedure based on the core hierarchy.

Preliminary studies indicate that a spectral placement does not lead to a satisfactory layout of the AS graph as a whole. However, the results improve for increasing core value. We therefore choose a spectral layout as initial placement for the core of the graph. Then, for the iterative addition of the other level of

---

Algorithm 1: Generic AS layout algorithm.

Input: graph $G = (V, E)$

let $k \leftarrow$ maximum coreness, $G_l \leftarrow$ the l-core, $C_l \leftarrow l$-core layer

calculate spectral layout for G_k

for $l \leftarrow k - 1, \ldots, 1$ **do**

    if $C_l \neq \emptyset$ **then**

        calculate barycentric layout for C_l in G_l, keeping G_{l+1} fixed

        calculate force-directed layout for C_l in G_l, keeping G_{l+1} fixed

        calculate force-directed layout for G_l

---

hierarchy, we first calculate a barycentric placement in which all new nodes are placed in the barycenter of their neighbors in this level. Unfortunately, barycentric layouts also have a number of drawbacks. Firstly, nodes that are structurally equivalent in the current subgraph are assigned to the same position. Secondly, all nodes are placed inside the convex hull of the already positioned nodes. In particular this means that the outermost placed nodes are those having highest coreness which is clearly contradictory to the intuition of importance. To overcome these difficulties, we use the barycentric layout as an initial placement for a subsequent force-directed refinement step, where only newly added nodes are displaced. In addition, a force-directed approach is applied for all nodes in order to relax the whole graph layout. However, the number of iterations and the maximal movement of the nodes is carefully restricted not to destroy the previously computed layout. A special feature of this relaxation step is the use of non-uniform natural spring lengths $l(u, v)$, where $l(u, v)$ scales with the smaller core value of the two incident nodes u and v. Thus, the effect of a barycentric layout is modeled, since edges between nodes of high coreness are longer than edges between nodes of low coreness. Accordingly, these springs prevent nodes with high coreness from drifting into the center of the layout.

Fitting the Parameters. Beside the choice of the hierarchical decomposition, the algorithm offers a few more degrees of freedom that allow an adjustment to a broad range of applications. Our choice of parameters are originated from the core structure of the AS graph. For the spectral layout we propose a modified Laplacian matrix $L' = 1/4 \cdot D - A$ [5]. Our experiments showed that the normalized adjacency matrix results in comparably good layouts while the standard Laplacian matrix performs significantly worse.

The force-directed placement is computed by a variant of the algorithm from [10]. Unlike the original algorithm, we calculate the displacement only for one vertex at a time and update its position immediately. Furthermore, we use the original forces but with non-uniform natural edge lengths $l(u, v)$ proportional to $\min\{\text{level}(u), \text{level}(v)\}^2$. For the local refinement step we perform at most 50 iterations and for the global roughly 20 iterations.

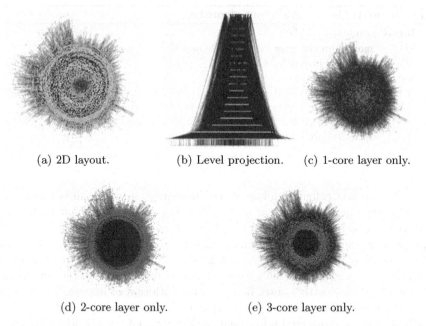

(a) 2D layout.   (b) Level projection.   (c) 1-core layer only.

(d) 2-core layer only.   (e) 3-core layer only.

Fig. 1. 2D layout and level projection of the AS graph (06/01/02).

3   Results

We illustrate the results of our method for real AS data sets as well as for
generated graphs. For a more detailed discussion, we also refer to [3]. The section
is concluded by techniques to aid the human perception.

Our real world data consist of three AS graphs collected by the Oregon
Routeview Project (http://www.routeviews.org) on different dates, i.e June,
1st 2001 (11,211 nodes, 23,689 edges, 19 levels), June, 1st 2002 (13,315 nodes,
27,703 edges, 20 levels), and June, 1st 2003 (15,415 nodes, 34,716 edges, 25
levels). In addition, we used INET 3.0 to generate artifical graphs that should
exhibit a similar topology. We discuss two different two-dimensional types of
figures, the 2D layout produced by Algorithm 1 and the projection of the 2.5D
layout into one of the full dimensions, also referred to as *level projection*. Nodes
are represented by ellipses of size decreasing according to the coreness and with
colors fading from black to white. Edges are always drawn as straight lines.

Real AS Graph. The 2D layouts are dominated by the nodes with small coreness leading to a huge periphery (Fig. 1(a)). On the other hand, most nodes with
higher coreness are contained in the convex hull of the core, which is apparent
in Figure 1(b) and documents the relation between importance and coreness. A
closer examination reveals three almost separated radial areas around the center.
The first one mainly contains the 3-core layer, while the 2-core layer forms the
second and third area that are distinguished by their density (see Fig. 1(c)–1(e)).

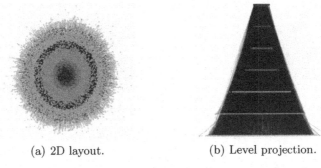

(a) 2D layout.   (b) Level projection.

Fig. 2. Layouts of the generated graph with 11,211 nodes.

This reflects the heterogenous importance distribution within these areas. In contrast, a large part of the 1-core layer is attracted to the central region. These properties can be observed for all three instances. The well-known growth of the AS graph affects especially the 2- and 3-core layers. We observe that the spatial distances of these two layers decreases over time.

Generated Graphs. There are significant differences of the generated graphs to the real AS graphs, e.g. in the number of edges (35,300 vs. 23,700) and core levels (8 vs. 19). An obvious difference of the generated graph is the more uniform distribution of cardinalities of the core layers (Fig 2). Accordingly, the separation of the different core layers is less visible in the layout.

Supporting Perception. There are several means for visual aid in 2.5D layouts, i.e. choice of perspective (in 3D), additional geometric objects emphasizing the levels of hierarchy, and colors. The choice of perspective is very powerful. We have already used this feature when presenting only the 2D layout and the level projection respectively. More general, a user can focus on individual aspects, i.e. a global oriented view, a hierarchical version, or a mixture of both. A beneficial consequence might be that unintended information is automatically masked out by the perspective. In order to simplify navigation in the three dimensional space, one can also introduce additional objects that mark the levels of hierarchy, i.e. rectangles, discs, or planes. Transparency or filters might even increase their effectiveness. Color can be used in various ways, to highlight nodes and edges of special interest, to code the levels of hierarchy, or to improve the overall perception. We used transparent rectangles that absorbed light to draw layers and colored the nodes accordingly to their coreness. The color of the edges are determined by a linear interpolation of their incident nodes' color (see Fig. 3).

4   Conclusion

Core based 2.5D visualizations of the AS graph support the recognition of its detailed hierarchy. Especially, it emphasizes the characteristics of the lower core

48   Michael Baur et al.

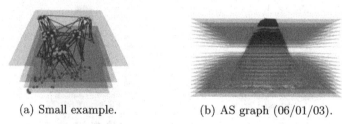

(a) Small example.   (b) AS graph (06/01/03).

Fig. 3. Visual support features.

layers and their connections with the highest layers. The evolution of the AS graph has an observable effect on the layout. Also there is a significant difference in the layouts of real AS graphs and generated ones.

References

1. V. Batagelj and M. Zaveršnik. Generalized cores. Preprint 799, University of Ljubljana, 2002.
2. G. Di Battista, F. Mariani, M. Patrignani, and M. Pizzonia. BGPlay: A System for Visualizing the Interdomain Routing Evolution. In *Proc. of Graph Drawing, GD'03*, volume 2912 of *Springer LNCS*, pages 295–306. Springer, 2004.
3. M. Baur, U. Brandes, M. Gaertler, and D. Wagner. Drawing the AS Graph in Two and a Half Dimensions. TR 2004-12, Informatics, University Karlsruhe, 2004.
4. U. Brandes and S. Corman. Visual unrolling of network evolution and the analysis of dynamic discourse. *Information Visualization*, 2(1):40–50, 2003.
5. U. Brandes and S. Cornelsen. Visual ranking of link structures. *Journal of Graph Algorithms and Applications*, 7(2):181–201, 2003.
6. U. Brandes, T. Dwyer, and F. Schreiber. Visual understanding of metabolic pathways across organisms using layout in two and a half dimensions. *Journal of Integrative Bioinformatics*, 0002, 2004.
7. CAIDA. Walrus – graph visualization tool, 2002.
8. CAIDA. Visualizing internet topology at a macroscopic scale, 2003.
9. P. Eades and Q. Feng. Multilevel visualization of clustered graphs. In *Proc. of Graph Drawing*, volume 1190 of *Springer LNCS*, pages 113–128. Springer, 1996.
10. T. Fruchtermann and E. Reingold. Graph drawing by force-directed placement. *Software – Practice and Experience*, 21(11):1129–1164, 1991.
11. M. Gaertler and M. Patrignani. Dynamic analysis of the autonomous system graph. In *IPS 2004 – Inter-Domain Performance and Simulation*, 2004.
12. C. Gkantsidis, M. Mihail, and E. Zegura. Spectral analysis of internet topologies. In *IEEE Infocom 2003*, 2003.
13. G. Sagie and A. Wool. A clustering approach for exploring the internet structure. In *Proc. 23rd IEEE Conv. of Electrical and Electronics Engineers in Israel*, 2004.
14. S. B. Seidman. Network structure and minimum degree. *Social Networks*, 5:269–287, 1983.
15. J. Winick and S. Jamin. Inet-3.0: Internet topology generator. Technical Report UM-CSE-TR-456-02, EECS, University of Michigan, 2002.

Boundary Labeling: Models and Efficient Algorithms for Rectangular Maps*

Michael A. Bekos[1], Michael Kaufmann[2],
Antonios Symvonis[1], and Alexander Wolff[3]

[1] National Technical University of Athens, Dept. of Mathematics,
Athens, Greece
{mikebekos,symvonis}@math.ntua.gr
[2] University of Tübingen, Institute for Informatics, Sand 13,
72076 Tübingen, Germany
mk@informatik.uni-tuebingen.de
[3] Faculty of Informatics, Karlsruhe University, P.O. Box 6980,
76128 Karlsruhe, Germany
http://i11www.ira.uka.de/people/awolff

Abstract. In this paper, we present *boundary labeling*, a new approach for labeling point sets with large labels. We first place disjoint labels around an axis-parallel rectangle that contains the points. Then we connect each label to its point such that no two connections intersect. Such an approach is common e.g. in technical drawings and medical atlases, but so far the problem has not been studied in the literature. The new problem is interesting in that it is a mixture of a label-placement and a graph-drawing problem.

1 Introduction

Label placement is one of the key tasks in the process of information visualization. In diagrams, maps, technical or graph drawings, features like points, lines, and polygons must be labeled to convey information. The interest in algorithms that automate this task has increased with the advance in type-setting technology and the amount of information to be visualized. Due to the computational complexity of the label-placement problem, which is NP-hard in general [5], cartographers, graph drawers, and computational geometers have suggested numerous approaches, such as expert systems, zero-one integer programming, approximation algorithms, simulated annealing, and force-driven algorithms to name only a few. An extensive bibliography about label placement can be found at [14]. The ACM Computational Geometry Impact Task Force report [3] denotes label placement as an important research area.

In this paper, we deal with labeling dense point sets with large labels. This is common e.g. in medical atlases where certain features of a drawing or photo are explained by blocks of text that are arranged around the drawing. Our model

* This work has partially been supported by the DFG grants Ka 512/8-2 and WO 758/4-1, by the German-Greek cooperation program GRC 01/048 and the EPEAEK program Pythagoras 89181(28).

is as follows: we assume that we are given a set $P = \{p_1, \ldots, p_n\}$ of points and an axis-parallel rectangle R that contains P. Each point, or *site*, p_i is associated with an axis-parallel rectangular open label. The labels have to be placed and connected to their corresponding sites by polygonal lines, so-called *leaders*, such that (a) no two labels intersect, (b) no two leaders intersect, and (c) the labels lie outside R but touch R. We investigate various constraints concerning the location of the labels and the type of leaders. More specifically we either allow to attach labels to one, two or all four sides of R, and we either use straight-line or rectilinear leaders. We propose efficient algorithms that find *some* non-intersecting leader-label placement, but we also consider two natural objectives: minimize the total length of the leaders and, if leaders are not straight lines, minimize the total number of bends over all leaders.

These new problems are combinations of label-placement and graph-drawing problems. Due to the complexity of either step there are still very few publications that combine graph drawing and label placement. Klau and Mutzel [11] have coined the term "graph labeling" for this discipline and have given a mixed-integer program for computing orthogonal graph layouts with node labels.

Leaders have so far only been used by Zoraster [15], Freeman et al. [6], and Fekete and Plaisant [4]. Zoraster [15] uses simulated annealing to label points and lines in seismic survey maps, while Freeman et al. [6] use an iterative raster-based method to determine positions for area labels in soil survey maps. Fekete and Plaisant [4] extend the *infotip paradigm* to cope with labeling dense point sets interactively. They draw a circle of fixed radius around the current cursor position, the so-called *focus circle*, and label only the sites that fall into the circle. Labels are left-aligned and placed in two stacks to the left and the right of the circle. To connect sites to their labels, Fekete and Plaisant use non-crossing leaders that consist of two or three line segments: one segment goes radially from the site to its projection on the focus circle and one or two axis-aligned segments go from there to the corresponding label.

Iturriaga and Lubiw [10] give an $O(n^4)$-time decision algorithm for attaching *elastic* labels to n points on the perimeter of a rectangle. An elastic label models a block of text of fixed area, but varying width and height. Iturriaga and Lubiw place their labels *inside* the rectangle. Iturriaga [9] also briefly investigates the inverse problem, where elastic labels must be attached to the sites *outside* the given rectangle R. She presents an algorithm that finds a label placement that uses the minimum-width strip around R. If n sites are given in order around R, her algorithm takes $O(n)$ time.

This paper is structured as follows. In Section 2 we model and define our problem. In Section 3 we are concerned with rectilinear leaders. We investigate algorithms for non-intersecting leader-label placement, for leader-bend and leader-length minimization. In Section 4 we consider straight-line leaders. For the one-side and the four-side case, we compute legal leader-label placements and we minimize (with a slower algorithm) the total leader length. We have implemented some of our algorithms. In Section 5 we give an example layout. A full version of this paper with more examples and proofs is available at [2].

2   Defining and Modeling the Problem

We consider the following problem. Given an axis-parallel rectangle $R = [l_R, r_R] \times [b_R, t_R]$ of width $W = r_R - l_R$ and height $H = t_R - b_R$, and a set $P \subset R$ of n points $p_i = (x_i, y_i)$, each associated with an axis-parallel rectangular open label l_i of width w_i and height h_i, our task is to find a *legal* or an *optimal* leader-label placement. Our criteria for a legal leader-label placement are the following:

1. Labels have to be disjoint.
2. Labels have to lie outside R but touch the boundary of R.
3. Leader c_i connects point p_i with label l_i for $1 \leq i \leq n$.
4. Intersections of leaders with other leaders, points or labels are not allowed.
5. The ports where leaders touch labels may be prescribed (the center of a label edge, say) or may be arbitrary.

In this paper we present algorithms that compute legal leader-label placements for various types of leaders defined below, but we also approach optimal placements according to the following two objective functions:

- short leaders (minimum total length) and
- simple leader layout (minimum number of bends).

These criteria have been adopted from the area of graph drawing since leaders do not play a significant role in the label-placement literature. We will evaluate the two criteria under two models for drawing leaders. In the first model we require that each leader is rectilinear, i.e. a connected sequence of orthogonal line segments. In the second model each leader is drawn straight-line. Clearly, minimizing the number of bends does not make sense for straight-line leaders.

A rectilinear leader consists of a sequence of axis-parallel segments that connects a site with its label. These segments are either parallel (p) or orthogonal (o) to the side of the bounding rectangle R to which the label is attached. This notation yields a classification scheme for rectilinear leaders: let a *type* be an alternating string over the alphabet $\{p, o\}$. Then a leader of type $t = t_1 \ldots t_k$ consists of an x- and y-monotone connected sequence $(e_1, \ldots, e_k)$ of segments from site to label, where each segment e_i has the direction that the letter t_i prescribes. In this paper we focus on leaders of the types opo and po, see Figures 1 and 2, respectively. We consider type-o leaders to be of type opo and of type po as well. We refer to straight-line leaders as type-s leaders.

In this paper we assume that input points are in general position, i.e. no three points lie on a line and no two points have the same x- or y-coordinate.

We start with a negative result. Assume that not all label heights are equal, that labels must be attached either to the right or left side of the rectangle R, and that the heights sum up to twice the height of R. Clearly the task of assigning the labels to the two sides corresponds to the well-known problem PARTITION, which is weakly NP-complete [7]. Due to this observation we first make some simplifying assumptions like uniform labels and then generalize our algorithms by adding more requirements.

3   Rectilinear Leaders

In this section we investigate different ways of drawing rectilinear leaders. We present algorithms for legal leader-label placement, leader-bend and leader-length minimization. We consider attaching labels to one, two, and four sides of the rectangle R and connecting sites to their labels with leaders of type *opo* and *po*, see Figures 1 and 2, respectively.

3.1   Leader-Bend Minimization

One-Side Labeling with Type-*opo* Leaders. We first consider the problem of attaching labels to one, say the right, side of the rectangle R. We assume that the sum of the label heights is at most the height of R and that the sites are sorted according to non-decreasing y-coordinate. If we use a slightly wider rectangle R' and leaders of type *opo*, then we can attach labels to the right side of R' and place non-crossing leaders in R' as follows. We first stack the labels on top of each other such that the lower left corner of l_1 is incident to the lower right corner of R'. Then we connect each site p_i by a horizontal segment $y_i \times [x_i, r_R]$ to the right side of R. Finally we use the gap between the right sides of R and R' to lay out the remaining parts of the leaders from the right side of R to the, say, midpoints of the left label edges, see Figure 1. This can be done with at most two bends per leader and without any crossing, since the vertical orders of sites and labels are identical and since we assume that no two sites have the same y-coordinate. Thus a legal one-side type-*opo* leader-label placement can be computed in $O(n \log n)$ time.

Clearly this approach is not optimal in terms of the total number of leader bends. Given the restriction to leaders of type *opo* and the trick with the extra space at the right side of R, routing the leaders is easy, and the remaining problem is a one-dimensional label-placement problem. There has been work on similar problems where labels are not restricted to a constant number of positions, but can slide. Our problem is new in that labels do not necessarily have to contain the point they label, but even if they do not (and thus contribute to the objective function in a negative way), they must be placed within an interval whose length is restricted (by the height of R).

Theorem 1. *A legal one-side type-opo leader-label placement with the minimum number of bends can be computed in $O(n^2)$ time and space.*

Proof. We use dynamic programming with a table T of size $n \times (n + 1)$. For $k \leq i$ the entry $T[i, k]$ will contain the minimum y-coordinate that is needed to accommodate the first i labels such that at least k of them use horizontal leaders. If it is impossible to connect k out of the first i labels with horizontal leaders, we set $T[i, k]$ to ∞. As usual, the table entries are computed bottom-up.

To compute a new entry $T[i, k]$, observe that there are only three interesting positions of the label l_i: (a) directly on top of l_{i-1} using a horizontal leader, (b) directly on top of l_{i-1} using a 2-bend leader, and (c) such that the top edge of l_i lies on the horizontal line through the i-th site. These cases and the case $T[i, k] = \infty$ can be distinguished in constant time. Thus T can be computed

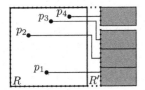
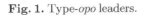

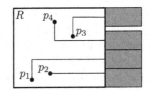

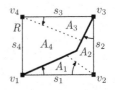

Fig. 1. Type-*opo* leaders.   **Fig. 2.** Type-*po* leaders.   **Fig. 3.** Partition into monotone regions.

in $O(n^2)$ time. Given T, the number of horizontal leaders is the largest k that fulfills $T[n,k] \leq t_R$. By using an extra table of the same size as T, label and leader positions of an optimal solution can be computed as well. □

3.2 Legal Leader-Label Placement

Four-Side Labeling with Type-*opo* Leaders. Our approach for attaching labels to all sides of the rectangle R is very simple. We partition R into four disjoint regions such that the algorithms from the previous subsection can be applied to each region separately. Points that lie on boundaries of our partition in the interior of R can be connected to a side of R via both incident regions. Thus we ignore the problem of how to distribute these boundaries.

We have two requirements for a region A in the partition of R: (a) A must be adjacent to a specific side s_A of R and (b) each point in A must see the point with the same x- or y-coordinate on s_A. Requirement (b) is a consequence of using type-*opo* leaders. If we manage to find a partition of R into four regions such that each region A contains the side s_A of R and A is monotone in the direction of s_A then obviously both requirements are fulfilled.

To avoid an NP-hard partition problem as discussed in Section 2 we assume that we know how many labels have to be attached to which side of R. To simplify the presentation, we assume uniform square labels. Let $n_1, \ldots, n_4$ be the number of labels that have to be attached to the respective sides and let $n = n_1 + \cdots + n_4$. We want to partition R into four regions $A_1, \ldots, A_4$ as described above, such that $|A_i \cap P| = n_i$ for $i = 1, \ldots, 4$. We do this by rotating rays around the rectangle corners until these conditions are fulfilled, see Figure 3.

The rotations can be implemented by sorting the sites according to the angles they enclose with the horizontal or vertical lines through the appropriate corners of R. Using the $O(n \log n)$-time algorithm of the previous subsection we have the following result:

Lemma 1. *Given a rectangle R of sufficient size, a set $P \subset R$ of n points in general position, square uniform labels, one per point, and numbers $n_1, \ldots, n_4$ that express how many labels are to be attached to which side of R, there is an $O(n \log n)$-time algorithm that attaches the labels to R and connects them to the corresponding points with non-intersecting <u>type-opo</u> leaders.*

One-Side Labeling with Type-*po* Leaders. In this subsection we describe how to compute a legal labeling with leaders of type-*po*, see Figure 2. We restrict ourselves to attaching labels to one side s of R. W.l.o.g., we assume that s is the right vertical side of R, and that the sites $p_1, \ldots, p_n$ are sorted according to increasing y-coordinate. We consider uniform labels. Since we do not attempt to minimize the number of bends, we simply stack labels to the right of s in the same vertical order as the corresponding sites.

Our algorithm is very simple: we go through the sites from bottom to top. Assume we have already placed non-intersecting leaders for the first $i - 1$ sites. Then we connect p_i to l_i by a leader c_i of type po, i.e. by a vertical segment (possibly of length zero) followed by a horizontal segment. Clearly c_i can be routed such that c_i does not contain any sites except p_i. Now we go through the sites $p_1, \ldots, p_{i-1}$ from right to left and test their leaders for intersection with c_i. Let p_j be the rightmost site p_j whose leader c_j intersects c_i. Then we reroute as in Figure 4: we connect p_j to l_i and p_i to l_j. We observe that the new leader c'_j of p_j does not intersect any other leader and that the new leader c'_i of p_i can only intersect leaders of sites to the left of p_j. For placing the leader of p_i we have to reroute at most $i - 1$ times, and after this process of rerouting no two leaders intersect any more. Thus we have:

Theorem 2. *A legal one-side type-po leader-label placement can be computed in $O(n^2)$ time given uniform labels.*

3.3   Leader-Length Minimization

In the remainder of this section we focus on obtaining label placements of minimum total leader length. We attach labels to the left and the right side of the rectangle R, and we treat uniform and non-uniform labels.

Type-*opo* Leaders and Uniform Labels. Labels are placed on opposite sides of the rectangle, say s_{left} and s_{right}, $n/2$ labels on each side. The labels are assumed to be uniform in the sense that they all are of identical height. The $n/2$ labels are of maximum height, covering the full length of the side of the rectangle they reside, and hence their position at each side is determined. We are given points $p_1, \ldots, p_n$ that have to be connected with leaders to labels on s_{left} and s_{right} so that the total leader length is minimized.

We consider type-*opo* leaders. Here we ignore the subproblem of routing. This can be done as for the one-side label placement in Section 3.1. Again we assume the existence of a slightly wider rectangle R'. The i-th point p which is assigned to s_{left} is connected to the i-th label of s_{left} with a type-*opo* leader. Since the location of each label is determined (and fixed) the length of the leader to the i-th label of s_{left} is defined. Call it $Left[p, i]$. We define $Right[p, i]$ analogously.

Theorem 3. *Given a rectangle R with $n/2$ uniform labels of maximum height on its left and on its right side, and a set $P \subset R$ of n points in general position, there is an $O(n^2)$-time algorithm that connects all points to their labels with non-intersecting type-opo leaders such that the total leader length is minimum.*

Proof. To compute a label placement of minimum total leader length, we use a dynamic programming algorithm. We assume that n is even. The case that n is odd is slightly more involved, see [2]. The algorithm constructs a table

$T[0 : n/2, 0 : n/2]$. Entry $T[l, r]$ contains the minimum total leader length for the $l + r$ lowest points where l of them have labels on s_{left} and r on s_{right}. It is easy to prove by induction that $T[l, r]$ satisfies the following recurrence relation for $l, r \le n/2$:

$$T[0, 0] = 0 \tag{1}$$
$$T[0, r] = T[0, r - 1] + Right[p_r, r] \tag{2}$$
$$T[l, 0] = T[l - 1, 0] + Left[p_l, l] \tag{3}$$
$$T[l, r] = \min\{ T[l, r - 1] + Right[p_{l+r}, r], \ T[l - 1, r] + Left[p_{l+r}, l] \} \tag{4}$$

Having computed table T, entry $T[n/2, n/2]$ corresponds to a label placement of minimum total leader length. The actual placement can be easily recovered by maintaining an additional table. The running time is obvious.   □

Type-*po* Leaders and Uniform Labels. Our next result also deals with two-side placement of uniform labels.

Theorem 4. *Given a rectangle R with $n/2$ uniform labels of maximum height on each of its left and right side, and a set $P \subset R$ of n points in general position, there is an $O(n^2)$-time algorithm that attaches each point to a label with non-intersecting type-po leaders such that the total leader length is minimum.*

Proof. We use the dynamic-programming algorithm of Theorem 3 for the case of type-*opo* leaders to get the label placement. It runs in $O(n^2)$ time. Observe that connecting a site to its label (at a fixed port) with a type-*opo* or a type-*po* leader requires the same leader length, namely, the Manhattan distance of site and port. So after obtaining the label placement (for type-*opo* leaders) we use type-*po* leaders routed in the way described in Section 3.2. Possible crossings of leaders to the same side are resolved as in Section 3.2 without changing the total length, while crossings of leaders that go to opposites sides cannot occur. This is due to the fact that swapping labels between a pair of points with crossing leaders would result in a solution with smaller total leader length.

Four-Side Labeling with Type-*opo* Leaders. We give a polynomial-time algorithm which finds type-*opo* leaders of minimum total length when the labels can be placed on all four sides of the boundary of the rectangle. We only assume that the labels have uniform size, the positions of the labels are disjoint, and the label ports are predefined. We have the following planarity result:

Lemma 2. *For any one-side solution of type-opo leaders with crossings there exists a crossing-free one-side solution of type-opo leaders which does not have a larger total leader length.*

Now we can use Vaidya's algorithm [13] for minimum-cost bipartite matching for points in the plane under the Manhattan metric. It runs in $O(n^2 \log^3 n)$ time and finds a matching between sites and ports that minimizes the total Manhattan distance of the matched pairs.

Theorem 5. *A crossing-free four-side solution of type-opo leaders with minimum total length can be computed in polynomial time.*

Proof. Assume now that the solution of the minimum-weight matching implies a crossing between two leaders. Clearly this crossing is between two segments inside of the rectangle. Replacing the crossing by an appropriate "knock-knee" [12] gives two leaders which might not be of type-*opo*. Rerouting the leaders in type-*opo* shape does not increase the leader lengths, and applying Lemma 2 to each of the two affected sides of the rectangle will provide a new solution of type-*opo* with at most the same total leader length. An argument similar to that used for the crossing resolution for type-*po* leaders shows that the process of crossing resolution terminates in polynomial time.   □

Type-*opo* Leaders and Non-uniform Labels. We focus on two-side label placement of type-*opo* leaders. We are given n points $p_i = (x_i, y_i), i = 1, 2, \ldots, n$, each associated with a label l_i of height h_i which can be placed on either the left side (s_{left}) or the right side (s_{right}) of rectangle R. Observe that the height of rectangle R must be large enough to accommodate the labels. In the event that the height of rectangle R is equal to half the sum of the label heights, managing to place the labels accounts to solving the partition problem. So, we cannot expect an algorithm that runs in polynomial time only to the number of points. Instead we get an algorithm that runs in polynomial time to the height of rectangle R, which can be considered to be the equivalent of the pseudo-polynomial solution to the partition problem.

Here we again ignore the routing of the type-*opo* leaders and assume the existence of a slightly wider rectangle R'.

Theorem 6. *Given a rectangle R of height H, a set $P \subset R$ of n points in general position where point p_i is associated with label l_i of height h_i, there is an $O(nH^2)$-time algorithm that places the labels to the sides of the rectangle and attaches the corresponding points with non-intersecting <u>type-opo</u> leaders such that the total leader length is minimum.*

Proof. We say that label l is *placed at height h* if its bottom edge has y-coordinate h. Assume that the i-th point p_i is connected to s_{left} and its label l_i is placed at height y then the length of the edge from p_i to l_i leftward is defined. Call it $Left[p_i, y]$. Similarly, we define $Right[p_i, y]$.

We denote by $T[i, y_L, y_R]$ the total length of the type-*opo* leaders of the i lowest points, where the left side of the rectangle is occupied up to y_L and the right side is occupied up to y_R. By $T_L[i, y_L, y_R]$ we denote the total leader length for the case where the i-th point has its label on the left side, the left side of the rectangle is occupied up to y_L (including label l_i) and the right side is occupied up to y_R. Similarly we define $T_R[i, y_L, y_R]$. Then, by induction we can show that the following recurrence relations hold (we omit the boundary conditions):

$$T[i, y_L, y_R] = \min\{T_L[i, y_L, y_R], T_R[i, y_L, y_R]\} \tag{5}$$

$$T_L[i, y_L + h_i, y_R] = T[i-1, y_L, y_R] + Left[p_i, y_L] \tag{6}$$

$$T_R[i, y_L, y_R + h_i] = T[i-1, y_L, y_R] + Right[p_i, y_R] \tag{7}$$

Based on them, we can compute table T by dynamic programming. After this computation, the minimum table entry of the form $T[n, a, b]$, where $0 < a, b \leq H$, gives the minimum total leader length. We can recover the label placement which realizes the computed total leader length by maintaining an additional table with dimensions equal to those of T. The dynamic programming algorithm will use $O(nH^2)$ time and space. □

4   Straight-Line Leaders

In this section we investigate straight-line or type-s leaders, i.e. we allow skewed lines but forbid bends. We first give a simple algorithm that computes a legal one-side labeling. Then we show how this algorithm can be improved either in terms of runtime or in terms of total leader length. Finally we sketch how it can be applied to four-side labeling.

One-Side Labeling. We adopt the scenario of Section 3.1. Let R be the bounding rectangle. We want to attach labels to the right side of R. We assume that labels are uniform and that their heights add up to the height of R. We also assume that the port m_i where the leader is connected to its label l_i is fixed, say m_i is in the middle of the left label edge. Thus the only task is to assign ports to points such that no two leaders intersect.

Let $M = \{m_1, \ldots, m_n\}$ be the ports sorted by y-coordinate from bottom to top. For $i = 1, \ldots, n$ we assign to m_i the first unlabeled point $p \in P$ that is hit by a ray r_i that emanates from m_i and is rotated around m_i in clockwise order. Initially r_i is pointing vertically downwards. The proof of correctness is trivial.

Clearly the algorithm can be implemented in $O(n^2)$ time, but we can do better. Let CH be the convex hull of $P \cup M$. Note that CH has an edge between the lowest port m_1 and the first point p reached by the rotating ray r_1. This edge is the first leader. Removing p and m_1 from CH yields the next leader and so on. Using a semi-dynamic convex-hull data structure [8] yields a total running time of $O(n \log n)$. This algorithm is correct since it mimics the slow one.

To compute an assignment that is minimum in terms of total leader length we proceed as described just before Theorem 5, except now we use *Euclidean* minimum-cost bipartite matching for the sets of sites and ports. This takes $O(n^{2+\delta})$ time [1], where $\delta > 0$ can be chosen arbitrarily small. For type-s leaders length minimization automatically ensures planarity. Thus we have:

Theorem 7. *A legal one-side type-s leader-label placement can be computed in* $O(n \log n)$ *time. Minimizing total leader length takes* $O(n^{2+\delta})$ *time for any* $\delta > 0$.

Four-Side Labeling. In this subsection, we partition the rectangle into convex polygons, such that the sites in each polygon can be connected to the labels on the boundary of the polygon using the one-side routing algorithm of the previous subsection. We assume uniform labels. Note that the only assumption we used about the relative position of sets P and M of sites and ports, respectively, was that M is contained in an edge of the convex hull of $P \cup M$. To make the one-side routing algorithm work, the convex polygons must be chosen such that they contain exactly as many sites as there are labels on their boundary. We construct in $O(n \log n)$ time eight polygons with this property by rotating ℓ, moving ℓ_{top} and ℓ_{bot}, and rotating ℓ_1 to ℓ_4 as indicated in Figure 5.

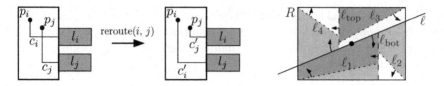

Fig. 4. Rerouting of crossing leaders.   **Fig. 5.** Partition for straight-line leaders.

As in the one-side case Euclidean minimum-cost bipartite matching yields a placement of minimum total leader-length. Thus we conclude:

Theorem 8. *A legal four-side type-s leader-label placement can be computed in $O(n \log n)$ time. Minimizing total leader length takes $O(n^{2+\delta})$ time for any $\delta > 0$.*

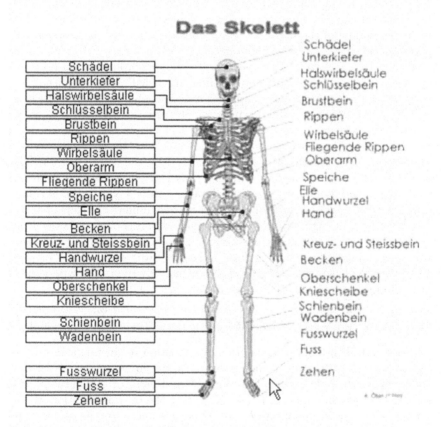

Fig. 6. A medical map with original labels and leaders (left) as well as labels and type-*opo* leaders computed by our algorithm (right). Drawing from http://www.vobs.at/bio/a-phys/pdf/a-skelett-a.jpg.

5 Examples

We have implemented some of the presented algorithms, but due to space constraints we can give only one example here. Figure 6 depicts a relatively small medical map of a skeleton. The original labels and leaders are on the right side of the drawing. We have mirrored the sites at the vertical line through the spine and have applied our algorithm for type-*opo* leaders such that labels were placed to the left of the drawing. For more examples, see [2].

References

1. P. K. Agarwal, A. Efrat, and M. Sharir. Vertical decomposition of shallow levels in 3-dimensional arrangements and its applications. In *Proc. 11th ACM Symp. Comp. Geom. (SoCG'95)*, pages 39–50, 1995.
2. M. A. Bekos, M. Kaufmann, A. Symvonis, and A. Wolff. Boundary labeling: Models and efficient algorithms for rectangular maps. Technical Report 2004-15, Fakultät für Informatik, Universität Karlsruhe, 2004. Available at `http://www.ubka.uni-karlsruhe.de/cgi-bin/psview?document=/ira/2004/15`.
3. B. Chazelle and 36 co-authors. The computational geometry impact task force report. In B. Chazelle, J. E. Goodman, and R. Pollack, editors, *Advances in Discrete and Computational Geometry*, vol. 223, pp. 407–463. AMS, 1999.
4. J.-D. Fekete and C. Plaisant. Excentric labeling: Dynamic neighborhood labeling for data visualization. In *Proc. Conference on Human Factors in Computer Systems (CHI'99)*, pages 512–519, 1999. ACM New York.
5. M. Formann and F. Wagner. A packing problem with applications to lettering of maps. In *Proc. 7th ACM Symp. Comp. Geom. (SoCG'91)*, pages 281–288, 1991.
6. H. Freeman, S. Marrinan, and H. Chitalia. Automated labeling of soil survey maps. In *Proc. ASPRS-ACSM Annual Convention, Baltimore*, vol. 1, pp. 51–59, 1996.
7. M. R. Garey and D. S. Johnson. *Computers and Intractability: A Guide to the Theory of NP-Completeness*. W. H. Freeman, New York, NY, 1979.
8. J. Hershberger and S. Suri. Applications of a semi-dynamic convex hull algorithm. *BIT*, 32:249–267, 1992.
9. C. Iturriaga. *Map Labeling Problems*. PhD thesis, University of Waterloo, 1999.
10. C. Iturriaga and A. Lubiw. Elastic labels around the perimeter of a map. *Journal of Algorithms*, 47(1):14–39, 2003.
11. G. W. Klau and P. Mutzel. Automatic layout and labelling of state diagrams. In W. Jäger and H.-J. Krebs, editors, *Mathematics – Key Technology for the Future*, pages 584–608. Springer-Verlag, Berlin, 2003.
12. T. Lengauer. *Combinatorial Algorithms for Integrated Circuit Layout*. B. G. Teubner, 1990.
13. P. M. Vaidya. Geometry helps in matching. *SIAM J. Comput.*, 18:1201–1225, 1989.
14. A. Wolff and T. Strijk. The Map-Labeling Bibliography. `http://i11www.ira.uka.de/map-labeling/bibliography/`, 1996.
15. S. Zoraster. Practical results using simulated annealing for point feature label placement. *Cartography and GIS*, 24(4):228–238, 1997.

Convex Drawings of 3-Connected Plane Graphs

Extended Abstract

Nicolas Bonichon[1], Stefan Felsner[2], and Mohamed Mosbah[1]

[1] LaBRI , Université Bordeaux-1,
351, cours de la Libération, 33405 Talence Cedex, France
{bonichon,mosbah}@labri.fr
[2] Technische Universität Berlin, Institut für Mathematik, MA 6-1,
Strasse des 17. Juni 136, 10623 Berlin, Germany
felsner@math.tu-berlin.de

Abstract. We use Schnyder woods of 3-connected planar graphs to produce convex straight line drawings on a grid of size $(n-2-\Delta)\times(n-2-\Delta)$. The parameter $\Delta \geq 0$ depends on the the Schnyder wood used for the drawing. This parameter is in the range $0 \leq \Delta \leq \frac{n}{2} - 2$.

1 Introduction

We investigate crossing-free straight-line drawings of planar graphs with the restriction that the vertices of the graph have to be located at integer grid points. The aim is to keep the area of an axis-aligned rectangle which covers the complete drawing as small as possible. It is known that a square of side-length $n - 2$, i.e., a $(n - 2) \times (n - 2)$ grid is enough to host every planar graph.

A drawing with the property that the boundary of every face (including the outer face) is a convex polygon is called a convex drawing. Convex drawings exist for every 3-connected planar graph. Again the aim is to keep the area of such a drawing as small as possible.

It is important to distinguish between convex drawings and strictly convex drawings. A drawing is strictly convex if every interior angle is less than 180° and every outer angle greater than 180°. In this paper we deal with convex drawings. The grid size for strictly convex drawings was recently studied by Rote [1], he proves that an $O(n^{7/3}) \times O(n^{7/3})$ grid is enough for strictly convex drawings of planar graphs with n vertices.

The question whether every planar graph has a straight line embedding on a grid of polynomial size was raised by Rosenstiehl and Tarjan [2]. Unaware of the problem Schnyder [3] constructed a barycentric representation which immediately translates to an embedding on the $(2n - 6) \times (2n - 6)$ grid. The first explicit answer to the question was given by de Fraysseix, Pach and Pollack [4, 5]. They construct straight line embeddings on an $(2n - 4) \times (n - 2)$ grid and show that the embedding can be computed in $O(n \log n)$. De Fraysseix et al. also observed a lower bound of $(\frac{2}{3}n - 1) \times (\frac{2}{3}n - 1)$ for grid embeddings of the n vertex graph containing a nested sequence of $n/3$ triangles. It is conjectured that this is the worst case, i.e., that every planar graph can be embedded on the

J. Pach (Ed.): GD 2004, LNCS 3383, pp. 60–70, 2004.
© Springer-Verlag Berlin Heidelberg 2004

$(\frac{2}{3}n-1) \times (\frac{2}{3}n-1)$ grid. 4-connected planar graphs with at least four vertices on the outer face can be drawn even more compactly. Work of He [6] and Miura et al. [7] shows that these graphs can be embedded on the $\frac{n}{2} \times \frac{n}{2}$ grid.

In his second paper Schnyder proves the existence of an embedding on the $(n-2) \times (n-2)$ grid which can be computed in $O(n)$ time. In general Schnyder's result from [8] is still unbeaten. Lately, Zhang and He [9] used the minimum Schnyder wood of a triangulation to prove a bound of $(n-1-\Delta^{\frown}) \times (n-1-\Delta^{\frown})$, where $\Delta^{\frown}$ is the number of cyclic faces in the minimum Schnyder wood.

Though it is implicitly contained in Steinitz's characterization of 3-connected planar graphs as the skeleton graphs of 3-dimensional polytopes the existence of convex drawings for these graphs is known as Tutte's theorem. The idea for Tutte's proof [10, 11] is known as *spring-embedding*. Technically the embedding is obtained as solution to a system of linear equations. Kant [12] has extended the approach of de Fraysseix et al. to construct convex drawings on the $(2n-4) \times (n-2)$ grid. The grid size was reduced to $(n-2) \times (n-2)$ by Chrobak and Kant [13]. Schnyder and Trotter [14] have worked on ideas of using Schnyder woods for convex grid embeddings. The basic approach was independently worked out by di Battista et al. [15] and Felsner [16] this results in convex grid drawings on the $(f-1) \times (f-1)$ grid, where f is the number of faces of the drawing. In this paper this basic algorithm is used but the size of the required grid is reduced by some new ideas. Loosely speaking some edges are eliminated which results in the reduction of f. This can be done until at most $n-\Delta$ faces remain. The eliminated edges can be reinserted in the resulting drawing on the $(n-1-\Delta) \times (n-1-\Delta)$ grid, with $\Delta \geq 0$. $\Delta \geq n-f$. The drawing procedure can be implemented to run in linear time. The algorithm has been implemented and integrated in PIGALE library[1].

In the next section we introduce Schnyder woods. It is shown how to use Schnyder woods to obtain convex drawings of 3-connected planar maps. The lattice of Schnyder woods is discussed and a new operation called *merge* is introduced as a tool for transforming Schnyder woods and their underlying graphs.

Section 3 contains the generic drawing algorithm. It is shown that this algorithm produces convex drawings and the size of the grid required for the drawing is analyzed. The main ingredient of this analysis is a bound on the number of merges applicable to a Schnyder wood. In particular it is shown that starting with the Schnyder wood of a triangulation a sequence of $n-4+\Delta^{\frown}-\Delta^{\oplus}$ merge operations is admissible.

Section 4 presents a technique to decrease of the side-length of the grid by one. This small reduction, however, is crucial to match Schnyder's $(n-2) \times (n-2)$ bound for planar triangulations.

2   Schnyder Woods

Schnyder defined special colorings and orientations of the inner edges of a triangulation. In [3] and [8] he applied these Schnyder woods to characterize planar

---
[1] http://pigale.sourceforge.net

graphs and to draw planar graphs on small grid sizes. Here we describe a generalization of Schnyder woods for 3-connected planar graphs. Such generalizations have been obtained in [15] and [16], in our exposition we follow [17].

A *planar map* M is a simple planar graph G together with a fixed planar embedding of G in the plane. A *suspension* M^σ of M is obtained as follows: Three different vertices from the outer face of M are specified and named a_1, a_2, a_3 in clockwise order. (For ease of visualization we identify the indices $1, 2, 3$ with colors red, green, blue). At each of the three special vertices a_i, called *suspension vertices*, a half-edge reaching into the outer face is attached.

Let M^σ be a suspension of a planar map. A *Schnyder wood* is an orientation and coloring of the edges of M^σ with the colors $1, 2, 3$ satisfying the following rules.

(W1) Every edge e is oriented by one or two opposite directions. The directions of edges are colored such that if e is bi-directed the two directions have distinct colors.

(W2) The half-edge at a_i is directed outwards and colored i.

(W3) Every vertex v has outdegree one in each color. The edges e_1, e_2, e_3 leaving v in colors 1,2,3 occur in clockwise order. Each edge entering v in color i enters v in the clockwise sector from e_{i+1} to e_{i-1}. See Figure 1.

(W4) There is no interior face whose boundary is a directed cycle in one color.

Fact 1 *There is a Schnyder wood for M^σ, if M^σ is the suspension M^σ of a 3-connected planar map. Actually, a Schnyder wood for M^σ exists under the weaker condition that the graph obtained by adding a new vertex v_∞ as the second endpoint for the three half-edges is planar and 3-connected.*

Given a Schnyder wood, let T_i be the set of edges colored i with the direction they have in this color. Since every inner vertex has outdegree one in T_i every v is the starting vertex of a unique i-path $P_i(v)$ in T_i.

Fact 2 *The digraph T_i is acyclic, even more, T_i is a tree with root a_i.*

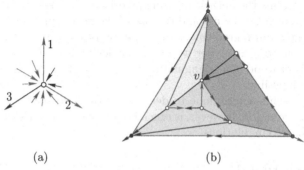

(a)   (b)

Fig. 1. (a) Edge colorings[2] and orientations at a vertex. (b) A Schnyder wood and the regions of vertex v.

---

[2] If you can't see the colors look up the colorful electronic versions at the authors homepages.

2.1   Convex Drawings via Face-Counting

Schnyder and Trotter [14] had some ideas of using Schnyder woods for convex grid embeddings. The approach has been worked out in [15] and [16]. We describe the technique omitting some details.

From the vertex condition (W3) it follows that for $i \neq j$ the paths $P_i(v)$ and $P_j(v)$ have v as the only common vertex. Therefore, $P_1(v), P_2(v), P_3(v)$ divide M into three regions $R_1(v)$, $R_2(v)$ and $R_3(v)$, where $R_i(v)$ denotes the region bounded by and including the two paths $P_{i-1}(v)$ and $P_{i+1}(v)$, see Fig. 1.

Fact 3 (a) $R_i(u) \subseteq R_i(v)$ iff $u \in R_i(v)$.
(b) $R_i(u) = R_i(v)$ iff there is a path of bicolored edges in colors $i - 1$ and $i + 1$ connecting u and v.
(c) For all u, v there are i and j with $R_i(u) \subset R_i(v)$ and $R_j(v) \subset R_j(u)$.

The *face-count* of a vertex v is the vector (v_1, v_2, v_3), where v_i is the number of faces of M contained in region $R_i(v)$.

Fact 4 For every edge $\{u, w\}$ and vertex $v \neq u, w$ there is a color i with $\{u, w\} \in R_i(v)$, hence, $u_i \leq v_i$ and $w_i \leq v_i$.

Inclusion properties of the three regions of adjacent vertices imply:

Fact 5 (a) If edge (u, v) is uni-directed in color i, then
$$u_i < v_i, \ u_{i-1} > v_{i-1} \text{ and } u_{i+1} > v_{i+1}.$$
(b) If (u, v) is directed in color $i - 1$ and (v, u) in color $i + 1$, then
$$u_i = v_i, \ u_{i-1} > v_{i-1} \text{ and } u_{i+1} < v_{i+1}.$$

Clearly, each vertex v has $v_1 + v_2 + v_3 = f - 1$, where f is the number of faces of M. Hence, we have a mapping of the vertices of the graph to the plane $T_f = \{(x_1, x_2, x_3) : x_1 + x_2 + x_3 = f - 1\}$ in $\mathbb{R}^3$. Connecting the points corresponding to adjacent vertices by the line segment between them yields a drawing $\mu(M)$ of M in the plane T_f.

Color and orientation of edges are nicely encoded in this drawing: Let v be a vertex with $\mu(v) = (v_1, v_2, v_3)$. The three lines $x_1 = v_1$, $x_2 = v_2$ and $x_3 = v_3$ partition the plane T_f into six wedges with apex $\mu(v)$. By Fact 5 the color and orientation of edges incident to v is determined by the wedge containing them, see Figure 2. In particular the bicolored edges are the edges supported by the lines defining the wedges.

Theorem 1. *The drawing $\mu(M)$ is a convex drawing of M in T_f. Dropping the third coordinate yields a convex drawing of M on the $(f - 1) \times (f - 1)$ grid.*

2.2   The Lattice of Schnyder Woods

In general the suspension M^σ of a 3-connected planar map will admit many Schnyder woods. Felsner [18] has shown that the set of all Schnyder woods of a given M^σ has the structure of a distributive lattice. As we will make use of some elements of this theory we recall some definitions and the main results.

Think of the three half-edges of M^σ as noncrossing infinite rays. These rays partition the outer face of M into three parts. The *suspension dual* M^{σ^*} of M^σ

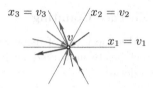

Fig. 2. Wedges and edges at a vertex v in the plane T_f.

is the dual of this map. Thus M^{σ^*} has a triangle b_1, b_2, b_3 corresponding to the unbounded face of M.

The *completion* $\widetilde{M^\sigma}$ of a plane suspension M^σ and its dual M^{σ^*} is obtained as follows: Superimpose M^σ and M^{σ^*} so that exactly the primal dual pairs of edges cross (the half edge at a_i has a crossing with the dual edge $\{b_j, b_k\}$, for $\{i, j, k\} = \{1, 2, 3\}$). At each crossing place a new vertex such that this new *edge vertex* is subdividing the two crossing edges.

The completion $\widetilde{M^\sigma}$ is planar, every edge-vertex has degree four and there are six half-edges reaching into the unbounded face.

A *3-orientation* of the completion $\widetilde{M^\sigma}$ of M^σ is an orientation of the edges of $\widetilde{M^\sigma}$ such that:

(O1) outdeg$(v) = 3$ for all primal- and dual-vertices v.
(O2) indeg$(v_e) = 3$ for all edge-vertices v_e (hence, outdeg$(v_e) = 1$).
(O3) All half-edges are out-edges of their vertex.

Theorem 2. *Let M^σ be a suspension of a 3-connected plane graph M. The following structures are in bijection: Schnyder woods of M^σ, Schnyder woods of the suspension dual M^{σ^*} and 3-orientations of the completion $\widetilde{M^\sigma}$.*

The bijections are illustrated in Figure 3.

The lattice structure of Schnyder woods is best understood by looking at 3-orientations: Let X be a 3-orientation and let C be directed cycle of a X. Reverting the orientation of all edges of C yields another 3-orientation X^C. If C is a simple cycle it has a connected interior an we can speak of the clockwise and the counterclockwise order of C. Define $X \succ X^C$ if C is a clockwise cycle in X. The transitive closure $\succ^*$ of this relation is an order relation on the set of 3-orientations.

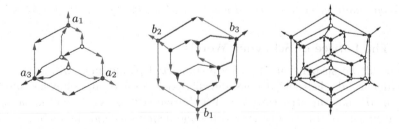

Fig. 3. The bijections for Theorem 2.

Theorem 3. *The relation $\succ^*$ is the order relation of a distributive lattice on the set of 3-orientations of the completion $\widehat{M^\sigma}$ of a suspension M^σ of a 3-connected planar map. The unique minimum 3-orientation contains no clockwise directed cycles.*

In view of Theorem 2 a suspension M^σ has unique minimum Schnyder wood S_{Min}. Figure 4 shows two sub-structures which are impossible in S_{Min}: an uni-directed edge incoming at v in color $i + 1$ such that the counterclockwise next edge is bi-directed, outgoing at v in color $i - 1$ and incoming in color i ; and a clockwise triangle of uni-directed edges, such a triangle must have colors i, $i+1$, $i + 2$ in this clockwise order.

Theorem 4. *([19]) Let G be a 3-connected plane graph. The minimal Schnyder Wood S_{Min} of G can be computed in linear time.*

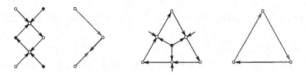

Fig. 4. Two types of clockwise cycles in 3-orientations and the corresponding substructures of Schnyder woods.

2.3   Merging and Splitting

The operations *merge* and *split* introduced in this section operate on Schnyder woods and the underlying graph. Merge and split can be seen as inverse operations, corresponding to the deletion and insertion of an edge.

Given a Schnyder wood, a *knee at vertex v* is a pair of uni-directed edges adjacent at an angle of v such that one of the edges is incoming and the other outgoing at v. Knees come in two kinds, if the in-edge of the knee is the clockwise neighbor of the out-edge at v we speak of a a *cw-knee*, otherwise, if the in-edge of the knee is the counterclockwise neighbor of the out-edge it is a *ccw-knee*.

Let (u, v), (v, w) be a knee at v. Suppose that the color of (v, w) is i by the vertex condition the color of (u, v) is $i - 1$ if it is a cw-knee and $i + 1$ if it is a ccw-knee. The *merge of the knee* consist of the deletion of the out-edge (v, w) while making (u, v) a bi-directed edge outgoing at v in color i and incoming in the same color as before. Depending on the type of the knee we distinguish between clockwise and counterclockwise merge operations. Figure 5 illustrates the definition.

Lemma 1. *Let S be a Schnyder wood, the coloring and orientation of edges after merging a knee is again a Schnyder wood.*

A *split of a bi-directed edge* is the inverse operation of a merge. In the context of this paper we only need one very specific type of split. The *short cw-split* is the inverse of a cw-merge with the additional property that (u, w) is an edge, i.e., u, v, w is a triangle.

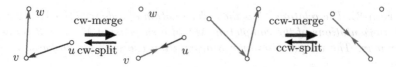

Fig. 5. Clockwise and counterclockwise merge and split.

3   The Drawing Algorithm

Let M be a 3-connected planar map with n vertices and f faces. The steps of the drawing algorithm with input M are the following:

(A1) Choose three vertices from the outer face for the suspension M^σ.
(A2) Compute the minimum Schnyder wood S_{Min} for M^σ and let $S_0 = S_{\text{Min}}$.
(A3) Compute a maximal cw-merge sequence $S_0 \to S_1 \to \ldots S_k$ of Schnyder woods, i.e., S_{i+1} is obtained from S_i by a cw-merge and S_k contains no cw-knee.
(A4) Use face-counting to draw S_k on the $(f - k - 1) \times (f - k - 1)$ grid.
(A5) Reinsert all edges which have been deleted by merge operations into the drawing from the previous step.

With Figure 6 we illustrate step A3 of the algorithm.

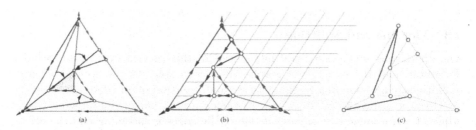

Fig. 6. (a) A Schnyder wood S_0, cw-knees are indicated by arcs. (b) The final Schnyder wood of a merge sequence. (c) The example graph with $n = f = 9$ drawn on the 6×6 grid.

3.1   The Drawing Is Convex

Theorem 5. *Reinserting all the edges which have been deleted by a sequence of cw-merge operations into the drawing of S_k obtained in A4 keeps the drawing planar and convex.*

The drawing steps of the algorithm (A4 and A5) are illustrated in Figure 6. Essential for the proof of the theorem is the following lemma:

Lemma 2. *Given a Schnyder wood of a suspended map M^σ and let F be an interior face. The orientation and color of edges around F obeys the following rule (see Figure 7):*

Fig. 7. The generic structure of a face as described by Lemma 2 and two concrete instances.

- *In clockwise order the types of edges at the boundary of the face can be described as follows (in case of bi-directed edges the clockwise color is noted first): One edge from the set {red-cw, blue-ccw, red-blue}, any number (may be 0) of edges green-blue, one edge from the set {green-cw, red-ccw, green-red}, any number of edges blue-red, one edge from the set {blue-cw, green-ccw, blue-green}, any number of edges red-green.*

3.2 The Number of Merges

Essential for the grid-size required for the drawing produced by the algorithm is the length k of the merge sequence computed in step A3. The main result in this subsection is a lower bound for k in terms of easily recognizable substructures of the initial Schnyder wood S computed in step A2 of the algorithm.

As a warm-up let us consider the case where M is a triangulation and S is an arbitrary Schnyder wood of M. Consider the $(2n - 4) - 4$ triangles of S which are bounded by three uni-directed edges. These triangles can be partitioned into two classes: Class one are those with at least two clockwise oriented edges on the boundary and class two are those with at least two counterclockwise edges on the boundary. Suppose that the number C_1 of triangles of class one is the larger one, i.e., $C_1 \geq n-4 \geq C_2$. In a triangle T of class one there is a knee of two consecutive clockwise edges of T, this knee is a candidate for a clockwise merge. Since every edge is clockwise only for one of its neighboring triangles these C_1 merges can be performed independently. It follows that starting from S there is a merge sequence of length $k \geq C_1 \geq n-4$. This estimate yields drawing of triangulations on grids of size at most $(f - (n - 4) - 1) \times (f - (n - 4) - 1) = (n - 1) \times (n - 1)$.

The following proposition gives a better result.

Proposition 1. *Let S be a Schnyder wood with $\Delta_S^{\oplus}$ clockwise and $\Delta_S^{\ominus}$ counterclockwise triangles. The number of cw-merges applicable in a merge sequence starting with S is at least $n - 4 - \Delta_S^{\oplus} + \Delta_S^{\ominus}$.*

To estimate the number of merges that can be applied to a Schnyder wood S of a non-triangulated map we need more terminology. Let $\Delta_S^{\ominus}$ be the number of faces, with a counterclockwise edge in each of the three colors and not adjacent to a suspension vertex. These edges do not need to be uni-directed.

$\Delta_S^{\wedge}$ counts the number of clockwise triangles of uni-directed edges together with patterns of the following type: an uni-directed edge incoming at v in color $i+1$ such that the counterclockwise next edge is bi-directed, outgoing at v in color $i-1$ and incoming in color i (see Figure 4).

Theorem 6. *Let S be a Schnyder wood of a 3-connected planar map. The number of cw-merges that can be applied to S is at least $f - n + \Delta^{\frown} - \Delta^{\wedge}$.*

Given an arbitrary Schnyder wood the contribution of $\Delta^{\frown} - \Delta^{\wedge}$ in the above formula may well be negative. However, the choice of $S = S_{\text{Min}}$ guarantees that $\Delta^{\wedge} = 0$. The findings of this section can be summarized as follows.

Theorem 7. *A 3-connected planar map M with n vertices has a convex drawing on a grid of size $(n-1-\Delta_{S_{\text{Min}}}^{\frown}) \times (n-1-\Delta_{S_{\text{Min}}}^{\frown})$, where $\Delta_{S_{\text{Min}}}^{\frown} \geq 0$ is the number of faces with a counterclockwise edge in each color in S_{Min}. Such drawing can be computed in linear time.*

4   Improvements and Limitations

Our ambition was to design an algorithm for convex drawings of 3-connected planar graphs which at least matches all known algorithms for this task. Theorem 7 shows that we are very close. Still, there is Schnyder's $(n-2) \times (n-2)$ bound for triangulations which is not completely matched by $(n-1-\Delta_{S_{\text{Min}}}^{\frown}) \times (n-1-\Delta_{S_{\text{Min}}}^{\frown})$ since there are triangulations with $\Delta_{S_{\text{Min}}}^{\frown} = 0$. An example of such a triangulation is shown in Figure 8.

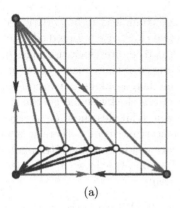   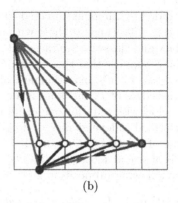

           (a)   (b)

Fig. 8. (a) A stacked triangulation on the $(n-1) \times (n-1)$ grid. (b) The same graph drawn with the improved method.

It is indeed the case that with any specialization of the algorithm from Section 3 the graph of Figure 8 (a) requires a grid of size $(n-1) \times (n-1)$.

4.1   From $n-1$ to $n-2$

In the standard algorithm, the face-count of the vertices a_1 is $(f-k-1,0,0)$, a_2 is $(0,f-k-1,0)$ and a_3 is $(0,0,f-k-1)$. In order to reduce the grid size, we change the face-count of these two vertices to the following ones: a_1 is $(f-k-2,0,1)$, a_2 is $(1,f-k-2,0)$ and a_3 is $(0,1,f-k-2)$.[3] The consequence on the final drawing is the following one: moving a_3 one unit to the left and one unit to the top, moving a_1 one unit down and moving a_3 one unit to the left. Figure 8 (b) gives and example of such drawing.

Using the technique of this section we then obtain:

Theorem 8. *A 3-connected planar map M with n vertices has a convex drawing on a grid of size $(n-2-\Delta_{S_{\mathsf{Min}}}^{\ominus}) \times (n-2-\Delta_{S_{\mathsf{Min}}}^{\ominus})$, where $\Delta_{S_{\mathsf{Min}}}^{\ominus} \geq 0$ is the number of faces with a counterclockwise edge in each color in S_{Min}.*

Concluding Remarks

As mentioned before, for some graphs, $\Delta_{S_{\mathsf{Min}}}^{\ominus}$ can be equal to zero. In [20] the asymptotic average value over n-vertices triangulations is given: $E(\Delta_{S_{\mathsf{Min}}}^{\ominus}) = n/8 + o(n)$. Hence the average grid size is significantly lower than the one of the existing algorithms.

References

1. Rote, G.: Strictly convex drawings of planar graphs (2004)
2. Rosenstiehl, P., Tarjan, R.E.: Rectilinear planar layouts and bipolar orientations of planar graphs. Discrete Comput. Geom. **1** (1986) 343–353
3. Schnyder, W.: Planar graphs and poset dimension. Order **5** (1989) 323–343
4. de Fraysseix, H., Pach, J., Pollack, R.: Small sets supporting Fary embeddings of planar graphs. In: Proc. 20th Annu. ACM Sympos. Theory Comput. (1988) 426–433
5. de Fraysseix, H., Pach, J., Pollack, R.: How to draw a planar graph on a grid. Combinatorica **10** (1990) 41–51
6. He, X.: Grid embeddings of 4-connected plane graphs. Discrete Comput. Geom. **17** (1997) 339–358
7. Miura, K., Nakano, S., Nishizeki, T.: Grid drawings of 4-connected plane graphs. Discrete Comput. Geom. **26** (2001) 73–87
8. Schnyder, W.: Embedding planar graphs on the grid. In: Proc. 1st ACM-SIAM Sympos. Discrete Algorithms. (1990) 138–148
9. Zhang, H., He, X.: Compact visibility representation and straight-line grid embedding of plane graphs. In: Proceedings WADS '03. Volume 2748 of Lecture Notes Comput. Sci., Springer-Verlag (2003) 493–504
10. Tutte, W.T.: Convex representations of graphs. Proceedings London Mathematical Society **10** (1960) 304–320

---

[3] The modification of the coordinates of a_3 do not change the grid size. It is done only to give the same role to each suspension vertex.

11. Tutte, W.T.: How to draw a graph. Proceedings London Mathematical Society **13** (1963) 743–768
12. Kant, G.: Drawing planar graphs using the canonical ordering. Algorithmica **16** (1996) 4–32
13. Chrobak, M., Kant, G.: Convex grid drawings of 3-connected planar graphs. Internat. J. Comput. Geom. Appl. **7** (1997) 211–223
14. Schnyder, W., Trotter, W.T.: Convex embeddings of 3-connected plane graphs. Abstracts of the AMS **13** (1992) 502
15. Di Battista, G., Tamassia, R., Vismara, L.: Output-sensitive reporting of disjoint paths. Algorithmica **23** (1999) 302–340
16. Felsner, S.: Convex drawings of planar graphs and the order dimension of 3-polytopes. Order (2001) 19–37
17. Felsner, S.: Geometric Graphs and Arrangements. Vieweg Verlag (2004)
18. Felsner, S.: Lattice structures from planar graphs. Electron. J. Comb. **11 R15** (2004) 24p.
19. Fusy, E., Poulalhon, D., G.Schaeffer: Coding, counting and sampling 3-connected planar graphs. In: 16th ACM-SIAM Sympos. Discrete Algorithms. (2005) to appear.
20. Bonichon, N., Gavoille, C., Hanusse, N., Poulalhon, D., Schaeffer, G.: Planar graphs, via well-orderly maps and trees. In: Proceedings WG '04. Lecture Notes Comput. Sci., Springer-Verlag (2004)

Partitions of Complete Geometric Graphs into Plane Trees[*]

Prosenjit Bose[1], Ferran Hurtado[2],
Eduardo Rivera-Campo[3], and David R. Wood[1,4]

[1] School of Computer Science, Carleton University,
Ottawa, Canada
{jit,davidw}@scs.carleton.ca
[2] Departament de Matemàtica Aplicada II, Universitat Politècnica de Catalunya,
Barcelona, Spain
Ferran.Hurtado@upc.es
[3] Departamento de Matemáticas, Universidad Autónoma Metropolitana,
Iztapalapa, México
erc@xanum.uam.mx
[4] Department of Applied Mathematics, Charles University,
Prague, Czech Republic

Abstract. Consider the following open problem: does every complete geometric graph K_{2n} have a partition of its edge set into n plane spanning trees? We approach this problem from three directions. First, we study the case of convex geometric graphs. It is well known that the complete convex graph K_{2n} has a partition into n plane spanning trees. We characterise all such partitions. Second, we give a sufficient condition, which generalises the convex case, for a complete geometric graph to have a partition into plane spanning trees. Finally, we consider a relaxation of the problem in which the trees of the partition are not necessarily spanning. We prove that every complete geometric graph K_n can be partitioned into at most $n - \sqrt{n/12}$ plane trees.

1 Introduction

A *geometric graph* G is a pair $(V(G), E(G))$ where $V(G)$ is a set of points in the plane in general position (that is, no three are collinear), and $E(G)$ is a set of closed segments with endpoints in $V(G)$. Elements of $V(G)$ are *vertices* and elements of $E(G)$ are *edges*. An edge with endpoints v and w is denoted by $\{v, w\}$ or vw when convenient. A geometric graph can be thought of as a straight-line drawing of its underlying (abstract) graph. A geometric graph is

---

[*] Research of all the authors was completed in the Departament de Matemàtica Aplicada II, Universitat Politècnica de Catalunya, Barcelona, Spain. Research of P. Bose supported by NSERC. Research of F. Hurtado supported by projects DURSI 2001SGR00224, MCYT-BFM2001-2340, MCYT-BFM2003-0368 and Gen. Cat 2001SGR00224. Research of E. Rivera-Campo supported by MECD, Spain and Conacyt, México. Research of D. R. Wood supported by NSERC and COMBSTRU.

J. Pach (Ed.): GD 2004, LNCS 3383, pp. 71–81, 2004.
© Springer-Verlag Berlin Heidelberg 2004

plane if no two edges cross. A *tree* is an acyclic connected graph. A subgraph H of a graph G is *spanning* if $V(H) = V(G)$. We are motivated by the following beautiful question.

Open Problem 1. Does every complete geometric graph with an even number of vertices have a partition of its edge set into plane spanning trees?

Since K_n, the complete graph on n vertices, has $\frac{1}{2}n(n-1)$ edges and a spanning tree has $n-1$ edges, there are $n/2$ trees in such a partition, and n is even. We approach this problem from three directions. In Section 2 we study the case of convex geometric graphs. We characterise the partitions of the complete convex graph into plane spanning trees. Section 3 describes a sufficient condition, which generalises the convex case, for a complete geometric graph to have a partition into plane spanning trees. In Section 4 we consider a relaxation of Open Problem 1 in which the trees of the partition are not necessarily spanning.

It is worth mentioning that decompositions of (abstract) graphs into trees have attracted much interest. In particular, Nash-Williams [5] obtained necessary and sufficient conditions for a graph to admit k edge-disjoint spanning trees, and Ringel's Conjecture and the Graceful Tree Conjecture about ways of decomposing complete graphs into trees are among the most outstanding open problems in the field. Nevertheless the non-crossing property that we require in our geometric setting changes the problems drastically.

2   Convex Graphs

A *convex graph* is a geometric graph with the vertices in convex position. An edge on the convex hull of a convex graph is called a *boundary edge*. Two convex graphs are *isomorphic* if the underlying graphs are isomorphic and the clockwise ordering of the vertices around the convex hull is preserved under this isomorphism. Suppose that G_1 and G_2 are isomorphic convex graphs. Then two edges cross in G_1 if and only if the corresponding edges in G_2 also cross. That is, in a convex graph, it is only the order of the vertices around the convex hull that determines edge crossings—the actual coordinates of the vertices are not important.

It is well known that Open Problem 1 has an affirmative solution in the case of convex complete graphs. That is, every convex complete graph K_{2n} can be partitioned into n plane trees, and since the book thickness of K_{2n} equals n, this bound is optimal even for partitions into plane subgraphs [2]. In this section we characterise the solutions to Open Problem 1 in the convex case. In other words, we characterise the book embeddings of the complete graph in which each page is a spanning tree.

First some well known definitions. A *leaf* of a tree is a vertex of degree at most one. A *leaf-edge* of a tree is an edge incident to a leaf. A tree has exactly one leaf if and only if it is a single vertex with no edges. Every tree with at least one edge has at least two leaves. A tree has exactly two leaves if and only if it is a path with at least one edge. Let T be a tree. Let T' be the tree obtained by deleting the leaves and leaf-edges from T. Let $\ell(T)$ be the number of leaves in T'. A *star* is a tree with at most one non-leaf vertex. Clearly a tree T is a star if and only if $\ell(T) \leq 1$. A *caterpillar* is a tree T such that T' is a path. The path T' is called the *spine* of the caterpillar. Clearly T is a caterpillar if and only if $\ell(T) \leq 2$. Observe that stars are the caterpillars whose spines consist of a single vertex.

We say a tree T is *symmetric* if there exists an edge vw of T such that if A and B are the components of $T \setminus vw$ with $v \in A$ and $w \in B$, then there exists a (graph-theoretic) isomorphism between A and B that maps v to w.

Theorem 1. *Let $T_1, T_2, \ldots, T_n$ be a partition of the edges of the convex complete graph K_{2n} into plane spanning convex trees. Then $T_1, T_2, \ldots, T_n$ are symmetric convex caterpillars that are pairwise isomorphic. Conversely, for any symmetric convex caterpillar T on $2n$ vertices, the edges of the convex complete graph K_{2n} can be partitioned into n plane spanning convex copies of T that are pairwise isomorphic.*

We will prove Theorem 1 by a series of lemmas. García *et al.* [4] proved:

Lemma 1 ([4]). *Let T be a tree with at least two edges. In every plane convex drawing of T there are at least $\max\{2, \ell(T)\}$ boundary edges, and there exists a plane convex drawing of T with exactly $\max\{2, \ell(T)\}$ boundary edges, such that if T is not a star then the boundary edges are pairwise non-consecutive.*

In what follows $\{0, 1, \ldots, 2n - 1\}$ are the vertices of a convex graph in clockwise order around the convex hull. In addition, all vertices are taken modulo $2n$. That is, vertex i refers to the vertex $j = i \bmod 2n$. Let G be a convex graph on $\{0, 1, \ldots, 2n - 1\}$. For all $0 \leq i, j \leq 2n - 1$, let $G[i, j]$ denote the subgraph of G induced by the vertices $\{i, i + 1, \ldots, j\}$.

Lemma 2. *For all $n \geq 2$, let $T_0, T_1, \ldots, T_{n-1}$ be a partition of the convex complete graph K_{2n} into plane spanning trees. Then (after relabelling the trees) for each $0 \leq i \leq n - 1$,*

(1) the edge $\{i, n + i\}$ is in T_i,
(2) T_i is a caterpillar with exactly two boundary edges, and
(3) for every non-boundary edge $\{a, b\}$ of T_i, there is exactly one boundary edge of T_i in each of $T_i[a, b]$ and $T_i[b, a]$.

Proof. The edges $\{\{i, n + i\} : 0 \leq i \leq n - 1\}$ are pairwise crossing. Thus each such edge is in a distinct tree. Label the trees such that each edge $\{i, n + i\}$ is in T_i. Since $n \geq 2$, each T_i has at least three edges, and by Lemma 1, has at least two boundary edges. There are $2n$ boundary edges in total and n trees. Thus each T_i has exactly two boundary edges, and by Lemma 1, $\ell(T_i) = 2$. For any tree T, $\ell(T) \leq 2$ if and only if T is a caterpillar. Thus each T_i is a caterpillar. Let $\{a, b\}$ be a non-boundary edge in some T_i. Then $T_i[a, b]$ has at least one boundary edge of T_i, as otherwise $T_i[a, b]$ would be a convex tree on at least three vertices with only one boundary edge (namely, $\{a, b\}$), which contradicts Lemma 1. Similarly $T_i[b, a]$ has at least one boundary edge of T_i. Thus each of $T_i[a, b]$ and $T_i[b, a]$ has exactly one boundary edge of T_i. □

Lemma 3. *Let $\{i,j\}$ be a non-boundary edge of a plane convex spanning caterpillar T such that $T[i,j]$ has exactly one boundary edge of T. Then exactly one of $\{i, j-1\}$ and $\{j, i+1\}$ is an edge of T.*

Proof. If both $\{i, j-1\}$ and $\{j, i+1\}$ are in T then they cross, unless $j-1 = i+1$ in which case T contains a 3-cycle. Thus at most one of $\{i, j-1\}$ and $\{j, i+1\}$ is in T.

Suppose, for the sake of contradiction, that neither $\{i, j-1\}$ nor $\{j, i+1\}$ are edges of T. Since T is spanning, there is an edge $\{i, a\}$ or $\{j, a\}$ in T for some vertex $i+1 < a < j-1$. Without loss of generality $\{i, a\}$ is this edge, as illustrated in Figure 1.

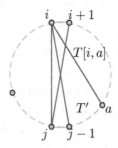

Fig. 1. One of $\{i, j-1\}$ and $\{j, i+1\}$ is an edge of T.

Since i, $i+1$ and a are distinct vertices of $T[i, a]$, the subtree $T[i, a]$ has at least three vertices, and by Lemma 1, has at least two boundary edges, one of which is $\{i, a\}$. Thus $T[i, a]$ has at least one boundary edge that is also a boundary edge of T. Now consider the subtree T' of T induced by $\{i\} \cup \{a, a+1, \ldots, j\}$. Then i, a, $j-1$ and j are distinct vertices of T', and T' has at least four vertices. Since $\{i, j-1\}$ is not an edge of T, and thus not an edge of T', the subtree T' is not a star. By Lemma 1, T' has at least two non-consecutive boundary edges, at most one of which is $\{i, j\}$ or $\{i, a\}$. Thus T' has at least one boundary edge that is also a boundary edge of T.

No boundary edge of T can be in both $T[i, a]$ and T'. Thus we have shown that $T[i, j]$ has at least two boundary edges of T, which is the desired contradiction. $\square$

In what follows we say an edge $e = \{i, j\}$ has *span*

$$\mathsf{span}(e) = \min\{(i-j) \bmod 2n, (j-i) \bmod 2n\} \ .$$

That is, $\mathsf{span}(e)$ is the number of edges in a shortest path between i and j that is contained in the convex hull.

Lemma 4. *Let $\{i, j\}$ be an edge of a plane convex spanning caterpillar T such that $1 \leq j-i \leq n$, and $T[i, j]$ has exactly one boundary edge of T. Then $T[i, j]$ has exactly one edge of span k for each $1 \leq k \leq j-i$. Moreover for each such $k \geq 2$ the edge of span k has an endpoint in common with the edge of span $k-1$, and the other two endpoints are consecutive on the convex hull.*

Proof. If $j - i = 1$ then $\{i, j\}$ is a boundary edge, and the result is trivial. Otherwise $\{i, j\}$ is not a boundary edge. By Lemma 3, exactly one of the edges $\{i, j - 1\}$ and $\{j, i + 1\}$ is in T. Without loss of generality $\{i, j - 1\}$ is in T. Thus the edge of span $j - i$ has an endpoint in common with the edge of span $j - i - 1$, and the other two endpoints are consecutive on the convex hull. The result follows by induction (on span) applied to the edge $\{i, j - 1\}$.   □

Theorem 2 below is the main theorem of this section, and its proof is illustrated in Figure 4. Let $e = \{a, b\}$ be an edge in the convex complete graph K_{2n}. Then $e + i$ denotes the edge $\{a + i, b + i\}$. For a set X of edges, $X + i = \{e + i : e \in X\}$, and $X^{(k)} = \{e \in X, \mathsf{span}(e) \geq k\}$.

Theorem 2. *Let $T_0, T_1, \ldots, T_{n-1}$ be a partition of the edges of the convex complete graph K_{2n} into plane spanning convex trees. Then $T_0, T_1, \ldots, T_{n-1}$ are pairwise isomorphic symmetric convex caterpillars.*

Proof. By Lemma 2, for each $0 \leq i \leq n - 1$, T_i is a caterpillar with two boundary edges, the edge $\{i, n + i\}$ is in T_i, and for every non-boundary edge $\{a, b\}$ of T_i, there is exactly one boundary edge of T_i in each of $T_i[a, b]$ and $T_i[b, a]$.

Let $H = T_0[0, n]$. Since $\{0, n\}$ is an edge of H, by Lemma 4, H has exactly one edge of span k for each $1 \leq k \leq n$. Furthermore, for each $1 \leq k \leq n - 1$, the edge of span k has an endpoint in common with the edge of span $k + 1$, and the other two endpoints are consecutive on the convex hull. Let $h_k = \{x_k, x_k + k\}$ denote the edge of span k in H. For each $1 \leq k \leq n - 1$, if $h_k \cap h_{k+1} = x_k + k$ ($= x_{k+1} + k + 1$) then we say the k-*direction* is 'clockwise'. Otherwise, $h_k \cap h_{k+1} = x_k$ ($= x_{k+1}$), and we say the k-*direction* is 'anticlockwise', as illustrated in Figure 2.

We will prove that H determines the structure of all the trees $T_0, T_1, \ldots, T_{n-1}$. We proceed by downwards induction on $k = n, n - 1, \ldots, 1$ with the hypothesis that for all $0 \leq i \leq n - 1$,

$$T_i^{(k)} = (H^{(k)} + i) \cup (H^{(k)} + n + i) \tag{1}$$

Consider the base case with $k = n$. The only edge in H of span n is $\{0, n\}$. Thus $H^{(n)} = \{0, n\}$, which implies that $H^{(n)} + i = \{i, n + i\}$, and $H^{(n)} + n + i =$

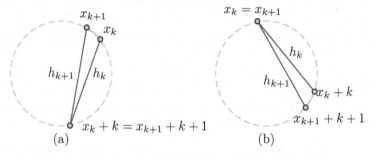

Fig. 2. k-direction is (a) clockwise and (b) anticlockwise.

$\{n+i, 2n+i\} = \{i, n+i\}$. Thus the right-hand side of (1) is $\{i, n+i\}$. The only edge in T_i of span n is $\{i, n+i\}$. Thus $T_i^{(n)} = \{i, n+i\}$, and (1) is satisfied for $k = n$.

Now suppose that (1) holds for some $k + 1 \geq 2$. We will prove that (1) holds for k. Suppose that the k-direction is clockwise. (The case in which the k-direction as anticlockwise is symmetric.) We proceed by induction on $j = 0, 1, \ldots, 2n - 1$ with the hypothesis:

$$\text{the edge } \{x_k + j, x_k + k + j\} \text{ is in the tree } T_{j \bmod n} . \tag{2}$$

The base case with $j = 0$ is immediate since by definition, $\{x_k, x_k + k\} \in E(T_0)$. Suppose that $\{x_k + j, x_k + k + j\} \in E(T_{j \bmod n})$ for some $0 \leq j < 2n - 1$. Consider the edge $e = \{x_k + j, x_k + k + j + 1\}$. Since the k-direction is clockwise, $x_k = x_{k+1} + 1$ and $x_k + k = x_{k+1} + k + 1$. Thus $e = \{x_{k+1} + 1 + j, x_{k+1} + k + 1 + j + 1\} = \{x_{k+1}, x_{k+1} + k + 1\} + j + 1 = h_k + j + 1$. Hence $e \in H + j + 1$, and since e has span $k + 1$, $e \in H^{(k+1)} + j + 1$. By induction from (1), $e \in T_{(j+1) \bmod n}^{(k+1)}$, as illustrated in Figure 3.

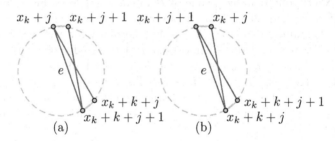

$$x_k + j \quad x_k + j + 1 \qquad x_k + j + 1 \quad x_k + j$$

$$e \qquad\qquad\qquad e$$

$$x_k + k + j \qquad\qquad x_k + k + j + 1$$
$$x_k + k + j + 1 \qquad\qquad x_k + k + j$$

$$\text{(a)} \qquad\qquad\qquad \text{(b)}$$

Fig. 3. k-direction is (a) clockwise and (b) anticlockwise.

By Lemma 3 applied to e, which is a non-boundary edge of $T_{(j+1) \bmod n}$, exactly one of $\{x_k + j, x_k + k + j\}$ and $\{x_k + j + 1, x_k + k + j + 1\}$ is an edge of $T_{(j+1) \bmod n}$. By induction from (2), $\{x_k + j, x_k + k + j\} \in T_{j \bmod n}$. Thus $\{x_k + j + 1, x_k + k + j + 1\} \in T_{(j+1) \bmod n}$. That is, (2) holds for $j + 1$. Therefore for all $0 \leq j \leq 2n - 1$, the edge $\{x_k + j, x_k + k + j\}$ is in $T_{j \bmod n}$. That is, $h_k + j$ is in $T_{j \bmod n}$. By (1) for $k + 1$ we have that (1) holds for k.

By (1) with $k = 1$, each tree T_i can be expressed as $T_i = (H+i) \cup (H+n+i)$. Clearly $H \cup (H+n)$ is a symmetric convex caterpillar. Thus each T_i is a translated copy of the same symmetric convex caterpillar. Therefore $T_0, T_1, \ldots, T_{n-1}$ are pairwise isomorphic symmetric convex caterpillars. □

Theorem 3. *For any symmetric convex caterpillar T on $2n$ vertices, the edges of the convex complete graph K_{2n} can be partitioned into n plane spanning pairwise isomorphic convex copies of T.*

Proof. Say $V(K_{2n}) = \{0, 1, \ldots, 2n - 1\}$ in clockwise order around the convex hull. Let $\{0, n\}$ be the edge of T such that after deleting $\{0, n\}$, A and B are

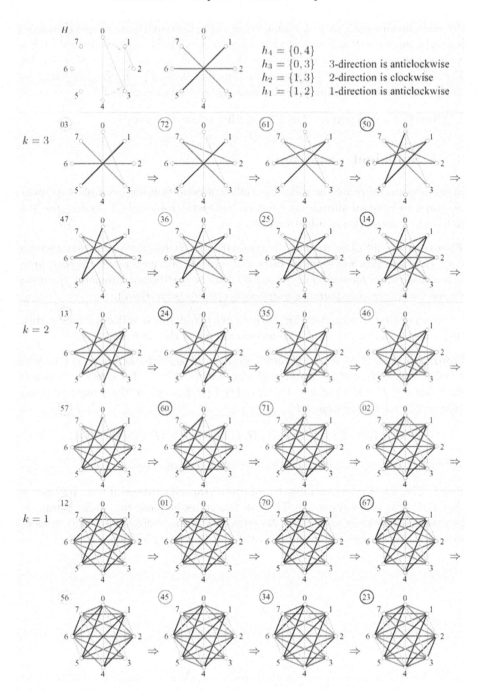

Fig. 4. Illustration for Theorem 2 with $n = 4$.

the components with $0 \in A$ and $n \in B$, and there exists a (graph-theoretic) isomorphism between A and B that maps 0 to n. It is easily seen that A has a plane representation on the vertices $\{0, 1, \ldots, n\}$. For each $0 \le i \le n - 1$, let $T_i = (A + i) \cup (A + n + i)$. Then as in Theorem 2, $T_0, T_1, \ldots, T_{n-1}$ is partition of K_{2n} into plane spanning pairwise isomorphic convex copies of T. $\qquad\square$

Observe that Theorems 2 and 3 together prove Theorem 1.

3   A Sufficient Condition

In this section we prove the following sufficient condition for a complete geometric graph to have an affirmative solution to Open Problem 1. A *double star* is a tree with at most two non-leaf vertices.

Theorem 4. *Let G be a complete geometric graph K_{2n}. Suppose that there is a set $\mathcal{L}$ of pairwise non-parallel lines with exactly one vertex of G in each open unbounded region formed by $\mathcal{L}$. Then $E(G)$ can be partitioned into plane spanning double stars (that are pairwise graph-theoretically isomorphic).*

Observe that in a double star, if there are two non-leaf vertices v and w then they must be adjacent, in which case we say vw is the *root edge*.

Lemma 5. *Let P be a set of points in general position. Let L be a line with $L \cap P = \emptyset$. Let H_1 and H_2 be the half-planes defined by L. Let v and w be points such that $v \in P \cap H_1$ and $w \in P \cap H_2$. Let $T(P, L, v, w)$ be the geometric graph with vertex set P and edge set*

$$\{vw\} \cup \{vx : x \in (P \setminus \{v\}) \cap H_1\} \cup \{wy : y \in (P \setminus \{w\}) \cap H_2\} \ .$$

Then $T(P, L, v, w)$ is a plane double star with root edge vw.

Proof. The set of edges incident to v form a star. Regardless of the point set, a geometric star is always plane. Thus no two edges incident to v cross. Similarly no two edges incident to w cross. No edge incident to v crosses an edge incident to w since such edges are separated by L, as illustrated in Figure 5. $\qquad\square$

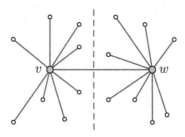

Fig. 5. A plane double star separated by a line.

Lemma 6. *Let P be a set of points in general position. Let L_1 and L_2 be non-parallel lines with $L_1 \cap P = L_2 \cap P = \emptyset$. Let v, w, x, y be points in P such that v, w, x, y are in distinct quarter-planes formed by L_1 and L_2, with each pair (v, w) and (x, y) in opposite quarter-planes. (Note that this does not imply that vw and xy cross.) Let T_1 and T_2 be the plane double stars $T_1 = T(P, L_1, v, w)$ and $T_2 = T(P, L_2, x, y)$. Then $E(T_1) \cap E(T_2) = \emptyset$.*

Proof. Suppose, for the sake of contradiction, that there is an edge $e \in E(T_1) \cap E(T_2)$. All edges of T_1 are incident to v or w, and all edges of T_2 are incident to x or y. Thus $e \in \{vx, vw, vy, xw, xy, wy\}$. By assumption, v, w, x, y are in distinct quarter-planes formed by L_1 and L_2, with each pair (v, w) and (x, y) in opposite quarter-planes. Thus e crosses at least one of L_1 and L_2. Without loss of generality e crosses L_1. Since $e \in E(T_1)$, and the only edge of T_1 that crosses L_1 is the root edge vw, we have $e = vw$. Since all edges of T_2 are incident to x or y and v, w, x, y are distinct, we have $e \notin E(T_2)$, which is the desired contradiction. Therefore $E(T_1) \cap E(T_2) = \emptyset$, as illustrated in Figure 6. □

Proof (of Theorem 4). As illustrated in Figure 7, let C be a circle such that the vertices of G and the intersection point of any two lines in $\mathcal{L}$ are in the interior of C. The intersection points of C and the lines in $\mathcal{L}$ partition C into $2n$ consecutive components $C_0, C_1, \ldots, C_{2n-1}$, each corresponding to a region containing a single vertex of G. Let i be the vertex in the region corresponding to C_i. Label the lines $L_0, L_1, \ldots, L_{n-1}$ so that for each $0 \leq i \leq n-1$, the components C_i and C_{i+n} run from $C \cap L_i$ to $C \cap L_{(i+1) \bmod n}$ in the clockwise direction.

For each $0 \leq i \leq n-1$, let T_i be the double star $T(V(G), L_i, i, i+n)$. By Lemma 5, each T_i is plane. Since $V(T_i) = V(G)$, T_i is a spanning tree of G. For all $0 \leq i < j \leq n-1$, the points $i, i+n, j, j+n$ are in distinct quarter-planes

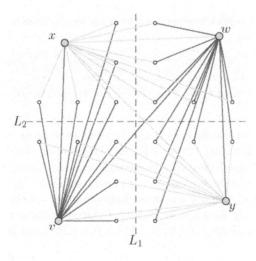

Fig. 6. Plane spanning double stars are edge-disjoint.

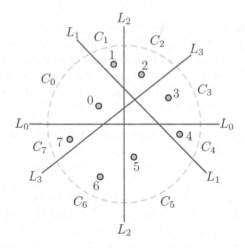

Fig. 7. Example of Theorem 4 with $n = 4$.

formed by L_i and L_j, with each pair $(i, i+n)$ and $(j, j+n)$ in opposite quarter-planes. Thus, by Lemma 6, $E(T_i) \cap E(T_j) = \emptyset$. Since each T_i has $2n - 1$ edges, and there are $n(2n - 1)$ edges in total, $T_0, T_1, \ldots, T_{n-1}$ is the desired partition of $E(G)$. $\square$

Note that each line in $\mathcal{L}$ in Theorem 4 is a halving line. Pach and Solymosi [6] proved a related result: a complete geometric graph on $2n$ vertices has n pairwise crossing edges if and only if it has precisely n halving lines.

4   Relaxations

We now drop the requirement that our plane trees be spanning. Thus we need not restrict ourselves to complete graphs with an even number of vertices. Theorem 4 generalises as follows.

Theorem 5. *Let G be a complete geometric graph K_n. Suppose that there is a set $\mathcal{L}$ of pairwise non-parallel lines with at least one vertex of G in each open unbounded region formed by $\mathcal{L}$. Then $E(G)$ can be partitioned into $n - |\mathcal{L}|$ plane trees.*

Proof. Let P be a set consisting of exactly one vertex in each open unbounded region formed by $\mathcal{L}$. Then $|P| = 2|\mathcal{L}|$. By Theorem 4, the induced subgraph $G[P]$ can be partitioned into $\frac{1}{2}|P|$ plane double stars. The edges incident to a vertex not in P can be covered by $n - |P|$ spanning stars, one rooted at each of the vertices not in P. Clearly a star is plane regardless of the vertex positions. Edges with both endpoints not in P can be placed in the star rooted at either endpoint. In total we have $\frac{1}{2}|P| + (n - |P|) = n - \frac{1}{2}|P| = n - |\mathcal{L}|$ plane trees. $\square$

Lemma 7. *Every complete geometric graph K_n with k pairwise crossing edges can be partitioned into $n - k$ plane trees.*

Proof. Let $E = \{e_i : 1 \leq i \leq k\}$ be a set of k pairwise crossing edges. For each $1 \leq i \leq k$, let L_i be the line obtained by extending the segment e_i, and rotating it about the midpoint of e_i by some angle of ϵ degrees. Clearly there exists an ϵ such that each edge e_i crosses every line L_j. Thus there is one endpoint of an edge in E in each open unbounded region formed by $L_1, L_2, \ldots, L_k$. The result follows from Theorem 5. □

Aronov *et al.* [1] proved that every complete geometric graph K_n has at least $\sqrt{n/12}$ pairwise crossing edges. Thus Lemma 7 implies:

Corollary 1. *Every complete geometric graph K_n can be partitioned into at most $n - \sqrt{n/12}$ plane trees.* □

We conclude with a seemingly easier problem than Open Problem 1.

Open Problem 2. Can the edges of every complete geometric graph K_n be partitioned into at most n/c plane subgraphs, for some constant $c > 1$?

Of course $c < 2$ since $n/2$ edges may be pairwise crossing. Dillencourt *et al.* [3] defined the *geometric thickness* of an (abstract) graph G to be the minimum k such that G has a representation as a geometric graph whose edges can be partitioned into k plane subgraphs. They proved that the geometric thickness of K_n is between $\lceil (n/5.646) + 0.342 \rceil$ and $\lceil n/4 \rceil$.

References

1. BORIS ARONOV, PAUL ERDŐS, WAYNE GODDARD, DANIEL J. KLEITMAN, MICHAEL KLUGERMAN, JÁNOS PACH, AND LEONARD J. SCHULMAN. Crossing families. *Combinatorica*, 14(2):127–134, 1994.
2. FRANK R. BERNHART AND PAUL C. KAINEN. The book thickness of a graph. *J. Combin. Theory Ser. B*, 27(3):320–331, 1979.
3. MICHAEL B. DILLENCOURT, DAVID EPPSTEIN, AND DANIEL S. HIRSCHBERG. Geometric thickness of complete graphs. *J. Graph Algorithms Appl.*, 4(3):5–17, 2000.
4. ALFREDO GARCÍA, CARMEN HERNANDO, FERRAN HURTADO, MARC NOY, AND JAVIER TEJEL. Packing trees into planar graphs. *J. Graph Theory*, 40(3):172–181, 2002.
5. C. ST. J. A. NASH-WILLIAMS. Decomposition of finite graphs into forests. *J. London Math. Soc.*, 39:12, 1964.
6. JÁNOS PACH AND JÓZSEF SOLYMOSI. Halving lines and perfect cross-matchings. In *Advances in discrete and computational geometry*, volume 223 of *Contemp. Math.*, pages 245–249. Amer. Math. Soc., 1999.

Additional PC-Tree Planarity Conditions

John M. Boyer

PureEdge Solutions Inc. Victoria, BC Canada
jboyer@acm.org, jboyer@PureEdge.com

Abstract. Recent research efforts have produced new algorithms for solving planarity-related problems. One such method performs vertex addition using the PC-tree data structure, which is similar to but simpler than the well-known PQ-tree. For each vertex, the PC-tree is first checked to see if the new vertex can be added without violating certain planarity conditions; if the conditions hold, the PC-tree is adjusted to add the new vertex and processing continues. The full set of planarity conditions are required for a PC-tree planarity tester to report only planar graphs as planar. This paper provides further analyses and new planarity conditions needed to produce a correct planarity algorithm with a PC-tree.

1 Introduction

The first linear-time planarity tests [1, 2] represent significant achievements but are also quite complex. Recent research has produced simpler linear-time planarity algorithms [3–5]. This paper discusses the planarity method of Shih and Hsu [5], which is based on a data structure called a PC-tree. The PC-tree method is a *vertex addition* method that adds each vertex to a *partial planar embedding* once it determines that planarity can be preserved while adding the vertex and all edges that connect it to other vertices in the partial embedding.

The PC-tree method processes the vertices in a post-order traversal of the depth first search (DFS) tree of the graph. Thus, there is a path of unprocessed vertices from every vertex to the root of the DFS-tree. If the graph is planar, then it must be possible to embed of the first k vertices so that all vertices with direct back edge connections to their unprocessed DFS ancestors are on the external face of the partial embedding. For each vertex i, the algorithm first checks the PC-tree for a number of defined planarity conditions. If all conditions are met, then a planarity reduction is applied to the PC-tree for vertex i.

If one or more planarity conditions were missing, then a planarity reduction would be applied when it should not be, ultimately causing a planar result to be reported on some non-planar graphs. The literature on PC-trees have not presented additional planarity conditions, instead focusing on the consecutive ones problem [6, 7] or on equating PQ-tree and PC-tree reductions [8]. A submitted book chapter [9] presents an alternate graph-theoretic view that shows the correctness of the general approach, but it uses constructs that are difficult to apply directly to a PC-tree. This paper presents the additional required

J. Pach (Ed.): GD 2004, LNCS 3383, pp. 82–88, 2004.
© Springer-Verlag Berlin Heidelberg 2004

planarity conditions that arise directly on the PC-tree, thus allowing more reasonable comparisons of complexity and empirical performance to be made with other planarity methods. In particular, although it is reasonable to assume that the 'batch' operations of vertex addition methods are more cumbersome to implement and less efficient than a finer grain edge addition method [4], proper comparisons cannot be done with only the planarity conditions in [5].

Section 2 provides some definitions and preliminary remarks. Sections 3, 4 and 5 present additional planarity conditions and further analyses for the PC-tree. Finally, Section 6 presents some concluding remarks.

2   Preliminaries

A PC-tree represents a partial planar embedding of a graph, with C-nodes representing all biconnected components and P-nodes representing cut vertices in the partial embedding and vertices with direct back edge connections that have not been embedded yet. Every P-node is associated with a vertex of the input graph. The neighbors of a C-node are P-nodes, which form the *representative bounding cycle* (RBC) of the C-node. The RBC corresponds to the external face cycle of the biconnected component represented by the C-node (for efficiency, nodes are removed from the RBC if they represent neither cut vertices in the partial embedding nor the endpoints of unembedded back edges). The P-nodes of the RBC are connected into a cycle. Traversal through a C-node occurs on one of the two paths along the RBC cycle between two neighbors of the C-node.

The PC-tree is denoted T, and T_r denotes a subtree of T rooted by node r. The current vertex being processed is denoted i. An *i-subtree* T_w is a PC-subtree of T_i that is rooted by the P-node for w with lowpoint(w) equal to i (i.e. the unembedded back edges from w and its descendants connect to i). An *i^*-subtree* T_x is a PC-subtree of T_i that is rooted by the P-node for x with lowpoint(x) $< i$ and that contains no vertex adjacent to i in the input graph (so, every unembedded back edge connects to an ancestor of i). To simplify discussion, the direct back edges to i and its ancestors are considered to be degenerate i-subtrees and i^*-subtrees. If the root node of an i-subtree or i^*-subtree is the child of a given PC-tree node, then we say that the i-subtree or i^*-subtree is a child of that node. A *terminal node* is a P-node or C-node of the PC-tree that has one or more i-subtree children, one or more i^*-subtree children, and no descendants in the PC-tree with both i-subtree and i^*-subtree children.

For each vertex i (in post-order of the DFS tree), the PC-tree is tested for planarity conditions before adding i to the partial embedding. Because three or more terminal nodes implies non-planarity, much of the discourse in [5] focuses on the one or two terminal node cases. Shih and Hsu present four necessary conditions for maintaining planarity in the one and two terminal node cases: "In Lemma 2.5, Corollary 2.6, [and] Lemmas 3.1 and 3.2 we made the assumption that graph G is planar in deriving at those conclusions. We shall show that if these conclusions hold at each iteration, then G must be planar by showing that these conditions imply a feasible internal embedding for each 2-connected component." [5, p. 188]. Then, the proof presented only describes how to perform the one and two terminal node planarity reductions, which does not prove that those reductions can always be performed if only the given planarity conditions are met. The remaining sections describe additional required planarity conditions and indicate how their violation implies non-planarity.

3   The i-i^* Subtree Patterns Around a Terminal C-Node

Lemma 3.2 of [5] seeks to characterize the allowable pattern of child i-subtrees and i^*-subtrees around a terminal C-node. It states that for the root j of any child i-subtree of a terminal C-node, one of the two RBC paths from j to the parent of the C-node must contain only i-subtrees. This condition is *necessary* but only *sufficient* in the one terminal node case when the terminal node has no proper ancestor with a child i^*-subtree. In the two terminal node case and the one terminal node case where the terminal node has a proper ancestor with a child i^*-subtree, it is possible to be compliant with the statement of the lemma yet still have a non-planarity condition. Theorem 1 states the additional restriction required on terminal C-nodes, and Figure 1 shows PC-trees that violate the restriction, along with the resulting $K_{3,3}$ minor.

Theorem 1. *If a terminal C-node c has a proper ancestor r with a child v such that T_v excludes c and is or contains an i^*-subtree, then c must have a child w for which an RBC path from w to the parent of c contains all of the i-subtree children of c.*

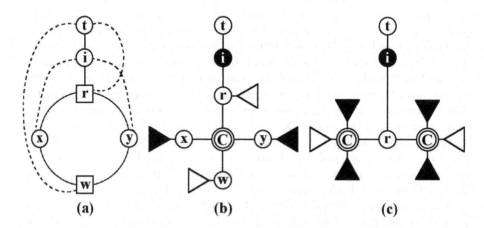

Fig. 1. (a) A $K_{3,3}$ non-planarity minor from [3]. (b) A corresponding PC-tree with one terminal C-node having the forbidden pattern of i-subtrees (dark triangles) and i^*-subtrees (light triangles). (c) An example with two terminal C-nodes, only one of which need be in the depicted state. Note the use of graph minors for simplification; in these examples, r could be any node on the path between i and the terminal C-node.

4   The i-i^* Subtree Patterns for an Intermediate C-Node

Given an intermediate C-node c along the path P between two terminal nodes, we consider the two RBC paths strictly between neighbors v and v' of c in P. The proof of Lemma 3.1 of [5] attempts to prove the following: neither RBC path of c can contain both an i-subtree and an i^*-subtree. It does not show the necessity of the broader planarity condition stated the lemma: of the two RBC paths strictly between v and v', one must contain only child i-subtrees and the other must contain only child i^*-subtrees. There are four issues. First, the proof of the simpler condition fails when the terminal node and the intermediate C-node are neighbors; the author has found other $K_{3,3}$ patterns (not depicted) for this case. Second, the proof is by contradiction but does not fully negate the lemma statement: the simpler condition (described above) can hold while still violating the lemma statement's planarity condition if both RBC paths contain only i^*-subtrees (see Figure 2(a)) or i-subtrees (reduces to Figure 1(c)). Third, stricter conditions are required if the intermediate C-node is m, the closest common ancestor of the terminal nodes, because it cannot be flipped. The graph is non-planar if an i-subtree appears below P on the RBC of m (see Figure 2(b)) or if an i^*-subtree appears on the RBC of m above P (resulting in a $K_{3,3}$ that edge contracts to the K_5 minor of Figure 3(a)). Theorem 2 states the required planarity conditions. A fourth problem is that analogous planarity conditions are required for one terminal node, producing the same non-planarity minors except the last case does not edge contract to a K_5 minor but still produces a $K_{3,3}$ (not depicted). Theorem 3 states the additional planarity conditions.

Theorem 2. *Let P denote the path between two terminal nodes u and u' with closest common ancestor m. Let c denote a C-node in $P - \{u, u'\}$ with neighbors v and v' in P. Of the two RBC paths strictly between v and v', one must contain no child i-subtrees and the other no child i^*-subtrees. Further, if $c = m$, then the RBC path containing the child i-subtrees must also contain the parent of c.*

Theorem 3. *Given one terminal node u, let P denote the path from u to the farthest ancestor u' with a child i^*-subtree. Let c be an intermediate C-node in path $P - \{u\}$. For $c \neq u'$, let v and v' denote the neighbors of c in P. For $c = u'$, let v denote the neighbor of c in P and let v' denote the closest child i^*-subtree along either RBC path from the parent p of c. The following conditions must hold: 1) The children of c in one RBC path strictly between v and v' must contain only child i-subtrees; 2) The opposing RBC path strictly between v and v' must contain only child i^*-subtrees; 3) If $c = u'$, then the RBC path containing the child i-subtrees must also contain p.*

5   Finding Non-planarity of $K_{3,3}$-Less Graphs

Consider extending Lemma 2.5 in [5] to a PC-tree that contains C-nodes. Specifically, suppose the closest common ancestor m of the two terminal nodes is a C-node whose parent has the only child i^*-subtree along the path P'. Figure 3

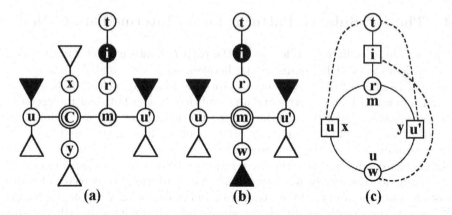

Fig. 2. (a) The intermediate C-node has child i^*-subtrees along both RBC paths between its parent (labelled m here) and the next node in path P (labelled u here). (b) The C-node labelled m has a child i-subtree below the path P between the terminal nodes. (c) The $K_{3,3}$ for these planarity condition violations; the labels m, x, y and u are the mapping for part (a), and the labels r, u, u' and w are for part (b).

depicts an example PC-tree and the corresponding K_5 minor pattern from [3]. In this case, the $K_{3,3}$ shown in the proof of Lemma 2.5 in [5] cannot be found, illustrating that the proof does not "go through for the case of general trees without any changes provided that the paths through a C-node are interpreted correctly" [5, p. 185]. Theorem 4 states the relevant planarity condition from Lemma 2.5 of [5], relying for its proof of necessity on both [5] and Figure 3.

Theorem 4. *Suppose there are two terminal nodes u and u' in T_r, and let m be their closest common ancestor. Let P' be the unique path from m to r. Every proper ancestor of m in T_r must have no child i^*-subtrees.*

This case is also important because it shows the method by which K_5 subdivisions and other $K_{3,3}$-less graphs are found by the PC-tree algorithm. In [5], the case of three terminal nodes is shown to produce a either a (subgraph homeomorphic to) $K_{3,3}$, or "we could have three terminal nodes being neighbors of a C-node, in which case we would get a subgraph homeomorphic to K_5 ..." Technically, the result is a K_5 minor, which could produce a subgraph homeomorphic to $K_{3,3}$ or K_5. Of greater importance, though, is the fact that a $K_{3,3}$ can also *always* be found in this case (though not the same one indicated for the three terminal node case in [5]). However, this case of three terminal node neighbors of a C-node is the only case mentioned in [5] for finding a K_5, yet there are many non-planar graphs that do not contain a $K_{3,3}$. Therefore, there must be *some other* condition that detects non-planarity for graphs that contain a K_5 but not a $K_{3,3}$ (e.g. all K_5 subdivisions). The K_5 in Figure 6 of [5] is equivalent to Figure 3(b). It does not result in three terminal nodes as stated in [5], but is instead discovered by violation of the planarity condition in Theorem 4.

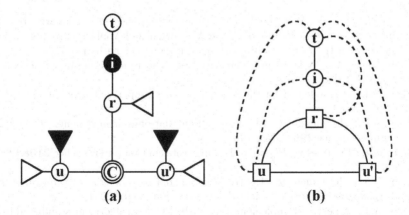

Fig. 3. (a) A PC-tree in which the closest common ancestor of terminal nodes u and u' is a C-node with a proper ancestor that has a child i^*-subtree. (b) The corresponding K_5 minor from [3]. Note: Due to the difference in definitions between graph minors and subgraph homeomorphism, this case implies a subgraph homeomorphic to $K_{3,3}$ or K_5.

6   Conclusion and Future Work

This paper presented the additional planarity conditions required to create a correct planarity algorithm using a PC-tree, allowing fair comparison with other recent approaches to planarity. While the August 2003 version of the implementation in [10] could not be empirically compared due to frequent incorrect results, Hsu also requested that a subsequent version with fixes not be empirically compared as he felt the implementation was only a proof of concept. However, there is strong evidence from [11] that a simplified vertex addition method can achieve far better performance than most prior methods, although those results also suggest that the edge addition methods in [4, 12] are faster. Future work must use the results of this paper to create correct, efficient PC-tree implementations for empirical comparisons, especially with [4, 12]. The results of this paper are also important for creating a Kuratowski subgraph isolator based on the PC-tree, the full exposition of which should translate from the graph minors used to express planarity condition violations to homeomorphic subgraphs.

References

1. Hopcroft, J., Tarjan, R.: Efficient planarity testing. Journal of the Association for Computing Machinery **21** (1974) 549–568
2. Booth, K.S., Lueker, G.S.: Testing for the consecutive ones property, interval graphs, and graph planarity using PQ–tree algorithms. Journal of Computer and Systems Sciences **13** (1976) 335–379
3. Boyer, J., Myrvold, W.: Stop minding your P's and Q's: A simplified $O(n)$ planar embedding algorithm. Proceedings of the Tenth Annual ACM-SIAM Symposium on Discrete Algorithms (1999) 140–146

4. Boyer, J., Myrvold, W.: On the cutting edge: Simplified $O(n)$ planarity by edge addition. Accepted to Journal of Graph Algorithms and Applications, August 2004 (Preprint at http://www.pacificcoast.net/~lightning/planarity.ps) 1–29
5. Shih, W.K., Hsu, W.L.: A new planarity test. Theoretical Computer Science **223** (1999) 179–191
6. Hsu, W.L.: A simple test for the consecutive ones property. Journal of Algorithms **42** (2002) 1–16
7. Hsu, W.L., McConnell, R.: PC-trees and circular-ones arrangements. Theoretical Computer Science **296** (2003) 59–74
8. Hsu, W.L.: PC-trees vs. PQ-trees. Lecture Notes in Computer Science **2108** (2001) 207–217
9. Hsu, W.L., McConnell, R.: PC-trees. submitted to *Handbook of Data Structures and Applications*, Dinesh P Mehta and Sartaj Sahni ed. (2004)
10. Hsu, W.L.: An efficient implementation of the PC-trees algorithm of shih and hsu's planarity test. Technical Report TR-IIS-03-015, Institute of Information Science, Academia Sinica (2003)
11. Boyer, J.M., Cortese, P.F., Patrignani, M., Di Battista, G.: Stop minding your P's and Q's: Implementing a fast and simple DFS-based planarity testing and embedding algorithm. In Liotta, G., ed.: Proceedings of the 11th International Symposium on Graph Drawing 2003. Volume 2912 of Lecture Notes in Computer Science., Springer-Verlag (2004) 25–36
12. de Fraysseix, H., Rosenstiehl, P.: A characterization of planar graphs by trémaux orders. Combinatorica **5** (1985) 127–135

GraphML Transformation*

Ulrik Brandes and Christian Pich**

Department of Computer & Information Science, University of Konstanz, Germany
Christian.Pich@uni-konstanz.de

Abstract. The efforts put into XML-related technologies have exciting consequences for XML-based graph data formats such as GraphML. We here give a systematic overview of the possibilities offered by XSLT style sheets for processing graph data, and illustrate that many basic tasks required for tools used in graph drawing can be implemented by means of style sheets, which are convenient to use, portable, and easy to customize.

1 Introduction

Among the multitude of software packages that process graphs, some are dedicated graph packages while others operate on graph structures implicitly. All of them have in common the need to input existing data and to output their computation results in files or streams. GraphML (Graph Markup Language) is an XML-based format for the description of graph structures, designed to improve tool interoperability and reduce communication overhead [1]. It is open to user-defined extensions for application-specific data. Thanks to its XML syntax, GraphML-aware applications can take advantage of a growing number of XML-related technologies and tools, such as parsers and validators.

It is straightforward to provide access to graphs represented in GraphML by adding input and output filters to an existing software application. However, we find that Extensible Stylesheet Language Transformations (XSLT) [7] offer a more natural way of utilizing XML formatted data, in particular when the resulting format of a computation is again based on XML. The mappings that transform input GraphML documents to output documents are defined in XSLT style sheets and can be used stand-alone, as components of larger systems, or in, say, web services.

This article is organized as follows. Section 2 provides some background on GraphML, XSLT and their combination. Basic means and concepts of transformations are outlined in Sect. 3, while different types of transformations are discussed in Sect. 4. The integration of XSLT extension mechanisms is described in Sect. 5, and results are discussed and summarized in Section 6.

---

* Research partially supported by DFG under grant Br 2158/1-2 and EU under grant IST-2001-33555 COSIN.
** Corresponding author.

2 Background

A key feature of GraphML is the separation into structural and data layer, both conceptually and syntactically; this enables applications to extend the standard GraphML vocabulary by custom data labels that are transparent to other applications not aware of the extension. Furthermore, applications are free to ignore unknown concepts appearing in the structural layer, such as <port>s, <hyperedge>s or nested <graph>s.

Thanks to its XML syntax, GraphML can be used in combination with other XML based formats: On the one hand, its own extension mechanism allows to attach <data> labels with complex content (possibly required to comply with other XML content models) to GraphML elements, such as Scalable Vector Graphics [5] describing the appearance of the nodes and edges in a drawing; on the other hand, GraphML can be integrated into other applications, e.g. in SOAP messages [6].

Since GraphML representations of graphs often need to be preprocessed or converted to other XML formats, it is convenient to transform them using XSLT, a language specifically designed for transforming XML documents; while originally created for formatting and presenting XML data, usually with HTML, it also allows general restructuring, analysis, and evaluation of XML documents. To reduce parsing overhead and to allow for XML output generation in a natural and embedded way, XSLT itself is in XML syntax.

Basically, the transformations are defined in style sheets (sometimes also called transformation sheets), which specify how an input XML document gets transformed into an output XML document in a recursive pattern matching process. The underlying data model for XML documents is the Document Object Model (DOM), a tree of DOM nodes representing the elements, attributes, text etc., which is held completely in memory. Fig. 1 shows the basic workflow of a transformation.

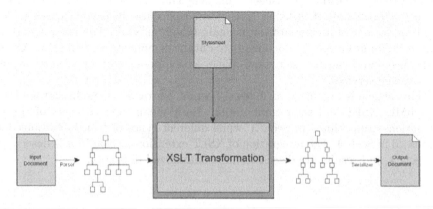

Fig. 1. Workflow of an XSLT transformation. First, XML data is converted to a tree representation, which is then used to build the result tree as specified in the style sheet. Eventually, the result tree is serialized as XML.

DOM trees can be navigated with the XPath language, a sublanguage of XSLT: It expresses paths in the document tree seen from a particular context node (similar to a directory tree of a file system) and serves to address sets of its nodes that satisfy given conditions. For example, if the context node is a `<graph>` element, all node identifiers can be addressed by `child::node/attribute::id`, or `node/@id` as shorthand. Predicates can be used to specify more precisely which parts of the DOM tree to select; for example, the XPath expression `edge[@source='n0']/data` selects only those `<data>` children of `<edge>`s starting from the `<node>` with the given identifier.

The transformation process can be roughly described as follows: A style sheet consists of a list of templates, each having an associated pattern and a template body containing the actions to be executed and the content to be written to the output. Beginning with the root, the processor performs a depth-first traversal (in document order) through the DOM tree. For each DOM node it encounters, it checks whether there is a template whose pattern it satisfies; if so, it selects one of the templates and executes the actions given in that template body (potentially with further recursive pattern matching for the subtrees), and does not do any further depth-first traversal for the DOM subtree rooted at that DOM node; else, it automatically continues the depth-first traversal recursively at each of its children.

3 Basic Means of Transformation

The expressivity and usefulness of XSLT transformations goes beyond their original purpose of only "adding style" to the input. The following is an overview of some important basic concepts of XSLT and how these concepts can particularly be employed in order to formulate advanced GraphML transformations that also take into account the underlying combinatorial structure of the graph instead of only the DOM tree. For some example style sheets, see Sect. 4.

3.1 Parameterization

Especially when integrated as component of a larger system, it is desirable or necessary to parameterize the transformations. Therefore, style sheets can be given an arbitrary number of global parameters `<xsl:param>` that serve as an interface to the outside world. When used autonomously, parameters are passed to the processor as command line parameters.

Such parameters are often used to determine which part of the source document is to be processed. For example, a GraphML file might contain multiple `<graph>`s; a parameter can express the unique identifier of a particular graph that is to be selected. Newer versions of XSLT even allow passing complex XML subtree structures to the transformation.

3.2 Recursion

In the pattern matching process described in Sect. 2, templates were instantiated and executed implicitly or explicitly, when a matching DOM node was encountered in the tree traversal. However, templates can also be given unique names and called like functions together with arbitrary scalar or complex arguments, independently from the tree traversal.

For implementation of more advanced computations, such as graph algorithms, templates may recursively call themselves, typically passing local parameters as function arguments. Similar to the global parameters (see Sect. 3.1), local parameters can represent both scalar values and complex tree fragments. With some limitations, XSLT can be considered a functional language, since templates (and the style sheet as a whole) define functions that are applied to subtrees of the input document and return fragments of the output document.

Due to the lack of assignment statements and side-effects, conventional imperative graph algorithms have to be formulated solely in terms of functions; states, data structures, and intermediate results must be expressed as parameters of function calls. For example, in a breadth-first search the set of all unvisited nodes is passed to a recursive incarnation of the BFS template, instead of becoming marked (see Sect. 4.3).

3.3   Modularization

To make transformations more flexible, they are not necessarily defined in one single file, but can be distributed over a set of modules. The main style sheet imports all templates from another style sheet with `<xsl:import>`, with its own templates having a higher priority, or includes them textually using an `<xsl:include>` tag. Alternatively, style sheets can be composed in advance instead of being imported and included at transformation runtime. Since XSLT is XML, it is even possible for style sheets to compose and transform other style sheets.

Another way of modularizing large transformations is to split them up into several smaller exchangeable style sheets that define successive steps of the transformation, each of which operates on the GraphML result produced in the previous step.

In effect, modularizing transformations facilitates implementing a family of general-purpose and specialized style sheets. Users are free to use specialized modules, or to design new custom templates that extend the general ones.

3.4   External Code

XSLT is designed to be an open, extensible system. While parameterization is one way of using an interface to the outside world when XSLT serves as a component, another even more powerful mechanism is the integration of extension functions into XSLT, i.e. code external to the style sheet. This is especially useful when pure XSLT implementations are inefficient to run or too complicated to use, especially when the input document is large, or when XSLT does not provide necessary functionality at all, e.g. when random numbers, mathematical operations, date functions, or complex string manipulations are needed.

It is important to note that extension functions and classes may violate the declarative, functional design idea of XSLT, since instance-level methods can provide information about mutable states, thus making side-effects possible because a template may now produce different output at different times of execution.

The mechanism is described in more detail in Sect. 5, where we present an extension to be used with GraphML.

4 Transformation Types

Since GraphML is designed as a general format not bound to a particular area of application, an abundance of XSLT use cases exist. However, we found that transformations can be filed into three major categories, depending on the actual purpose of transformation. Note that transformations may correspond to more than one type.

4.1 Internal

While one of GraphML's design goals is to require a well-defined interpretation for all GraphML files, there is no uniqueness the other way round, i.e. there are various GraphML representations for a graph; for example, its <node>s and <edge>s may appear in arbitrary order. However, applications may require their GraphML input to satisfy certain preconditions, such as the appearance of all <node>s before any <edge> in order to set up a graph in memory on-the-fly while reading the input stream.

Generally, some frequently arising transformations include

- pre- and postprocessing the GraphML file to make it satisfy given conditions, such as rearranging the markup elements or generating unique identifiers,
- inserting default values where there is no explicit entry, e.g. edge directions or default values for <data> tags,
- resolving XLink references in distributed graphs,
- filtering out unneeded <data> tags that are not relevant for further processing and can be dropped to reduce communication or memory cost, and
- converting between graph classes, for example eliminating hyperedges, expanding nested graphs, or removing multiedges.

For such GraphML-to-GraphML transformations that operate on the syntactical representation rather than on the underlying combinatorial structure, XSLT style sheets are a very useful and lightweight tool. Often, the source code fits on one single page. See, e.g., Fig. 2 and Fig. 3.

4.2 Format Conversion

Although in recent years GraphML and similar formats like GXL [9] became increasingly used in various areas of interest, there are still many applications and services not (yet) capable of processing them. To be compatible, formats need to be translatable to each other, preserving as much information as possible.

In doing so, it is essential to take into account possible structural mismatch in terms of both the graph models and concepts that can be expressed by the

involved formats, and their support for additional data. Of course, the closer
the conceptual relatedness between source and target format is, the simpler the
style sheets typically are.

```xsl
<xsl:stylesheet version="2.0"
    xmlns:xsl="http://www.w3.org/1999/XSL/Transform">
  <xsl:output method="xml" indent="yes" encoding="iso-8859-1"/>

  <xsl:template match="data|desc|key|default"/> <!-- empty template-->

  <xsl:template match="/graphml">
    <graphml>
      <xsl:copy-of select="key|desc|@*"/>
      <xsl:apply-templates match="graph"/> <!-- process graph(s) -->
    </graphml>
  </xsl:template>

  <xsl:template match="graph">  <!-- override template -->
    <graph>
      <xsl:copy-of select="key|desc|@*"/>
      <xsl:copy-of select="node"/> <!-- nodes first -->
      <xsl:copy-of select="edge"/> <!-- then edges -->
    </graph>
  </xsl:template>
</xsl:stylesheet>
```

Fig. 2. This transformation rearranges the graph so that the nodes appear before the
edges. All subtrees related to data extensions (data and key tags) are omitted.

```xsl
<xsl:stylesheet version="2.0"
    xmlns:xsl="http://www.w3.org/1999/XSL/Transform">
  <xsl:output method="xml" indent="yes" encoding="iso-8859-1"/>

  <xsl:import href="rearrange.xsl"/> <!-- import templates -->

  <xsl:template match="graph">
    <graph>
      <xsl:copy-of select="key|desc|@*"/>
      <xsl:copy-of select="node"/>
      <xsl:apply-templates match="edge"/>
    </graph>
  </xsl:template>

  <xsl:template match="edge"> <!-- new template rule for edges-->
    <xsl:copy>
      <xsl:copy-of select="@*[name()!='id']|*"/>
      <xsl:attribute name="id">       <!-- create new ID attribute -->
        <xsl:value-of select="generate-id()"/> <!-- XPath-generated ID -->
      </xsl:attribute>
    </xsl:copy>
  </xsl:template>
</xsl:stylesheet>
```

Fig. 3. The transformation in Fig. 2 is extended by importing its templates and over-
riding the template for graphs, as described in Sect. 3.3. Edges are copied to the output
document, except for their identifiers, which are generated anew.

While conversion will be necessary in various settings, two use cases appear to be of particular importance:

- *Conversion into another graph format:* We expect GraphML to be used in many applications to archive attributed graph data and in Web services to transmit aspects of a graph. While it is easy to output GraphML, style sheets can be used to convert GraphML into other graph formats and can thus be utilized in translation services like GraphEx [3]. Converting between GraphML and GXL is discussed in [2].
- *Export to some graphics format:* Of course, graph-based tools in general and graph drawing tools in particular will have to export graphs in graphics formats for visualization purposes. In fact, this is the most natural use of style sheets, and we give an example tranformation to SVG (see Appendix A).

The transformation need not be applied to a filed document, but can also be carried out in memory by applications that ought to be able to export in some target format. Note that, even though XSLT is typically used for mapping between XML documents, it can also be utilized to generate non-XML output.

4.3   Algorithmic

Algorithmic style sheets appear in transformations which create fragments in the output document that do not directly correspond to fragments in the input document, i.e. when there is structure in the source document that is not explicit in the markup. This is typical for GraphML data: For example, it is not possible to determine whether or not a given `<graph>` contains cycles by just looking at the markup; some algorithm has to be applied to the represented graph.

To get a feel for the potential of algorithmic style sheets, we implemented some basic graph algorithms using XSLT, and with recursive templates outlined in Sect. 3.2, it proved powerful enough to formulate even more advanced algorithms. For example, a style sheet can be used to compute the distances from a single source to all other nodes or execute a layout algorithm, and then attach the results to `<node>`s in `<data>` labels. See Fig. 4 and Appendix A.

5   Java Language Binding

We found that pure XSLT functionality is expressive enough to solve even more advanced GraphML related problems. However, it suffers from some general drawbacks:

- With growing problem complexity, the style sheets tend to become disproportionately verbose.
- Algorithms must be reformulated in terms of recursive templates, and there is no way to use existing implementations.
- Computations may perform poorly, especially for large input. This is often due to excessive DOM tree traversal and overhead generated by template instantiation internal to the XSLT processor.
- There is no direct way of accessing system services, such as date functions or data base connectivity.

```xsl
<xsl:stylesheet version="2.0" xmlns:xsl="http://www.w3.org/1999/XSL/Transform">
  <xsl:output method="xml" indent="yes" encoding="iso-8859-1"/>

  <xsl:param name="source">s</xsl:param>   <!-- global parameter -->

  <xsl:template match="data|desc|key"/>

  <xsl:template match="/graphml/graph">
    <graphml>
      <graph>
        <xsl:copy-of select="@*|*[name()!='node']"/>
        <key for="node" name="distance"/>
        <xsl:variable name="bfsnodes">
          <xsl:call-template name="bfs">
            <xsl:with-param name="V" select="node[@id!=$source]"/>
            <xsl:with-param name="W" select="node[@id=$source]"/>
            <xsl:with-param name="dist" select="number(0)"/>
          </xsl:call-template>
        </xsl:variable>
        <xsl:copy-of select="$bfsnodes/node"/>
        <xsl:for-each select="node[not(@id=$bfsnodes/node/@id)]">
          <xsl:copy>
            <xsl:copy-of select="*|@*"/>
            <data key="distance">-1</data>   <!-- not reachable -->
          </xsl:copy>
        </xsl:for-each>
      </graph>
    </graphml>
  </xsl:template>

  <xsl:template name="bfs">
    <xsl:param name="dist"/> <!-- current distance to source -->
    <xsl:param name="V"/>    <!-- unvisited nodes -->
    <xsl:param name="W"/>    <!-- BFS front nodes -->
    <xsl:for-each select="$W">
      <xsl:copy>
        <xsl:copy-of select="*|@*"/>
        <data key="distance"><xsl:value-of select="$dist"/></data>
      </xsl:copy>
    </xsl:for-each>
    <xsl:variable name="new" select="$V[@id=../edge[@source=$W/@id]/@target]"/>
    <xsl:if test="$new"> <!-- newly visited nodes? -->
      <xsl:call-template name="bfs"> <!-- start BFS from them -->
        <xsl:with-param name="V" select="$V[count(.|$new)!=count($new)]"/>
        <xsl:with-param name="W" select="$new"/>
        <xsl:with-param name="dist" select="$dist+1"/>
      </xsl:call-template>
    </xsl:if>
  </xsl:template>
</xsl:stylesheet>
```

Fig. 4. An algorithmic style sheet that starts a breath-first search from a source specified in a global parameter. The computed distances from that source node are attached to the nodes as data tags with a newly introduced key.

Therefore, most XSLT processors allow the integration of extension functions implemented in XSLT or some other programming language. Usually, they support at least their native language. For example, Saxon [4] can access and use external Java classes since itself is written entirely in Java. In this case, extension functions are methods of Java classes available on the class path when the transformation is being executed, and get invoked within XPath expressions. Usually, they are static methods, thus staying compliant with XSLT's design idea of declarative style and freeness of side-effects. However, XSLT allows to create objects and to call their instance-level methods by binding the created objects to XPath variables.

Fig. 5 shows the architecture of a transformation integrating external classes. See Appendix A for a style sheet that makes use of extension functions for random graph generation.

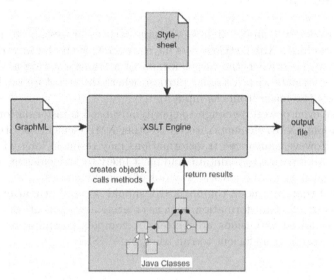

Fig. 5. Extending a transformation with Extension Functions. The box around the Java classes may represent a wrapper class.

In particular, this technique enables developers to implement extensions for graph algorithms. They can either implement extension functions from scratch, or make use of already existing off-the-shelf graph libraries. We implemented a prototype extension for GraphML that basically consists of three layers:

- Java classes for graph data structures and algorithms.
- A wrapper class (the actual XSLT extension) that converts GraphML markup to a wrapped graph object, and provides computation results.
- The style sheet that instantiates the wrapper and communicates with it.

Thus, the wrapper acts as a mediator between the graph object and the style sheet. The wrapper instantiates a graph object corresponding to the GraphML markup, and, for instance, applies a graph drawing algorithm to it. In turn, it provides the resulting coordinates and other layout data in order for the style sheet to insert it into the XML (probably GraphML) result of the transformation, or to do further computations.

The approach presented here is only one of many ways of mapping an external graph description file to an internal graph representation. A stand-alone application could integrate a GraphML parser, build up its graph representation in memory apart from XSLT, execute a transformation, and serialize the result as GraphML output. However, the intrinsic advantage of using XSLT is that it generates output in a natural and embedded way, and that the output generation process can be customized easily.

6   Discussion

We have presented a simple, lightweight approach to processing graphs represented in GraphML. XSLT style sheets have proven to be useful in various areas of application, both when the target format of a transformation is GraphML, and in other formats with a similar purpose where the structure of the output does not vary too much from the input.

They are even powerful enough to specify advanced transformations that go beyond mapping XML elements directly to other XML elements or other simple text units. However, advanced transformations may result in long-winded style sheets that are intricate to maintain, and most likely to be inefficient. Extension functions appear to be the natural way out of such difficulties.

We found that, as rule-of-thumb, XSLT should be used primarily to do the structural parts of a transformation, such as creating new elements or attributes, whereas specialized extensions are better for complex computations that are difficult to express or inefficient to run using pure XSLT.

References

1. U. Brandes, M. Eiglsperger, I. Herman, M. Himsolt, and M. S. Marshall. GraphML progress report: Structural layer proposal. *Proc. 9th Intl. Symp. Graph Drawing (GD '01)*, LNCS 2265:501–512. Springer, 2001.
2. U. Brandes, J. Lerner, and C. Pich: GXL to GraphML and vice versa with XSLT. *Proc. 2nd Intl. Workshop Graph-Based Tools (GraBaTs '04)*. To appear.
3. S. Bridgeman. GraphEx: An improved graph translation service. *Proc. 11th Intl. Symp. Graph Drawing (GD '03)*, LNCS 2912:307–313. Springer, 2004.
4. M. Kay. *SAXON*. http://saxon.sourceforge.net/.
5. W3C. *Scalable Vector Graphics*. http://www.w3.org/TR/SVG.
6. W3C. *SOAP*. http://www.w3.org/TR/soap12-part0/.
7. W3C. *XSL Transformations*. http://www.w3.org/TR/xslt.
8. R. Wiese, M. Eiglsperger, and M. Kaufmann. yFiles: Visualization and automatic layout of graphs. *Proc. 9th Intl. Symp. Graph Drawing (GD '01)*, LNCS 2265:453–454. Springer, 2001.
9. A. Winter. Exchanging Graphs with GXL. *Proc. 9th Intl. Symp. Graph Drawing (GD '01)*, LNCS 2265:485–500. Springer, 2001.

A   Supplement

Due to space limitations, we do not give extensive examples for style sheets. The following examples are referred to in this paper and can be obtained from the GraphML homepage (graphml.graphdrawing.org):

GraphML → SVG. An example for conversion into an XML-based graphics format (requires coordinates).

Spring Embedder. A computational style sheet computing coordinates using a popular layout algorithm.

Random Graph Generator. Generates random graphs in the Erdős-Rényi model by calling an external random number generator (Java language binding).

Clustering Cycles into Cycles of Clusters*

Extended Abstract

Pier Francesco Cortese, Giuseppe Di Battista,
Maurizio Patrignani, and Maurizio Pizzonia

Dipartimento di Informatica e Automazione – Università Roma Tre, Italy
{cortese,gdb,patrigna,pizzonia}@dia.uniroma3.it

Abstract. In this paper we study the clustered graphs whose underlying graph is a cycle. This is a simple family of clustered graphs that are "highly non connected". We start by studying 3-cluster cycles, that are clustered graphs such that the underlying graph is a simple cycle and there are three clusters all at the same level. We show that in this case testing the c-planarity can be done efficiently and give an efficient drawing algorithm. Also, we characterize 3-cluster cycles in terms of formal grammars. Finally, we generalize the results on 3-cluster cycles considering clustered graphs that at each level of the inclusion tree have a cycle structure. Even in this case we show efficient c-planarity testing and drawing algorithms.

1 Introduction

Consider the following problem. A cycle is given where each vertex has a label. Is it possible to add new edges so that: (i) the new graph (i.e. cycle plus new edges) is planar and (ii) for each label, the subgraph induced by the vertices with that label is connected? An example is in Fig. 1.a. In this case the problem admits a solution, depicted in Fig. 1.b.

In this paper we tackle problems of the above type. Such kind of problems arise in the field of *clustered planarity* [9, 8]. Given a graph, a *cluster* is a non empty subset of its vertices. A *clustered graph* consists of a graph G and a rooted tree T such that the leaves of T are the vertices of G. Each node ν of T corresponds to the cluster $V(\nu)$ of G whose vertices are the leaves of the subtree rooted at ν. The subgraph of G induced by $V(\nu)$ is denoted as $G(\nu)$. An edge e between a vertex of $V(\nu)$ and a vertex of $V - V(\nu)$ is said to be *incident* on ν. Graph G and tree T are called *underlying graph* and *inclusion tree*, respectively. A clustered graph is *connected* if for each node ν of T we have that $G(\nu)$ is connected.

In a *drawing* of a clustered graph each vertex of G is a point and each edge is a simple curve between its end-vertices. For each node ν of T, $G(\nu)$ is drawn inside a

* Work partially supported by European Commission – Fet Open project COSIN – COevolution and Self-organisation In dynamical Networks – IST-2001-33555, by European Commission – Fet Open project DELIS – Dynamically Evolving Large Scale Information Systems – Contract no 001907, by "Progetto ALINWEB: Algoritmica per Internet e per il Web", MIUR Programmi di Ricerca Scientifica di Rilevante Interesse Nazionale, and by "The Multichannel Adaptive Information Systems (MAIS) Project", MIUR–FIRB.

J. Pach (Ed.): GD 2004, LNCS 3383, pp. 100–110, 2004.

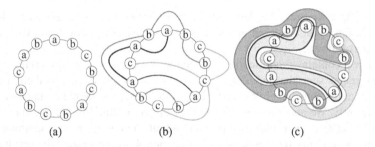

Fig. 1. (a) An example of a cycle with labels in $\{a, b, c\}$. (b) The cycle with extra edges. (c) The corresponding clustered drawing of the cycle.

simple closed region $R(\nu)$ such that: (i) for each node μ of T that is neither an ancestor nor a descendant of ν, $R(\mu)$ is completely contained in the exterior of $R(\nu)$; (ii) an edge e incident on ν crosses the boundary of $R(\nu)$ exactly once. We say that edge e and region R have an *edge-region crossing* if both endpoints of e are outside R and e crosses the boundary of R. A drawing of a clustered graph is *c-planar* if it does not have edge crossings and edge-region crossings. A clustered graph is *c-planar* if it has a c-planar drawing.

Consider again the example of Fig. 1 according to the above definitions. The cycle is the underlying graph of a clustered graph. Vertices with the same label are in the same cluster. The inclusion tree consists of a root with three children, denoted a, b, and c. The children of node x are the vertices labeled x. The edges added to the cycle are used to "simulate" the closed regions containing the clusters (See Fig. 1.c). In this paper we call *saturator* such set of edges. The clustered graph of the example is c-planar. Further, the problem of adding extra edges to a labeled cycle admits a solution iff the corresponding clustered graph is c-planar. Observe that the clustered graph of the example is not connected.

Clustered planarity, because of its practical impact and because of its theoretical appeal, attracted many research contributions. Feng, Cohen, and Eades devised the first polynomial time c-planarity testing algorithm for connected clustered graphs [9]. A planarization algorithm for connected clustered graph is shown in [5]. However, the complexity of the problem for a non connected clustered graph is still unknown.

A contribution on this topic has been given by Gutwenger et al. that presented a polynomial time algorithm for c-planarity testing for *almost connected* clustered graphs [10]. In almost connected clustered graphs either all nodes corresponding to non connected clusters are in the same path in T starting at the root of T, or for each non connected cluster its parent and all its siblings are connected. Also, the works in [1, 2] by Biedl, Kaufmann, and Mutzel can be interpreted as a linear time c-planarity test for non connected clustered graphs with exactly two clusters at the same level.

Another contribution studying the interplay between c-planarity and connectivity has been presented in [3] by Cornelsen and Wagner. They show that a *completely connected* clustered graph is c-planar iff its underlying graph is planar. A completely connected clustered graph is so that not only each cluster is connected but also its complement is connected.

In this paper we study the clustered graphs whose underlying graph is a cycle. This is a simple family of clustered graphs that are "highly non connected". The paper is organized as follows.

Section 2 contains preliminaries. In Section 3 we study *3-cluster cycles*, that are clustered graphs such that the underlying graph is a simple cycle and there are three clusters all at the same level. We show that, in this case, testing the c-planarity can be done efficiently. We also give an efficient drawing algorithm. Further, we show that in this specific case if the c-planarity problem admits a solution then a saturator exists that is composed only by simple paths. In Section 4 we generalize the results on 3-cluster cycles considering clustered graphs that on each level of the inclusion tree have a cycle structure. Even in this case we show efficient c-planarity testing and drawing algorithms. Section 5 contains conclusions and open problems.

2   Preliminaries

We assume familiarity with connectivity and planarity of graphs [7, 6]. We also assume familiarity with formal grammars [11].

Given a c-planar non connected clustered graph $C(G, T)$, a *saturator* of C is a set of edges that can be added to the underlying graph G so that C becomes connected without loosing its c-planarity. Finding a saturator of a clustered graph is important since it allows to apply to C the same drawing techniques that have been devised for connected clustered graphs.

We call *3-cluster cycle* a clustered graph such that the underlying graph is a simple cycle and there are exactly three clusters all at the same level (plus the root cluster). In a 3-cluster cycle the inclusion tree consists of a root node with three children and each vertex of the underlying cycle is a child of one of these three nodes. Given a 3-cluster cycle, we associate a label in $\{a, b, c\}$ to each of the three clusters.

Consider a 3-cluster cycle and arbitrarily select a starting vertex and a direction. We can visit the cycle and denote it by the sequence σ of labels associated with the clusters encountered during the visit. The same clustered cycle is also denoted by any cyclic permutation of σ and by any reverse sequence of such permutations. We use Greek letters to denote general sequences and Roman letters to identify single-character sequences. Given a sequence σ, we denote with $\overline{\sigma}$ its reverse sequence.

A non c-planar c-cluster cycle is $abcabc$, while a c-planar one is $abcbac$.

It is easy to see that repeated consecutive labels can be collapsed into a single label without affecting the c-planarity property of a 3-cluster cycle. Hence, in the following we consider only 3-cluster cycles where consecutive vertices belong to distinct clusters. Also, since clusters can not be empty, in a 3-cluster cycle at least one occurrence of each label can be found.

We assign a cyclic order to the clusters so that $a \prec b$, $b \prec c$, and $c \prec a$. A sequence σ is *monotonic increasing* (*decreasing*) if for each pair x, y of consecutive labels of σ $x \prec y$ ($y \prec x$). A sequence is *cyclically increasing* (*decreasing*) *monotonic* if all its cyclic permutations are increasing (decreasing) monotonic.

Given a 3-cluster cycle σ, $Balance(\sigma)$ is a number defined as follows. Select a starting vertex and a direction. Set counter c to zero. Visit σ adding (subtracting) one unit to c when passing from x to y, where $x \prec y$ ($y \prec x$). Observe that, when the starting vertex is reached again, c is a multiple of 3 that can be positive, negative, or zero. If we selected a different starting vertex, while preserving the direction, we would obtain the same value. On the contrary, if σ was visited in the opposite direction the opposite value would be obtained for c. $Balance(\sigma) = |c|$. For example, $Balance(ababc) = 3$ and $Balance(cbacba) = 6$.

Observe that, when representing a 3-cluster cycle with a sequence of labels, by reading the sequence from left to right, we implicitly choose a direction for visiting the cycle. For simplicity, we adopt the convention of representing a 3-cluster cycle with a sequence σ such that, when the vertices of the cycle are visited according to the order induced by σ, a non negative value for c is obtained.

3   Cycles with Three Clusters

In this section we address the problem of testing the c-planarity of a 3-cluster cycle. The following lemma introduces transformations that can be used to simplify 3-cluster cycles without affecting their c-planarity properties.

Lemma 1. *Let* $\sigma = \sigma_1 x \alpha y \overline{\alpha} x \alpha y \sigma_2$ *be a 3-cluster cycle such that* σ_1, σ_2, *and* α *are possibly empty and* $x \alpha y$ *is monotonic. The 3-cluster cycle* $\sigma' = \sigma_1 x \alpha y \sigma_2$ *is c-planar if and only if* σ *is c-planar.* $Balance(\sigma) = Balance(\sigma')$.

Proof Sketch: Suppose there exists a c-planar drawing of σ'. The black line in Fig. 2.a shows an example of such a drawing for the portion concerning subsequence $x \alpha y$. Such a drawing can be modified by replacing the edge between y and the first vertex of σ_2 with the sequence $\overline{\alpha} x \alpha y$. Such sequence can be drawn arbitrarily close to $x \alpha y$ preserving c-planarity. Finally, the just added instance of y may be connected to the first vertex of σ_2. The result is shown in Fig. 2.a where the added part is drawn gray.

Now, suppose that there exists a c-planar drawing of σ. Fig. 2.b shows an example of such a drawing for the part concerning subsequence $x \alpha y \overline{\alpha} x \alpha y$. The inlet formed by $x \alpha y \overline{\alpha} x$ may contain parts of σ that are denoted by Q in Fig. 2.b. The parts of σ that are contained in the inlet formed by $y \overline{\alpha} x \alpha y$ are denoted by P. The embedding of P and Q may be rearranged preserving c-planarity as in Fig. 2.c. Path $\overline{\alpha} x \alpha y$ can now

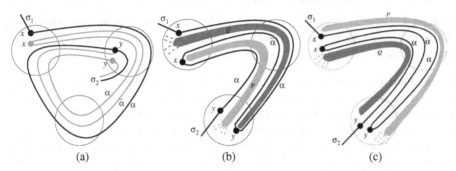

(a)   (b)   (c)

Fig. 2. Illustration of the proof of Lemma 1. (a) Necessary condition. (b) and (c) Sufficient condition.

be deleted and substituted by an edge connecting vertex y with the first vertex of σ_2. Finally, observe that, since we have removed from σ two monotonic sub-sequences, one increasing and one decreasing, with the same length, $Balance(\sigma') = Balance(\sigma)$. □

For example, Lemma 1 allows to study the c-planarity of $cabcab$ instead of the c-planarity of $cabcacbabcab$ (by taking $\sigma_1 = c$, $x = a$, $\alpha = bc$, $y = a$, and $\sigma_2 = b$).

Lemma 2. *Let σ be a 3-cluster cycle. There exists a 3-cluster cycle σ' such that: $Balance(\sigma') = Balance(\sigma)$, σ' is c-planar iff σ is c-planar, and either σ' is cyclically monotonic or $\sigma' = x\alpha y\beta$, where*

1. *α and β are non empty,*
2. *$x\alpha y$ is maximal monotonic increasing, and*
3. *$y\beta x$ is maximal monotonic decreasing.*

Because of space limitation, the proof for this Lemma is omitted. This proof is based on the main idea of repeatedly applying Lemma 1 starting from the shortest monotonic subsequences [4].

The following two lemmas (Lemma 3 and Lemma 4) study the c-planarity of the simple families of 3-cluster cycles cited in Lemma 2.

Lemma 3. *A 3-cluster cycle σ such that σ is cyclically monotonic is c-planar if and only if $Balance(\sigma) = 3$.*

Proof Sketch: Since σ is monotonic we have that $Balance(\sigma) \neq 0$. Recall that $Balance(\sigma)$ is a multiple of 3. If $Balance(\sigma) = 3$, then it can only be the case that $\sigma = abc$ or $\sigma = bca$ or $\sigma = cab$ and it is trivial to see that σ is c-planar.

Suppose that $Balance(\sigma) \geq 6$. We show that σ is not c-planar. Suppose by contradiction that there exists a c-planar drawing Γ_σ of σ. Consider the vertices v_1, v_2, v_3, v_4, v_5 and v_6 of σ as drawn in Γ_σ (see Fig. 3.b). The two edges incident to v_4 separate v_1 from the rest of the vertices of its cluster. Thus, it is possible to add an edge (v_1, v_4) preserving the planarity of the drawing. For similar reasons, it is possible to add the edges (v_2, v_5) and (v_3, v_6). A contradiction arises from the fact that a subdivision of a $K_{3,3}$ can be found in the drawing. Consider, the vertices v_1, v_2, v_3, v_4, v_5 and v_6. Vertex v_1 is connected to v_6 with a path in σ and it is directly connected to v_2 and v_4. Vertices v_3 and v_5 are directly connected to v_2, v_4, and v_6.   □

Lemma 4. *Let $\sigma = x\alpha y\beta$ be a 3-cluster cycle, where α and β are non empty, $x\alpha y$ is maximal monotonic increasing, and $y\beta x$ is maximal monotonic decreasing. We have that σ is c-planar iff $Balance(\sigma)$ is in $\{0, +3\}$.*

Proof Sketch: Let $Balance(\sigma) = 3k$, with k non negative integer. Suppose k is equal to 0 or 1. A c-planar drawing of σ can be constructed by placing the vertices on three half-lines as in the examples shown in Fig. 4.a and 4.b, respectively. The vertices of each half-line can be enclosed into a region representing their cluster.

Suppose that $k > 1$. We show that σ is not c-planar. Suppose for a contradiction that σ is c-planar and let $\Gamma(\sigma)$ be a c-planar drawing of σ. Denote with $v_1, \ldots, v_n$ the vertices of σ starting from the first vertex of α and suppose, without loss of generality, that the length of α is greater or equal than the length of β.

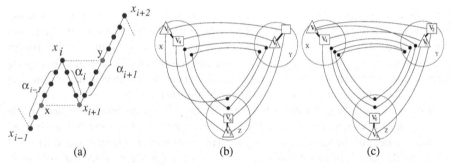

Fig. 3. Illustrations for the proofs of Lemma 2 (a), Lemma 3 (b) and of Lemma 4 (c).

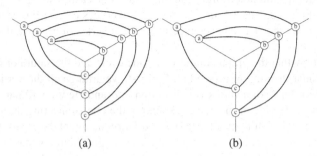

Fig. 4. The construction of a c-planar drawing for a cycle σ when $Balance(\sigma) = 0$ (a) and when $Balance(\sigma) = 3$ (b).

Consider the relative position of v_1 and v_4 in Γ in their cluster X (see Fig. 3.c). We have that the two edges incident on v_4 separate v_1 from the rest of the vertices of X. Thus, it is possible to join v_1 and v_4 with an edge (v_1, v_4) that is entirely contained into the cluster X and that preserves the planarity of the drawing. Analogously, it is possible to join vertices v_2 and v_5 in cluster Y with the edge (v_2, v_5) and vertices v_3 and v_6 in cluster Z with the edge (v_3, v_6).

A contradiction arises since a subgraph that is a subdivision of $K_{3,3}$ can be found in the drawing. In fact, exploiting the edges of σ and the edges introduced above, each vertex in $\{v_1, v_3, v_5\}$ is connected to all vertices in $\{v_2, v_4, v_6\}$. Vertex v_1 is directly connected to v_2 and to v_6 with edges of σ, while it is connected to v_4 with edge (v_1, v_4); vertex v_3 is directly connected to v_2 and to v_4 with edges of σ, while it is connected to v_n with edge (v_3, v_n); finally, vertex v_5 is directly connected to v_4 with an edge of σ, it is connected to v_n with a path in σ, and it is connected to v_2 with edge (v_2, v_5). □

Because of Lemma 2, Lemma 3, and Lemma 4, the problem of testing whether a 3-cluster cycle σ is c-planar can be reduced to the problem of computing $Balance(\sigma)$. Since it is easy to compute $Balance(\sigma)$ in linear time (see Section 2), the following theorem holds.

Theorem 1. *Given an n-vertex 3-cluster cycle, there exists an algorithm to test if it is c-planar in $O(n)$ time.*

In what follows we introduce a simple algorithm which guarantees the computation of a c-planar drawing of a 3-cluster cycle, if it admits one, in linear time. Consider a

3-cluster cycle σ with $Balance(\sigma) \in \{0, +3\}$. Set a counter to zero. Visit σ starting from the first vertex and adding (subtracting) one unit to the counter when passing from x to y, where $x \prec y$ ($y \prec x$). Without loss of generality we will assume that the counter never reaches a negative value. Otherwise, we can replace σ with an equivalent cyclic permutation of it that has the above property and that can be obtained in linear time. Let K be the maximum value assumed by the counter during the visit.

We say that a vertex of σ *belongs to the k-th level* iff the counter has value k when reaching such a vertex. The first vertex of σ belongs to level 0. Note that each level contains vertices of the same cluster. Also, vertices belonging to level k and level $k + 3$ belong to the same cluster. We denote with $\sigma|_k$ the sequence σ *restricted to level* k, obtained from σ by deleting all the vertices not belonging to the k-th level.

We construct a saturator in the following way. For each level $k \in \{0, \ldots, K\}$, we connect with an edge each pair of consecutive vertices of $\sigma|_k$. For each level $k \in \{0, \ldots, K-3\}$, we insert an edge connecting the first vertex of $\sigma|_k$ with the last vertex of $\sigma|_{k+3}$.

Now we show that the graph composed by the cycle and the saturator is planar by providing a planar drawing of it (see Fig. 5). First, we arrange all the vertices of σ on a grid: the x-coordinate of a vertex is its position in σ and the y-coordinate is its level. Then, we draw each edge of the cycle (excluding the one connecting the first and the last vertex of σ) with a straight segment without introducing intersections. Second, for each level $k \in \{0, \ldots, K\}$, we draw those edges of the saturator that connect pairs of consecutive vertices of $\sigma|_k$ with straight segments without introducing intersections. Note that, the sequence of the clusters at levels $0, \ldots, K - 3$ is the same sequence as that of the clusters at levels $3, \ldots, K$. Also, at this point of the construction, for each $k \in \{0, \ldots, K\}$ the first and the last vertices of $\sigma|_k$ are on the external face. Hence, the drawing can be completed without intersections by adding, for each level $k \in \{0, \ldots, K - 3\}$, the edge of the saturator connecting the first vertex of $\sigma|_k$ with

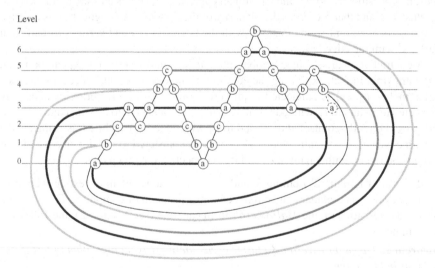

Fig. 5. The construction of a c-planar drawing of a 3-cluster cycle σ in the case in which $Balance(\sigma) = 3$.

the last vertex of $\sigma|_{k+3}$ as shown in the example of Fig. 5. Finally, since the first and the last vertex of σ are on the same face, they can be connected with a curve contained into such a face without introducing intersections. To explicitly represents clusters as simple closed regions starting from the saturator we select a region of the plane at small distance (strictly greater than zero) from each saturator edge and delete the saturator.

It is easy to implement the above algorithm to work in linear time by building the lists of vertices for each level while visiting σ. Notice that K is bounded by the number of the vertices of the cycle.

Hence, we can state the following result.

Theorem 2. *Given an n-vertex c-planar 3-cluster cycle σ, there exists an algorithm that computes a c-planar drawing of σ in $O(n)$ time.*

From the above construction we also have the following.

Theorem 3. *A c-planar 3-cluster cycle admits a saturator that is the collection of three disjoint paths.*

If we consider the representation of 3-cluster cycles as strings, it is possible to show, in terms of formal grammars, that the set of 3-cluster cycles is a regular set, while the set of c-planar 3-cluster cycles is generated by a context-free grammar [4].

4   Cycles in Cycles of Clusters

In this section we present a generalization of the results of Section 3. First, we generalize the results on 3-cluster cycles to the case of clusters that form a cycle whose length is greater than three. Second, we tackle the general problem of testing the c-planarity of a cycle that is clustered into a cycle of clusters that is in turn clustered into another cycle of clusters, and so on. Fig. 6.a shows c-planar clustered graph whose underlying graph is a cycle for which two levels of clusters are defined. Fig. 6.b puts in evidence the inclusion relationship between clusters of a given level and clusters of the level directly above it. The same figure shows also that the clusters of each level form a cycle.

We start by introducing preliminary assumptions and definitions. We consider clustered graphs $C(G, T)$ in which all the leaves of the inclusion tree T have the same distance from the root (we call *depth* that distance). A clustered graph which has not this property can be easily reduced to this case by inserting "dummy" nodes in T. Hence, from now on we consider only inclusion trees whose leaves are all at the same depth. We define as $G^l(V^l, E^l)$ the graph whose vertices are the nodes of T at distance l from its root, and an edge (μ, ν) exists if and only if an edge of G exists incident to both μ and ν.

For example, G^0 has only one vertex and G^L, where L is the depth of the tree, is the underlying graph G of $C(G, T)$. We label each vertex ν of G^l with the cluster (corresponding to a vertex of G^{l-1}) which ν belongs to. If G^l is a cycle, then it is possible to identify G^l with the cyclic sequence of the labels of its vertices. If also G^{l-1} is a cycle, we consider the labels of G^l cyclically ordered according to the order they appear in G^{l-1}. At this point, $Balance(G^l)$ can be defined as in Section 3 and can assume values $0, k, 2k, 3k, \ldots$ where k is the length of G^{l-1}.

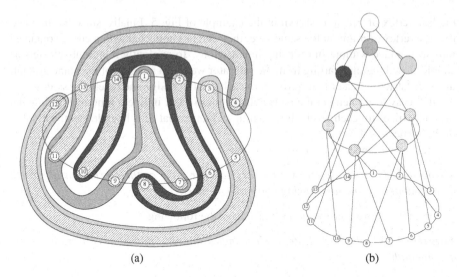

(a)   (b)

Fig. 6. A clustered graph where at each level of the inclusion tree the nodes form a cycle. (a) A c-planar drawing. (b) The inclusion tree augmented with edges that put in evidence the adjacencies between nodes at the same level.

According to the above definitions a 3-cluster cycle is a clustered graph where T as depth 2, G^2 is a cycle and G^1 is a cycle of length 3. In fact, the results of Section 3 can be extended to the case in which G^1 is a cycle of an arbitrary length.

Theorem 4. *Given an n-vertex clustered graph $C(G,T)$, such that T has depth 2 and G^1 and G^2 are cycles, there exists an algorithm to test if C is c-planar in $O(n)$ time. If C is c-planar, a c-planar drawing of C can be computed in $O(n)$ time.*

Proof Sketch: The proof exploits the same considerations and constructions of Theorems 1 and 2. If the length of G^1 is k then C is c-planar iff $Balance(G^2) \in \{0, k\}$. In order to find a c-planar drawing of C, if it exists, the same strategy described in Section 3 can be applied, where, since in the construction depicted in Fig. 5 vertices belonging to level j and level $j + k$ belong to the same cluster, an edge of the saturator is added between the first vertex of level j and the last vertex of level $j + k$ instead of between the first vertex of level j and the last vertex of level $j + 3$.   □

Let $C(G,T)$ be a clustered graph and l be an integer between 1 and L, where L is the depth of T. Clustered graph $C^l(G, T^l)$ is obtained from C by replacing T with a tree T^l obtained from T by connecting all the nodes at depth l with the root and deleting all the nodes having depth greater than zero and less than l. The c-planarity of C^l can be used to study the c-planarity of C^{l-1}, as is shown in the following lemma.

Lemma 5. *Let $C(G,T)$ be a clustered graph and l be an integer between 1 and L, where L is the depth of T. Let C^l be c-planar, G^l be a cycle, and G^{l-1} be a cycle of length k. C^{l-1} is c-planar iff $Balance(G^l) \in \{0, k\}$.*

Proof Sketch: First, we prove that if $Balance(G^l) \in \{0, k\}$, then C^{l-1} is c-planar. Since $Balance(G^l) \in \{0, k\}$, then it exists a planar drawing of G^l augmented with the edges of a saturator connecting vertices of G^l with the same label. Those edges can be added to the internal or external face of cycle G^l according to the output of the algorithm described in Section 3. Let Γ_{C^l} be a c-planar drawing of C^l. Since G^l is a cycle, there exist in Γ_{C^l} exactly two faces containing vertices belonging to all the clusters corresponding to vertices of G^{l-1}. Call such faces *internal* and *external* face arbitrarily. A c-planar drawing $\Gamma_{C^{l-1}}$ can be constructed by adding to Γ_{C^l} an edge for each edge of the saturator of G^l in such a way to place on the internal (external) face of Γ_{C^l} the edges of the saturator that are added to the internal (external) face of G^l.

The second part of the proof shows that if $Balance(G^l)$ is not in $\{0, k\}$ then then C^{l-1} is not c-planar. Assume for a contradiction that $Balance(G^l)$ is not in $\{0, k\}$ and a c-planar drawing of $\Gamma_{C^{l-1}}$ exists.

By using similar arguments as in the proofs of Lemmas 3 and 4, a subdivision of a $K_{3,3}$ can be found where the vertices of the subdivision are actually vertices of G^l, that is, clusters of C. $\square$

Lemma 6. *Let $C = (G, T)$ be a clustered graph and let be l an integer between 1 and L, where L is the depth of T. If C^l is not c-planar, then $C^0 = C$ is not c-planar.*

Proof Sketch: If C^l is not c-planar, there is a subdivision of $K_{3,3}$ or K_5 in the graph G augmented with the edges of the saturator of C^l. The same obstruction can be found in the graph G augmented with the edges of saturator of C^0; hence C^0 can not be c-planar.
$\square$

Theorem 5. *Given an n-vertex clustered graph $C(G, T)$, such that T has depth L and, for $l > 0$, G^l is a cycle, there exists an algorithm to test if C is c-planar in $O(Ln)$ time.*

Proof Sketch: The proof is based on iteratively applying, level by level, Lemma 5 to the clustered graphs C^l for $l = L, L-1, \ldots, 2$. Since each test can be performed in $O(n)$ time, the statement follows. $\square$

Theorem 6. *Given an n-vertex clustered graph $C(G, T)$, such that T has depth L and, for $l > 0$, G^l is a cycle, if C is c-planar there exists an algorithm to compute a c-planar drawing of C in $O(Ln)$ time.*

Proof Sketch: The proof of Lemma 5 is a constructive one. Thus, by applying, level by level, Lemma 5 starting from level L to level 1, a c-planar drawing of C can be obtained. Since each step may be performed in $O(n)$ the statement follows. $\square$

5   Conclusions and Open Problems

In this paper we studied a peculiar family of non-connected clustered graphs. Namely, we studied clustered graphs whose underlying graph is a simple cycle. Besides the general problem of stating the complexity of the c-planarity testing of non-connected clustered graphs, several other problems remain open.

Are there other families of non-connected clustered graphs whose c-planarity can be efficiently assessed and whose underlying graph has a simple structure? For example, what happens if the underlying graph is a tree? It is easy to show that a two-level clustered graph whose underlying graph G^2 is a path and such that graph G^1 is a cycle, is c-planar. It is also easy to find an example of a two level clustered graph whose underlying graph G^2 is a tree, such that G^1 is a cycle and that is not c-planar.

Suppose that the underlying graph has a fixed embedding. Can this hypothesis simplify the c-planarity testing?

Can the techniques introduced in this paper be combined with techniques known in the literature for devising tools able to handle the c-planarity testing and embedding problem for more complex families of clustered graphs?

References

1. T. Biedl, M. Kaufmann, and P. Mutzel. Drawing planar partitions II: HH-Drawings. In *Workshop on Graph-Theoretic Concepts in Computer Science (WG'98)*, volume 1517, pages 124–136. Springer-Verlag, 1998.
2. T. C. Biedl. Drawing planar partitions III: Two constrained embedding problems. Tech. Report RRR 13-98, RUTCOR Rutgen University, 1998.
3. S. Cornelsen and D. Wagner. Completely connected clustered graphs. In *Proc. 29th Intl. Workshop on Graph-Theoretic Concepts in Computer Science (WG 2003)*, volume 2880 of *LNCS*, pages 168–179. Springer-Verlag, 2003.
4. P. F. Cortese, G. Di Battista, M. Patrignani, and M. Pizzonia. Clustering cycles into cycles of clusters. Technical Report RT-DIA-91-2004, Dipartimento di Informatica e Automazione, Università di Roma Tre, Rome, Italy, 2004.
5. G. Di Battista, W. Didimo, and A. Marcandalli. Planarization of clustered graphs. In *Proc. Graph Drawing 2001 (GD'01)*, LNCS, pages 60–74. Springer-Verlag, 2001.
6. G. Di Battista, P. Eades, R. Tamassia, and I. G. Tollis. *Graph Drawing*. Prentice Hall, Upper Saddle River, NJ, 1999.
7. S. Even. *Graph Algorithms*. Computer Science Press, Potomac, Maryland, 1979.
8. Q. W. Feng, R. F. Cohen, and P. Eades. How to draw a planar clustered graph. In Ding-Zhu Du and Ming Li, editors, *Proc. COCOON'95*, volume 959 of *LNCS*, pages 21–30. Springer-Verlag, 1995.
9. Q. W. Feng, R. F. Cohen, and P. Eades. Planarity for clustered graphs. In P. Spirakis, editor, *Symposium on Algorithms (Proc. ESA '95)*, volume 979 of *LNCS*, pages 213–226. Springer-Verlag, 1995.
10. C. Gutwenger, M. Jünger, S. Leipert, P. Mutzel, M. Percan, and R. Weiskircher. Advances in C-planarity testing of clustered graphs. In Stephen G. Kobourov and Michael T. Goodrich, editors, *Proc. Graph Drawing 2002 (GD'02)*, volume 2528 of *LNCS*, pages 220–235. Springer-Verlag, 2002.
11. J. E. Hopcroft and J. D. Ullman. *Introduction to Automata Theory, Languages, and Computation*. Addison-Wesley, 1979.

Unit Bar-Visibility Layouts
of Triangulated Polygons
Extended Abstract

Alice M. Dean[1], Ellen Gethner[2], and Joan P. Hutchinson[3]

[1] Skidmore College, Saratoga Springs, NY 12866
adean@skidmore.edu
[2] University of Colorado at Denver, Denver CO 80217
ellen.gethner@cudenver.edu
[3] Macalester College, St. Paul, MN 55105
hutchinson@macalester.edu

Abstract. A *triangulated polygon* is a 2-connected maximal outerplanar graph. A *unit bar-visibility graph* (UBVG for short) is a graph whose vertices can be represented by disjoint, horizontal, unit-length bars in the plane so that two vertices are adjacent if and only if there is a non-degenerate, unobstructed, vertical band of visibility between the corresponding bars. We give combinatorial and geometric characterizations of the triangulated polygons that are UBVGs. To each triangulated polygon G we assign a character string with the property that G is a UBVG if and only if the string satisfies a certain regular expression. Given a string that satisfies this condition, we describe a linear-time algorithm that uses it to produce a UBV layout of G.

1   Introduction

A *bar-visibility layout* of a graph G is a representation of G in the plane by disjoint horizontal line segments ('bars') in which each vertex corresponds to a bar and two vertices are adjacent if and only if there is an unobstructed, non-degenerate vertical *visibility band* between the corresponding bars. If G has such a layout it is called a *bar-visibility graph* (BVG for short). A BVG layout induces a plane embedding of G in a natural way, by placing each vertex on its corresponding bar and drawing edges between pairs of vertices whose bars have vertical visibility. A BVG and its corresponding layout are shown in Fig. 1. The original motivation for studying BVGs was the design of electronic circuits; another application is the display of data, using bars 'fattened' into rectangles that hold labels, with relations between data items represented by visibility bands.

Bar-visibility graphs were fully characterized in the mid-1980s [9, 12, 13] as those planar graphs having a planar embedding with all cutpoints on a common face, and linear-time recognition and layout algorithms were given. Generalizations of bar-visibility graphs have also been studied, including visibility representations using different objects like rectangles and with different rules for visibility between objects [1, 5, 6, 8, 10, 11].

J. Pach (Ed.): GD 2004, LNCS 3383, pp. 111–121, 2004.

Fig. 1. A triangulated polygon and its UBV layout.

The usefulness of bar-visibility layouts diminishes when the relative lengths of bars vary widely. The simplest way to restrict the relative lengths of bars is to require all bars to have equal length; such a graph is called a *unit bar-visibility graph* (or UBVG). Fundamental results concerning these graphs appear in [7]; however, in contrast to BVGs, no full characterization of these graphs has been found. We characterize a significant subclass of UBVGs, the *triangulated polygons*.

A triangulated polygon is a 2-connected, maximal outerplanar graph; in other words, a graph with a plane embedding as a simple, closed curve whose interior is subdivided by diagonals into triangles. The graph in Fig. 1 is a triangulated polygon, and the layout is a UBV layout. To each triangulated polygon G we associate a character string called the *internal spine string* that encodes enough information about G to determine whether or not G is a UBVG and, if it is, to produce a UBV layout of G. The layout algorithm runs in linear time.

In Section 2 we define the maximal and internal spine strings corresponding to triangulated polygon. In Section 3 we state a series of necessary conditions on the maximal and internal spine strings, leading to our main theorem characterizing those triangulated polygons that are UBVGs, and we outline the proof of necessity. In Section 4 we use the internal spine string to give a linear-time algorithm that produces a UBV layout of the corresponding triangulated polygon.

2   Spine Strings and Clumps

If G is a plane graph, we call the unbounded face of G the *external* face, and the other faces are called *internal*. G^* denotes the *dual* of G, in which the vertices are the faces of G, and two vertices are adjacent if and only if the corresponding faces of G share an edge. The *internal dual* of G, denoted G_I^*, is the subgraph of G^* induced by the internal faces of G. A graph G is *outerplanar* if it has a plane embedding in which all vertices lie on the external face; such an embedded graph is called *outerplane*. A straightforward but key observation is that a 2-connected graph is outerplane if and only if its internal dual G_I^* is a tree. Lastly a *maximal* outerplanar graph is one in which each internal face is a triangle, hence the internal dual of such a graph has maximum degree at most 3. If a maximal outerplanar graph is 2-connected, then it has a *unique* outerplane embedding as a *triangulated polygon*, and we generally do not distinguish between the graph and its outerplane embedding.

A *caterpillar* is a tree containing a path P, called a *spine*, such that all vertices have distance at most 1 from P. A *subdivided* caterpillar, in which each edge is replaced by a path, has a path P, also called a spine, that contains all vertices of degree 3 or more. It follows from results of [7] that if a triangulated polygon G is induced by a UBV layout, then its internal dual G is a subdivided caterpillar. This condition is necessary but not sufficient. Given a triangulated polygon G whose internal dual is a subdivided caterpillar, we define below a character string that encodes key aspects of the embedding of G. The central result of this paper is that this string encodes necessary and sufficient information to determine if G is a UBVG.

Definition 1.   *1. Let G be a triangulated polygon G whose internal dual G_I^* is a subdivided caterpillar (necessarily of maximum degree 3). Choose a maximal spine of G_I^*, $P^* = F_0, F_1, \ldots, F_k, F_{k+1}$. As P^* is traversed in order of increasing i, each face F_i, $i = 1, \ldots, k$, shares one edge with F_{i-1} and another with F_{i+1}. Denote the vertex incident with these two edges by v_i, and denote the third edge of F_i, which is not incident with any other face on P^*, by e_i. If P^* is oriented left-to-right in order of traversal, then e_i lies either above or below v_i; we say briefly that e_i lies above (resp., below) P^*. Define a string S_M of length k, composed of the four symbols A, N_A, B, N_B, as follows. If e_i lies above P^*, then the ith character of S_M is either A or N_A, depending on whether F_i does or does not have a leg-neighbor above P^*. Similarly, if e_i lies below P^*, then the ith character of S_M is either B or N_B, depending on whether F_i does or does not have a leg-neighbor below P^*. The string S_M is called a* maximal spine string *for G.*
2. *Given any string composed of the symbols A, N_A, B, N_B, an A-clump (resp. B-clump) is a maximal length substring using only the symbols A and N_A (resp., B and N_B). A trivial clump is an A-clump or B-clump comprised entirely of N_A or N_B terms.*
3. *If S_M is a maximal spine string, then the* internal spine string S_I *is the substring obtained by deleting all symbols including and preceding those in the first non-trivial clump of S_M, and also all symbols including and following those in the last non-trivial clump of S_M. It is possible that S_I is the empty string.*

The triangulated polygon in Fig. 1 has (non-unique) maximal spine string $S_M = N_B B N_B N_A A N_A A N_A N_B B N_B$, comprising three clumps. The corresponding internal spine string is $S_I = N_A A N_A A N_A$. In what follows we write an arbitrary maximal spine string S_M as a string of clumps, $S_M = T_0 C_1 C_2 \ldots C_k T_{k+1}, k \geq 0$, where T_0 is the union of all trivial clumps at the beginning of S_M, T_{k+1} is the union of all trivial clumps at the end of S_M, and $C_1, \ldots, C_k$ are the remaining clumps of S_M, where C_1 and C_k are necessarily non-trivial. The corresponding internal spine string is $S_I = C_2 \ldots C_{k-1}$.

3   Necessity and the Characterization Theorem

Given a triangulated polygon G whose internal dual is a subdivided caterpillar, we choose a maximal spine string S_M and divide it into clumps, $S_M =

$T_0C_1\ldots C_kT_{k+1}$, as described in Def. 1. Certain graphs can be eliminated immediately if their clumps have too many non-trivial terms, or if two non-trivial terms in a single clump are too far apart, as given below in Thm. 2. For the remaining, 'feasible' graphs, additional parsing of the clumps is required, as given in Thm. 6. Analysis of this parsing applied to the *internal* spine string determines whether a UBV layout exists; Thm. 8 gives the full characterization in terms of valid maximal and internal spine strings.

Theorem 2. *Let G be a triangulated polygon whose internal dual is a subdivided caterpillar with maximal spine $S_M = T_0C_1\ldots C_kT_{k+1}$, as described in Def. 1. Each of the following conditions is necessary for G to be a UBVG.*

1. *If $k = 1$, then C_1 contains at most four A- or B-terms.*
2. *If $k \geq 2$, then C_1 and C_k each contain at most three A- or B-terms, and C_i, for $i = 2,\ldots,k-1$, contains at most two A- or B-terms.*
3. *If $k \geq 3$, then no C_i, $2 \leq i \leq k-1$, contains any substring of the form $AN_A^{++}A$ or $BN_B^{++}B$, where the notation $++$ indicates an exponent that is at least two.*

A triangulated polygon G that satisfies the conditions of Thm. 2 is called *UBVG-feasible* or *feasible*. Having eliminated all 'infeasible' graphs from consideration, we do a further parsing of the clumps, leading to an analysis of the internal spine string that characterizes those feasible graphs having UBV layouts.

The relation of the spine string to a UBV layout of the triangulated polygon G comes from the fact that in both settings there are notions of the directions *left, right, up,* and *down*. For the spine string the directions are defined relative to a traversal of the spine. For a UBV layout the directions indicate relative positions of bars for adjacent faces, as defined below. G is a UBVG if and only if these two notions of direction are compatible.

Definition 3. *Suppose that the triangulated polygon G is a UBVG with UBV layout $U(G)$. We assume henceforth that each bar in a UBV layout has length 1 and is at a unique vertical level, usually at integer heights.*

1. *We denote the height of a bar b by $y(b)$, and its left x-coordinate by $x(b)$ (thus its right x-coordinate is $x(b)+1$). Two bars in a UBV layout are called collinear if a common x-value is shared by an endpoint of each bar; if the two bars have the same left x-coordinate (and hence also the same right x-coordinate), then they are called flush.*
2. *If B is any set of bars of $U(G)$, we define the rectangle $Rec(B)$ to be the smallest rectangle containing all the bars of B. The left and right x-coordinates of $Rec(B)$ are denoted $x_1(B)$ and $x_2(B)$, and its bottom and top y-coordinates are denoted $y_1(B)$ and $y_2(B)$. $Cor(B)$ denotes the two-way infinite vertical corridor bounded by the lines $x = x_1(B)$ and $x = x_2(B)$.*
3. *Let f and f' be internal faces of G, and let f be a neighbor of f' in G_I^* (i.e., the two faces share an edge). If $x_1(f) < x_1(f')$ (resp., $x_2(f) > x_2(f')$), we call f a left-neighbor (resp. right-neighbor) of f'.*

4. If f is a neighbor of f', but is neither a left- nor right-neighbor, then either $y_1(f) < y_1(f')$ or $y_2(f) > y_2(f')$, but not both, since G is outerplanar. In the former case we call f a down-neighbor of f', and in the latter case we call it an up-neighbor of f'. In both cases either $x_1(f) = x_1(f')$ or $x_2(f) = x_2(f')$, and we call f a left-flush or right-flush neighbor of f' accordingly. Note that, if f is a left neighbor of f', then f' cannot be a left neighbor of f, although it could be an up-, down-, or right-neighbor of f.

Two important geometric lemmas follow from results of [6]. Lemma 4 says that no path of faces in the internal dual of a triangulated polygon can have a UBV layout that proceeds left-to-right ('increases') and then later proceeds right-to-left ('decreases'), or vice-versa; we refer to this as the 'No U-Turn' property.

Lemma 4 (No U-Turn Lemma). *Let G be a triangulated polygon induced by a UBV layout, and let $P^* = F_1, \ldots, F_k$ be a path in G_I^*. Then the sequence $\{x_1(F_i)\}$ comprises a (monotone) decreasing subsequence followed by a (monotone) increasing subsequence, either of which may be empty. Similarly the sequence $\{x_2(F_i)\}$ comprises an increasing subsequence followed by a decreasing subsequence.*

Applying the No-U-Turn Lemma to the spine S_M and the legs incident with faces of S_M, we see that at most one leg may protrude to left of its spine neighbor, and at most one may protrude to the right. In [4] it is shown that the first two conditions in Thm. 2 guarantee that the beginning and ending clumps can always be laid out if S_M is feasible. The remaining clumps, contained in the *internal* spine string, must have legs composed entirely of up-neighbors or down-neighbors, when traversed starting at the face on the spine. The question then becomes whether there is space enough, using only bars of unit length, to lay out multiple legs on the internal spine.

As we move along a path P^* in the maximal spine, in order of increasing i, there is a path of vertices *below* P^* that we denote $a_0, a_1, \ldots$, and a path of vertices *above* P^*, denoted $b_0, b_1, \ldots$. A single clump C in P^* comprises a path of faces all incident with a common vertex; assume, without loss of generality, that C is an A-clump, and that this vertex is a_j, for some j. The opposite edges of the triangles in C form a path of b-vertices, $b_0, \ldots, b_k$, so that the ith triangle of C, $i = 0, \ldots, k-1$, has vertices a_j, b_i, b_{i+1}. If the ith face is *non-trivial*, then vertices b_i and b_{i+1} are incident with another vertex, c_i, so that the three vertices b_i, b_{i+1}, c_i form the initial triangle on a leg of the subdivided caterpillar G_I^*. We always use c_i for the first leg-vertex off an A-triangle and d_i for the first leg-vertex off a B-triangle. This labeling is used in Fig. 1.

In [4] it shown that, if P^* is a path in the *internal* spine, then we may make additional assumptions, without loss of generality, about the bars representing the paths of a-,b-,c-, and d-vertices in any UBV layout of G:

1. For each of the paths of a-,b-,c-, and d-vertices, the left x-coordinates of the corresponding bars form a *strictly increasing* sequence.
2. The set of d-bars lies *fully below* the set of a-bars, the set of a-bars is fully below the set of b-bars, and the set of b-bars is fully below the set of c-bars.

The second geometric lemma gives further restrictions on the paths of a-, b-, and c-bars in a single A-clump (and by symmetry, the paths of a-, b-, and d-vertices in a B-clump) in the layout of the internal spine. In particular, the heights of the path of b-vertices in a single A-clump form a sequence with a single relative maximum; we refer to this as the 'one extremum property.'

Lemma 5 (One Extremum Lemma). *Suppose $b_0, b_1, \ldots, b_k$ is a path of bars in a UBV layout of a triangulated polygon, all visible to a single bar a_0, such that $x(b_{i-1}) < x(b_i)$ and $y(b_i) > y(a_0)$ for all i. Assume, as usual, that the bars representing the b_i-vertices are all at distinct heights $y(b_i)$.*

1. *There is a single value m, $0 \le m \le k$, such that the sequence of heights $y(b_i)$ increases for $0 \le i \le m$ and decreases for $m \le i \le k$.*
2. *For $0 \le i \le k - 1$, let $X_{A,i}$ denote the triangle $\{b_i, b_{i+1}, a_0\}$, and suppose for some i that $X_{A,i}$ has an up-neighbor U_i. If $\{y(b_i), y(b_{i+1})\}$ is increasing, then $Cor(b_i)$ does not intersect $Cor(b_{i+2})$. If $\{y(b_i), y(b_{i+1})\}$ is decreasing, then $Cor(b_{i+1})$ does not intersect $Cor(b_{i-1})$.*
3. *If $X_{A,i}$ has an up-neighbor, then it is incident with one of the vertices b_{m-1}, b_m, b_{m+1}; in other words, $i \in \{m - 2, m - 1, m, m + 1\}$.*

Theorem 6. *Let G be a triangulated polygon with a UBV layout. Let S_I be an internal spine string, and let C be a clump in S_I. If C is an A-clump (resp., B-clump), let $y(b_0), \ldots, y(b_k)$ (resp., $y(a_0), \ldots, y(a_k)$) be the sequence of heights of the b-bars (resp., a-bars) of C in the UBV layout of S_I. If C is an A-clump (resp., B-clump), then it follows from the One Extremum Lemma that C has a unique relative maximum b_m (resp., unique relative minimum a_m). The position of b_m (or a_m) in the sequence is determined by which of the following classes C belongs to. Below the exponents $*, \#, +,$ and $++$, respectively, represent integer powers that are at least 0, equal to 0 or 1, at least 1, and at least 2.*

1. *$ForcedMax = \{N_A^{++} A N_A^{++}, \ N_A^* A N_A^\# A N_A^*\}$: The sequence $\{y(b_i)\}$ is neither strictly increasing nor strictly decreasing. The value b_m is a maximum that does not occur at $m = 0$ or $m = k$. In other words, $1 \le m \le k - 1$. An analogous statement holds for the class $ForcedMin = \{N_B^{++} B N_B^{++},$ $N_B^* B N_B^\# B N_B^*\}$: The sequence $\{y(b_i)\}$ is neither strictly increasing nor strictly decreasing. The value b_m is a minimum that does not occur at $m = 0$ or $m = k$. In other words, $1 \le m \le k - 1$.*
2. *$MaxOrIncrease = \{N_A^{++} A N_A^\#\}$: The sequence $\{y(b_i)\}$ is not strictly decreasing. The value b_m is a maximum that does not occur at $m = 0$. In other words, $1 \le m \le k$. Analogously, $MinOrDecrease = \{N_B^\# B N_B^{++}\}$.*
3. *$MaxOrDecrease = \{N_A^\# A N_A^{++}\}$: The sequence $\{y(b_i)\}$ is not strictly increasing. The value b_m is a maximum that does not occur at $m = k$. In other words, $0 \le m \le k - 1$. Analogously, $MinOrIncrease = \{N_B^{++} B N_B^\#\}$.*
4. *$Wild_A = \{N_A^\# A N_A^\#, N_A^+\}$: The sequence $\{y(b_i)\}$ may be strictly increasing, strictly decreasing, or increasing followed by decreasing. The value b_m is a maximum that may occur anywhere in the sequence. Analogously, $Wild_B = \{N_B^\# B N_B^\#, N_B^+\}$. Elements of $Wild_A$ and $Wild_B$ are called wildcards. The class of A-singletons, $S_A = \{A, N_A\}$, is a subset of $Wild_A$. Analogously, the B-singletons comprise a subset of $Wild_B$.*

Based on simple principles of calculus, e.g., that two consecutive relative maxima on the graph of a continuous function must have between them at least one relative minimum, it might appear that Thm. 6 eliminates some feasible triangulated polygons as candidates for being UBVGs. A closer examination of the classes reveals that the feasibility conditions have already eliminated these cases. However, there is a subtle condition that two successive 'ForcedMax' clumps, or other such 'special needs' pairs of clumps, must satisfy to provide sufficient space to lay out the clumps between. Filling in that final condition yields the main result of the paper, given in Thm. 8 below.

Definition 7. *Let G be a feasible triangulated polygon with internal spine string S_I, parsed into clumps that are then classified as in Thm. 6. Let C_i and C_j, $i < j$, be two* successive, non-wildcard *clumps in S_I. In other words, neither C_i nor C_j is a wildcard, but every clump between C_i and C_j is a wildcard. The ordered pair (C_i, C_j) is called a* special needs *pair if it is one of the following pairs of A-clumps or the corresponding B-clump twin: (ForcedMax, ForcedMax), (ForcedMax, MaxOrIncrease), (MaxOrDecrease, ForcedMax), (MaxOrDecrease, MaxOrIncrease).*

For example, the string $N_A^4 A N_A^2 N_B N_A^{100} B N_A^5 A$ is parsed as (*ForcedMax*, S_B, *Wild$_A$*, S_B, *MaxOrIncrease*), and it contains the special needs pair (*ForcedMax*, *MaxOrIncrease*).

Theorem 8 (Main Theorem). *Let G be a feasible triangulated polygon with internal spine string S_I. G is a UBVG if and only if the following condition holds (or its equivalent with the roles of A and B interchanged): between every special needs pair (C_i, C_j), there is either at least one wildcard A-clump that is the singleton clump N_A or at least one wildcard B-clump with two or more terms, namely $N_B^+ B N_B^\#, N_B^\# B N_B^+,$ or N_B^{++}.*

Section 4 outlines the sufficiency proof and layout algorithm for graphs satisfying the conditions of Thm. 8.

4   Sufficiency and the Layout Algorithm

In this section we outline an efficient algorithm that accepts as input any feasible spine string whose internal spine string has at least two non-wildcard clumps and satisfies Thm. 8, and that produces as output a set of coordinate pairs that are the left endpoints of a corresponding UBV layout. We assume for simplicity that the legs of the caterpillar are not subdivided. The more general case in which the internal spine string satisfies the remaining conditions in Thm. 8, and the caterpillar legs may be subdivided departs only slightly from the upcoming treatment: complete details are included in [4]. The proof of the sufficiency of Thm. 8 contains three main components: (a) parsing and labeling the clumps of S_I in accordance with Thm. 6, which, under the conditions of Thm. 8, leads in a natural way to (b) a description of a UBV layout algorithm, and finally (c) verification that the resulting UBV layout corresponds to the original input spine string. We outline these three ideas next.

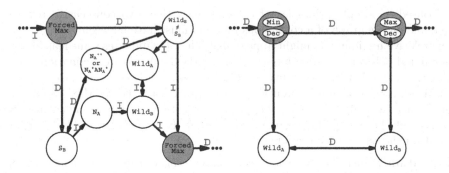

Fig. 2. (a) *ForcedMax → ForcedMax* (b) *MinOrDecrease → MaxOrDecrease*.

Parsing and Labeling the Internal Spine String. We begin by parsing the internal spine string S_I into a sequence of alternating A- and B-clumps and assigning class labels to the clumps as described in Thm. 6. Denote the resulting sequence by $C(S_I)$ (for *clumped spine string*). The conditions in Thm. 8 guide the layout of the bars corresponding to G one clump at a time in between and including special needs pairs of clumps. The layout between successive pairs of non-wildcard clumps that are not special needs pairs is simpler to accomplish (see Fig. 2(b)). All layouts between successive non-wildcard clumps are captured in a set of *clump labelers*: we represent a clump labeler between each pair of successive non-wildcard clumps as a directed graph, where each node is labeled by the name of a clump and each directed edge is labeled with either **I** (for *increase*) or **D** (for *decrease*). Fig. 2 shows two of the 36 possible clump labelers. Let S be a substring of $C(S_I)$ that begins and ends on non-wildcard clumps and has only wildcard clumps in between. Feed string S one clump at a time from left to right into the appropriate clump labeler. After traversing the clump labeler, each clump in S is marked with exactly one of *Max, Min, Increase* or *Decrease*. A trace of $S = N_A^{100} A N_A A N_A^3 N_B A B N_A B N_B A A$ fed through the clump labeler in Fig. 2(a) is shown in Table 1.

After the entire string $C(S_I)$ has visited the appropriate clump labelers, all clumps have been marked with one of *Max, Min, Increase,* and *Decrease* so that

Table 1. Trace for $S = N_A^{100} A N_A A N_A^3 N_B A B N_A B N_B A A$ using Fig. 2(a).

	Current Clump	In Node	Next Clump	Exit Direction	Next Node	Clump Label
1	$N_A^{100} A N_A A N_A^3$	ForcedMax	N_B	D	S_B	Max
2	N_B	S_B	A	D	N_A^{++} or $N_A^* A N_A^*$	Decrease
3	A	N_A^{++} or $N_A^* A N_A^*$	B	D	S_B	Decrease
4	B	S_B	N_A	I	N_A	Min
5	N_A	N_A	$B N_B$	I	$Wild_B$	Increase
6	$B N_B$	$Wild_B$	$A A$	I	ForcedMax	Increase
7	$A A$	ForcedMax		D		Max

the clumped spine string is an alternation of *Max*, *Decrease^i*, *Min*, *Increase^j*, or *Min*, *Increase^j*, *Max*, *Decrease^i*, or simply *Increase^i* or *Decrease^i*, where $i, j \geq 0$. With this information in hand, the left endpoint coordinates of the bars corresponding to each clump are computed. We make use of Thm. 8 to sketch the UBV layout algorithm.

UBV LAYOUT ALGORITHM

Input: A feasible internal spine string S_I with at least two non-wildcard clumps
Output: A set of coordinate pairs, each of which represents the left endpoint of a unit bar
Initialization: *coordinatePairs*= $\varnothing$; **input** clump labelers
   Step 1: Relabel S_I as $C(S_I)$ and extract the sequence of *non-wildcard* clumps $C_1, \ldots, C_u$.
   Step 2: for $i = 1$ **to** $u - 1$ **do**
   follow the clump labeler from C_i to C_{i+1}; mark each clump including and between C_i and C_{i+1} with one of *Max*, *Min*, *Increase*, or *Decrease*.
   Step 3: Compute left endpoints of coordinates of bars before and during each direction change and store in *coordinatePairs*; **return** *coordinatePairs*.

Coordinates of Bars. Suppose $C(S_I)$ contains a total of ℓ clumps and let $k = \min\{\frac{1}{2\ell-2}, \frac{1}{12}\}$. We construct a generic increasing sequence of wildcard clumps, where the left endpoint coordinates of the associated bars are each a function of k; the construction can then be modified to accommodate any increasing sequence of (not necessarily wildcard) clumps and subsequently translated to any location in the plane. We then construct the layout of clump $N_A A N_A A N_A$; any *Max* that is not of the form S_A can be modified from the latter construction. The layouts for a generic decreasing sequence and any *Min* not of the form S_B are accomplished by laying out the twin of the previous two constructions. The constructions lend themselves to interlocking any combination of sequences. Thus, the locations of bars in each clump are computed in the algorithm after the assignments of *Max*, *Min*, *Increase*, and *Decrease* to all of the clumps in $C(S_I)$. The parameter k is chosen to guarantee sufficient room to lay out the bars corresponding to the legs of the caterpillar.

Generic Increasing Sequence of Clumps. Any increasing sequence of clumps consists of (a) an alternation of elements from S_A and S_B, (b) an alternation of elements from *MinOrIncrease* and *MaxOrIncrease*, (c) the same as (a) with one element from *MinOrIncrease* or *MaxOrIncrease* in the interior of the sequence, (d) the same as (b) with one singleton clump in the interior of the sequence, (e) an alternation of elements from *MinOrIncrease* and S_A, (f) an alternation of elements from *MaxOrIncrease* and S_B, or finally (g) any combination of concatenations of (a)-(f). The constructions of the layouts in (a)-(f) are similar, and can all be modified from alternations of $N_A A N_A$ and $N_B B N_B$ laid out as an increasing sequence; the modifications consist of adding or removing bars to each clump (left to right) and translating subsets of the bars as required to maintain or create the needed visibilities. As such, for this note, we illustrate the layout for alternations of $N_A A N_A$ and $N_B B N_B$ in an increasing sequence.

The left endpoints of the bars for such a sequence of m clumps are given by the union of $aBars = \{(\lfloor \frac{i}{3} \rfloor + (\lfloor \frac{i}{3} \rfloor + (i \mod 3))k, i) : i = 0, 1, \ldots, 3m - 1\}$, $bBars = \{(\lfloor \frac{i}{3} \rfloor + 1 + (\lfloor \frac{i}{3} \rfloor - 3 + (i \mod 3))k, i - 3) : i = 0, 1, \ldots, 3m - 1\}$, $cBars = \{\lfloor \frac{i+1}{3} \rfloor + (\lfloor \frac{i+1}{3} \rfloor + (i + 1 \mod 3)k, i + 6) : i = 0, 3, \ldots, 3m - 3\}$, and $dBars = \{\lfloor \frac{i+1}{3} \rfloor + 1 + (\lfloor \frac{i+1}{3} \rfloor - 3 + (i + 1 \mod 3)k, i - 7) : i = 0, 3, \ldots, 3m - 3\}$, where $\lfloor x \rfloor$ denotes the floor of x.

By construction, c_{3i} is flush with a_{3i} and is visible only to a_{3i} and a_{3i+1}; similarly, d_{3i} is flush with b_{3i+1} and is visible only to b_{3i} and b_{3i+1}.

Generic Max. Any clump that is not of the form S_A and that is to be laid out as a *Max* can be modified from $N_A A N_A A N_A \in ForcedMax$ and then translated to any location in the plane. The following set of left endpoints, each of which is a function of k, represents such a *Max*: $maxBars = \{(1 - k, -1), (0, 0), (k, 1), (2k, 2), (1 + 3k, 3), (1 + k, 4), (2 - 4k, 1), (k, 5), (1 + 2k, 6)\}$. Note that there is room on the left side to attach an incoming increasing sequence and room on the right side to attach an outgoing decreasing sequence.

Singleton Min. Finally, at times the singleton S_B must be laid out as a *Min* and the singleton S_A must be laid out as a *Max* (see Thm. 6 and Thm. 8). We give a generic construction, parameterized by k, for the layout of $N_A A N_A A N_A S_B N_A$ $S_B N_A A N_A A N_A$, which shows the layout of S_B as a *Min* in between two *Forced-Maxes* (note the occurrence of N_A contiguous with S_B). Let $singetonMin = \{(2 - 2k, 0), (1 - k, 1), (3 - 3k, 1), (0, 2), (4 - 4k, 2), (k, 3), (2 - 4k, 3), (4 - 5k, 3), (2k, 4), (3 - 8k, 4), (4 - 6k, 4), (1 + 3k, 5), (3 - 7k, 5), (1 + k, 6), (3 - 5k, 6), (k, 7), (1 + 2k, 7), (3 - 6k, 7), (4 - 5k, 7)\}$. Any layout that requires an S_B to be used as a *Min* can be modified from this construction; similarly, the twin gives the construction for laying out S_A as a *Max*.

Example. Fig. 3 illustrates the ideas from this abstract by showing a UBV layout for a triangulated polygon with 67 vertices whose spine string is BN_B $(N_A A N_A N_B B N_B)^2 N_A A N_A N_B B N_B N_A A B A B N_A A N_A A N_A N_B A B N_A N_B N_A A$ $N_A A N_A N_B N_A N_B A B N_A A$.

Proving the sufficiency of Thm. 8 is equivalent to proving the correctness of the UBV Layout Algorithm. It is easy to see that the UBV Layout algorithm takes $O(n)$-time, where n is the length of the input spine string.

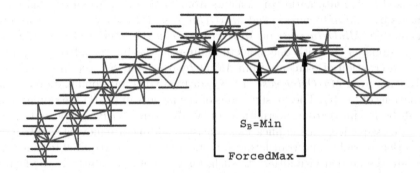

Fig. 3. UBV layout of a triangulated polygon with 67 vertices.

We conclude by noting that these techniques should also be useful for characterizing outerplanar near-triangulations (not 2-connected), near-triangulations (not outerplanar), and outerplanar near-quadrangulations. Other questions of interest include determining the computational complexity of UBVG testing, classification results for layouts in which bars are permitted to have two or more distinct lengths [2], and layouts in which visibility is permitted to extend past a fixed number of obstructing bars [3].

References

1. Bose, P., Dean, A., Hutchinson, J., Shermer, T. On rectangle visibility graphs. In: Lecture Notes in Computer Science 1190: Graph Drawing 1996. S. North (ed.). Springer-Verlag, Berlin (1997), 25–44.
2. Chen, G. and Keating, K. Bar Visibility Graphs with Bounded Length. Preprint.
3. Dean, A., Evans, W., Gethner, E., Laison, J., Safari, M. and Trotter, T. Bar k-Visibility Graphs. In preparation (2004).
4. Dean, A., Gethner, E., Hutchinson, J. A Characterization of Triangulated Polygons that are Unit Bar-Visibility Graphs. In preparation (2004).
5. Dean, A., Hutchinson, J. Rectangle-visibility layouts of unions and products of trees. J. Graph Alg. and App. **2** (1998), 1–21.
6. Dean, A., Hutchinson, J. Rectangle-visibility representations of bipartite graphs. Disc. App. Math. **75** (1997), 9–25.
7. Dean, A. Veytsel, N. Unit bar-visibility graphs. Congressus Numerantium **160** (2003), 161–175.
8. Hutchinson, J., Shermer, T., Vince, A. On representations of some thickness two graphs. Computational Geometry, Theory and Applications, **13** (1999) 161–171.
9. Rosenstiehl, P. and Tarjan,R. Rectilinear planar layouts and bipolar orientations of planar graphs. Discrete Comput. Geom. **1** (1986), 343–353.
10. Shermer, T. On rectangle visibility graphs III, external visibility and complexity. Proc. 8th Canadian Conf. on Computational Geometry (1996) 234–239.
11. Streinu, I. and Whitesides, W. Rectangle Visibility Graphs: Characterization, Construction, and Compaction. Lecture Notes in Computer Science #2607. H. Alt and M. Habib (eds.). Springer-Verlag, 2003, 26–37.
12. Tamassia, R., Tollis, I. A Unified Approach to Visibility Representations of Planar Graphs. J. of Discrete and Computational Geometry. **1** (1986), 321–341.
13. Wismath, S. Characterizing Bar Line-of-Sight Graphs. Proc. 1st ACM Symp. Computational Geometry (1985). 147–152.

Really Straight Graph Drawings*

Vida Dujmović[1,2], Matthew Suderman[1], and David R. Wood[2,3]

[1] School of Computer Science, McGill University, Montréal, Canada
{vida,suderman}@cs.mcgill.ca
[2] School of Computer Science, Carleton University, Ottawa, Canada
davidw@scs.carleton.ca
[3] Department of Applied Mathematics, Charles University, Prague, Czech Republic

Abstract. We study straight-line drawings of graphs with few segments and few slopes. Optimal results are obtained for all trees. Tight bounds are obtained for outerplanar graphs, 2-trees, and planar 3-trees. We prove that every 3-connected plane graph on n vertices has a plane drawing with at most $5n/2$ segments and at most $2n$ slopes, and that every cubic 3-connected plane graph has a plane drawing with three slopes (and three bends on the outerface). Drawings of non-planar graphs with few slopes are also considered. For example, it is proved that graphs of bounded degree and bounded treewidth have drawings with $\mathcal{O}(\log n)$ slopes.

1 Introduction

A common requirement for an aesthetically pleasing drawing of graph is that the edges are straight. This paper studies the following additional requirements of straight-line graph drawings:

1. minimise the number of segments in the drawing
2. minimise the number of distinct edge slopes in the drawing

First we formalise these notions. Consider a mapping of the vertices of a graph to distinct points in the plane. Now represent each edge by the closed line segment between its endpoints. Such a mapping is a *(straight-line) drawing* if each edge does not intersect any vertex, except for its own endpoints. By a *segment* in a drawing, we mean a maximal set of edges that form a line segment. The *slope* of a line L is the angle swept from the X-axis in an anticlockwise direction to L (and is thus in $[0, \pi)$). The *slope* of an edge or segment is the slope of the line that extends it. A *crossing* in a drawing is a pair of edges that intersect at some point other than a common endpoint. A drawing is *plane* if it has no crossings. A *plane graph* is a planar graph with a fixed combinatorial embedding and a specified outerface. We emphasise that a plane drawing of a plane graph must preserve the embedding and outerface. That every plane graph has a plane drawing is a classical result independently due to Wagner and Fáry.

---

* Research initiated at the International Workshop on Fixed Parameter Tractability in Geometry and Games, organised by Sue Whitesides; Bellairs Research Institute of McGill University, Holetown, Barbados, Feb. 7-13, 2004. Research supported by NSERC and COMBSTRU.

It is easily seen that a graph has a (plane) drawing on two slopes if and only if it has a (plane) drawing on any two slopes [3]. Garg and Tamassia [8] proved that it is $\mathcal{NP}$-complete to decide whether a graph has a rectilinear planar drawing (that is, with vertical and horizontal edges). Thus it is $\mathcal{NP}$-complete to decide whether a graph has a plane drawing with two slopes.

Our results include lower and upper bounds on the minimum number of segments and slopes in plane drawings of graphs, as summarised in Table 1. Due to space limitations, a number of auxiliary results and most proofs are omitted from this paper; see [3] for all the details. We refer the reader to the survey of Bodlaender [1] for the definition of treewidth, pathwidth, and k-tree.

First observe that the minimum number of slopes in a drawing of (plane) graph G is at most the minimum number of segments in a drawing of G. Upper bounds for plane graphs are stronger than for planar graphs, since for planar graphs one has the freedom to choose the embedding and outerface. On the other hand, lower bounds for planar graphs are stronger than for plane graphs. For example, consider the n-vertex planar triangulation illustrated in Figure 1. It has at least $n + 2$ slopes in every plane drawing. Now fix the outerface to that illustrated in (a). Then there are at least $2n - 2$ slopes in every plane drawing. However, using the embedding shown in (b), there is a plane drawing with only $\lceil 3n/2 \rceil$ slopes.

Section 2 studies plane drawings of 3-connected plane and planar graphs. In the case of slope-minimisation for plane graphs we obtain a bound that is tight in the worst case. However, our lower bound examples have linear maximum degree. In Section 3 we (drastically) improve this result in the case of cubic graphs, by proving that every 3-connected plane cubic graph has a plane drawing with three slopes, except for three edges on the outerface that have their own slope. As a corollary we prove that every 3-connected plane cubic graph has a plane 'drawing' with three slopes and three bends on the outerface. Section 4 considers non-plane drawings of arbitrary graphs with few slopes. For example, we prove that every graph with bounded degree and bounded treewidth has a drawing with $\mathcal{O}(\log n)$ slopes.

Before continuing, we outline some related research from the literature.

- Eppstein [6] characterised those planar graphs that have plane drawings with a segment between every pair of vertices. In some sense, these are the plane drawings with the least number of slopes.
- The *geometric thickness* of a graph G is the minimum k such that G has a drawing in which every edge receives one of k colours, and monochromatic

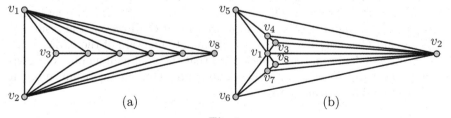

Fig. 1.

Table 1. Summary of results (ignoring additive constants). Here n is the number of vertices, η is the number of vertices of odd degree, and Δ is the maximum degree. The lower bounds are existential, except for trees, for which the lower bounds are universal.

graph family	# segments		# slopes	
	$\geq$	$\leq$	$\geq$	$\leq$
trees	$\eta/2$	$\eta/2$	$\lceil\Delta/2\rceil$	$\lceil\Delta/2\rceil$
maximal outerplanar	n	n	-	n
plane 2-trees	$2n$	$2n$	$2n$	$2n$
plane 3-trees	$2n$	$2n$	$2n$	$2n$
plane 2-connected	$5n/2$	-	$2n$	-
planar 2-connected	$2n$	-	n	-
plane 3-connected	$2n$	$5n/2$	$2n$	$2n$
planar 3-connected	$2n$	$5n/2$	n	$2n$
plane 3-connected cubic	-	$n+2$	3	3

edges do not cross (see [5, 7]). In any drawing, edges with the same slope do not cross. Thus the geometric thickness of G is a lower bound on the minimum number of slopes in a drawing of G.

- A drawing is *convex* if all the vertices are on the convex hull, and no three vertices are collinear. The *book thickness* of a graph (also called *pagenumber* and *stacknumber*) is the same as geometric thickness except that the drawing must be convex (see [4] for numerous references). Since edges with the same slope do not cross, the book thickness of G is a lower bound on the minimum number of slopes in a convex drawing of G.
- Plane orthogonal drawings with two slopes (and few bends) have been extensively studied (see [12]). For example, Ungar [14] proved that every cyclically 4-edge-connected plane cubic graph has a plane drawing with two slopes and four bends on the outerface. Thus our above-mentioned result for 3-connected plane cubic graphs nicely complements this theorem of Ungar.
- A drawing of the complete graph K_n is defined by a set of n points with no three collinear. Jamison [9] proved that the minimum number of slopes in a drawing of K_n is n. The upper bound is obtained by positioning the vertices of K_n on the vertices of a regular n-gon, as illustrated in Figure 2(a) and (b). In fact, Jamison [9] proved that every drawing of K_n with exactly n slopes is affinely equivalent to a regular n-gon. In [3] we study drawings of complete multi-partite graphs. For example, we prove that the minimum number of slopes in a convex drawing of $K_{n,n}$ is n, as illustrated in Figure 2(c).
- Wade and Chu [15] recognised that drawing arbitrary graphs with few slopes is an interesting problem. They defined the *slope-number* of a graph G to be the minimum number of slopes in a drawing of G. However, the results of Wade and Chu only pertain to K_n. Seemingly unaware of the earlier work of Scott and Jamison, they rediscovered that the minimum number of slopes in a drawing of K_n is n. In addition, they presented an algorithm to test if K_n can be drawn using a given set of slopes.

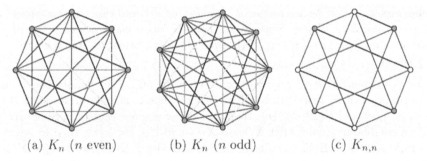

(a) K_n (n even)   (b) K_n (n odd)   (c) $K_{n,n}$

Fig. 2. Drawings of K_n and $K_{n,n}$ with n slopes.

2   3-Connected Plane Graphs

Theorem 1. *Every 3-connected plane graph with n vertices has a plane drawing with at most $5n/2 - 3$ segments and at most $2n - 10$ slopes.*

The proof of Theorem 1 is based on the canonical ordering of Kant [10]. Let G be a 3-connected plane graph. Kant [10] proved that G has a canonical ordering defined as follows. Let $\sigma = (V_1, V_2, \ldots, V_K)$ be an ordered partition of $V(G)$. That is, $V_1 \cup V_2 \cup \cdots \cup V_K = V(G)$ and $V_i \cap V_j = \emptyset$ for all $i \neq j$. Define G_i to be the plane subgraph of G induced by $V_1 \cup V_2 \cup \cdots \cup V_i$. Let C_i be the subgraph of G induced by the edges on the boundary of the outerface of G_i. Then σ is a *canonical ordering* of G if:

- $V_1 = \{v_1, v_2\}$, where v_1 and v_2 lie on the outerface and $v_1v_2 \in E(G)$.
- $V_K = \{v_n\}$, where v_n lies on the outerface, $v_1v_n \in E(G)$, and $v_n \neq v_2$.
- Each C_i ($i > 1$) is a cycle containing v_1v_2.
- Each G_i is biconnected and internally 3-connected; that is, removing any two interior vertices of G_i does not disconnect it.
- For each $i \in \{2, 3, \ldots, K-1\}$, one of the following condition holds:
    1. $V_i = \{v_i\}$ where v_i is a vertex of C_i with at least three neighbours in C_{i-1}, and v_i has at least one neighbour in $G \setminus G_i$.
    2. $V_i = (s_1, s_2, \ldots, s_\ell, v_i)$, $\ell \geq 0$, is a path in C_i, where each vertex in V_i has at least one neighbour in $G \setminus G_i$. Furthermore, the first and the last vertex in V_i have one neighbour in C_{i-1}, and these are the only two edges between V_i and G_{i-1}.

The vertex v_i is called the *representative* vertex of V_i, $2 \leq i \leq K$. The vertices $\{s_1, s_2, \ldots, s_\ell\} \subseteq V_i$ are called *division* vertices. Let $S \subset V(G)$ be the set of all division vertices. A vertex u is a *successor* of a vertex $w \in V_i$ if uw is an edge and $u \in G \setminus G_i$, and u is a *predecessor* of $w \in V_i$ if uw is an edge and $u \in V_j$ for some $j < i$. We also say that u is a predecessor of V_i. Let $P(V_i) = (p_1, p_2, \ldots, p_q)$ be the set of predecessors of V_i ordered by the path from v_1 to v_2 in $C_{i-1} \setminus v_1v_2$. Vertex p_1 and p_q are the *left* and *right* predecessors of V_i respectively, and vertices $p_2, p_3, \ldots p_{q-1}$ are called *middle* predecessors of V_i.

Theorem 2. *Let σ be a canonical ordering of an n-vertex m-edge plane 3-connected graph G. Define S as above. Then G has a plane drawing D with at most $m - \max\{\lceil n/2 \rceil - |S| - 3, |S|\}$ segments, and at most $m - \max\{n - |S| - 4, |S|\}$ slopes.*

Proof Construction. For every vertex v, let $X(v)$ and $Y(v)$ denote the x and y coordinates of v, respectively. If a vertex v has a neighbour w, such that $X(w) < X(v)$ and $Y(w) < Y(v)$, then we say vw is a *left edge* of v. Similarly, if v has a neighbour w, such that $X(w) > X(v)$ and $Y(w) < Y(v)$, then we say vw is a *right edge* of v. If vw is an edge such that $X(v) = X(w)$ and $Y(v) < Y(w)$, than we say vw is a *vertical edge above* v and *below* w.

We define D inductively on $\sigma = (V_1, V_2, \ldots, V_K)$ as follows. Let D_i denote a drawing of G_i. A vertex v is a *peak in D_i*, if each neighbour w of v has $Y(w) \leq Y(v)$ in D_i. We say that a point p in the plane is *visible in D_i* from vertex $v \in D_i$, if the segment $\overline{pv}$ does not intersect D_i except at v. At the ith induction step, $2 \leq i \leq K$, D_i will satisfy the following invariants:

Invariant 1: $C_i \setminus v_1 v_2$ is *strictly X-monotone*; that is, the path from v_1 to v_2 in $C_i \setminus v_1 v_2$ has (strictly) increasing X-coordinates.

Invariant 2: Every peak in D_i, $i < K$, has a successor.

Invariant 3: Every representative vertex $v_j \in V_j$, $2 \leq j \leq i$ has a left and a right edge. Moreover, if $|P(V_j)| \geq 3$ then there is a vertical edge below v_j.

Invariant 4: D_i has no edge crossings.

For the base case $i = 2$, position the vertices v_1, v_2 and v_3 at the corners of an equilateral triangle so that $X(v_1) < X(v_3) < X(v_2)$ and $Y(v_1) < Y(v_2) < Y(v_3)$. Draw the division vertices of V_2 on the segment $v_1 v_3$. This drawing of D_2 satisfies all four invariants. Now suppose that we have a drawing of D_{i-1} that satisfies the invariants. There are two cases to consider in the construction of D_i, corresponding to the two cases in the definition of the canonical ordering.

Case 1. $|P(V_i)| \geq 3$: If v_i has a middle predecessor v_j with $|P(V_j)| \geq 3$, let $w = v_j$. Otherwise let w be any middle predecessor of v_i. Let L be the open ray $\{(X(w), y) : y > Y(w)\}$. By invariant 1 for D_{i-1}, there is a point in L that is visible in D_{i-1} from every predecessor of v_i. Represent v_i by such a point, and draw segments between v_i and each of its predecessors. That the resulting drawing D_i satisfies the four invariants can be immediately verified.

Case 2. $|P(V_i)| = 2$: Suppose that $P(V_i) = \{w, u\}$, where w and u are the left and the right predecessors of V_i, respectively. Suppose $Y(w) \geq Y(u)$. (The other case is symmetric.) Let P be the path between w and u on $C_{i-1} \setminus v_1 v_2$. As illustrated in Figure 3, let A_i be the region $\{(x, y) : y > Y(w) \text{ and } X(w) \leq x \leq X(u)\}$. Assume on the contrary that $D_{i-1} \cap A_i \neq \emptyset$. By the monotonicity of D_{i-1}, $P \cap A_i \neq \emptyset$. Let $p \in P \cap A_i$. Since $Y(p) > Y(w) \geq Y(u)$, P is X-monotone and thus has a vertex between w and u that is a peak. By the definition of the canonical ordering σ, the addition of V_i creates a face of G, since V_i is added in the outerface of G_{i-1}. Therefore, each vertex between w and u on P has no successor, and is thus not a peak in D_{i-1} by invariant 2, which is the desired contradiction. Therefore $D_{i-1} \cap A_i = \emptyset$.

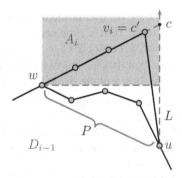

Fig. 3. Illustration for Case 2.

Let L be the open ray $\{(X(u), y) : y > Y(u)\}$. If $w \notin S$, then by invariant 3, w has a left and a right edge in D_{i-1}. Let c be the point of intersection between L and the line extending the left edge at w. If $w \in S$, then let c be any point in A_i on L. By invariant 1, there is a point $c' \notin \{c, w\}$ on $\overline{wc}$ such that c' is visible in D_{i-1} from u. Represent v_i by c', and draw two segments $\overline{v_i u}$ and $\overline{v_i w}$. These two segments do not intersect any part of D_{i-1} (and neither is horizontal). Represent any division vertices in V_i by arbitrary points on the open segment $\overline{w v_i} \cap A_i$. Therefore, in the resulting drawing D_i, there are no crossings and the remaining three invariants are maintained. This completes the construction of D. The analysis for the number of segments and slopes is in [3]. □

Proof (of Theorem 1). Whenever a set V_i is added to G_{i-1}, at least $|V_i| - 1$ edges that are not in G can be added so that the resulting graph is planar. Thus $|S| = \sum_i (|V_i| - 1) \leq 3n - 6 - m$. Hence Theorem 2 implies that G has a plane drawing with at most $m - n/2 + |S| + 3 \leq 5n/2 - 3$ segments, and at most $m - n + |S| - 4 \leq 2n - 10$ slopes. □

Since deleting an edge from a drawing cannot increase the number of slopes, and every plane graph can be triangulated to a 3-connected plane graph, Theorem 1 implies that every n-vertex plane graph has a plane drawing with at most $2n - 10$ slopes. Note that we cannot draw the same conclusion for segments, since deleting an edge in a drawing may increase the number of segments. The famous 'nested-triangles' planar graph leads to the following lower bound.

Lemma 1. *For all $n \equiv 0 \pmod 3$, there is an n-vertex planar triangulation with maximum degree six that has at least $2n - 6$ segments in every plane drawing, regardless of the choice of outerface.*

3   Cubic 3-Connected Plane Graphs

A graph in which every vertex has degree three is *cubic*.

Theorem 3. *Every cubic 3-connected plane graph has a plane drawing in which every edge has slope in $\{\pi/4, \pi/2, 3\pi/4\}$, except for three edges on the outerface.*

Proof. Let $\sigma = (V_1, V_2, \ldots, V_K)$ be a canonical ordering of G. We re-use the notation from Theorem 2, except that a representative vertex of V_i may be the first or last vertex in V_i. Since G is cubic, $|P(V_i)| = 2$ for all $1 < i < K$, and every vertex not in $\{v_1, v_2, v_n\}$ has exactly one successor. We proceed by induction on i with the hypothesis that G_i has a plane drawing D_i that satisfies:

Invariant 1: $C_i \backslash v_1 v_2$ is X-*monotone*; that is, the path from v_1 to v_2 in $C_i \backslash v_1 v_2$ has non-decreasing X-coordinates.

Invariant 2: Every peak in D_i, $i < K$, has a successor.

Invariant 3: If there is a vertical edge above v in D_i, then all the edges of G that are incident to v are in G_i.

Invariant 4: D_i has no edge crossings.

Let D_2 be the drawing of G_2 constructed as follows. Draw $v_1 v_2$ horizontally with $X(v_1) < X(v_2)$. This accounts for one edge whose slope is not in $\{\pi/4, \pi/2, 3\pi/4\}$. Now draw $v_1 v_3$ with slope $\pi/4$, and draw $v_2 v_3$ with slope $3\pi/4$. Add any division vertices on the segment $v_1 v_3$. Now v_3 is the only peak in D_2, and it has a successor by the definition of the canonical ordering. Thus all the invariants are satisfied for the base case D_2.

Now suppose that $2 < i < K$ and we have a drawing of D_{i-1} that satisfies the invariants. Suppose that $P(V_i) = \{u, w\}$, where u and w are the left and the right predecessors of V_i, respectively. Without loss of generality, $Y(w) \leq Y(u)$. Let the representative vertex v_i be last vertex in V_i. Position v_i at the intersection of a vertical segment above w, and a segment of slope $\pi/4$ from u, and add any division vertices on $\overline{uv_i}$, as illustrated in Figure 4(a). Note that there is no vertical edge above w by invariant 3 for D_{i-1}. (For the case in which $Y(u) < Y(w)$, we take the representative vertex v_i to be the first vertex in V_i, and the edge wv_i has slope $3\pi/4$, as illustrated in Figure 4(b).)

Clearly the resulting drawing D_i is X-monotone. Thus invariant 1 is maintained. The vertex v_i is the only peak in D_i that is not a peak in D_{i-1}. Since v_i has a successor by the definition of the canonical ordering, invariant 2 is maintained. The vertical edge wv_i satisfies invariant 3, since v_i is the sole successor of w. Thus invariant 3 is maintained. No vertex between u and w (on the path from u to w in $C_{i-1} \backslash v_1 v_2$) is higher than the higher of u and w. Otherwise there would be a peak, not equal to v_n, with no successor, and thus violating invariant 2 for D_{i-1}. Thus the edges in $D_i \backslash D_{i-1}$ do not cross any edges in D_i. In particular, there is no edge ux in D_{i-1} with slope $\pi/4$ and $Y(x) > Y(u)$. The vertex v_n can be easily added to the drawing to complete the construction. □

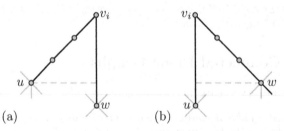

(a)   (b)

Fig. 4. Construction of a 3-slope drawing of a cubic 3-connected plane graph.

It is easily seen that the bound of six on the number of slopes in Theorem 3 is optimal for any 3-connected cubic plane graph whose outerface is a triangle. An easy variation on the algorithm in Theorem 3 gives:

Corollary 1. *Every cubic 3-connected plane graph has a plane 'drawing' with three slopes and three bends on the outerface.*

4   Drawings of General Graphs with Few Slopes

This section is motivated by the following fundamental open problem: Is there a function f such that every graph with maximum degree Δ has a drawing with at most $f(\Delta)$ slopes? This is open even for $\Delta = 3$. Note that:

- The best lower bound that we are aware of is $\Delta + 1$ for the complete graph.
- There is no such function f for convex drawings. Malitz [11] proved that there are Δ-regular n-vertex graphs with book thickness $\Omega(\sqrt{\Delta}n^{1/2-1/\Delta})$. Since book thickness is a lower bound on the number of slopes in a convex drawing, every convex drawing of such a graph has $\Omega(\sqrt{\Delta}n^{1/2-1/\Delta})$ slopes.
- An affirmative solution to this problem would imply that geometric thickness is bounded by maximum degree, which is an open problem due to Eppstein [7]. Duncan *et al.* [5] recently proved that graphs with maximum degree at most four have geometric thickness at most two.

Let H be a *(host)* graph. The vertices of H are called *nodes*. An H-*partition* of a graph G is a function $f : V(G) \to V(H)$ such that for every edge $vw \in E(G)$ we have $f(v) = f(w)$ or $f(v)f(w) \in E(H)$. In the latter case, we say vw is *mapped* to the edge $f(v)f(w)$. The *width* of f is the maximum of $|f^{-1}(x)|$, taken over all nodes $x \in V(H)$, where $f^{-1}(x) = \{v \in V(G) : f(v) = x\}$. In the following result, we describe how to produce a drawing of a graph G given an H-partition of G and a drawing D of H. The general approach is to scale D appropriately, and then replace each node of H by a copy of the drawing of K_k on a regular k-gon. The only difficulty is to scale D so that we obtain a valid drawing of G.

Lemma 2 ([3]). *Let H be a graph admitting a drawing D with s distinct slopes and ℓ distinct edge lengths. Let G be a graph admitting an H-partition of width k. Then G has a drawing with $ks\ell(k-1) + k + s$ slopes.*

Lemma 2 suggests looking at host graphs that admit drawings with few slopes and few edge lengths. Obviously a path has a drawing with one slope and one edge length. Based on this idea, we prove that every graph with bandwidth b has a drawing with at most $\frac{1}{2}b(b+1)+1$ slopes. Based on results from the literature that bound bandwidth in terms of maximum degree Δ, we conclude:

- Every interval graph has a drawing with at most $\frac{1}{2}\Delta(\Delta+1)+1$ slopes.
- Every co-comparability graph (which includes the permutation graphs) has a drawing with at most $\Delta(2\Delta-1)+1$ slopes.
- Every AT-free graph has a drawing with at most $\frac{3}{2}\Delta(3\Delta+1)+1$ slopes.

Lemma 2 motivates the study of drawings of trees with few slopes and few distinct edge lengths.

Lemma 3. *Every tree T with pathwidth $k \geq 1$ has a plane drawing with $\max\{\Delta(T) - 1, 1\}$ slopes and $2k - 1$ distinct edge lengths.*

Lemma 4 ([13]). *Every tree T has a path P, called a "backbone", such that $T \setminus V(P)$ has smaller pathwidth than T, and the endpoints of P are leaves of T.*

Proof (of Lemma 3). We refer to T as T_0. Let n_0 be the number of vertices in T_0, and let $\Delta_0 = \Delta(T_0)$. The result holds trivially for $\Delta_0 \leq 2$. Now assume that $\Delta_0 \geq 3$. Let S be the set of slopes $S = \{\frac{\pi}{2}(1 + \frac{i}{\Delta_0 - 2}) : 0 \leq i \leq \Delta_0 - 2\}$. We proceed by induction on n with the hypothesis: "There is a real number $\ell = \ell(n_0, \Delta_0)$, such that for every tree T with $n \leq n_0$ vertices, maximum degree at most Δ_0, and pathwidth $k \geq 1$, and for every vertex r of T with degree less than Δ_0, T has a plane drawing D in which:

 - r is at the top of D (that is, no point in D has greater Y-coordinate than r),
 - every edge of T has slope in S,
 - every edge of T has length in $\{\ell^0, \ell^1, \ldots, \ell^{2k-1}\}$, and
 - if r is contained in some backbone of T, then every edge of T has length in $\{\ell^0, \ell^1, \ldots, \ell^{2k-2}\}$."

The result follows from the induction hypothesis, since we can take r to be the endpoint of a backbone of T_0, in which case $\deg(r) = 1 < \Delta_0$, and thus every edge of T_0 has length in $\{\ell^0, \ell^1, \ldots, \ell^{2k-2}\}$.

The base case with $n = 1$ is trivial. Now suppose that the hypothesis is true for trees on less than n vertices, and we are given a tree T with n vertices and pathwidth k, and r is a vertex of T with degree less than Δ_0.

If r is contained in some backbone B of T, then let $P = B$. Otherwise, let P be a path from r to an endpoint of a backbone B of T. Note that P has at least one edge. As illustrated in Figure 5, draw P horizontally with unit-length edges. Every vertex in P has at most $\Delta_0 - 2$ neighbours in $T \setminus V(P)$, since r has degree less than Δ_0 and the endpoints of a backbone are leaves. At each vertex $x \in P$, the children $\{y_0, y_1, \ldots, y_{\Delta_0 - 3}\}$ of x are positioned below P and on the unit-circle centred at x, so that each edge xy_j has slope $\frac{\pi}{2}(1 + j/(\Delta_0 - 2)) \in S$.

Every connected component T' of $T \setminus V(P)$ is a tree rooted at some vertex r' adjacent to a vertex in P. Thus r' has already been positioned in the drawing of T. If T' is a single vertex, then we no longer need to consider this T'.

We consider two types of subtrees T', depending on whether the pathwidth of T' is less than k. Suppose that the pathwidth of T' is k (it cannot be more). Then $T' \cap B \neq \emptyset$ since B is a backbone of T. Thus $T' \cap B$ is a backbone of T' containing r'. Thus we can apply the stronger induction hypothesis in this case.

Every T' has less vertices than T, and every r' has degree less than Δ_0 in T'. Thus by induction, every T' has a drawing with r' at the top, and every edge of T' has slope in S. Furthermore, if the pathwidth of T' is less than k, then every edge of T' has length in $\{\ell^0, \ell^1, \ldots, \ell^{2k-3}\}$. Otherwise r' is in a backbone of T', and every edge of T' has length in $\{\ell^0, \ell^1, \ldots, \ell^{2k-2}\}$.

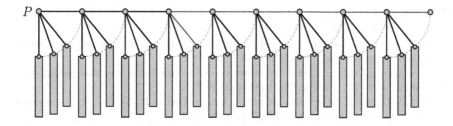

Fig. 5. Drawing of T with few slopes and few edge lengths.

There exists a scale factor $\ell < 1$, depending only on n_0 and Δ_0, so that by scaling the drawings of every T' by ℓ, the widths of the drawings are small enough so that there is no crossings when the drawings are positioned with each r' at its already chosen location. (Note that ℓ is the same value at every level of the induction.) Scaling preserves the slopes of the edges. An edge in any T' that had length ℓ^i before scaling, now has length ℓ^{i+1}.

Case 1. r is contained in some backbone B of T: By construction, $P = B$. So every T' has pathwidth at most $k - 1$, and thus every edge of T' has length in $\{\ell^1, \ell^2, \ldots, \ell^{2k-2}\}$. All the other edges of T have unit-length. Thus we have a plane drawing of T with edge lengths $\{\ell^0, \ell^1, \ldots, \ell^{2k-2}\}$, as claimed.

Case 2. r is not contained in any backbone of T: Every edge in every T' has length in $\{\ell^1, \ell^2, \ldots, \ell^{2k-1}\}$. All the other edges of T have unit-length. Thus we have a plane drawing of T with edge lengths $\{\ell^0, \ell^1, \ldots, \ell^{2k-1}\}$, as claimed. $\square$

Theorem 4. *Let G be a graph with n vertices, maximum degree Δ, and tree-width k. Then G has a drawing with $\mathcal{O}(k^3 \Delta^4 \log n)$ slopes.*

Proof. Ding and Oporowski [2] proved that for some tree T, G has a T-partition of width at most $\max\{24k\Delta, 1\}$. Let $w = \max\{24k\Delta, 1\}$. For each node $x \in V(T)$, there are at most $w\Delta$ edges of G incident to vertices mapped to x. Hence we can assume that T is a forest with maximum degree at most $w\Delta$, as otherwise there is an edge of T with no edge of G mapped to it, in which case the edge of T can be deleted. Similarly, T has at most n vertices. Now, T has pathwidth at most $\log(2n+1)$ (see [1]). By Lemma 3, T has a drawing with at most $w\Delta - 1$ slopes and at most $2 \log(2n+1) - 1$ distinct edge lengths. By Lemma 2, G has a drawing in which the number of slopes is at most $w(w\Delta - 1)(2 \log(2n + 1) - 1)(w - 1) + (w\Delta - 1) + w \in \mathcal{O}(w^3 \Delta \log n) \subseteq \mathcal{O}(k^3 \Delta^4 \log n)$. $\square$

Corollary 2. *Every n-vertex graph with bounded degree and bounded treewidth has a drawing with $\mathcal{O}(\log n)$ slopes.* $\square$

Acknowledgements

Thanks to all of the participants of the Bellairs workshop for creating a stimulating working environment. Special thanks to Mike Fellows for suggesting the problem.

References

1. HANS L. BODLAENDER. A partial k-arboretum of graphs with bounded treewidth. *Theoret. Comput. Sci.*, 209(1-2):1–45, 1998.
2. GUOLI DING AND BOGDAN OPOROWSKI. Some results on tree decomposition of graphs. *J. Graph Theory*, 20(4):481–499, 1995.
3. VIDA DUJMOVIĆ, MATTHEW SUDERMAN, AND DAVID R. WOOD. Really straight graph drawings. arXiv.org:cs.DM/0405112, 2004.
4. VIDA DUJMOVIĆ AND DAVID R. WOOD. On linear layouts of graphs. *Discrete Math. Theor. Comput. Sci.*, 6(2):339–358, 2004.
5. CHRISTIAN A. DUNCAN, DAVID EPPSTEIN, AND STEPHEN G. KOBOUROV. The geometric thickness of low degree graphs. In *Proc. 20th ACM Symp. on Computational Geometry* (SoCG '04), pp. 340–346. ACM Press, 2004.
6. DAVID EPPSTEIN. Dilation-free planar graphs.
   http://www.ics.uci.edu/~eppstein/junkyard/dilation-free/, 1997.
7. DAVID EPPSTEIN. Separating thickness from geometric thickness. In JÁNOS PACH, editor, *Towards a Theory of Geometric Graphs*, vol. 342 of *Contemporary Mathematics*, pp. 75–86. Amer. Math. Soc., 2004.
8. ASHIM GARG AND ROBERTO TAMASSIA. On the computational complexity of upward and rectilinear planarity testing. *SIAM J. Comput.*, 31(2):601–625, 2001.
9. ROBERT E. JAMISON. Few slopes without collinearity. *Discrete Math.*, 60:199–206, 1986.
10. GOOS KANT. Drawing planar graphs using the canonical ordering. *Algorithmica*, 16(1):4–32, 1996.
11. SETH M. MALITZ. Graphs with E edges have pagenumber $O(\sqrt{E})$. *J. Algorithms*, 17(1):71–84, 1994.
12. MD. SAIDUR RAHMAN, TAKAO NISHIZEKI, AND SHUBHASHIS GHOSH. Rectangular drawings of planar graphs. *J. Algorithms*, 50:62–78, 2004.
13. MATTHEW SUDERMAN. Pathwidth and layered drawings of trees. *Internat. J. Comput. Geom. Appl.*, 14(3):203–225, 2004.
14. PETER UNGAR. On diagrams representing maps. *J. London Math. Soc.*, 28:336–342, 1953.
15. GREG A. WADE AND JIANG-HSING CHU. Drawability of complete graphs using a minimal slope set. *The Computer Journal*, 37(2):139–142, 1994.

Layouts of Graph Subdivisions[*]

Vida Dujmović[1,2] and David R. Wood[2,3]

[1] School of Computer Science, McGill University, Montréal, Canada
vida@cs.mcgill.ca
[2] School of Computer Science, Carleton University, Ottawa, Canada
davidw@scs.carleton.ca
[3] Department of Applied Mathematics, Charles University, Prague, Czech Republic

Abstract. A *k-stack layout* (respectively, *k-queue layout*) of a graph consists of a total order of the vertices, and a partition of the edges into k sets of non-crossing (non-nested) edges with respect to the vertex ordering. A *k-track layout* of a graph consists of a vertex k-colouring, and a total order of each vertex colour class, such that between each pair of colour classes no two edges cross. The *stack-number* (respectively, *queue-number*, *track-number*) of a graph G, denoted by $\mathsf{sn}(G)$ ($\mathsf{qn}(G)$, $\mathsf{tn}(G)$), is the minimum k such that G has a k-stack (k-queue, k-track) layout. This paper studies stack, queue, and track layouts of graph subdivisions. It is known that every graph has a 3-stack subdivision. The best known upper bound on the number of division vertices per edge in a 3-stack subdivision of an n-vertex graph G is improved from $\mathcal{O}(\log n)$ to $\mathcal{O}(\log \min\{\mathsf{sn}(G), \mathsf{qn}(G)\})$. This result reduces the question of whether queue-number is bounded by stack-number to whether 3-stack graphs have bounded queue number. It is proved that every graph has a 2-queue subdivision, a 4-track subdivision, and a mixed 1-stack 1-queue subdivision. All these values are optimal for every non-planar graph. In addition, we characterise those graphs with k-stack, k-queue, and k-track subdivisions, for all values of k. The number of division vertices per edge in the case of 2-queue and 4-track subdivisions, namely $\mathcal{O}(\log \mathsf{qn}(G))$, is optimal to within a constant factor, for every graph G. Applications to 3D polyline grid drawings are presented. For example, it is proved that every graph G has a 3D polyline grid drawing with the vertices on a rectangular prism, and with $\mathcal{O}(\log \mathsf{qn}(G))$ bends per edge.

1 Introduction

This paper studies stack, queue and track layouts of subdivisions of graphs. The contributions of this paper are three-fold. First, we characterise those graphs admitting k-stack, k-queue or k-track subdivisions, for all k. In addition, we prove bounds on the number of division vertices per edge that are asymptotically tight in a number of cases. These results are presented in Section 3. Second, we use these subdivision layouts to reduce two of the major open problems in the theory of stack and queue layouts to certain special cases. These results, along with relationships amongst various thickness parameters, are presented in

---

[*] Research supported by NSERC and COMBSTRU.

Section 4. As the third contribution, we apply our results concerning track layouts of subdivisions to the study of three-dimensional polyline graph drawings. These results are presented in Section 5. Due to space limitations, many proofs and some references are omitted – see [7] for all the details. All logarithms are base 2 unless stated otherwise.

2   Preliminaries

We consider undirected, finite, and simple graphs G with vertex set $V(G)$ and edge set $E(G)$. The number of vertices and edges of G are respectively denoted by $n = |V(G)|$ and $m = |E(G)|$. A *subdivision* of G is a graph obtained from G by replacing each edge $vw \in E(G)$ by a path with at least one edge whose endpoints are v and w. Internal vertices on this path are called *division* vertices. Let G' be the subdivision of G with one division vertex per edge.

A *graph parameter* is a function α that assigns to every graph G a nonnegative integer $\alpha(G)$. Let $\mathcal{G}$ be a class of graphs. By $\alpha(\mathcal{G})$ we denote the function $f : \mathbb{N} \to \mathbb{N}$, where $f(n)$ is the maximum of $\alpha(G)$, taken over all n-vertex graphs $G \in \mathcal{G}$. We say $\mathcal{G}$ has *bounded* α if $\alpha(\mathcal{G}) \in \mathcal{O}(1)$. A graph parameter α is *bounded by* a graph parameter β (for some class $\mathcal{G}$), if there exists a *binding* function g such that $\alpha(G) \leq g(\beta(G))$ for every graph G (in $\mathcal{G}$). If α is bounded by β (in $\mathcal{G}$) and β is bounded by α (in $\mathcal{G}$) then α and β are *tied* (in $\mathcal{G}$).

A *vertex ordering* of a graph G is a total order σ of the vertex set $V(G)$. Let $L(e)$ and $R(e)$ denote the endpoints of each edge $e \in E(G)$ such that $L(e) <_\sigma R(e)$. Consider two edges $e, f \in E(G)$ with no common endpoint such that $L(e) <_\sigma L(f)$. If $L(e) <_\sigma L(f) <_\sigma R(e) <_\sigma R(f)$ then e and f *cross*, and if $L(e) <_\sigma L(f) <_\sigma R(f) <_\sigma R(e)$ then e and f *nest*. A *stack* (respectively, *queue*) is a set of edges $E' \subseteq E(G)$ such that no two edges in E' cross (nest). Observe that when traversing the vertex ordering, edges in a stack (queue) appear in LIFO (FIFO) order – hence the names. A *k-stack* (*queue*) *layout* of G consists of a vertex ordering σ of G and a partition $\{E_\ell : 1 \leq \ell \leq k\}$ of $E(G)$, such that each E_ℓ is a *stack* (*queue*) in σ. A graph admitting a k-stack (queue) layout is called a *k-stack* (*queue*) *graph*. The *stack-number* of a graph G, denoted by $\mathsf{sn}(G)$, is the minimum k such that G is a k-stack graph. The *queue-number* of a graph G, denoted by $\mathsf{qn}(G)$, is the minimum k such that G is a k-queue graph. For a summary of results regarding stack and queue layouts see [8].

A *vertex t-colouring* of a graph G is a partition $\{V_i : 1 \leq i \leq t\}$ of $V(G)$ such that for every edge $vw \in E(G)$, if $v \in V_i$ and $w \in V_j$ then $i \neq j$. Suppose that $<_i$ is a total order of each colour class V_i. Then the pair $(V_i, <_i)$ is called a *track*, and $\{(V_i, <_i) : 1 \leq i \leq t\}$ is a *t-track assignment* of G. We denote track assignments by $\{V_i : 1 \leq i \leq t\}$ when the ordering on each colour class is implicit. An *X-crossing* in a track assignment consists of two edges vw and xy such that $v <_i x$ and $y <_j w$, for distinct colours i and j. A *(k, t)-track layout* of G consists of a t-track assignment of G and a (non-proper) edge k-colouring of G with no monochromatic X-crossing. $(1, t)$-track layouts (that is, with no X-crossing) are of particular interest due to applications in three-dimensional graph drawing (see Section 5). A $(1, t)$-track layout is called a *t-track layout*. A graph admitting a t-track layout is called a *t-track graph*. The *track-number* of G, denoted by $\mathsf{tn}(G)$, is the minimum t such that G is a t-track graph. For a summary of bounds on the track-number see [6].

3   Layouts of Subdivisions

Stack and queue layouts of graph subdivisions are a central topic of this paper. That every graph has a 3-stack subdivision has been observed by many authors [10, 17, 11, 1]. Note that 3-stack layouts are important in complexity theory, and 3-stack layouts of knots and links, so called *Dynnikov digrams*, have also recently been considered (see the references in [7]). It is interesting to determine the minimum number of division vertices in a 3-stack subdivision of a given graph. The previously best known bounds are due to Enomoto and Miyauchi [10], who proved that every graph has a 3-stack subdivision with $\mathcal{O}(\log n)$ division vertices per edge. Moreover, Enomoto *et al.* [12] proved that this bound is tight up to a constant factor for K_n (and some slightly more general families). Thus Enomoto *et al.* [12] claimed that the $\mathcal{O}(\log n)$ upper bound is 'essentially best possible'. We prove the following refinement of the upper bound of Enomoto and Miyauchi [10], in which the number of division vertices per edge depends on the stack-number or queue-number of the given graph. Moreover, we characterise those graphs admitting k-stack subdivisions for all k.

Theorem 1. (a) *Every graph G has a 3-stack subdivision with*
   $\mathcal{O}(\log \min\{\mathsf{sn}(G), \mathsf{qn}(G)\})$ *division vertices per edge.*
(b) *A graph has a 2-stack subdivision if and only if it is planar. Every planar graph has a 2-stack subdivision with at most one division vertex per edge.*
(c) *A graph has a 1-stack subdivision if and only if it is outerplanar. Every outerplanar graph has a 1-stack layout (with no division vertices).*

Proof Outline. Let H be the subdivision of G with $2\lceil \log \mathsf{sn}(G) \rceil - 2$ division vertices per edge. As illustrated in Figure 1, we now prove that H has a 3-stack subdivision. Consider a $\mathsf{sn}(G)$-stack layout of G. Let T be the complete binary tree of height $\lceil \log \mathsf{sn}(G) \rceil$. Consider each stack of G to correspond to a distinct leaf of T. Now define a mapping of the vertices of H into the nodes of T such that adjacent vertices of H are mapped to adjacent nodes of T or to the same leaf of T. In particular, the original vertices of G are mapped to the root, and each subdivided edge e is mapped to a walk from the root to the leaf corresponding to the stack containing e, and then back to the root. A depth-first ordering of $V(T)$ gives a 3-stack layout of T in which edges with a common endpoint are in distinct stacks. From this layout of T we can obtain the desired 3-stack layout of H by appropriately ordering the vertices of H that are mapped to a single node of T, and by assigning each edge e of H to the same stack as the edge of T that e is mapped to. The proof that G has a 3-stack subdivision with $\mathcal{O}(\log \mathsf{qn}(G))$ division vertices per edge is similar. Parts (b) and (c) are easy extensions of known results.   □

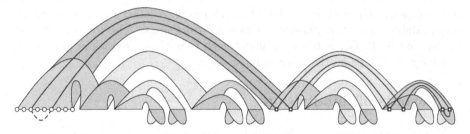

Fig. 1. 3-stack subdivision of a 16-stack graph; one edge is indicated.

Since $\mathsf{sn}(G)$ and $\mathsf{qn}(G)$ are both no more than n, our bound in Theorem 1(a) is at most the $\mathcal{O}(\log n)$ bound of Enomoto and Miyauchi [10] (ignoring constant factors). We prove the following analogous result for queue layouts, in which, additionally, the number of division vertices per edge is optimal.

Theorem 2. (a) *Every graph G has a 2-queue subdivision with $\mathcal{O}(\log \mathsf{qn}(G))$ division vertices per edge, and every 2-queue subdivision of G has an edge with $\Omega(\log \mathsf{qn}(G))$ division vertices per edge.*
(b) *A graph has a 1-queue subdivision if and only if it is planar.*

Thus, at least for the representation of graph subdivisions, two queues suffice rather than three stacks. In this sense, queues are more powerful than stacks. We have the following analogous result for track layouts.

Theorem 3. (a) *Every graph G has a 4-track subdivision with $\mathcal{O}(\log \mathsf{qn}(G))$ division vertices per edge, and every 4-track subdivision of G has an edge with $\Omega(\log \mathsf{qn}(G))$ division vertices.*
(b) *A graph has a 3-track subdivision if and only if it is planar.*
(c) *A graph has a 2-track subdivision if and only if it is a forest of caterpillars.*

A trade-off between the number of stacks and the number of division vertices in 3-stack subdivisions was observed by Enomoto and Miyauchi [11], who proved that for all $s \geq 3$, every graph has an s-stack subdivision with $\mathcal{O}(\log_{s-1} n)$ division vertices per edge. Again Enomoto *et al.* [12] proved that this bound is tight up to a constant factor for K_n. As described in Table 1, our results for 3-stack subdivisions, 2-queue subdivisions, and 4-track subdivisions generalise in a similar fashion to the result of Enomoto and Miyauchi [11]. Moreover, we

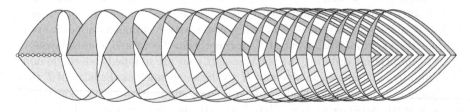

Fig. 2. A 2-queue subdivision of an 8-queue graph.

Table 1. Layouts of a subdivision of a graph G.

graph	type of layout		# division vertices per edge
arbitrary	s-stack	$(s \geq 3)$	$\mathcal{O}(\log_{s-1} \mathsf{sn}(G))$
arbitrary	s-stack	$(s \geq 3)$	$\mathcal{O}(\log_{s-1} \mathsf{qn}(G))$
planar	2-stack		1
arbitrary	q-queue	$(q \geq 2)$	$\Theta(\log_q \mathsf{qn}(G))$
planar	1-queue		$n-2$
arbitrary	s-stack q-queue	$(s \geq 1, q \geq 1)$	$\mathcal{O}(\log_{(s+q)q} \mathsf{sn}(G))$
arbitrary	s-stack q-queue	$(s \geq 1, q \geq 1)$	$\mathcal{O}(\log_{(s+q)q} \mathsf{qn}(G))$
planar	1-stack 1-queue		4
arbitrary	$(d+1,2)$-track	$(d \geq 2)$	$\Theta(\log_d \mathsf{qn}(G))$
arbitrary	$(d,3)$-track	$(d \geq 2)$	$\Theta(\log_d \mathsf{qn}(G))$
arbitrary	$(d+2)$-track	$(d \geq 2)$	$\Theta(\log_d \mathsf{qn}(G))$
planar	3-track		$n-2$

generalise stack and queue layouts through the notion of a *mixed* layout. Here each edge is assigned to a stack or to a queue, defined with respect to a common vertex ordering. We speak of an *s-stack q-queue mixed layout* and an *s-stack q-queue graph*. Part of the motivation for studying mixed stack and queue layouts is that they model the double-ended queue (dequeue) data structure, since a dequeue may be simulated by two stacks and one queue.

4   Relationships

The following lemma highlights the fundamental relationship between track layouts, and queue and stack layouts. Its proof follows immediately from the definitions, and is illustrated in Figure 3 for $k = 1$.

Lemma 1. *Let $\{A, B\}$ be a track assignment of a bipartite graph G. Then the following are equivalent:*

(a) *$\{A, B\}$ admits a $(k, 2)$-track layout of G,*
(b) *the vertex ordering with A followed by B admits a k-queue layout of G, and*
(c) *the vertex ordering with A followed by the reversal of B admits a k-stack layout of G.*

The relationship between queue and track layouts in Lemma 1 was extended by Dujmović *et al.* [6] who proved that queue-number and track-number are tied. Despite a wealth of research on stack and queue layouts, the following fundamental questions of Heath *et al.* [15] remain unanswered[1].

---

[1] Heath *et al.* [15], in their study of the relationship between stack- and queue-number, restricted themselves to linear binding functions. For example, for stack-number to be bounded by queue-number meant that $\mathsf{sn}(G) \in \mathcal{O}(\mathsf{qn}(G))$ for every graph G. Thus Heath *et al.* [15] considered Open Problem 1 to be solved in the negative by displaying an infinite class of graphs $\mathcal{G}$, such that $\mathsf{sn}(\mathcal{G}) \in \Omega(3^{\mathsf{qn}(\mathcal{G})})$. In our more liberal definition of a binding function, this result merely provides a lower bound on a potential binding function.

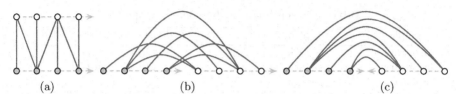

Fig. 3. Layouts of a caterpillar: (a) 2-track, (b) 1-queue, (c) 1-stack.

Open Problem 1. [15] Is stack-number bounded by queue-number?

Open Problem 2. [15] Is queue-number bounded by stack-number?

Suppose that stack-number is bounded by queue-number, but queue-number is not bounded by stack-number. This would happen, for example, if there exists a constant s such that for every q there exists an s-stack graph with no q-queue layout. Then we would consider stacks to be more 'powerful' than queues. In the remainder of this section we show that the study of stack, queue and track layouts of subdivisions provides insights into these open problems.

Let α be a graph parameter. Let sub-α be the graph parameter defined by sub-$\alpha(G) = \alpha(G')$ for every graph G. We say α is *topological* if α and sub-α are tied. For example, chromatic number is not topological since G' is bipartite. On the other hand tree-width is topological. In fact, the tree-width of G equals the tree-width of every subdivision of G. Similarly crossing number is topological.

The *thickness* of a graph G, denoted by $\theta(G)$, is the minimum number of sub-graphs in a partition of $E(G)$ into planar subgraphs. Thickness is not topological since it is easily seen that $\theta(G') \leq 2$. The *geometric thickness* of a graph G, denoted by $\bar{\theta}(G)$, is the minimum number of colours such that G can be drawn in the plane with edges as coloured straight-line segments, such that monochromatic edges do not cross. Eppstein [13] proved that $\bar{\theta}(G') \leq 2$ for every graph G. Thus geometric thickness is not topological.

Stack-number (or book-thickness) is equivalent to geometric thickness with the additional requirement that the vertices are in convex position. Thus

$$\forall \text{ graph } G, \ \theta(G) \leq \bar{\theta}(G) \leq \mathsf{sn}(G) \ . \tag{1}$$

Blankenship and Oporowski [1], Enomoto and Miyauchi [10], and Eppstein [13] independently proved that $\mathsf{sn}(K_n)$ is bounded by $\mathsf{sn}(K_n')$. The proofs by Blankenship and Oporowski [1] and Eppstein [13] use essentially the same Ramsey-theoretic argument. Since $\bar{\theta}(K_n') = 2$, Eppstein [13] observed that stack-number is not bounded by geometric thickness. Using a more elaborate Ramsey-theoretic argument, Eppstein [13] proved that geometric thickness is not bounded by thickness. In particular, for every t there exists a graph with thickness three and geometric thickness at least t. Blankenship and Oporowski [1] conjecture that their result for complete graphs extends to all graphs.

Conjecture 1. [1] There exists a function f, such that for every graph G and every subdivision H of G with at most one division vertex per edge, we have $\mathsf{sn}(G) \leq f(\mathsf{sn}(H))$.

We now prove that Conjecture 1 is related to Open Problem 1.

Theorem 4. *If Conjecture 1 is true then stack-number is topological, and stack-number is bounded by queue-number.*

Proof Outline. Conjecture 1 would imply that sn is bounded by sub-sn, which would imply that stack-number is topological since it is easily seen that $\mathsf{sn}(G') \leq \mathsf{sn}(G) + 1$. It follows from Conjecture 1 that there exists a function f^* such that for any s-stack subdivision of a graph G with k division vertices per edge, G has a $f^*(s, k)$-stack layout. By Theorem 1(a), every graph G has a 3-stack subdivision with $\mathcal{O}(\log \mathsf{qn}(G))$ division vertices per edge. Thus $\mathsf{sn}(G) \leq f^*(3, \mathcal{O}(\log \mathsf{qn}(G)))$, and stack-number is bounded by queue-number.   □

We now turn our attention to the question of whether queue-number is topological. The next lemma is proved by repeated application of the Erdös-Szekeres Theorem regarding monotone subsequences.

Lemma 2. *If a q-queue subdivision of a graph G has at most k division vertices per edge, then $\mathsf{qn}(G) \in \mathcal{O}(q^{2k})$.*

Lemma 2 is used to prove the lower bounds on the number of division vertices per edge in Theorem 2(a) and Theorem 3(a). It follows from Lemma 2 that:

Theorem 5. *Queue-number is topological (for all graphs), and track-number is topological for any proper minor-closed graph family.*

We now relate queue-number to a new thickness parameter. Let the 2-*track thickness* of a bipartite graph G, denoted by $\theta_2(G)$, be the minimum k such that G has a $(k, 2)$-track layout. By (1) and Lemma 1(c),

$$\forall \text{ bipartite graphs } G, \ \theta(G) \leq \overline{\theta}(G) \leq \mathsf{sn}(G) \leq \theta_2(G) \ .$$

Let the 2-*track sub-thickness* of a graph G, denoted by sub-$\theta_2(G)$, be the 2-track thickness of G'. This is well-defined since G' is bipartite.

Theorem 6. *Queue-number is tied to 2-track thickness for bipartite graphs, and queue-number is tied to 2-track sub-thickness (for all graphs).*

Theorem 6 is somewhat counterintuitive since, at first glance, queue layouts may have many crossings, as opposed to the various thickness parameters. The immediate implication for Open Problem 1 is that stack-number is bounded by queue-number if and only if stack-number is bounded by 2-track sub-thickness. While it is an open problem whether stack number is bounded by track-number or by queue-number, in [6] we prove the weaker result that geometric thickness is bounded by track-number, which implies that geometric thickness is bounded by queue-number. We have the following reductions for Open Problem 2.

Theorem 7. *The following are equivalent:*

(a) *queue-number is bounded by stack-number,*
(b) *bipartite 3-stack graphs have bounded queue-number,*
(c) *bipartite 3-stack graphs have bounded 2-track thickness.*

Moreover, if queue-number is bounded by stack-number then queue-number is bounded by a polynomial function of stack-number.

Proof Outline. That (a) implies (b) is immediate. Theorem 6 proves that (b) and (c) are equivalent. It remains to prove that (b) implies (a). Suppose that every bipartite 3-stack graph has queue-number at most some constant q. Consider an arbitrary graph G. An easy extension of Theorem 1(a) proves that G has a 3-stack bipartite subdivision D with $\mathcal{O}(\log \mathsf{sn}(G))$ division vertices per edge. By assumption, $\mathsf{qn}(D) \leq q$. By Lemma 2, and with an abuse of $\mathcal{O}()$ notation, $\mathsf{qn}(G) \in \mathcal{O}(q^{\mathcal{O}(\log \mathsf{sn}(G))}) \in \mathcal{O}(\mathsf{sn}(G)^{\mathcal{O}(q)})$. Thus queue-number is bounded by a polynomial function of stack-number.   □

For Theorem 7 to hold, it is essential that the number of division vertices per edge in Theorem 1(a) is some function of $\mathsf{sn}(G)$, thus emphasising the significance of our bound in comparison with previous results.

5   Three-Dimensional Polyline Drawings

A *3D polyline drawing* of a graph represents the vertices by distinct points in $\mathbb{Z}^3$ (called *gridpoints*), and represents each edge as a polyline between its endpoints with bends (if any) also at gridpoints, such that distinct edges only intersect at common endpoints, and each edge only intersects a vertex that is an endpoint of that edge. A 3D polyline drawing with at most b bends per edge is called a *3D b-bend drawing*. A 3D 0-bend drawing is called a *3D straight-line drawing*. Of course, a 3D b-bend drawing of a graph G is precisely a 3D straight-line drawing of a subdivision of G with at most b division vertices per edge. The *bounding box* of a 3D polyline drawing is the minimum axis-aligned box containing the drawing. If the bounding box has side lengths $X - 1$, $Y - 1$ and $Z - 1$, then we speak of an $X \times Y \times Z$ polyline drawing with *volume* $X \cdot Y \cdot Z$. That is, the volume of a 3D drawing is the number of gridpoints in the bounding box.

This paper initiates the study of upper bounds on the volume and number of bends per edge in arbitrary 3D polyline drawings. The volume of 3D straight-line drawings has been widely studied [4, 3, 14, 19, 2]. Table 2 summarises the best known upper bounds on the volume and bends per edge, including those established in this paper. Our upper bound of $\mathcal{O}(m \log q)$ is within a factor of $\mathcal{O}(\log q)$ of being optimal for all q-queue graphs, since Bose *et al.* [2] proved that 3D polyline drawings have at least $\frac{1}{8}(n + m)$ volume.

Track layouts have previously been used to produce 3D drawings with small volume (see [5]). The principle idea is to position the vertices in a single track on a vertical 'rod'. Since there are no X-crossings in the track layout, no edges between the same pair of tracks can cross.

Theorem 8. [9, 5] *Let G be a c-colourable t-track graph. Then*

(a) *G has a $\mathcal{O}(t) \times \mathcal{O}(t) \times \mathcal{O}(n)$ straight-line drawing with $\mathcal{O}(t^2 n)$ volume, and*
(b) *G has a $\mathcal{O}(c) \times \mathcal{O}(c^2 t) \times \mathcal{O}(c^4 n)$ straight-line drawing with $\mathcal{O}(c^7 tn)$ volume.*

Moreover, if G has an $X \times Y \times Z$ straight-line drawing then G has track-number $\mathsf{tn}(G) \leq 2XY$.

By Theorem 3(a), every graph has a 4-track subdivision with $\mathcal{O}(\log n)$ division vertices per edge, and hence a 3D polyline drawing with $\mathcal{O}(n + m \log n)$ volume by Theorem 8(a). We have the following specific results.

Table 2. Volume of 3D polyline drawings of graphs with n vertices and $m \geq n$ edges.

graph family	bends per edge	volume	reference
arbitrary	0	$\mathcal{O}(n^3)$	Cohen *et al.* [3]
arbitrary	0	$\mathcal{O}(m^{4/3}n)$	Dujmović and Wood [9]
maximum degree Δ	0	$\mathcal{O}(\Delta mn)$	Dujmović and Wood [9]
bounded maximum degree	0	$\mathcal{O}(m^{1/2}n)$	Dujmović and Wood [9]
bounded chromatic number	0	$\mathcal{O}(n^2)$	Pach *et al.* [19]
bounded chromatic number	0	$\mathcal{O}(m^{2/3}n)$	Dujmović and Wood [9]
H-minor free (H fixed)	0	$\mathcal{O}(n^{3/2})$	Dujmović and Wood [9]
bounded tree-width	0	$\mathcal{O}(n)$	Dujmović *et al.* [5]
c-colourable q-queue	1	$\mathcal{O}(cqm)$	Theorem 9(a)
arbitrary	1	$\mathcal{O}(nm)$	Theorem 9(b)
q-queue	2	$\mathcal{O}(qn)$	Theorem 9(c)
q-queue (constant $\epsilon > 0$)	$\mathcal{O}(1)$	$\mathcal{O}(mq^\epsilon)$	Theorem 10
q-queue	$\mathcal{O}(\log q)$	$\mathcal{O}(m \log q)$	Theorem 12

Theorem 9. *Every c-colourable q-queue graph has: (a) a $2 \times c(q+1) \times (n+m)$ polyline 1-bend drawing, (b) an $n \times m \times 2$ polyline 1-bend drawing, and (c) a $2 \times 2q \times (2n-3)$ polyline 2-bend drawing.*

The next result highlights the apparent trade-off between few bends and small volume.

Theorem 10. *For every $\epsilon > 0$, every q-queue graph has a $2 \times \mathcal{O}(q^\epsilon) \times \mathcal{O}(n+m/\epsilon)$ polyline drawing with $\mathcal{O}(1/\epsilon)$ bends per edge.*

Felsner *et al.* [14] introduced 3D straight-line graph drawings with the vertices positioned on the edges of a triangular or rectangular prism.

Theorem 11. *Every planar graph has a $2 \times 2 \times \mathcal{O}(n^2)$ polyline drawing on a triangular prism with at most $n - 2$ bends per edge. Only planar graphs have polyline drawings on a triangular prism.*

Theorem 12. *Every q-queue graph G has a $2 \times 2 \times \mathcal{O}(n + m \log q)$ polyline drawing on a rectangular prism with $\mathcal{O}(\log q)$ bends per edge.*

Proof. By Theorem 3(a), G has a 4-track subdivision D with $\mathcal{O}(\log q)$ division vertices per edge. The number of vertices of D is $\mathcal{O}(n + m \log q)$. Let $\{V_1, V_2, V_3, V_4\}$ be the tracks. Let $n' = \max\{|V_1|, |V_2|, |V_3|, |V_4|\}$. Position the ith vertex in V_1 at $(0,0,2i)$. Position the ith vertex in V_2 at $(1,0,2i)$. Position the ith vertex in V_3 at $(0,1,2i)$. Position the ith vertex in V_4 at $(1,1,2i+1)$. Clearly the only possible crossing is between edges vw and xy with $v \in V_1$, $w \in V_4$, $x \in V_2$, and $y \in V_3$. Such a crossing point is on the line $L = \{(\frac{1}{2}, \frac{1}{2}, z) : z \in \mathbb{R}\}$. However, vw intersects L at $(\frac{1}{2}, \frac{1}{2}, \alpha + \frac{1}{2})$ for some integer α, and xy intersects L at $(\frac{1}{2}, \frac{1}{2}, \beta)$ for some integer β. Thus vw and xy do not intersect. The bounding box is $2 \times 2 \times 2n'$, which is $2 \times 2 \times \mathcal{O}(n + m \log q)$. □

Note that Di Giacomo and Meijer [4] proved that a 4-track graph has a $2 \times 2 \times n$ drawing. When $n' < \frac{n}{2}$ the above construction has less volume.

6 Planar Graphs

Felsner *et al.* [14] asked the following question (in their conference paper).

Open Problem 3. [14] Does every n-vertex planar graph have a 3D straight-line drawing with $\mathcal{O}(n)$ volume?

By Theorem 8, this question has an affirmative answer if planar graphs have bounded track-number. Whether planar graphs have bounded track-number is an open problem due to Hubert de Fraysseix [private communication, 2000], and since queue-number is tied to track-number for planar graphs [5, 6], is equivalent to the following open problem due to Heath *et al.* [15]. Note that the best known upper bound on the queue-number of planar graphs is $\mathcal{O}(\sqrt{n})$.

Open Problem 4. [15] Do planar graphs have bounded queue-number?

We make the following contribution to the study of this problem, which is analogous to Theorem 7, since 2-stack graphs are precisely the subgraphs of Hamiltonian planar graphs.

Theorem 13. *Let $\mathcal{F}(n)$ be the family of functions $\mathcal{O}(1)$ or $\mathcal{O}(\text{polylog } n)$. The following are equivalent:*

(a) *n-vertex planar graphs have queue-number in $\mathcal{F}(n)$,*
(b) *n-vertex bipartite Hamiltonian planar graphs have queue-number in $\mathcal{F}(n)$,*
(c) *n-vertex bipartite Hamiltonian planar graphs have 2-track thickness in $\mathcal{F}(n)$.*
(d) *n-vertex planar graphs have $\mathcal{O}(1) \times \mathcal{O}(1) \times \mathcal{O}(n)$ polyline $\mathcal{O}(1)$-bend drawings.*

Acknowledgements

Thanks to Stefan Langerman for stimulating discussions on 3D polyline drawings. Thanks to Franz Brandenburg and Ulrik Brandes for pointing out the connection to double-ended queues. Thanks to Ferran Hurtado and Prosenjit Bose for graciously hosting the second author, whose research was partially completed at the Departament de Matemàtica Aplicada II, Universitat Politècnica de Catalunya, Barcelona, Spain.

References

1. ROBIN BLANKENSHIP AND BOGDAN OPOROWSKI. Drawing subdivisions of complete and complete bipartite graphs on books. Technical Report 1999-4, Department of Mathematics, Louisiana State University, 1999.
2. PROSENJIT BOSE, JUREK CZYZOWICZ, PAT MORIN, AND DAVID R. WOOD. The maximum number of edges in a three-dimensional grid-drawing. *J. Graph Algorithms Appl.*, 8(1):21–26, 2004.
3. ROBERT F. COHEN, PETER EADES, TAO LIN, AND FRANK RUSKEY. Three-dimensional graph drawing. *Algorithmica*, 17(2):199–208, 1996.
4. EMILIO DI GIACOMO AND HENK MEIJER. Track drawings of graphs with constant queue number. In LIOTTA [16], pages 214–225.

5. VIDA DUJMOVIĆ, PAT MORIN, AND DAVID R. WOOD. Layout of graphs with bounded tree-width. *SIAM J. Comput.*, to appear.
6. VIDA DUJMOVIĆ, ATTILA PÓR, AND DAVID R. WOOD. Track layouts of graphs. Submitted; see arXiv:cs.DM/0407033, 2004.
7. VIDA DUJMOVIĆ AND DAVID R. WOOD. Stacks, queues and tracks: Layouts of graph subdivisions. Submitted; see Tech. Rep. TR-2003-08, School of Computer Science, Carleton University, Ottawa, Canada, 2003.
8. VIDA DUJMOVIĆ AND DAVID R. WOOD. On linear layouts of graphs. *Discrete Math. Theor. Comput. Sci.*, 6(2):339–358, 2004.
9. VIDA DUJMOVIĆ AND DAVID R. WOOD. Three-dimensional grid drawings with sub-quadratic volume. In PACH [18], pages 55–66.
10. HIKOE ENOMOTO AND MIKI SHIMABARA MIYAUCHI. Embedding graphs into a three page book with $O(M \log N)$ crossings of edges over the spine. *SIAM J. Discrete Math.*, 12(3):337–341, 1999.
11. HIKOE ENOMOTO AND MIKI SHIMABARA MIYAUCHI. Embedding a graph into a $d + 1$-page book with $\lceil m \log_d n \rceil$ edge-crossings over the spine. *IPSJ SIGNotes ALgorithms*, 051, 2001. Abstract No. 008.
12. HIKOE ENOMOTO, MIKI SHIMABARA MIYAUCHI, AND KATSUHIRO OTA. Lower bounds for the number of edge-crossings over the spine in a topological book embedding of a graph. *Discrete Appl. Math.*, 92(2-3):149–155, 1999.
13. DAVID EPPSTEIN. Separating thickness from geometric thickness. In PACH [18], pages 75–86.
14. STEFAN FELSNER, GIUSSEPE LIOTTA, AND STEPHEN WISMATH. Straight-line drawings on restricted integer grids in two and three dimensions. *J. Graph Algorithms Appl.*, 7(4):363–398, 2003.
15. LENWOOD S. HEATH, FRANK THOMSON LEIGHTON, AND ARNOLD L. ROSENBERG. Comparing queues and stacks as mechanisms for laying out graphs. *SIAM J. Discrete Math.*, 5(3):398–412, 1992.
16. GUISEPPE LIOTTA, editor. *Proc. 11th International Symp. on Graph Drawing (GD '03)*, volume 2912 of *Lecture Notes in Comput. Sci.* Springer, 2004.
17. MIKI SHIMABARA MIYAUCHI. An $O(nm)$ algorithm for embedding graphs into a 3-page book. *Trans. IEICE*, E77-A(3):521–526, 1994.
18. JÁNOS PACH, editor. *Towards a Theory of Geometric Graphs*, volume 342 of *Contemporary Mathematics*. Amer. Math. Soc., 2004.
19. JÁNOS PACH, TORSTEN THIELE, AND GÉZA TÓTH. Three-dimensional grid drawings of graphs. In BERNARD CHAZELLE, JACOB E. GOODMAN, AND RICHARD POLLACK, editors, *Advances in discrete and computational geometry*, volume 223 of *Contemporary Mathematics*, pages 251–255. Amer. Math. Soc., 1999.

Label Number Maximization in the Slider Model
Extended Abstract

Dietmar Ebner, Gunnar W. Klau, and René Weiskircher

Institute of Computer Graphics and Algorithms, Vienna University of Technology
{ebner,gunnar,weiskircher}@ads.tuwien.ac.at
http://www.ads.tuwien.ac.at

Abstract. We consider the *NP*-hard label number maximization problem LNM: Given a set of rectangular labels, each of which belongs to a point feature in the plane, the task is to find a *labeling* for a largest subset of the labels. A labeling is a placement such that none of the labels overlap and each is placed so that its boundary touches the corresponding point feature. The purpose of this paper is twofold: We present a new force-based simulated annealing algorithm to heuristically solve the problem and we provide the results of a very thorough experimental comparison of the best known labeling methods on widely used benchmark sets. The design of our new method has been guided by the goal to produce labelings that are similar to the results of an experienced human performing the same task. So we are not only looking for a labeling where the number of labels placed is high but also where the distribution of the placed labels is good.

Our experimental results show that the new algorithm outperforms the other methods in terms of quality while still being reasonably fast and confirm that the simulated annealing method is well-suited for map labeling problems.

1 Introduction

The growing amount of data for which informational graphics have to be produced leads to an increasing need for automatic labeling procedures.

Several criteria have been developed that characterize a high-quality labeling:

(C1) On a good map the placement of labels is unambiguous. This implies that labels are close to the point features they belong to.
(C2) The information of the labels is legible.
(C3) No or only a few labels overlap. Obviously, overlaps decrease the legibility of a map.
(C4) The number of omitted labels is low.

The cartographic literature contains more rules, see, *e.g.*, the papers by Imhof [6] and Yoeli [14]. Yet, the overall aim in automatic map labeling is to devise algorithms that produce labelings of maximum legibility.

J. Pach (Ed.): GD 2004, LNCS 3383, pp. 144–154, 2004.
© Springer-Verlag Berlin Heidelberg 2004

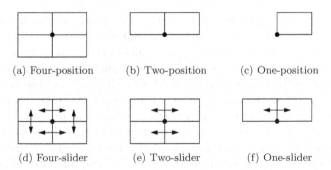

(a) Four-position   (b) Two-position   (c) One-position

(d) Four-slider   (e) Two-slider   (f) One-slider

Fig. 1. Axis-parallel rectangular labeling models. A label can be placed in any of the positions indicated by the rectangles and can slide in the directions of the arrows.

An instance of a labeling problem consists of a set of point features, information about the label sizes, and a mapping from labels to point features. In general it is not possible to place all the given labels in their original size without any overlap. The literature suggests several possibilities to deal with this problem; among these are decreasing the size of the labels to allow a placement of all labels without any overlap (*label size maximization*), and keeping the sizes of the labels fix while looking for the maximum number of labels that can be placed (*label number maximization problem*, LNM).

Research in automated map labeling has mainly focused on the six *labeling models* shown in Figure 1, the most popular of which are the four-position and the four-slider model. The dots in the figure represent the point feature to be labeled.

Definition 1 (LNM in the 4-slider model). *Given a set $\Lambda = \{\lambda_1, \ldots, \lambda_k\}$ (the labels), two functions $w, h : L \to \mathbb{R}$, and a function $a : \Lambda \to \mathbb{R}^2$, find a subset $\Lambda' \subseteq \Lambda$ of largest cardinality and a function $\rho : \Lambda' \to \mathbf{R}$, where $\mathbf{R}$ is the set of axis-parallel rectangles in the plane, so that the following conditions hold:*

(L1) Rectangle $\rho(\lambda)$ has width $w(\lambda)$ and height $h(\lambda)$ for every $\lambda \in \Lambda'$.
(L2) Point $a(\lambda)$ lies on the boundary of $\rho(\lambda)$ for all $\lambda \in \Lambda'$.
(L3) The open intersection $\rho(\lambda) \cap \rho(\mu)$ is empty for all $\lambda, \mu \in \Lambda', \lambda \neq \mu$.

An assignment of labels to rectangles that satisfies the three properties (L1)–(L3) is called a *labeling*. Properties (L1) and (L2) make sure that each label λ is drawn with the given size in the 4-slider model. Property (L3) forbids overlaps between the labels.

Force-directed methods have originally been developed for drawing graphs. In practice, these techniques often perform remarkably well on medium-sized instances and are easy to implement. Further, the resulting drawings typically capture symmetries while avoiding the expensive computations to look for them explicitly.

These algorithms, going back to Eades [3] and Kruskal and Seery [10], view the input graph as a system of objects with forces acting between them. Configurations of the objects with low energy correspond to aesthetically pleasing layouts of the graph. Algorithms for this task are mostly variations of iterative gradient-based methods such as the Newton-Raphson method.

Davidson and Harel consider in [2] the number of edge crossings in a drawing as an additional, discrete term in the objective function and can therefore not apply gradient methods to find an equilibrium. The authors propose the *simulated annealing* approach. This approach defines for each configuration a finite set of neighboring configurations and tries one of them at random. New configurations are always accepted if they decrease the energy of the system but even if they increase the energy, they are accepted with a probability that decreases with time. As we will point out in Section 2, we will use simulated annealing for similar reasons as Davidson and Harel.

Van Kreveld, Strijk, and Wolff [13] show *NP*-completeness of the decision problem in the four-slider model (independently, Marks and Shieber have shown this in [11]). The main result in [13] is a $\frac{1}{2}$-approximation algorithm that is able to find a solution of LNM in any of the slider-models with unit height rectangles. The algorithm is a $\Theta(n \log n)$-time greedy sweep-line algorithm. For the same models, the authors develop a polynomial time approximation scheme. Strijk and van Kreveld extend the above mentioned $\frac{1}{2}$-approximation algorithm for the slider models in [12] to labels with different heights. If r denotes the number of different label heights, the running time of the algorithm is $O(rn \log n)$. The algorithm is based on the simple greedy strategy of iteratively placing the *leftmost label* until no more points can be labeled without intersections. The *leftmost label* is defined to be the label, whose right edge is leftmost among all label candidates, which are those labels that have not been placed yet minus a set of labels that are already known to be unplaceable in the current configuration.

Klau and Mutzel present in [9] an exact algorithm for the label number maximization problem that works in any of the labeling models. The method is based on a pair of so-called constraint graphs that code horizontal and vertical positioning relations. The key idea is to link the two graphs by a set of additional constraints, thus characterizing all feasible solutions of LNM. This combinatorial description enables the formulation of a zero-one integer linear program whose solution leads to an optimal labeling.

The paper [1] by Christensen, Marks, and Shieber contains an extensive computational study of labeling methods in the four-position model. The authors also present a simulated annealing method for this problem that is the clear winner of the study in terms of labeling quality while still being reasonably fast. Furthermore, they propose a procedure for randomly creating labeling instances. We use this benchmark generator, which has become a widely used tool in map labeling research, for our computational experiments in Section 3.

Already in 1982, Hirsch introduced a model that is similar to the four-slider model and proposed an algorithmic labeling method that can be interpreted as a force-directed approach. The algorithm starts with an initial label placement and tests for overlaps. Based on the amount of intersecting area, overlap vectors are computed for labels involved in an overlap conflict. For each label, the summation of these vectors helps in heuristically deciding where to move the label. Successive movements of all labels in conflict, which Hirsch calls a *map sweep*, is done by using one of the following two methods: (a) Moving labels in the direction of the computed vectors with sequential stops at preferred positions. This method allows the label to be placed at any possible position. (b) Performing a discrete jump to a position indicated by the vector angle. Here, the primary aim is to solve an overlap situation where the first method fails. Hirsch does not consider the number maximization problem explicitly. Also, although his overlap vectors resemble the intersection-proportional component within our force system, he does not consider distance-related forces and suggests a different method for finding an equilibrium of minimum energy. His approach can be seen as a gradient-driven heuristic.

2   Force-Directed Map Labeling

In this section we describe our force-based simulated annealing algorithm for the label number maximization problem. Our approach uses repulsive forces between labels, which are used to compute a force vector for each label. The length and direction of these vectors gives us an idea of where to place individual labels and how to solve potential conflicts between two or more labels. As a side-effect we achieve another important benefit, which makes the method usable for practical applications: Our forces are defined to grow super linearly with decreasing distance between two labels. Therefore, labels are not placed close to each other if possible and the method achieves a good distribution of the labels in the available space. This improves the readability of the labels and results in an aesthetically pleasing arrangement. To avoid being trapped in local minima of the energy function, we combine the purely force directed method with the simulated annealing approach.

Every force-directed algorithm consists of two major parts: (a) a force-system between the objects and (b) a method that seeks an equilibrium of minimum energy. In our case a low energy equilibrium configuration should correspond to a pleasing labeling. In contrast to applications in graph drawing labels are bound to their point feature and may not be positioned freely in the available space. We only allow intermediate positions that satisfy at least the first condition, hence we do not need any attractive forces between points and labels. Furthermore we restrict the computation of forces to pairs of labels that might intersect. We call the set of those labels for each label λ the *neighborhood*

$$N(\lambda) = \{\mu \in \Lambda \mid w(\lambda) + w(\mu) \geq |x_\lambda - x_\mu| \wedge h(\lambda) + h(\mu) \geq |y_\lambda - y_\mu|\} \ .$$

Our main goal is to place as many labels as possible in the available space without any intersections. Therefore the decisive factor in our force system is the amount of intersection between two labels. We call this force the *intersection-proportional* component. The amount of the second force acting in our model, the so-called *distance-related* part, depends on the distance between two labels and grows, if two rectangles are placed close to each other. If labels overlap a very small area ϵ the intersection-proportional component can become arbitrary small. Thus we add a constant value to the force function if and only if two labels overlap to punish overlaps stronger. The distance-related part is not the significant value in our model, its only purpose is to guide the algorithm to a well distributed labeling.

For every two labels $\lambda, \mu \in \Lambda$, we define $d_{\min} : \Lambda \times \Lambda \to \mathbb{R}$ as

$$d_{\min}(\lambda, \mu) = \begin{cases} 0 & \text{if } \lambda \text{ and } \mu \text{ overlap} \\ \min\{\|p, q\| \mid p \in \lambda, q \in \mu\} & \text{otherwise .} \end{cases}$$

The function $\|p, q\| : \mathbb{R}^2 \times \mathbb{R}^2 \to \mathbb{R}$ denotes the Euclidean distance between the two points p and q in the Euclidean plane $\mathbb{R}^2$.

We can now define the force function $f = (f_x, f_y)$ for each label in the following way: For each label $\lambda \in \Lambda$ with center point $c_\lambda = (x_\lambda, y_\lambda)$, the x-component of the force function $f_x : \Lambda \to \mathbb{R}$ is defined as

$$f_x(\lambda) = \sum_{\mu \in N(\lambda)} \left(f_i(\lambda, \mu) + f_a(\lambda, \mu) + f_d(\lambda, \mu) \right) (x_\lambda - x_\mu) / (\|c_\lambda, c_\mu\|) \ ,$$

$$\text{where} \qquad f_i(\lambda, \mu) = \delta_1 \, i_x(\lambda, \mu) \, i_y(\lambda, \mu) \ ,$$

$$f_d(\lambda, \mu) = \frac{\delta_2}{\max(\varepsilon, d_{\min}(\lambda, \mu))^2} \ , \quad \text{and} \quad f_a(\lambda, \mu) = \begin{cases} \delta_3 & \text{if } \lambda \text{ and } \mu \text{ overlap} \\ 0 & \text{otherwise .} \end{cases}$$

The y-component f_y is defined analogously. The constants $\delta_1, \delta_2, \delta_3 \in \mathbb{R}$ control the influence of the particular term on the force function f. Note that the direction of the force between two labels is defined by the location of their center points and that ε limits the amount of f_d to a value of δ_2 / ε^2.

"Force has no place where there is need of skill." [4]

A purely force directed method performs poorly if the labels take a significant fraction of the available drawing area. There is only little space for manoeuvre when seeking an equilibrium, especially if incremental methods are used. Often, real-world labeling instances contain dense areas that do not leave much space for moving labels around without producing new intersections. The problem is aggravated by the fact that we only allow horizontal and vertical moves around the label's border. The same observation holds for the algorithm proposed by Hirsch in [5] and is well described by Christensen, Marks, and Shieber in [1].

Figure 2(a) shows an example of a bad local minimum that is difficult to escape from by using incremental moves. It is not possible to transform the bad labeling on the left continuously into the good labeling on the right without a temporary increase of overlaps and thus of the overall energy of the system.

Another problem arises from the direction of the forces. Since labels have non uniform size and they are bound to their point features, the direction of our resulting force vector does not always indicate a solution for the conflict. Figure 2(b) shows a very simple example consisting of just two point features. Any algorithm that strictly follows the direction of the force vector is not able to resolve the shown configuration, even though the optimal solution is self-evident.

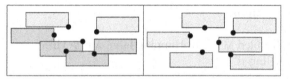  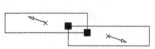

(a) Bad local minimum vs. optimal labeling.   (b) Forces cannot alwyas re-
                                                    solve overlaps.

Fig. 2. Problems with forces.

Therefore, we need to accept worse intermediate configurations to be able to
escape local minima and we propose to use the simulated annealing method for
this purpose.

Simulated annealing is a very flexible optimization method and can be used
in a wide range of combinatorial optimization problems. It has been proposed
in [7] and is derived from the following observation: When cooling down a liquid
rapidly to a crystal form, the system results in amorphous structures with a high
energy while slow cooling results in a crystal structure with lower energy.

The general simulated annealing procedure applies a series of sequential
moves while simultaneously decreasing the temperature. The main idea is that
the probability with which the change from a state with energy E_1 to a state
with energy E_2 will be accepted is $e^{-\frac{E_2 - E_1}{kT}}$, where k is a positive constant.
Thus the probability for moves that increase the energy decreases with a falling
temperature.

The hybrid force-based simulated annealing algorithm for the label number
maximization problem works as follows:

1: compute random initial labeling σ in the eight-pos. model
2: initialize temperature T and cooling rate α
3: compute forces for current conf. and init. set of active labels Φ
4: $M_{\max} \leftarrow 30|\Lambda|$; taken $\leftarrow$ rejected $\leftarrow 0$;
5: **repeat**
6:    $\hat{\sigma} \leftarrow \sigma$;
7:    choose random candidate $\lambda \in \Phi$
8:    **if** $|f_x(\lambda)| > F_{\min} \vee |f_y(\lambda)| > F_{\min}$ **then**
9:        change σ by moving λ in the direction indicated by its force vector
10:       **if** $|f_x(\lambda)| > F_{\min} \vee |f_y(\lambda)| > F_{\min}$ **then**
11:           change σ by moving λ to a random position
12:   **else**
13:       change σ by moving λ to a random position
14:   **if** force$(\sigma) <$ force$(\hat{\sigma}) \vee$ random $r \in [0 \ldots 1] < e^{\frac{\text{force}(\sigma) - \text{force}(\hat{\sigma})}{T}}$ **then**
15:       taken $\leftarrow$ taken $+ 1$
16:       update set of active labels Φ
17:   **else**
18:       rejected $\leftarrow$ rejected $+ 1$
19:       $\sigma \leftarrow \hat{\sigma}$;

20:    **if** taken + rejected $\geq M_{max}$ **then**
21:      **if** taken = 0 **then**
22:        **if** point selection is disabled $\vee \neg\exists$ overlapping label $\hat{\lambda} \in \Phi$ **then**
23:          **return** current labeling σ
24:        **else**
25:          $\sigma \leftarrow \sigma \setminus \hat{\lambda}$
26:          update set of active labels Φ
27:      $T = \alpha T$
28:      $M_{\mathrm{max}} = \max(|\Lambda|, \min(10|\Lambda|, 50|\Phi|));$ taken $\leftarrow$ rejected $\leftarrow 0$
29: **until** $|\Phi| = 0$
30: **return** current labeling σ

The algorithm performs a series of temperature stages. After each stage the temperature is decreased by a constant precomputed factor, which decreases the probability of accepting moves that lead to a higher energy state. To speed up convergence we compute a set of *active labels* Φ, which either intersect at least one other label or their associated force vector indicates movement to a new position with lower energy. The algorithm returns the current solution if $|\Phi| = 0$ and chooses the label with the most overlaps for removal if no move has been accepted for a full temperature stage.

In each iteration we randomly choose a label $\lambda \in \Phi$ and try to move it according to its force vector. If this move does not lead to an equilibrium or the force vector does not indicate movement even though the label is involved in an overlap, we move the label to a random position in the eight-position model instead. The new position is always accepted if it decreases the energy and may be accepted if it does not increase the energy by more than the current temperature allows.

At each temperature stage we perform M_{max} moves. We initialize this value with $30|\Lambda|$ and perform $\max(|\Lambda|, \min(10|\Lambda|, 50|\Phi|))$ moves in all subsequent stages. The initial temperature is chosen such that we accept an increase in the overall force of f_{avg} with a probability of 30%, where f_{avg} represents the amount of force for an overlap of 50% of two average sized labels. The cooling rate α is chosen such that the temperature T becomes less than 1 after 15 stages. The parameter α should be changed to adjust the trade-off between quality and speed. The above settings yield high-quality labelings in reasonable computation time.

Whenever we move a label λ to a different position or remove it from the labeling in line 25, the forces on all labels $\lambda' \in N(\lambda)$ change. Since a simple approach takes time $O(n^2)$ in the worst case we store the forces between each pair of labels in a quadratic matrix. This enables us to update the forces in linear time by recalculating only the change of the particular addend for each neighbor $N(\lambda)$. Furthermore we have to update the set of active labels Φ, since some labels $\lambda' \in N(\lambda)$ may have to be added to or removed from this set.

Since labels have to be placed according to the four-slider model, moving a label alongside its force vector becomes more difficult than moving, *e.g.*, zero-sized nodes in a graph drawing application. A position that corresponds to an equilibrium of the forces is not always valid with respect to the point. Furthermore our forces depend on a combination of the overlapping area and the distance between two labels, which are both defined differently depending on the specific domain, and are thus not continuous. Thus we can not apply numerical algorithms like the Newton-Raphson method or similar techniques, since they require at least the first derivation of the function. In place of this we start moving the label by 20% of the remaining width/height in the particular direction and halve the amount of movement if the indicated direction changes until we achieve an equilibrium or the maximum number of moves has been performed.

We perform at least $|\Lambda|$ moves before removing a heuristically chosen label. Thus the running time depends to a great extent on the number of labels that the algorithm cannot place. Most problem instances in our test suits of real world labeling problems do not contain many of these unplaceable labels. Therefore, our method performs well on these problems. However, if running time is a critical criterion, this step can be replaced through a faster cleanup heuristic.

3   Computational Study

In this section we report on the extensive computational experiments we have performed to evaluate quality and resource requirements of our new method in comparison to the best-known algorithms for label number maximization. We want to emphasize that both the data we used and our implementation of the evaluated algorithms are publicly available under the Gnu General Public License at http://www.ads.tuwien.ac.at/research/labeling.

We have implemented all major map labeling algorithms that we found in the literature on point feature map labeling in the slider model. All computations were done on a Pentium 4 with 2.8GHz and 2GB of RAM. For each run, we set a limit of 30 minutes computation time.

- The algorithm RANDOM, which places labels randomly, if possible, has been incorporated into the study only for comparative reasons.
- Christensen, Marks and Shieber present in [1] a simulated annealing approach that beats most other algorithms in both speed and quality. Since their implementation uses the four-position model, in general, the quality of their solutions cannot be as good as those of algorithms for the four-slider model. Nevertheless we decided to include this algorithm in our computational study to compare one of the best known labeling methods in the four-position model to the remaining algorithms. We isolated configuration changes to either obstructed or deleted labels, since this causes the algorithm to converge much faster.
- We followed the suggestion of Christensen, Marks, and Shieber in [1] and reduced the radius of the circle in Hirsch's algorithm (HIRSCH) to zero. Furthermore we neglect any cartographic preferences.
- APPROX is our implementation of the algorithm described in [13, 12] that runs in $O(n^2)$, does not rely on unique label heights, and is quite fast in practice.

- We computed optimal labelings in the 4-slider model using OPT, an implementation of the algorithm presented in [9]. Note that, due to the running
  time limit, only instances up to maximally 850 labels could be computed.
- Finally, FDL is our JAVA implementation of the new force-directed method.

We ran the implementations on different data sets. Among them are (a)
instances generated with the widely used benchmark generator by Christensen,
Marks, and Shieber and (b) instances derived from real world data giving the
positions of ground water drill holes in Munich.

We generated 25 random problem instances of type (a) for each instance size
in $\{100, 150, \ldots, 1450, 1500\}$ labels, resulting in 685 instances, as in the study
on the four-position model [1]. The numbers of labels in the real-world problem
set (b) are in the set $\{250, 500, 750, \ldots, 2750, 3000\}$ and there are 30 instances
for each number of labels.

Figure 3(a) illustrates the performance of the evaluated algorithms in terms
of quality, whereas Figure 3(b) displays their running time behavior. Of course,
OPT performs best in terms of quality but also needs the largest amount of
resources. Among the heuristic methods, our new algorithm produces the best
scores but also takes more time to compute them – especially for large instances.
We want to remark, however, that the random instances larger than 1000 labels
do not resemble real-world instances since they get very dense (see the discussion
on the real-world Munich drill hole instances below). The plots also reveal that
the approximation algorithm performs surprisingly well in terms of quality (for
very large instances it becomes as good as FDL) with the advantage that its
running time does not explode.

We then compared the heuristic methods on the easier real-world instances.
Figures 4(a) and 4(b) show the results. It can be seen that all methods apart
from RANDOM have quite good results with FDL being the winner. In fact, these
instances have been generated so that always 100% of the labels can be placed
– even in the four-position model. FDL is the only method that achieved the
perfect score on all instances. As already mentioned, the running time of FDL
depends heavily on the number of labels that cannot be placed. As this number

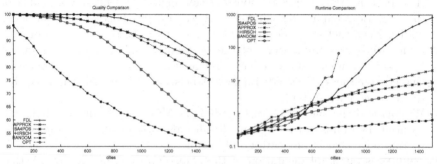

(a) Percentage of labeled point features. (b) Runtime of the algorithms in seconds.

Fig. 3. Results for the random benchmark set.

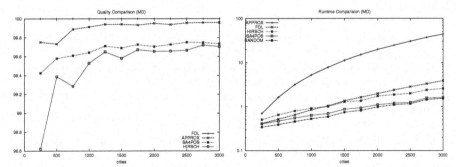

(a) Percentage of labeled point features (b) Runtime of the algorithms in seconds. (RANDOM is far behind and thus not shown).

Fig. 4. Results for the real-world benchmark set.

is zero for these instances, the running time behavior is very good for FDL as for all other methods apart from the approximation algorithm.

Our computational results confirm the outcome of the 1995 study [1]: simulated annealing is very well-suited for labeling problems and outperforms other methods in terms of quality.

4   Conclusions

We have presented a new hybrid heuristical approach for the label number maximization problem. Our algorithm uses an underlying force system that serves two purposes. First, a minimum energy configuration of this system corresponds to placements with evenly distributed labels that is appealing to a human observer. The second task of the force system is to determine which labels should be left out to obtain a labeling without overlaps. We combine this with a simulated annealing algorithm to escape local minima.

Our extensive computational experiments on widely used benchmark data show that our algorithm finds labelings that are close to optimality in a short amount of computing time. We find that our results often look similar to those of a human cartographer.

Future lines of research might include to adapt the approach to line and area labeling. We will also investigate how to combine force-based graph drawing with our approach to attack the combined drawing and labeling problem. Further, we want to integrate the approach into the Human-Guided Search (HuGS) system, see [8], to allow for human interaction.

Acknowledgments

We thank Tycho Strijk for useful discussions regarding the implementation APPROX and one of the anonymous referees of a previous version for his constructive feedback.

154   Dietmar Ebner, Gunnar W. Klau, and René Weiskircher

References

1. J. Christensen, J. Marks, and S. Shieber. An empirical study of algorithms for point-feature label placement. *ACM Trans. Graph.*, 14(3):203–232, 1995.
2. R. Davidson and D. Harel. Drawing graphs nicely using simulated annealing. *ACM Transactions on Graphics*, 15(4):301–331, 1996.
3. P. Eades. A heuristic for graph drawing. *Congressus Numerantium*, 42:149–160, 1984.
4. Herodotus. *The History of Herodotus.* 440 B.C.
5. S. A. Hirsch. An algorithm for automatic name placement around point data. *The American Cartographer*, 9:5–17, 1982.
6. E. Imhof. Die Anordnung der Namen in der Karte. *International Yearbook of Cartography*, 2:93–129, 1962.
7. S. Kirkpatrick, Jr. C. D. Gelatt, and M. P. Vecchi. Optimization by simulated annealing. *Science*, 220:671–680, 1983.
8. G. W. Klau, N. Lesh, J. Marks, M. Mitzenmacher, and G. T. Schafer. The HuGS platform: A toolkit for interactive optimization. In *Proc. of AVI 2002 (International Working Conference on Advanced Visual Interfaces)*, 2002.
9. G. W. Klau and P. Mutzel. Optimal labelling of point features in rectangular labelling models. *Mathematical Programming*, 94(2-3):435–458, 2003.
10. J. Kruskal and J. Seery. Designing network diagrams. *First General Conf. on Social Graphics*, pages 22–50, 1980.
11. J. Marks and S. Shieber. The computational complexity of cartographic label placement. Technical Report TR-05-91, Harvard University, Cambridge, MA, U.S.A., 1991.
12. T. Strijk and M. van Kreveld. Practical extensions of point labeling in the slider model. In *Proc. 7th ACM Symp. Adv. Geogr. Inform. Syst.*, pages 47–52, 1999.
13. M. van Kreveld, T. Strijk, and A. Wolff. Point labeling with sliding labels. *Computational Geometry: Theory and Applications*, 13:21–47, 1999.
14. P. Yoeli. The logic of automated map lettering. *The Cartographic Journal*, 9:99–108, 1972.

An Efficient Implementation
of Sugiyama's Algorithm
for Layered Graph Drawing*

Markus Eiglsperger[1], Martin Siebenhaller[2], and Michael Kaufmann[2]

[1] Universität Konstanz, Fakultät für Informationswissenschaften,
78457 Konstanz, Germany
markus.eiglsperger@uni-konstanz.de
[2] Universität Tübingen, WSI für Informatik, Sand 13,
72076 Tübingen, Germany
{siebenha,mk}@informatik.uni-tuebingen.de

Abstract. Sugiyama's algorithmic framework for layered graph drawing is commonly used in practical software. The extensive use of dummy vertices to break long edges between non-adjacent layers often leads to unsatisfactorial performance. The worst-case running-time of Sugiyama's approach is $O(|V||E|\log|E|)$ requiring $O(|V||E|)$ memory, which makes it unusable for the visualization of large graphs. By a conceptually simple new technique we are able to keep the number of dummy vertices and edges linear in the size of the graph and hence reduce the worst-case time complexity of Sugiyama's approach by an order of magnitude to $O((|V|+|E|)\log|E|)$ requiring $O(|V|+|E|)$ space.

1 Introduction

Most approaches for drawing directed graphs used in practice follow the same framework developed by Sugiyama et al. [17], which produces layered layouts [3]. This framework consists of four phases: In the first phase, called *Cycle Removal*, the directed input graph $G = (V, E)$ is made acyclic by reversing appropriate edges. During the second phase, called *Layer Assignment*, the vertices are assigned to horizontal layers. Before the third phase starts, long edges between vertices of non-adjacent layers are replaced by chains of dummy vertices and edges between the corresponding adjacent layers. Hence in the third phase, called *Crossing Reduction*, an ordering of the vertices within a layer is computed such that the number of edge crossings is reduced. Finally, the fourth phase, called *Horizontal Coordinate Assignment*, calculates an x-coordinate for each vertex. Now the dummy vertices introduced after the layer assignment are removed and replaced by bends.

Unfortunately, almost all problems occuring during the single phases of this approach are NP-hard: Feedback-arc set [12], Precedence Constrained Multi-processor Scheduling [5], 2-layer crossing minimization [8], etc. Nevertheless, for

---

* This work has partially been supported by the DFG-grant Ka512/8-2. It has been performed when the first author was with the Universität Tübingen.

all these problems appropriate heuristics have been developed and nearly all practical graph drawing software use this approach, mostly enriched by modifications required in practice like large vertices, same-layer-edges, clustering, etc.

In the following, we review Sugiyama's framework for drawing directed graphs in more detail and give the necessary definitions and results. Then we use this as basis for our new approach. In the rest of this work we assume that the input graph is already acyclic.

1.1   Layer Assignment and Normalization

Let $L_1,..,L_h$ be a partition of V with $L_i \subset V$, $1 \leq i \leq h$ and $\bigcup_{i=1}^{h} L_i = V$ (h denotes the number of layers). Such a partition is called a layering of G if for all $e = (v, w)$ with $v \in L_i$ and $w \in L_j$ holds $i < j$. The number of vertices in a layer L_i is denoted with n_i. The span of edge e is $j - i$. In a layered drawing, all vertices $v \in L_i$ are drawn on a horizontal line (same y-coordinate). We call the layering proper if $span(e) = 1$ for all edges $e \in E$. In most applications the layers of the vertices can be assigned arbitrarily and, in some cases, the layer assignment is even part of the input.

For edges $e = (u, v)$ with $span(e) > 1$ and for which the endpoints u and v lie on layers L_i and L_j, we replace edge e by a chain of dummy vertices $u = d_i, d_{i-1}, \ldots, d_{j+1}, d_j = v$ where any two consecutive dummy vertices are connected by a dummy edge. Vertex d_k for $i \leq k \leq j$ is placed on layer L_k. This process is called *normalization* and the result the *normalized graph* $G_N = (V_N, E_N)$. With this construction, the next phase starts with a proper layering.

Gansner et al. [10] presented an algorithm, which calculates a layer assignment of the vertices such that the total number of dummy vertices is minimized. The algorithm for minimizing the number of dummy vertices is a network simplex method and no polynomial time bound has been proven for it, but several linear time heuristics for this problem work well in practice [14, 15]. In the worst case $|V_N| = O(|V||E|)$ and $|E_N| = O(|V||E|)$.

After the final layout of the modified graph, we replace the chains of dummy edges by polygonal chains in which the former dummy vertices become bends.

1.2   Crossing Reduction

The vertices within each layer L_i are stored in an ordered list, which gives the left-to-right order of the vertices on the corresponding horizontal line. Such an ordering is called a *layer ordering*. We will often identify the layer with the corresponding list L_i. The ordering of the vertices within adjacent layers L_{i-1} and L_i determines the edge crossings with endpoints on both layers.

Crossing reduction is usually done by a layer-by-layer sweep where each step minimizes the number of edge crossings for a pair of adjacent layers. This layer-by-layer sweep is performed as follows: We start by choosing an arbitrary vertex order for the first layer L_1 (we number the layers from top to bottom). Then iteratively, while the vertex ordering of layer L_{i-1} is kept fixed, the vertices of

L_i are put in an order that minimizes crossings. This step is called one-sided two-layer crossing minimization and is repeated for $i = 2, .., h$. Then the sweep direction is reversed and repeated until no further crossings can be saved.

Many heuristics have been proposed to attack the one-sided two-layer crossing minimization problem [3, 6]. Most important are the *median* and the *barycenter heuristic*, where the new position of each vertex v in list L_i is chosen relative to the position of the adjacent vertices from list L_{i-1}.

To decide whether we improved the number of crossings by a sweep, we have to count this number. This important subproblem, called the *bilayer cross counting* problem, has to be solved in each of the steps. The naive sweep-line algorithm needs time $O(|E'| + |C'|)$ where $|E'|$ is the number of edges between the two layers and $|C'|$ the number of crossings between these edges [15]. It has recently been improved to $O(|E'| \log |V'|)$ by Waddle [19] and Barth et al. [2].

The algorithm reduces the bilayer cross counting problem to the problem of counting the inversions in the vertex sequences of layers L_{i+1} and L_i respectively. The number of inversions are counted by means of an efficient data structure, called the accumulator tree T.

1.3   Horizontal Coordinate Assignment

The horizontal coordinate assignment computes the x-coordinate for each vertex with respect to the layer ordering computed by the crossing reduction phase. There are two objectives to consider to get nice drawings. First the drawings should be compact and second the edges should be as vertical as possible.

Gansner et al. [10] model this problem as a linear program:

$$\min \sum_{(v,w) \in E} \Omega(v, w) \cdot |x(v) - x(w)|$$

$$s.t. \quad x(b) - x(a) \geq \delta(a, b) \quad a, b \text{ consecutive in } L_i, \ 1 \leq i \leq h$$

where $\Omega(v, w)$ denotes the priority to draw edge (v, w) vertical and $\delta(a, b)$ denotes the minimum distance of consecutive vertices a and b. This linear program can be interpreted as a rank assignment problem on a compaction graph $G_a = (V, \{(a, b) : a, b \text{ consecutive in } L_i, \ 1 \leq i \leq h\})$ with length function δ. Each valid rank assignment corresponds to a valid drawing. The above objective function can be modeled by adding vertices and edges to G_a [10].

The drawback of the above approach is, that edges can have as many bends as dummy vertices. This creates sometimes a "spaghetti" effect which reduces the readability. To avoid this negative behaviour the *linear segments model* was proposed, where each edge is drawn as polyline with at most three segments. The middle segment is always drawn vertical. In general, linear segment drawings have less bends but need more area than drawings in other models. There have been a number of algorithms proposed for this model [4, 15]. The approach of Brandes and Köpf [4] produces pleasing results in linear time.

1.4   Drawbacks

The complexity of algorithms in the Sugiyama framework heavily depends on the number of dummy vertices inserted. Although this number can be minimized efficiently, it may still be in the order of $O(|V||E|)$ [9]. Assume we use an algorithm based on the Sugiyama framework which uses the fastest available algorithms for each phase. Then this algorithm has running time $O(|V||E|\log|E|)$ and uses $O(|V||E|)$ memory.

To improve the running time and space complexity we avoid introducing dummy vertices for each layer that an edge spans. We rather split edges only in a limited number of segments. As a result, there may be edges which traverse layers without having a dummy vertex in it. We will extend the existing crossing reduction and coordinate assignment algorithms to handle this case.

A similar idea is used in the Tulip-software described in [1]. Unfortunately, no details are given. However, in this approach, only the proper edges are considered in the crossing reduction phase and the long edges are ignored. This leads to drawings which have many more crossings than drawings using the traditional Sugiyama approach. In contrast, we will show that our approach yields the same results as the methods traditionally used in practice.

2   The New Approach

The basic idea of our new approach is the following: Since in the linear segments model each edge consists of at most two bends, all corresponding dummy vertices in the middle layers have the same x-coordinate. We combine them into one *segment* and therefore reduce the size of the normalized graph dramatically. More precisely, if edge $e = (v, w)$ spans between layers L_i and L_j with $|j - i| > 2$, we introduce only two dummy vertices: p_e at layer L_{i+1} (called p-vertex) and q_e at layer L_{j-1} (called q-vertex), as well as three edges: (v, p_e), $s_e = (p_e, q_e)$, and (q_e, w). The first and the last edge are proper while s_e, called the *segment* of e, is not necessarily proper. If $|j - i| = 2$ we insert a single dummy vertex r_e. We call this transformation *sparse normalization* and the result the *sparse normalized graph* $G_S = (V_S, E_S)$. The size of the sparse normalized graph is linear with respect to the size of the input graph.

A layer L of a sparse normalized graph contains vertices and segments. A layer ordering of a sparse normalized graph is a linear ordering of the vertices and segments in a layer and is called a *sparse layer ordering*. For a graph G, there is a one-to-one correspondence between layer orderings of the normalized graph G_N and sparse layer orderings of the sparse normalized graph G_S.

Let us look at the layer orderings of normalized graphs: instead of storing the layer ordering in lists, we can store it in a directed graph D. This graph has an edge between vertices v and w if and only if these two vertices are in the same layer i and are consecutive in L_i. The ordering $<$ defined as $v < w$ if and only if there is a directed path from v to w in D, is a complete ordering for the vertices of a layer, i.e., either $v < w$ or $w < v$ for $v, w \in L_i$. In fact D is the compaction graph G_a mentioned in the preceding section. The

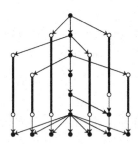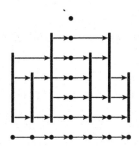

Fig. 1. In the left figure a sparse normalized graph is shown. Thick lines denote the segments. The right figure shows the corresponding compaction graph.

graph D has $|V_N|$ vertices and $O(|V_N|)$ edges, which results in a worst case size of $O(|V||E|)$.

We want to reduce the size of D to $O(|V| + |E|)$ without losing the property that $<$ defines a total layer ordering. The key observation therefor is that the edges between two segments in D can be omitted if no two segments cross.

Given a layer L_i, we partition the layer in the following way:

$$S_{i_0}, v_{i_0}, S_{i_1}, v_{i_1}, S_{i_2}, v_{i_2}, \ldots, S_{i_{n_{i-1}}}, v_{i_{n_{i-1}}}, S_{i_{n_i}}.$$

The list S_{i_k} contains the segments which are between vertices $v_{i_{k-1}}$ and v_{i_k} for $1 \leq k \leq n_i - 1$, S_{i_0} contains the segments before v_{i_0} and $S_{i_{n_i}}$ the segments after $v_{i_{n_{i-1}}}$. We denote the first element of a non-empty list S as $head(S)$ and the last element as $tail(S)$. Furthermore, let v be a vertex in V_S. We denote with $s(v)$ the segment to which v is incident if v is a p- or q-vertex, otherwise $s(v) = v$.

Definition 1. *Given a directed acyclic graph $G = (V, E)$ and a sparse layer ordering in which no two segments cross. The sparse compaction graph (N, A) of the sparse normalized graph $G_S = (V_S, E_S)$ of G is defined as:*

$$N = \{V_S \setminus \{v : v \text{ is } p\text{- or } q\text{-vertex}\}\} \cup \{s_e : s_e \text{ is segment of } e \in E\}$$
$$A = \{(s(v_{i_{j-1}}), s(v_{i_j})) : 1 \leq i \leq h, \ 1 \leq j \leq n_i - 1, \ S_{i_j} = \emptyset\} \cup$$
$$\{(s(tail(S_{i_j})), s(v_{i_j})) : 1 \leq i \leq h, \ 0 \leq j \leq n_i - 1, \ S_{i_j} \neq \emptyset\} \cup$$
$$\{(s(v_{i_{j-1}}), s(head(S_{i_j}))) : 1 \leq i \leq h, \ 1 \leq j \leq n_i, \ S_{i_j} \neq \emptyset\}$$

If we look at two consecutive layers L_n and L_s of a sparse normalized graph we have the following properties:

P1: A segment s_e in L_n is either also in L_s or the adjacent q-vertex q_e is in L_s.
P2: A segment s_e in L_s is either also in L_n or the adjacent p-vertex p_e is in L_n.

Theorem 1. *The ordering $<$ induced by the sparse compaction graph (N, A) of a sparse normalized graph $G_S = (V_S, E_S)$ defines a sparse layer ordering. The compaction graph (N, A) has linear size with respect to G.*

Our new approach is now as follows: In the first phase we create a sparse normalization of the input graph. In the second phase we perform crossing minimization on the sparse normalization. In the third phase we take the resulting sparse compaction graph and perform a coordinate assignment in linear time using an approach similar to the one described in [4]. It remains to show how we can perform crossing minimization on a sparse normalization efficiently.

3   Efficient Crossing Reduction

In this section we present an algorithm which performs crossing minimization using the barycenter or median heuristic on a sparse normalization. The output is a sparse compaction graph which induces a sparse layer ordering with the same number of crossings as these heuristics would produce for a normalization. For our algorithm it is not important which strategy we choose as long as it conforms to some rules.

Definition 2. *A measure m defines for each vertex v in a layer L_{i+1} a non-negative value $m(v)$. If v has only one neighbor w in L_i, then $m(v) = pos(w)$, where $pos(w)$ is the position of w in layer L_i.*

Clearly the barycenter and median heuristic define such a measure.

Lemma 1. *Using such a measure m there are no segments crossing each other.*

Proof. A segment represents a chain of dummy vertices. Each dummy vertex v on a layer L_i has exactly one neighbor w in layer L_{i-1}. Hence when we use a measure m then $m(v) = pos(w)$. Thus two segments never change their relative ordering and thus never produce a crossing with each other.   □

3.1   2-Layer Crossing Minimization

The input of our two-layer crossing minimization algorithm is an *alternating layer L_i* and the sparse compaction graph for the layers $L_1, \ldots, L_i$. An alternating layer consists of an alternating sequence of vertices and containers, where each container represents a maximal sequence of segments. The output is an alternating layer L_{i+1} and the sparse compaction graph for $L_1, \ldots, L_{i+1}$, in which the vertices and segments are ordered by some measure. Note that the representation of layer L_i will be lost, since the containers are reused for layer L_{i+1}.

The containers correspond to the lists S of the previous section. The segments in the container are ordered. The data structure implementing the container must support the following operations:

- $\mathbf{S} = \mathbf{create}()$: Creates an empty container S.
- $\mathbf{append(S, s)}$: Appends segment s to the end of container S.
- $\mathbf{join(S_1, S_2)}$: Appends all elements of container S_2 to container S_1.
- $(\mathbf{S_1, S_2}) = \mathbf{split(S, s)}$: Split container S at segment s into containers S_1 and S_2. All elements less than s are in container S_1 and those who are greater than s in S_2. Element s is neither in S_1 nor S_2.
- $(\mathbf{S_1, S_2}) = \mathbf{split(S, k)}$: Split container S at position k. The first k elements in container S are in S_1 and the remainder in S_2.
- $\mathbf{size(S)}$: Returns the number of elements in container S.

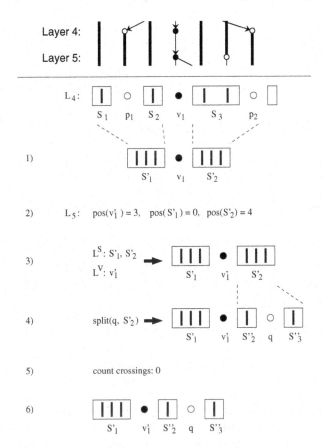

Fig. 2. The six steps applied to layers 4 and 5 from figure 1.

Our algorithm *Crossing_Minimization(L_i, L_{i+1})* consists of six steps:

- In the first step we append the segment $s(v)$ for each p-vertex v in layer L_i to the container preceding v. Then we join this container with the succeeding container. The result is again an alternating layer (p-vertices are omitted).
- In the second step we compute the measure values for the elements in L_{i+1}. First we assign a position value $pos(v_{i_j})$ to all vertices v_{i_j} in L_i. $pos(v_{i_0}) = size(S_{i_0})$ and $pos(v_{i_j}) = pos(v_{i_{j-1}}) + size(S_{i_j}) + 1$. Note that the pos values are the same as they would be in the median or barycenter heuristic if each segment was represented as dummy vertex. Each non-empty container S_{i_j} has pos value $pos(v_{i_j-1}) + 1$. If container S_{i_0} is non-empty it has pos value 0. Now we assign the measure to all non-q-vertices and containers in L_{i+1}. Recall that the measure of a container is its old position.
- In the third step we calculate an initial ordering of L_{i+1}. We sort all non-q-vertices in L_{i+1} according to their measure in a list L^V. We do the same for the containers and store them in a list L^S. Then we merge these two sorted lists in the following way:

> **if** $m(head(L^V)) \leq pos(head(L^S))$ **then** $v = pop(L^V)$, $append(L_{i+1}, v)$
> **if** $m(head(L^V)) \geq (pos(head(L^S)) + size(head(L^S))) - 1)$
>   **then** $S = pop(L^S)$, $append(L_{i+1}, S)$
> **else** $S = pop(L^S)$, $v = pop(L^V)$, $k = \lceil m(v) - pos(S) \rceil$, $(S_1, S_2) = split(S, k)$,
> $append(L_{i+1}, S_1)$, $append(L_{i+1}, v)$, $pos(S_2) = pos(S) + k$, $push(L^S, S_2)$.

- In the fourth step we place the q-vertices according to the position of their segment. We do this by calling $split(s(v))$ for all q vertices v in layer L_{i+1}.
- In the fifth step we perform cross counting according to the scheme proposed by Barth et al. Using the $size(S)$ operation, we put appropriate weights on the container S, such that the number of segments in the container can be taken into account without any loss of performance.
- In the sixth step we perform a scan on L_{i+1} and insert empty containers between two consecutive vertices, and call $join(S_1, S_2)$ on two consecutive containers in the list. This ensures that L_{i+1} is an alternating layer.

Finally we create the edges in the sparse compaction graph for layer L_{i+1}.

3.2   The Overall Algorithm

The first and the last layer never contain segments because of property P1 and P2. Therefore when we perform a sweep or reverse sweep it is easy to create the initial alternating layer. During the reverse sweeps we simply have to take the former p-vertices as q-vertices and vice versa and apply the 2-layer crossing minimization algorithm of the previous section.

There are no other changes to the original Sugiyama approach except for the different calculation of the measure m for all vertices in a layer, the normalization of the layer lists such that the lists are alternating, and the modified counting scheme for crossings. We summarize this section in the following theorem.

Theorem 2. *The approach described above is equivalent to traditional crossing reduction.*

4   An Efficient Data Structure

Let n denote the maximal number of elements in a container. To be competitive, we need a data structure that supports append, split, join and size operations in $O(\log n)$. Thus we use splay trees, a data structure developed by Sleator und Tarjan [16]. Splay trees are self-adjusting binary search trees, which are easy to implement because the tree is allowed to become unbalanced and we need not keep balance information. Nevertheless we can perform all required operations in $O(\log n)$ amortized time. A single operation might cost $O(n)$ but k consecutive operations starting from an empty tree take $O(k \log n)$ time.

The basic operation on a splay tree is called a 'splay'. Splaying node x makes x the root of the tree by a series of special rotations. We use splay trees to represent containers. So we have to implement the container operations.

- **append(S, s):** We search the rightmost element in the tree (last element in the container) by going from the root down taking always the right child. Now, we insert s as the right child of the rightmost element and then splay s. The append operation is performed once for each p-vertex.
- **join(S$_1$, S$_2$):** To join two containers, we search the rightmost element of S_1, splay it and then make S_2 to the right child of it. This operation can only be invoked by an append operation or during the normalization of a layer list. Thus, it is invoked $O(|V| + |E|)$ times.
- **size(S):** While performing the rotations we have to update the size information. Therefore each node knows the size of the subtree rooted by it. So we can maintain the correct size at no extra cost.
- **split(S, s):** First we have to search s in the container. We can not perform a conventional tree search because the elements have only an implicit ordering (their container position) which is not stored by the element. To avoid a search operation, we store a pointer to s in the corresponding p-vertex (this split operation is only used when we are processing the q-vertex layer and the q-vertex knows its corresponding p-vertex). So we just have to splay s and then take its left and its right child as root for the resulting lists. The split operation is performed once for each q-vertex.
- **split(S, k):** First we have to search the element at position k. We use a conventional binary tree search. Let $p(x)$ denote the parent of x and $l(x)$ $(r(x))$ the left (right) child of x. The positions are computed by the following formula: $pos(x) = pos(p(x)) + size(l(x)) + 1$, if x is a right child and $pos(x) = pos(p(x)) - size(r(x)) - 1$ if x is a left child. If x is the root then $pos(x) = size(l(x)) + 1$. After we have found the element at position k, we just splay it and then take its right child as root for the second list. This split operation is performed at most once for each common vertex.

Theorem 3. *[16] A sequence of k arbitrary update operations on a collection of initially empty splay trees takes $O(k + \sum_{j=1}^{k} \log n_j)$ time, where n_j is the number of items in the tree or trees involved in operation j.*

The update operations include insert, join and split operations; 'append' is a special case of the insert operation and the size operation does not change the data structure. Each new iteration starts with empty containers and there are at most $O(|E|)$ elements. Thus we have an overall cost of $O((|V| + |E|) \log |E|)$.

5   Conclusion: Complexity and Practical Behaviour

We have given a new technique that leads to a drastic reduction of the complexity of the important algorithm of Sugiyama for automatic graph drawing. We close with some remarks on the complexity of the algorithm. We first do the normalization of the graph by introducing at most $O(|E|)$ new vertices and edges. Then we perform the layer-by-layer sweep with the modified two-layer crossing minimization procedure. Using the splay-tree data structure as well as the cross-counting scheme by Barth et al., we can ensure that each crossing minimization step can be executed in time $O(n \log n)$ where n denotes the number of vertices and edges involved in this step. Summed up over all layers, the complexity remains $O((|V| + |E|) \log |E|)$. The coordinate assignment is performed in time $O(|V| + |E|)$ using a variant of the algorithm of Brandes and Köpf [4]. Our approach favourably compares to the previous implementations of Sugiyama's algorithm where the complexity might be quadratic in the size of the graph.

We implemented our approach in Java using the yFiles library[20]. We made some preliminary tests and compared our approach to the results achieved with other layout tools using Sugiyama's algorithm. All experiments have been performed on a Pentium IV System with 1.5 GHz and 512 MB main memory running Redhat Linux 9. For our measurement we used the following types of graphs:

- **Long Edge Graphs:** These graphs have many long edges. They have $n/2$ vertical vertices $v_1, \ldots, v_{n/2}$ and $n/2$ horizontal vertices $h_1, \ldots, h_{n/2}$. The vertical vertices are connected by edges (v_i, v_{i+1}) for $1 \leq i \leq n/2 - 1$. The graph also have edges (v_i, h_j) for $1 \leq i, j \leq n/2$.
- **Random Graphs:** They have n vertices and $2.5n$ random edges.

We run the experiments for VCG [18], Dot [11] and our new approach. We also added an algorithm 'Traditional' which uses the same code as our new approach but insert the traditional dummy vertices. Table 1 shows the time taken by the cross counting step, which is given in milliseconds/iteration as well as the number of dummy vertices in the normalized graph, when applying the network simplex for layer assignment. The network simplex gives a solution which minimize the edge length. So the results for other methods are even worse.

Our approach achieved significant improvements in running time for both graph types. This is due to the enormous increase of the number of dummy vertices in the common approach. The results show that our improvements are

Table 1. Experimental results for the long edge graphs and the random graphs.

Size (n)	Time (ms/iter)*				#Dummy vertices	
(long edges)	VCG	Dot	Traditional	New	Common	New
60	146	499	116	19	13050	1710
80	455	2852	306	42	31200	3080
100	1040	13346	658	69	61250	4850
120	2060	42414	1219	98	106200	7020
140	3702	103327	2020	158	169050	9590

Size (n)	Time (ms/iter)*				#Dummy vertices	
(random)	VCG	Dot	Traditional	New	Common	New
100	11	33	16	4	2725	295
200	40	275	60	9	9486	596
500	311	4404	416	29	49203	1485
1000	2978	60783	2643	72	233486	3001
2000	14419	n/a**	n/a**	190	796653	6019

* results are averaged over 10 passes   ** not enough memory

also relevant for practice, even if the number of dummy vertices is usually far less than $|V| \cdot |E|$ there. The number of crossings in our new approach is comparable with the number computed by the other tools. The slight differences are based on the fact, that each implementation has its own refinements (e.g. how to handle nodes having the same median weight). Only Dot has noticeable less crossings but is therefor very slow. This is possibly due to an additional optimization method. Our improvements made it possible to layout graphs for which this was formerly not possible because of the enormous memory consumption of Sugiyama's algorithm. Our approach has just a linear memory consumption.

References

1. D. Auber: *Tulip – A Huge Graph Visualization Framework*. In: Jünger, Mutzel (eds.): Graph Drawing Software, Springer-Verlag, pp. 105–126, 2003.
2. W. Barth, M. Jünger and P. Mutzel: *Simple and Efficient Bilayer Cross Counting*. In: Proceedings of Graph Drawing 2002, Springer LNCS 2528, pp. 130–141, 2002.
3. O. Bastert and C. Matuszewski: *Layered drawings of digraphs*. In: Kaufmann, Wagner (eds.): Drawing Graphs: Methods and Models, Springer LNCS 2025, pp. 104–139, 2001.
4. U. Brandes and B. Köpf: *Fast and Simple Horizontal Coordinate Assignment*. In: Proceedings of Graph Drawing 2001, Springer LNCS 2265, pp. 31–44, 2001.
5. E. Coffman and R. Graham: *Optimal scheduling for two processor systems*. Acta Informatica, 1: 200–213, 1972.
6. G. Di Battista, P. Eades, R. Tamassia and I. G. Tollis: *Graph Drawing: Algorithms for the Visualization of Graphs*. Prentice Hall, 1999.
7. P. Eades and D. Kelly: *Heuristics for Reducing Crossings in 2-Layered Networks*. Ars Combin., 21.A: 89–98, 1986.
8. P. Eades and N. Wormald: *Edge crossings in drawings of bipartite graphs*. Algorithmica, 11(4): 379–403, 1994.
9. A. Frick: *Upper bounds on the number of hidden nodes in Sugiyama's algorithm*. In: Proceedings of Graph Drawing 1996, Springer LNCS 1190, pp. 169–183, 1996.
10. E. Gansner, E. Koutsofios, S. North and K. Vo: *A technique for drawing directed graphs*. In: IEEE Transactions on Software Engineering, 19(3): 214–229, 1993.
11. Graphviz – open source graph drawing software: *http://www.research.att.com/sw/tools/graphviz/*.
12. R. M. Karp: *Reducibility among Combinatorial Problems*. In: Miller R. E., Thatcher J. W. (eds.): Complexity of Computer Computations, Plenum Press, New York, pp. 85–103, 1972.
13. C. Matuszewski, R. Schönfeld and P. Molitor: *Using sifting of k-layer straightline crossing minimization*. In: Proceedings of the 7th Symposium on Graph Drawing (GD'99), Springer LNCS 1731, pp. 217–224, 1999.
14. N. Nikolov and P. Healy: *How to layer a directed Acyclic Graph*. In: Proceedings of the 9th Symposium on Graph Drawing (GD'01), Springer LNCS 2265, pp. 16–30, 2002.
15. G. Sander: *Graph layout through the VCG tool*. In: Proceedings of Graph Drawing 1994, Springer LNCS 894, pp. 194–205, 1995.
16. D. Sleator and R. E. Tarjan: *Self-Adjusting Binary Search Trees*. In: Journal of the ACM, 3: 652–686, 1985.

17. K. Sugiyama, S. Tagawa and M. Toda: *Methods for visual understanding of hierarchical system structures*. In: IEEE Transactions on Systems, Man and Cybernetics, SMC-11(2): 109–125, 1981.
18. VCG – Visualization of Compiler Graphs:
    http://rw4.cs.uni-sb.de/users/sander/ html/gsvcg1.html.
19. V. Waddle and A. Malhotra: *An E log E line crossing algorithm for levelled graphs*. In: Proceedings of the 7th Symposium on Graph Drawing (GD'99), Springer LNCS 1731, pp. 59–70, 1999.
20. yFiles – a Java Graph Layout and Visualization Library: *http://www.yworks.com.*

Random Geometric Graph Diameter
in the Unit Disk with ℓ_p Metric
Extended Abstract

Robert B. Ellis[1,*], Jeremy L. Martin[2,**], and Catherine Yan[1,***]

[1] Department of Mathematics, Texas A&M University,
College Station, TX 77843-3368, USA
{rellis,cyan}@math.tamu.edu
[2] School of Mathematics, University of Minnesota,
206 Church St. SE, Minneapolis, MN 55455, USA
martin@math.umn.edu

Abstract. Let n be a positive integer, $\lambda > 0$ a real number, and $1 \leq p \leq \infty$. We study the *unit disk random geometric graph* $G_p(\lambda, n)$, defined to be the random graph on n vertices, independently distributed uniformly in the standard unit disk in $\mathbb{R}^2$, with two vertices adjacent if and only if their ℓ_p-distance is at most λ. Let $\lambda = c\sqrt{\ln n/n}$, and let a_p be the ratio of the (Lebesgue) areas of the ℓ_p- and ℓ_2-unit disks. Almost always, $G_p(\lambda, n)$ has no isolated vertices and is also connected if $c > a_p^{-1/2}$, and has $n^{1-a_pc^2}(1 + o(1))$ isolated vertices if $c < a_p^{-1/2}$. Furthermore, we find upper bounds (involving λ but independent of p) for the diameter of $G_p(\lambda, n)$, building on a method originally due to M. Penrose.

1 Introduction

Let D be the Euclidean unit disk in $\mathbb{R}^2$ and n a positive integer. Let V_n be a set of n points in D, distributed independently and uniformly with respect to the usual Lebesgue measure on $\mathbb{R}^2$. For $p \in [1, \infty]$, the ℓ_p *metric* on $\mathbb{R}^2$ is defined by

$$d_p((x_1, y_1), (x_2, y_2)) = \begin{cases} (|x_2 - x_1|^p + |y_2 - y_1|^p)^{1/p} & \text{when } p \in [1, \infty) \ , \\ \max\{|x_2 - x_1|, |y_2 - y_1|\} & \text{when } p = \infty \ . \end{cases}$$

For $\lambda \in (0, \infty)$, the *unit disk random geometric graph* $G_p(\lambda, n)$ on the vertex set V_n is defined by declaring two vertices $u, v \in V_n$ to be adjacent if and only if $d_p(u, v) \leq \lambda$. In addition to their theoretical interest, random geometric graphs have important applications to wireless communication networks; see, e.g., [1–3].

Together with X. Jia, the first and third authors studied the case $p = 2$ in [4]. In this extended abstract, we generalize to arbitrary p those results of [4]

* Partially supported by NSF grant DMS-9977354.
** Partially supported by an NSF Postdoctoral Fellowship.
*** Partially supported by NSF grants DMS-0245526 and DMS-0308827 and a Sloan Fellowship. The author is also affiliated with Dalian University of Technology.

J. Pach (Ed.): GD 2004, LNCS 3383, pp. 167–172, 2004.
© Springer-Verlag Berlin Heidelberg 2004

concerning connectedness and graph diameter. Complete results with proofs will be included in a forthcoming paper.

We will say that $G_p(\lambda, n)$ has a property P *almost always* if

$$\lim_{n \to \infty} \Pr\left[G_p(\lambda, n) \text{ has the property } P\right] = 1 \ .$$

Denote by $B_p(u, r)$ the ℓ_p-ball of radius r with center $u \in \mathbb{R}^2$. It is not hard to show that the area of $B_p(u, r)$ is $4r^2 \Gamma((p+1)/p)^2 / \Gamma((p+2)/p)$, where $\Gamma(\cdot)$ is the usual gamma function. We omit the calculation, which uses the beta function; see [5, §12.4]. An important quantity in our work will be the ratio

$$a_p := \frac{\text{Area}(B_p(u, r))}{\text{Area}(B_2(u, r))} = \frac{4\Gamma\left(\frac{p+1}{p}\right)^2}{\pi\Gamma\left(\frac{p+2}{p}\right)} \ .$$

By another elementary calculation, the ℓ_p-diameter of the unit disk D is

$$\text{diam}_p(D) := \max_{u,v \in D}\{d_p(u, v)\} = \begin{cases} 2^{1/2+1/p} & \text{when } 1 \le p \le 2 \ , \\ 2 & \text{when } 2 \le p \le \infty \ . \end{cases}$$

The diameter is achieved by the points $(\sqrt{2}/2, \sqrt{2}/2)$ and $(-\sqrt{2}/2, -\sqrt{2}/2)$ when $1 \le p \le 2$, and by $(0, 1)$ and $(0, -1)$ when $2 \le p \le \infty$.

Let $\lambda = c\sqrt{\ln n / n}$. In Sect. 2, we show that almost always, $G_p(\lambda, n)$ has $n^{1-a_p c^2}(1 + o(1))$ isolated vertices when $c < a_p^{-1/2}$ and no isolated vertices when $c > a_p^{-1/2}$. Penrose [6] has shown that, almost always, $G_p(\lambda, n)$ is connected when it has no isolated points; combining this with our result, it follows that when $c > a_p^{-1/2}$, the graph $G_p(\lambda, n)$ is almost always connected.

The *diameter* of a graph G, denoted $\text{diam}(G)$, is defined as the maximum distance in G between any two of its vertices. This graph-theoretic quantity should not be confused with the diameter of a geometric object with respect to the ℓ_p-metric; we will always denote the latter by diam_p. In Sect. 3, we show that if $c > a_p^{-1/2}$, then almost always $\text{diam}(G_p(\lambda, n)) \le K/\lambda$, where $K \approx 387.17\ldots$ is a constant independent of p. In Sect. 4, we show that when c is larger than a constant depending only on p, we have almost always $\text{diam}(G_p(\lambda, n)) \le 2 \cdot \text{diam}_p(D)(1 + o(1))/\lambda$. In fact, there is a function $c_p(\delta) > 0$ with the following property: if $c > c_p(\delta)$, then almost always $\text{diam}(G_p(\lambda, n)) \le \text{diam}_p(D)(1 + \delta + o(1))/\lambda$.

2   Isolated Vertices

Theorem 1. *Let $1 \le p \le \infty$, let $\lambda = c\sqrt{\ln n / n}$, and let X be the number of isolated vertices in $G_p(\lambda, n)$. Then, almost always,*

$$X = \begin{cases} 0 & \text{when } c > a_p^{-1/2}, \\ n^{1-a_p c^2}(1 + o(1)) & \text{when } 0 < c < a_p^{-1/2} \ . \end{cases}$$

We sketch the proof, which uses the *second moment method* [7] to show that the expected number of isolated vertices is $\mathbb{E}[X] = n^{1-a_p c^2}$, and that the variance is $\mathrm{Var}[X] = o(\mathbb{E}[X]^2)$. When $c < a_p^{-1/2}$, an application of Chebyshev's inequality yields $X = n^{1-a_p c^2}(1 + o(1))$. Let A_i be the event that vertex v_i has degree 0. Then

$$\frac{a_p}{2}\pi\lambda^2(1 + O(\lambda)) \le \mathrm{Area}\left(B_p(v_i, \lambda) \cap D\right) \le a_p\pi\lambda^2 \ ,$$

where the upper (resp. lower) bound is achieved when $B_p(v_i, \lambda) \subseteq D$ (resp. $B_p(v_i, \lambda) \not\subseteq D$). Conditioning on the event that $B_p(v_i, \lambda) \subseteq D$, we have

$$(1 - a_p\lambda^2)^{n-1} \le \Pr[A_i] \le \Pr[B_p(v_i, \lambda) \subseteq D](1 - a_p\lambda^2)^{n-1}$$
$$+ \Pr[B_p(v_i, \lambda) \not\subseteq D]\left(1 - \frac{a_p}{2}\lambda^2(1 + O(\lambda))\right)^{n-1} \ .$$

By linearity of expectation, $\mathbb{E}[X] = n \cdot \Pr[A_i] = n^{1-a_p c^2}(1 + o(1))$. The variance is $\mathrm{Var}[X] = O\left(n^{\frac{3}{2} - \frac{3}{2}a_p c^2}\sqrt{\ln n}\right)$, computed via $\Pr[A_i \wedge A_j]$, conditioned on $d_p(v_i, v_j)$. The rest of the proof is a straightforward computation.

Penrose [6, Thm. 1.1] showed that for every $t \ge 0$, the d-dimensional unit-*cube* random geometric graph simultaneously becomes $(t + 1)$-connected and achieves minimum degree $t + 1$. Penrose's proof remains valid for the unit disk. The precise statement is as follows: for $t \ge 0$ and $1 < p \le \infty$, almost always,

$$\min\{\lambda \mid G_p(\lambda, n) \text{ is } (t + 1)\text{-connected}\}$$
$$= \min\{\lambda \mid G_p(\lambda, n) \text{ has minimum degree } t + 1\} \ .$$

Penrose's proof also works for $p = 1$ in dimension 2, though not for arbitrary dimension d. Combining Penrose's theorem for $t = 0$ with Theorem 1 yields the following.

Theorem 2. *Let $1 \le p \le \infty$ and $\lambda = c\sqrt{\ln n/n}$. Suppose that $c > a_p^{-1/2}$. Then, almost always, the unit disk random geometric graph $G_p(\lambda, n)$ is connected.*

3   Diameter of $G_p(\lambda, n)$ near the Connectivity Threshold

Suppose that $G_p(\lambda, n)$ is connected by virtue of Theorem 2. Usually, $G_p(\lambda, n)$ will contain two vertices whose ℓ_p-distance is close to $\mathrm{diam}_p(D)$, so that the graph has diameter at least $\mathrm{diam}_p(D)/\lambda$. It appears to be much more difficult to obtain an *upper* bound on diameter. However, there is an upper bound which is a constant multiple of the lower bound, as we now explain.

Theorem 3. *Let $1 \le p \le \infty$ and $\lambda = c\sqrt{\ln n/n}$, where $c > a_p^{-1/2}$. Suppose that $K > 256\sqrt{2} + 8\pi \approx 387.17\ldots$. Then, almost always, $\mathrm{diam}(G_p(\lambda, n)) < K/\lambda$.*

We sketch the proof of this theorem. For any two points $u, v \in D$, define

$$T_{u,v}(k) := (\text{convex hull of } B_2(u, k\lambda) \cup B_2(v, k\lambda)) \cap D \ .$$

We impose upon this lozenge-shaped region a grid composed of squares with side length proportional to λ. Let $A_n(k)$ be the event that there exist two points $u, v \in V_n$ such that

(i) at least one of u, v lies in $B_2(O, 1 - (k + \sqrt{2})\lambda)$, and
(ii) there is no path in $G_p(\lambda, n)$ joining u to v that lies entirely inside $T_{u,v}(k)$.

We claim that

$$\text{if } k > 128/(\pi\sqrt{2}) \approx 28.180\ldots, \text{ then } \lim_{n \to \infty} \Pr[A_n(k)] = 0. \tag{1}$$

Indeed, if the event $A_n(k)$ occurs, then by Penrose's argument [6, p. 162], there exists a curve L separating u and v which intersects a large number of grid squares, none of which contains any vertex of V_n (see Fig. 1). Combining this fact with a Peierls argument, as in [8, Lemma 3], leads to the bound on k given in (1).

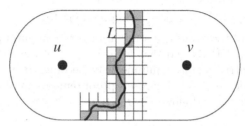

Fig. 1. Two vertices $u, v \in V_n$ which are not connected by any path in $T_{u,v}(k)$, and the "frontier" L separating them.

Let $u, v \in V_n$. If k is large enough, then (1) guarantees the existence of a path from u to v inside $T_{u,v}(k)$. Comparing the total area of $T_{u,v}(k)$ to the area of the ℓ_p-balls around the vertices in a shortest path from u to v inside $T_{u,v}(k)$, one obtains the desired diameter bound on $G_p(\lambda, n)$, completing the proof. (Minor adjustments are needed if u or v is close to the boundary of D.)

Corollary 1. *Let* $1 \leq p \leq \infty$ *and* $\lambda = c\sqrt{\ln n/n}$, *where* $c > a_p^{-1/2}$. *Suppose that* $K > 256\sqrt{2} + 8\pi \approx 387.17\ldots$ *. Then, almost always, every two vertices* u, v *in the unit disk random geometric graph* $G_p(\lambda, n)$ *are joined by a path of length at most* $K d_p(u, v)/\lambda$ *in* $G_p(\lambda, n)$.

4 Diameter of $G_p(\lambda, n)$ for Larger c

By means of a "spoke overlay" construction, we improve the upper bound in Theorem 3 by increasing the constant c slightly and reducing the constant K substantially. Roughly, a spoke consists of a number of evenly spaced, overlapping ℓ_p-balls whose centers lie on a diameter L of the Euclidean unit disk D. We superimpose several spokes on D so that the regions of intersection of the ℓ_p-balls are distributed fairly evenly around D. The idea is that if the constant

c is large enough, then, almost always, every region of intersection contains at least one vertex of V_n, so that $G_p(\lambda, n)$ contains a path joining vertices near the antipodes of D on L. The lengths of such paths, which may be calculated geometrically, give an upper bound for the diameter of $G_p(\lambda, n)$.

Definition 1 (Spoke construction). *Fix $1 \le p \le \infty$, $\theta \in (-\pi/2, \pi/2]$, and $r > 0$. Let D be the Euclidean unit disk. For $m \in \mathbb{Z}$, put*

$$u_m = u_m(r, \theta) = ((r/2 + rm)\cos\theta, \, (r/2 + rm)\sin\theta) \in \mathbb{R}^2 \ .$$

The corresponding spoke *is defined to be the point set $U_{p,\theta}(r) = \{u_m\} \cap D$, together with a collection of ℓ_p-balls of radius $\lambda/2$, one centered at each point $u_m \in U_{p,\theta}(r)$.*

The points u_m lie on the line L_θ through O at angle θ, and the Euclidean distance $d_2(u_m, u_{m'})$ equals $r|m - m'|$. By choosing r sufficiently small, we can ensure that each pair of adjacent ℓ_p-balls intersects in a set with positive area (the shaded rectangles in Fig. 2). Thus the two outermost points on each spoke are joined by a segmented path of Euclidean length approximately 2, which has approximately $2 \cdot \text{diam}_p(D)/\lambda$ edges when $r = \min\{\lambda 2^{-1/2-1/p}, \lambda/2\}$.

Define $A_p^*(r, \lambda/2)$ to be the minimum area of intersection between two ℓ_p-balls in $\mathbb{R}^2$ of radius $\lambda/2$ whose centers are at Euclidean distance r. The general formula for this quantity seems to involve integrals that cannot be evaluated exactly, except for very special cases such as $p = 1, 2, \infty$. However, for fixed r, it is certainly true that $A_p^*(r, \lambda/2) = \Theta(\lambda^2)$.

Theorem 4. *Let $1 \le p \le \infty$, $\lambda = c\sqrt{\ln n/n}$, and $r = \min\{\lambda 2^{-1/2-1/p}, \lambda/2\}$. Suppose that*

$$c > \sqrt{\pi\lambda^2/(2A_p^*(r, \lambda/2))} \ . \tag{2}$$

Then, almost always, as $n \to \infty$,

$$\text{diam}(G_p(\lambda, n)) \le (2 \cdot \text{diam}_p(D) + o(1))/\lambda.$$

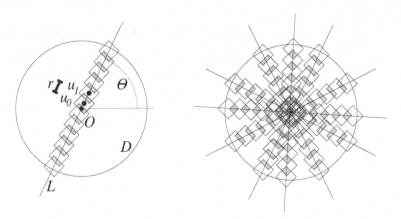

Fig. 2. The spoke overlay construction with $p = 1$, in the unit disk D. The left-hand figure shows a single spoke with parameters r, L, θ. The right-hand figure shows how spokes at different angles are superimposed on D.

Since $A_p^*(r, \lambda/2) = \Theta(\lambda^2)$, the lower bound (2) for c depends only on p.

We sketch the proof of Theorem 4. The spoke construction uses approximately $\ln n$ spokes $U_{p,\theta}(r)$, at evenly spaced angles. Almost always, for each spoke, every intersection of two consecutive ℓ_p-balls of radius $\lambda/2$ contains at least one vertex of V_n, provided that the bound (2) holds.

Let $v_1, v_2 \in V_n$. For $i = 1, 2$, by Corollary 1, there is a vertex $v_i' \in V_n$ lying inside some spoke U_i, connected to v_i by a path in $G_p(\lambda, n)$ of length $o(1/\lambda)$. Moreover, v_i' is connected to a vertex near the origin by a path consisting of vertices in $U_i \cap V_n$, lying in successive ℓ_p-balls of the spoke. Thus each of these two paths contains at most $\mathrm{diam}_p(D)/\lambda$ vertices, and concatenating these paths gives the desired upper bound on the diameter of $G_p(\lambda, n)$.

We can make the average Euclidean distance covered in a path from v_i' to v_j' larger by increasing r. This change decreases the area of intersection of consecutive ℓ_p-balls, which in turn requires an increase in c in order to guarantee a vertex of V_n in every region of intersection. This leads to the following corollary.

Corollary 2. *Let $1 \leq p \leq \infty$ and let $\lambda = c\sqrt{\ln n/n}$. For every $\delta \in (0, 1]$, there exists $c_p(\delta) > 0$ such that if $c > c_p(\delta)$, then $G_p(\lambda, n)$ is almost always connected, and has diameter at most $\mathrm{diam}_p(D)(1 + \delta + o(1))/\lambda$ as $n \to \infty$.*

References

1. Chen, X., Jia, X.: Package routing algorithms in mobile ad-hoc wireless networks. In: 2001 International Conference on Parallel Processing Workshops. (2001) 485–490
2. Stojmenovic, I., Seddigh, M., Zunic, J.: Dominating sets and neighbor elimination-based broadcasting algorithms in wireless networks. IEEE Trans. Parallel Distrib. Syst. **13** (2002) 14–25
3. Wu, J., Li, H.: A dominating-set-based routing scheme in ad hoc wireless networks. Telecommunication Systems **18** (2001) 13–36
4. Ellis, R.B., Jia, X., Yan, C.H.: On random points in the unit disk. (preprint)
5. Whittaker, E.T., Watson, G.N.: A course of modern analysis. Cambridge University Press (1996)
6. Penrose, M.D.: On k-connectivity for a geometric random graph. Random Structures Algorithms **15** (1999) 145–164
7. Alon, N., Spencer, J.H.: The probabilistic method. 2nd edn. John Wiley (2000)
8. Klarner, D.A.: Cell growth problems. Canad. J. Math. **19** (1967) 851–863

Algorithms for Drawing Media

David Eppstein[*]

Computer Science Department,
School of Information & Computer Science,
University of California, Irvine
eppstein@uci.edu

Abstract. We describe algorithms for drawing media, systems of states, tokens and actions that have state transition graphs in the form of partial cubes. Our algorithms are based on two principles: embedding the state transition graph in a low-dimensional integer lattice and projecting the lattice onto the plane, or drawing the medium as a planar graph with centrally symmetric faces.

1 Introduction

Media [7, 8] are systems of states, tokens, and actions of tokens on states that arise in political choice theory and that can also be used to represent many familiar geometric and combinatorial systems such as hyperplane arrangements, permutations, partial orders, and phylogenetic trees. In view of their importance in modeling social and combinatorial systems, we would like to have efficient algorithms for drawing media as state-transition graphs in a way that makes the action of each token apparent. In this paper we describe several such algorithms.

Formally, a *medium* consists of a finite set of *states* transformed by the actions of a set of *tokens*. A string of tokens is called a *message*; we use upper case letters to denote states, and lower case letters to denote tokens and messages, so Sw denotes the state formed by applying the tokens in message w to state S. Token t is *effective* for S if $St \neq S$, and message w is *stepwise effective* for S if each successive token in the sequence of transformations of S by w is effective. A message is *consistent* if it does not contain the reverse of any of its tokens. A set of states and tokens forms a medium if it satisfies the following axioms:

1. Each token t has a unique *reverse* $\tilde{t}$ such that, for any states $S \neq Q$, $St = Q$ iff $Q\tilde{t} = S$.
2. For any states $S \neq Q$, there exists a consistent message w with $Sw = Q$.
3. If message w is stepwise effective for S, then $Sw = S$ if and only if the number of copies of t in w equals the number of copies of $\tilde{t}$ for each token t.
4. If $Sw = Qz$, w is stepwise effective for S, z is stepwise effective for Q, and both w and z are consistent, then wz is consistent.

The states and state transitions of a medium can also be viewed as a graph, and it can be shown that these graphs are *partial cubes* [12]: that is, their vertices can be mapped to a hypercube $\{0, 1\}^d$ in such a way that graph distance equals L_1 distance in

[*] Supported in part by NSF grant CCR-9912338.

J. Pach (Ed.): GD 2004, LNCS 3383, pp. 173–183, 2004.
© Springer-Verlag Berlin Heidelberg 2004

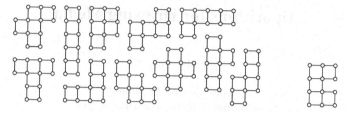

Fig. 1. 11 of the 12 pentominos represent isometric lattice embeddings of media. The twelfth, the U pentomino, does not, because a pair of vertices that are three edges apart in the graph have placements that are only one unit apart.

the hypercube. For media, we can find such a mapping by choosing arbitrarily state S, and assigning any state S' a coordinate per token t that is 1 when a consistent path from S to S' contains t and 0 otherwise. Conversely, any d-dimensional partial cube gives rise to a medium with its vertices as states and with $2d$ tokens; the action of any token is to change one of the partial cube coordinates to a zero or to a one, if it does not already have that value and if such a change would produce another vertex of the partial cube.

We assume throughout, as in [7], that we are given as input an explicit description of the states, tokens, and actions of a medium. However, our algorithms are equally applicable to any partial cube or family of partial cube graphs such as the median graphs. If a partial cube representation is not given, it can be found (and the corresponding medium constructed) in time $O(mn)$ via known algorithms [1, 11, 12, 15].

2   Lattice Dimension

As we have seen, media can be embedded *isometrically* (that is, in a distance-preserving way) into hypercubes $\{0,1\}^d$ (with L_1 distance), and hypercubes can be embedded isometrically into integer lattices $\mathbb{Z}^d$, so by transitivity media can be embedded isometrically onto integer lattices. Conversely any finite isometric subset of an integer lattice forms a partial cube and corresponds as described above to a medium.

If the dimension of the lattice in which a medium is embedded is low, we may be able to use the embedding as part of an effective drawing algorithm. For instance, if a medium M can be embedded isometrically onto the planar integer lattice $\mathbb{Z}^2$, then we can use the lattice positions as vertex coordinates of a drawing in which each edge is a vertical or horizontal unit segment (Figure 1). If M can be embedded isometrically onto the cubic lattice $\mathbb{Z}^3$, in such a way that the projection onto a plane perpendicular to the vector $(1,1,1)$ projects different vertices to distinct positions in the plane, then this projection produces a planar graph drawing in which the edges are unit vectors at $60°$ and $120°$ angles (Figure 10, center).

Recently, we showed that the *lattice dimension* of a medium or partial cube, that is, the minimum dimension of a lattice $\mathbb{Z}^d$ into which it may be isometrically embedded, may be determined in polynomial time [6]. We now briefly our algorithm for finding low-dimensional lattice embeddings.

Suppose we are given an undirected graph G and an isometry $\mu: G \mapsto \{0,1\}^\tau$ from G to the hypercube $\{0,1\}^\tau$ of dimension τ. Let $\mu_i: G \mapsto \{0,1\}$ map each vertex v of

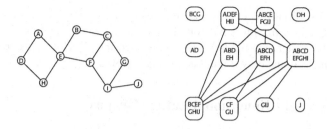

Fig. 2. A medium, left, and its semicube graph, right. From [6].

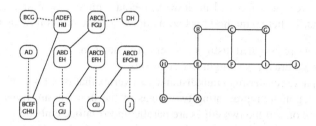

Fig. 3. A matching in the semicube graph (left, solid edges) completed to a set of paths by adding edges from each semicube to its complement (left, dashed edges), and the corresponding lattice embedding of the original medium (right). From [6].

G to the ith coordinate of $\mu(v)$, and assume that each coordinate μ_i takes on both value 0 and 1 for at least one point . From G and μ we can define 2τ distinct *semicubes* $S_{i,\chi} = \{v \in V(G) \mid \mu_i(v) = \chi\}$, for any pair i,χ with $0 \leq i < \tau$ and $\chi \in \{0,1\}$. We now construct a new graph $\mathrm{Sc}(G)$, which we call the *semicube graph* of G. We include in $\mathrm{Sc}(G)$ a set of 2τ vertices $u_{i,\chi}$, $0 \leq i < \tau$ and $\chi \in \{0,1\}$. We include an edge in $\mathrm{Sc}(G)$ between $u_{a,b}$ and $u_{c,d}$ whenever $S_{a,b} \cup S_{c,d} = V(G)$ and $S_{a,b} \cap S_{c,d} \neq \emptyset$; that is, whenever the corresponding two semicubes cover all the vertices of G non-disjointly. Although defined from some particular isometry μ, the semicube graph turns out to be independent of the choice of μ. An example of a partial cube G and its semicube graph $\mathrm{Sc}(G)$ is shown in Figure 2. The main result of [6] is that the lattice dimension of G can be determined from the cardinality of a maximum matching in $\mathrm{Sc}(G)$:

Theorem 1 (Eppstein [6]). *If G is a partial cube with isometric dimension τ, then the lattice dimension of G is $d = \tau - |M|$ where M is any maximum matching in $\mathrm{Sc}(G)$.*

More specifically, we can extend a matching in $\mathrm{Sc}(G)$ to a collection of d paths by adding to the matching an edge from each semicube to its complement. The dth coordinate of a vertex in the lattice embedding equals the number of semicubes that contain the vertex in even positions along the dth path.

We can use this result as part of a graph drawing system, by embedding our input medium in the lattice of the lowest possible dimension and then projecting that lattice onto the plane. For two-dimensional lattices, no projection is needed, and we have already discussed projection of certain three-dimensional integer lattices onto two-dimensional triangular lattices. We discuss more general techniques for lattice projection in the next section. It is essential for this result that we require the embedding to be isometric. Even for trees it is NP-complete to find an embedding into $\mathbb{Z}^2$ with unit length edges that is not required to be distance-preserving [2]. However a tree embeds isometrically in $\mathbb{Z}^2$ if and only if it has at most four leaves [14].

3   Drawing High-Dimensional Lattice Graphs

Two-dimensional lattice embeddings of media, and some three-dimensional embeddings, lead to planar graph drawings with all edges short and well separated by angles. However, we are also interested in drawing media without low dimensional embeddings. We describe here a method for finding drawings with the following properties:

1. All vertices are assigned distinct integer coordinates in $\mathbb{Z}^2$.
2. All edges are drawn as straight line segments.
3. No edge passes closer than unit distance to a vertex that is not one of its endpoints.
4. The line segments representing two edges of the drawing are translates of each other if and only if the two edges are parallel in the lattice embedding.
5. The medium corresponding to a Cartesian product of intervals $[a_0, b_0] \times [a_1, b_1] \times \cdots [a_{d-1}, b_{d-1}]$ is drawn in area $O(n^2)$, where n is the number of its states.

Because of property 4, the lattice embedding and hence the medium structure of the state transition graph can be read from the drawing. To achieve these properties, we map $\mathbb{Z}^d$ to $\mathbb{Z}^2$ linearly, by choosing wo vectors X and $Y \in \mathbb{Z}^d$, and mapping any point $p \in \mathbb{Z}^d$ to the point $(X \cdot p, Y \cdot p) \in \mathbb{Z}^2$. We now describe how these vectors X and Y are chosen. If $L \subset \mathbb{Z}^d$ is the set of vertex placements in the lattice embedding of our input medium, define a *slice* $L_{i,j} = \{p \in L \mid p_i = j\}$ to be the subset of vertices having ith coordinate equal to j. We choose the coordinates X_i sequentially, from smaller i to larger i, so that all slices $L_{i,j}$ are separated from each other in the range of x-coordinates they are placed in. Specifically, set $X_0 = 0$. Then, for $i > 0$, define

$$X_i = \max_j \left(\min_{p \in L_{i,j}} \sum_{k=0}^{i-1} X_k p_k - \max_{q \in L_{i,j-1}} \sum_{k=0}^{i-1} X_k q_k \right),$$

where the outer maximization is over all j such that $L_{i,j}$ and $L_{i,j-1}$ are both nonempty. We define Y similarly, but we choose its coordinates in the opposite order, from larger i to smaller: $Y_{d-1} = 0$, and

$$Y_i = \max_j \left(\min_{p \in L_{i,j}} \sum_{k=i+1}^{d-1} X_k p_k - \max_{q \in L_{i,j-1}} \sum_{k=i+1}^{d-1} X_k q_k \right).$$

Theorem 2. *The projection method described above satisfies the properties 1–5 listed above. The method's running time on a medium with n states and τ tokens is $O(n\tau^2)$.*

Proof. Property 2 and property 4 follow immediately from the fact that we our drawing is formed by projecting $\mathbb{Z}^d$ linearly onto $\mathbb{Z}^2$, and from the fact that the formulas used to calculate X and Y assign different values to different coordinates of these vectors.

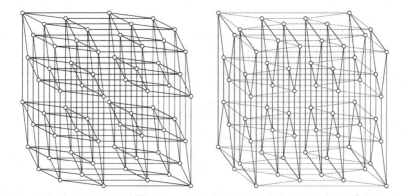

Fig. 4. Left: lattice drawing of six-dimensional hypercube; right: a drawing with geometric thickness two is possible, but the vertex placement is less regular and edges formed by actions of the same token are not all drawn parallel.

All vertices are assigned distinct coordinates (property 1): for, if vertices p and q differ in the ith coordinates of their lattice embeddings, they belong to different slices $L_{i,j}$ and $L_{i,j'}$ and are assigned X coordinates that differ by at least X_i (unless $i = X_i = 0$ in which case their Y coordinates differ by at least Y_i).

The separation between vertices and edges (property 3) is almost equally easy to verify: consider the case of three vertices p, q, and r, with an edge pq to be separated from r. Since p and q are connected by an edge, their lattice embeddings must differ in only a single coordinate i. If r differs from p and q only in the same coordinate, it is separated from edge pq by a multiple of (X_i, Y_i). Otherwise, there is some coordinate $i' \neq i$ in which r differs from both p and q. If $i' > i$, the construction ensures that the slice $L_{i',j}$ containing pq is well separated in the x-coordinate from the slice $L_{i',j'}$ containing r, and if $i' < i$ these slices are well separated in the y coordinate.

Finally, we consider property 5. For Cartesian products of intervals, in the formula for X_i, the value for the subexpression $\min_{p \in L_{i,j}} \sum_{k=0}^{i-1} X_k p_k$ is the same for all j considered in the outer maximization, and the value for the subexpression $\max_{q \in L_{i,j-1}} \sum_{k=0}^{i-1} X_k q_k$ is also the same for all j considered in the outer maximization, because the slices are all just translates of each other. Therefore, there is no gap in x-coordinates between vertex placements of each successive slice of the medium. Since our drawings of these media have vertices occupying contiguous integer x coordinates and (by a symmetric argument) y coordinates, the total area is at most n^2.

The time for implementing this method is dominated by that for finding a minimum-dimension lattice embedding of the input graph, which can be done in the stated time bound [6].   □

When applied to a hypercube, the coordinates X_i become powers of two, and this vertex placement algorithm produces a uniform placement of vertices (Figure 4, left) closely related to the Hammersley point set commonly used in numerical computation and computer graphics for its low discrepancy properties [16]. Other examples of drawings produced by this method can be seen in Figures 6, 9, and 10(left).

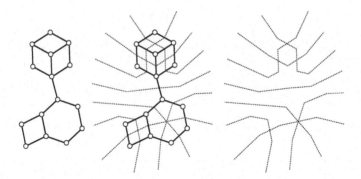

Fig. 5. Left: a graph with a face-symmetric planar drawing; center: connecting opposite pairs of edge midpoints produces a weak pseudoline arrangement; right: the arrangement.

4 Face-Symmetric Planar Drawings

Our two-dimensional and projected three-dimensional lattice drawings are planar (no two edges cross) and each internal face is symmetric (squares for two-dimensional lattices, 60°-120° rhombi and regular hexagons for projected three-dimensional lattices). We now describe a different type of drawing of the state-transition graphs of media as planar graphs, generalizing this symmetry property. Specifically, we seek straight-line planar drawings in which each internal face is strictly convex and centrally symmetric; we call such a drawing a *face-symmetric planar drawing*.

A *weak arrangement of pseudolines* [9] is a collection of curves in the plane, each homeomorphic to a line, such that any pair of curves in the collection has at most one point of intersection, and such that if any two curves intersect then they cross properly at their intersection point. Weak arrangements of pseudolines generalize pseudoline arrangements [10] and hyperbolic line arrangements, and are a special case of the *extendible pseudosegment arrangements* defined by Chan [3]. Any weak pseudoline arrangement with n pseudolines partitions the plane into at least $n + 1$ and at most $n(n + 1)/2 + 1$ *cells*, connected components of the set of points that do not belong to any pseudoline. Any pseudoline in the arrangement can be partitioned into *nodes* (crossing points) and *arcs* (connected components of the complement of the crossing points); we use this terminology to avoid confusion with the vertices and edges of the medium state-transition graphs we hope to draw. Each arc is adjacent to two cells and two nodes. We define the *dual* of a weak pseudoline arrangement to be the graph having a vertex for each cell of the arrangement and an edge connecting the vertices dual to any two cells that share a common arc; this duality places the graph's vertices in one-to-one correspondence with the arrangement's cells, and the graph's edges in one-to-one correspondence with the arrangement's arcs.

Lemma 1. *If G has a face-symmetric planar drawing, then G is the dual of a weak pseudoline arrangement.*

Lemma 2. *If G is the dual of a weak pseudoline arrangement, then G is the state transition graph of a medium.*

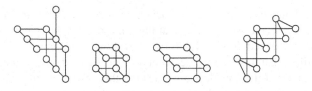

Fig. 6. Media with planar state-transition graphs but with no face-symmetric planar drawing.

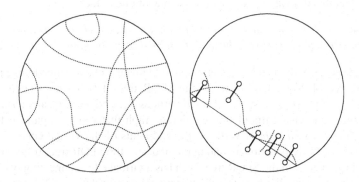

Fig. 7. Converting a weak pseudoline arrangement into a face-symmetric planar drawing. Left: arrangement drawn inside a circle O such that crossings with O are equally spaced around the circle. Right: edges dual to arcs of ℓ_i are drawn as unit length and perpendicular to the chord through the points where ℓ_i crosses O.

By these lemmas (the proofs of which we omit due to lack of space), every face-symmetric planar drawing represents the state transition graph of a medium. However, not every medium, and not even every medium with a planar state transition graph, has such a drawing; see for instance Figure 6, the medium in Figure 9(right), and the permutahedron in Figure 10(left) for media that have planar state transition graphs but no face-symmetric planar drawing.

Lemma 3. *If G is the dual of a weak pseudoline arrangement, then G has a face-symmetric planar drawing.*

Proof. Let G be dual to a weak pseudoline arrangement $\mathcal{A}$; the duality fixes a choice of planar embedding of G as well as determining which faces of that embedding are internal and external. Denote by $|\mathcal{A}|$ the number of pseudolines in $\mathcal{A}$. Let O be a circle (the size and placement of the circle within the plane being irrelevant to our construction), and deform $\mathcal{A}$ as necessary so that each pseudoline crosses O, with all nodes interior to O, and so that the $2|\mathcal{A}|$ points where pseudolines cross O are spaced at equal distances around the perimeter of O (Figure 7, left). Then, for each pseudoline ℓ_i of $\mathcal{A}$, let c_i be the chord of O connecting the two points where ℓ_i crosses O. We will draw G in such a way that the edges of G that are dual to arcs of ℓ_i are drawn as unit length segments perpendicular to c_i (Figure 7, right). To do so, choose an arbitrary starting vertex v_0 of G, and place it arbitrarily within the plane. Then, the placement of any other vertex v_i of G can be found by following a path from v_0 to v_i in G, and for each edge of the path moving unit distance (starting from the location of v_0) in the direction determined for that edge as described above, placing v_i at the point reached by this motion when the end of the path is reached. It is straightforward to show from Lemma 2 and the axioms defining a medium that this vertex placement does not depend on the choice of the path from v_0 to v_i, and that if all vertices are placed in this way then all edges of G will be unit length and perpendicular to their corresponding chords c_i. Thus, we have a drawing of G, in which we can identify sets of edges corresponding to the faces of G. We omit the proof that this drawing is face-symmetric planar due to lack of space.   □

Lemma 4. *If G is biconnected, at most one planar embedding of G is dual to a weak pseudoline arrangement. This embedding (if it exists) can be found in time $O(n)$.*

Proof. We use a standard technique in graph drawing and planar embedding problems, the SPQR tree [4, 13]. Each node v in the SPQR tree of G has associated with it a multigraph G_v consisting of some subset of vertices of G, edges of G, and *virtual edges* representing contracted parts of the remaining graph that can be separated from the edges of G_v by a *split pair* of vertices (the endpoints of the virtual edge). The non-virtual edges of G are partitioned among the nodes of the SPQR tree. If two nodes are connected by an edge in the SPQR tree, each has a virtual edge connecting two vertices shared by both nodes. We root the SPQR tree arbitrarily; let (s_v, t_v) denote the split pair connecting a non-root node v to its parent, and let H_v denote the graph represented by the SPQR subtree rooted at v. We work bottom up in the rooted tree, showing by induction on tree size that the following properties hold for each node of the tree:

1. Each graph H_v has at most one planar embedding that can be part of an embedding of G dual to a weak pseudoline arrangement.
2. If v is a non-root node, and G is dual to a weak pseudoline arrangement, then edge $s_v t_v$ belongs to the outer face of the embedding of H_v.
3. If v is a non-root node, form the path p_v by removing virtual edge $s_v t_v$ from the outer face of H_v. Then p_v must lie along the outer face of any embedding of G dual to a weak pseudoline arrangement.

SPQR trees are divided into four different cases (represented by the initials S, P, Q, and R) and our proof follows the same case analysis, in each case showing that the properties at each node follow from the same properties at the descendant nodes. We omit the details of each case due to lack of space.   □

Theorem 3. *Given an input graph G, we can determine whether G is the dual of a weak pseudoline arrangement, and if so construct a face-symmetric planar drawing of G, in linear time.*

Proof. If G is biconnected, we choose a planar embedding of G by Lemma 4. Otherwise, each articulation point of G must be on the outer face of any embedding. We find biconnected components of G, embed each component by Lemma 4, and verify that these embeddings place the articulation points on the outer faces of each component. We then connect the embeddings together into a single embedding having as its outer face the edges that are outer in each biconnected component; the choice of this embedding may not be unique but does not affect the correctness of our algorithm.

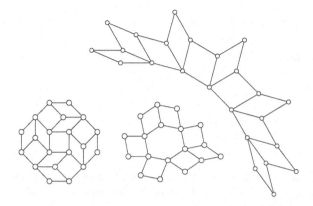

Fig. 8. Face-symmetric planar drawings of three irregular media.

Once we have an embedding of G, we must verify that we have the dual of a weak pseudoline arrangement and construct a face-symmetric planar drawing. We first make sure all faces of G are even, and apply Lemma 1 to construct an arrangement of curves $\mathcal{A}$ dual to G. We test that $\mathcal{A}$ has no closed curves, then apply the construction of Lemma 3 to produce vertex placements for a drawing of G, test for each edge of G that the endpoints of that edge are placed at unit distance apart with the expected slope, and test that each internal face of G is drawn as a correctly oriented strictly convex polygon. If our input passes all these tests we have determined that it is the dual of a weak pseudoline arrangement and found a face-symmetric planar drawing.   □

Our actual implementation is based on a simpler but less efficient algorithm that uses the known medium structure of the input to construct the dual weak pseudoline arrangement one curve at a time, before applying the construction of Lemma 3 to produce a face-symmetric planar drawing from the weak pseudoline arrangement. Examples of drawings produced by our face-symmetric planar drawing code are shown in Figure 8.

5   Implementation and Examples

We implemented our algorithms in Python, with drawings output in SVG format. Our code allows various standard combinatorial media (such as the collection of permutations on n items) to be specified on the command line; irregular media may be loaded from a file containing hypercube or lattice coordinates of each state. We have seen already examples of our implementation's output in Figures 4, 6, 8, and 9. Figure 10 provides additional examples. All figures identified as output of our code have been left unretouched, with the exception that we have decolorized them for better printing.

6   Conclusions and Open Problems

We have shown several methods for drawing the state transition graphs of media. There are several interesting directions future research in this area could take.

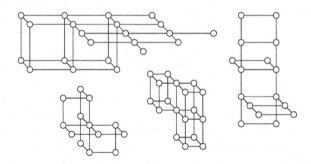

Fig. 9. Lattice drawings of four irregular media with three-dimensional lattice embeddings, from [6]. The bottom left drawing is of a medium isomorphic to the weak ordering medium shown in Figure 10(right).

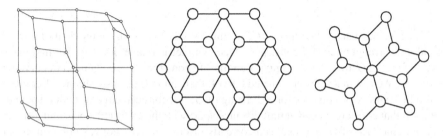

Fig. 10. Media defined by orderings of n-item sets. Left: Lattice drawing of total orderings (permutations) on four items. Center: Projected three-dimensional lattice drawing of partial orderings on three items. Right: Face-symmetric planar drawing of weak orderings on three items.

– If a three-dimensional lattice embedding can be projected perpendicularly to the vector $(1, 1, 1)$ (or more generally $(\pm 1, \pm 1, \pm 1)$) without placing two vertices in the same point, the projection produces a planar drawing with all edges having equal lengths and angles that are multiples of $60°$ (e.g., Figure 10, center). Our lattice dimension algorithm can find a three-dimensional embedding, if one exists, and it is trivial to test the projection property. However, a medium may have more than one three-dimensional embedding, some of which have the projection property and some of which don't. For instance, the medium in the lower right of Figure 9 is the same weak ordering medium as the one in Figure 10(right), however the former drawing is from a lattice embedding without the projection property. Is it possible to efficiently find a projectable three-dimensional lattice embedding, when one exists? More generally, given an arbitrary dimension lattice embedding of a medium, can we find a planar projection when one exists?
– Hypercubes have projected lattice drawings with $O(n^2)$ area and unit separation between vertices and nonadjacent edges. Can similar area and separation bounds be achieved for projected lattice drawings of more general classes of media?
– Our lattice and face-symmetric planar drawings have several desirable qualities; for instance, all edges corresponding to a single token are drawn as line segments with the same slope and length, and our lattice drawings have good vertex-vertex

and vertex-edge separation. However, we have not seriously examined the extent to which other important graph drawing properties may be achieved. For instance, d-dimensional hypercubes (and therefore also media with up to $2d$ tokens) may be drawn with geometric thickness [5] at most $\lceil d/3 \rceil$ (Figure 4, right) however our lattice projection methods achieve geometric thickness only $\lceil d/2 \rceil$ while the only way we know how to achieve the better $\lceil d/3 \rceil$ bound is to use a more irregular drawing in which edges coming from the same token are no longer parallel.

References

1. F. Aurenhammer and J. Hagauer. Recognizing binary Hamming graphs in $O(n^2 \log n)$ time. *Mathematical Systems Theory* 28:387–395, 1995.
2. S. Bhatt and S. Cosmodakis. The complexity of minimizing wire lengths in VLSI layouts. *Inform. Proc. Lett.* 25:263–267, 1987.
3. T. M. Chan. On levels in arrangements of curves. *Discrete & Comput. Geom.* 29(3):375–393, April 2003.
4. G. Di Battista and R. Tamassia. Incremental planarity testing. *Proc. 30th IEEE Symp. Foundations of Computer Science (FOCS 1989)*, pp. 436–441, 1989.
5. D. Eppstein. Layered graph drawing. http://www.ics.uci.edu/~eppstein/junkyard/thickness/.
6. D. Eppstein. The lattice dimension of a graph. To appear in *Eur. J. Combinatorics*, arXiv:cs.DS/0402028.
7. D. Eppstein and J.-C. Falmagne. Algorithms for media. ACM Computing Research Repository, June 2002, arXiv:cs.DS/0206033.
8. J.-C. Falmagne and S. Ovchinnikov. Media theory. *Discrete Applied Mathematics* 121(1–3): 103–118, September 2002.
9. H. de Fraysseix and P. Ossona de Mendez. Stretching of Jordan arc contact systems. *Proc. 11th Int. Symp. Graph Drawing (GD 2003)*, pp. 71–85. Springer-Verlag, Lecture Notes in Computer Science 2912, 2003.
10. J. E. Goodman and R. Pollack. Allowable sequences and order types in discrete and computational geometry. *New Trends in Discrete and Computational Geometry*, chapter V, pp. 103–134. Springer-Verlag, Algorithms and Combinatorics 10, 1993.
11. W. Imrich and S. Klavžar. On the complexity of recognizing Hamming graphs and related classes of graphs. *Eur. J. Combinatorics* 17:209–221, 1996.
12. W. Imrich and S. Klavžar. *Product Graphs*. John Wiley & Sons, 2000.
13. P. Mutzel. The SPQR-tree data structure in graph drawing. *Proc. 30th Int. Coll. Automata, Languages and Computation (ICALP 2003)*, pp. 34–46. Springer-Verlag, Lecture Notes in Computer Science 2719, June 2003.
14. S. Ovchinnikov. The lattice dimension of a tree. arXiv.org, February 2004, arXiv:math.CO/0402246.
15. P. Winkler. Isometric embeddings in products of complete graphs. *Discrete Applied Mathematics* 7:221–225, 1984.
16. T.-T. Wong, W.-S. Luk, and P.-A. Heng. Sampling with Hammersley and Halton points. *J. Graphics Tools* 2(2):9–24, 1997.

Confluent Layered Drawings*

David Eppstein, Michael T. Goodrich, and Jeremy Yu Meng

School of Information and Computer Science,
University of California, Irvine,
Irvine, CA 92697, USA
{eppstein,goodrich,ymeng}@ics.uci.edu

Abstract. We combine the idea of confluent drawings with Sugiyama style drawings, in order to reduce the edge crossings in the resultant drawings. Furthermore, it is easier to understand the structures of graphs from the mixed style drawings. The basic idea is to cover a layered graph by complete bipartite subgraphs (bicliques), then replace bicliques with tree-like structures. The biclique cover problem is reduced to a special edge coloring problem and solved by heuristic coloring algorithms. Our method can be extended to obtain multi-depth confluent layered drawings.

1 Introduction

Layered drawings visualize hierarchical graphs in a way such that vertices are arranged in layers and edges are drawn as straight lines or curves connecting these layers. A common method was introduced by Sugiyama, Tagawa and Toda [25] and by Carpano [4]. Several closely related methods were proposed later (see e.g. [12, 19, 15, 22, 6, 20, 11].)

Crossing reduction is one of the most important objectives in layered drawings. But it is well known that for two-layer graphs the straight-line crossing minimization problem is NP-complete [14]. The problem remains NP-complete even when one layer is fixed. Jünger and Mutzel [16] present exact algorithms for this problem, and perform experimental comparison of their results with various heuristic methods. Recently new methods related to crossing reduction ([26, 1, 8, 23, 10]) have been proposed.

However when the given two-layer graph is dense, even in an optimum solution, there are still a large number of crossings. Then the resulting straight-line drawing will be hard to read, since edge-crossing minimization is one of the most important aesthetic criteria for visualizing graphs [24]. This give us a motivation for exploring new approaches to reduce the crossings in a drawing other than the traditional methods.

In addition, it is sometime of interest to find the bicliques between two layers. For example in the drawing of a call graph, it is interesting to find out which set

* Work by the first author is supported by NSF grant CCR-9912338. Work by the second and the third author is supported by NSF grants CCR-0098068, CCR-0225642, and DUE-0231467.

of modules are calling a common set of functions and what are those common functions. Call graphs are usually visualized as layered drawings. However it is hard to learn this information from layered drawings by traditional Sugiyama-style approaches, especially when the input graphs are dense.

Our previous work [5] introduces the concept of confluent drawings. In [5] we talk about the confluent drawability of several classes of graphs and give a heuristic for finding confluent drawings of graphs with bounded arboricity. In this paper we experiment with an implementation of confluent drawings for the layered graphs. However we relax the constraint of planarity and allow crossings in the drawings, while it is not allowed to have crossings in a confluent drawing in our previous definitions.

We are aware of the Edge Concentration method by Newbery [21]. Edge Concentration and our method share a same idea of covering by bicliques. But in Newbery's method, dummy nodes (edge concentrators) are explicit in the drawing and treated equally as original nodes, which causes the nodes' original levels to change. In our method dummy nodes are implicit in the curve representation of edges and the original levels are preserved. Furthermore, our method uses a very different algorithm to compute the biclique covers.

2   Definitions

In this section we give definitions for confluent layered drawings. The definitions almost remain the same as in our previous confluent drawing paper, except that the planarity constraints are dropped. Fig. 1 gives an idea of confluent layered drawings. Edges in the drawing are represented as smooth curves.

A curve is *locally monotone* if it contains no sharp turns, that is, it contains no point with left and right tangents that forms a angle less than or equal to 90

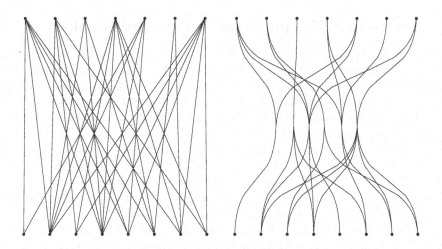

Fig. 1. An example confluent layered drawing.

degrees. Intuitively, a locally-monotone curve is like a single train track, which can make no sharp turns. *Tracks* are the union of locally-monotone curves. They are formed by merging edges together.

A drawing A formed by a collection of tracks on the plane is called a *confluent drawing* for an undirected graph G if and only if

- There is a one-to-one mapping between the vertices in G and A, so that, for each vertex $v \in V(G)$, there is a corresponding vertex $v' \in A$, and all vertices of G are assigned to distinct points in the plane.
- There is an edge (v_i, v_j) in $E(G)$ if and only if there is a locally-monotone curve e' connecting v'_i and v'_j in along tracks in A.

The directed version of a confluent drawing is defined similarly, except that in such a drawing the locally-monotone curves are directed and in every track formed by the union of directed curves, the curves must be oriented consistently.

Self loops and parallel edges of G are not allowed in our definitions, although multiple ways of realizing the same edge are allowed. Namely, for an edge in the original graph, there could be more than one locally monotone path in the drawing corresponding to this edge.

We apply the idea of confluent drawings on layered graphs. Particularly, in the resultant confluent drawing, we replace bicliques in a biclique cover of a two-layer graph $G = (U, L, E)$ by tree-like structures and draw them with smooth curves. As we can see in Fig. 1, our method can greatly reduce the crossings in the drawings of dense bipartite graphs. Additionally, nodes of a biclique can be easily identified by following the smooth curve paths.

Since it is valid to have more than one confluent path between two nodes u and l in the confluent drawing when $(u, l) \in E$, as defined above, it is straightforward that a confluent layered drawing can be obtained by computing a biclique cover C of G, then visualizing each biclique in C as a tree-like structure. We show how to compute a biclique cover of G in the next section.

3   Computing Biclique Cover of Bipartite Graphs

Fishburn and Hammer [9] show that the biclique cover problem is equivalent to a restricted edge coloring problem. This coloring is not much useful for general graphs. However, it has a nice result for triangle-free graphs, and since bipartite graphs belong to the class of triangle-free graphs, an immediate result is that this type of edge coloring can be used to find a biclique cover of a bipartite graph. This result is useful in layered drawing because the edges between any two layers in such a drawing induce a bipartite subgraph.

An edge coloring $c: E \leftarrow \{1, 2, \ldots, k\}$ for $G = (V, E)$ is *simply-restricted* if no induced K_3 is monochromatic and the vertex-disjoint edges in an induced P_4 or C_4^c have different colors. Fig. 2 shows the conditions that such induced subgraphs of a simply-restricted edge coloring must satisfy.

Let $d(G)$ denote the bipartite dimension of G, which is the minimum cardinality of a biclique cover of G. Let $\chi_s(G)$ be the chromatic number of a simply-restricted edge coloring of G. $\chi_s(G)$ is 0 if $E = \emptyset$; otherwise, it is the minimum k

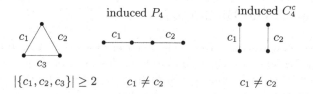

Fig. 2. The required conditions of induced subgraphs of a simply edge coloring.

for which G has a simply-restricted coloring $c\colon E \leftarrow \{1, 2, \ldots, k\}$. The following theorem states the equivalence of $d(G)$ and $\chi_s(G)$ for triangle-free graphs.

Theorem 1 *Fishburn and Hammer [9])*
$d(G) = \chi_s(G)$ *for every triangle-free graph.*

Let E_j be the set of edges with color j in a simply-restricted edge coloring for a triangle-free graph G. As we can see in the second part of the proof of Theorem 1 (omitted here, included in the full version of this paper), E_j is included in the edge set of a biclique subgraph of G. Therefore, every edge set of a single color induces a biclique subgraph of G. By computing a simply restricted edge coloring we can get a biclique cover of G.

Because it is known that the problem of COVERING BY COMPLETE BIPARTITE SUBGRAPHS is NP-hard (Garey and Johnson [13] GT18), it is unlikely to have efficient optimization algorithms for finding the minimum biclique cover of a bipartite graph. Thus we only focus on fast heuristics for computing a near-optimal biclique cover.

The simply-restricted edge coloring problem can be transformed into a vertex coloring problem. So, instead of devising a special algorithm for the simply-restricted edge coloring, we can choose to use one of the existing vertex coloring algorithms. Well known heuristic algorithms for vertex coloring include Recursive Largest First (RLF) algorithm of Leighton [18], DSATUR algorithm of Brélaz [2]. For more about heuristics on graph coloring, see Campers et al. [3].

The above method of computing a biclique cover by coloring doesn't distinguish between two kinds of bicliques: $K_{p,1}$ and $K_{1,r}$, where $p, q, r > 1$. So if we are more interested in finding out the set of common callers and callees, we would need to give higher priority to $K_{p,q}$ than $K_{1,r}$ when covering the edges.

After the biclique cover of the two-layer bipartite graph is computed, each biclique in the cover is drawn as a tree-like structure in the final drawing. Doing this repeatedly for every two adjacent layers, we can get the drawings for multi-layer graphs.

The time complexity of the algorithm depends on the coloring heuristic subroutines. For a graph with a set of vertices V, both the RLF algorithm and the DSATUR algorithm run in worst case $O(|V|^3)$ time. There are some other faster coloring heuristics with $O(|V|^2)$ time, but their output qualities are worse. Suppose we have a two-layer bipartite graph $G = (V, E)$. The transformation from the simply-restricted edge coloring into vertex coloring version takes $O(|E|^2)$ time. Using RLF or DSATUR costs $O(|E|^3)$, thus the total time is $O(|E|^3)$.

4   Layout of the Bicliques

We described how to compute a biclique cover of a two-layer bipartite graph in the previous section. Now it is time to show how the bicliques are laid out. In the confluent layered drawings, each biclique in the biclique cover is visualized as a tree-like structure, as in Fig. 1. Now here are the questions. What are the best positions to place the centers of the tree-like structures? How to arrange the curves so that they form confluent tracks defined in Section 2?

4.1   Barycenter Method to Place Centers

In the case where the positions of nodes in the upper level and lower level are fixed, one would like to put the center of a tree to the center of the nodes belonging to the corresponding biclique. For example, in Fig. 3, the drawing on the left is visually better than the drawing on the right. Firstly it has better angular resolution and better edge separation. Secondly it is easier for people to see the biclique as a whole. Then the next question is: what does the center of those nodes mean? In our method, the natural candidate position for a center of the tree-like structure is the barycenter, i.e., the average position, of all the nodes in this biclique.

It looks bad too if these tree centers stay very close to each other. So we need to specify a minimum separation between two centers.

The above requirements can be formulated into constraints:

1. A tree center stays within the range of its leaves.

$$\min_j x_{ij} \leq x_i \leq \max_j x_{ij},$$

    where x_i is the x-coordinate of the ith tree center c_i, and x_{ij} is the x-coordinate of the jth leaf of c_i.
2. The distance between any two centers is greater than or equal to the minimum separation.

$$\forall i \neq j, \qquad |x_i - x_j| \geq \delta$$

    where δ is some pre-specified minimum distance.

Under these constraints, we want a tree center to stay as close as possible to the barycenter of all its leaf nodes. i.e., we want to minimize $\sum_i (x_i - \text{avg}_j(x_{ij}))^2$,

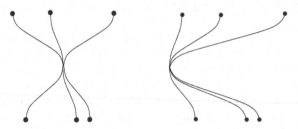

Fig. 3. Good-looking tree and bad-looking tree with centers placed differently.

subject to the above constraints. This is a Quadratic Programming problem, and unfortunately it is NP-hard (Garey and Johnson [13], MP2).

Since it is unlikely to have efficient algorithms for solving this optimization problem, and a small deviation of a tree center from the perfect position won't cause too much displeasure, we use instead a very simple heuristic method to place the tree centers. We first assign to each tree center the x-coordinate of the barycenter of its leaves. Then we sort tree centers by their x-value. The third step is to try to place these tree centers at their x-coordinates one by one. Assume we have k centers to place. Start from the jth center, where $j = \lfloor \frac{k}{2} \rfloor$. Place center j at its barycenter, then try to place centers one by one in the following order: $j-1, j-2, \ldots, 1$. If constraint 2 is violated, the violating center is placed the minimum distance away from the previous placed center. Tree centers to the right of center j are placed similarly in the order of $j+1, j+2, \ldots, k$. It is easy to see that the running time of the barycenter method is dominated by the sorting of the tree centers.

4.2   Placing Tree Centers to Reduce Crossings

Alternatively, one might want to place these centers on positions such that the total number of edge crossings is as few as possible, especially in the case where nodes of upper level and lower level are not fixed. If this is the main concern, we can place the tree centers in another way in order to reduce the edge crossings.

After the biclique cover of a two-layer graph $G = (U, L, E)$ is computed, we construct a new three-layer graph G'. We treat these tree centers as nodes of a middle layer. The set of vertices includes three levels: an upper layer $U' = U$, a middle layer M consisting of tree centers, and a lower layer $L' = L$. The edges of G' are added as follows: for each biclique B_i in the biclique cover, add one edge between the tree center node m_i and each node $u \in U$ that belongs to B_i. Similarly add one edge between m_i and each node $l \in L$ that belongs to B_i.

Now a two-layer graph of the original problem is transformed into a three-layer graph G'. Straight-line crossing reduction algorithms can be applied on G'. After the crossing reduction, we obtain the ordering of nodes in each of the three layers. The orderings will be used to compute the positions of nodes and tree centers in the final drawings. Note that when crossing reduction method is used to place tree centers, it is not always true that a tree center always stays within the x-range of its leaves, i.e., bad centers like the one in Fig. 3 could appear.

Here we are using straight-line edge crossing reduction algorithms for our confluent layered drawings with curve edges. Readers may suspect the equality of the crossing number in the straight-line drawing for the new three-layer graph G' and the crossing number of our curve edge drawings. We will confirm this equality after we describe the generation of curves in the next section.

4.3   Curves

After the positions of tree centers (and the positions of nodes if not given) are computed, we are now ready to place the confluent tracks for the edges.

We use Bézier curves to draw the curve edges in confluent drawings. Given a set of control points $P_1, P_2, \ldots, P_n$, the corresponding Bézier curve is given by

$$C(u) = \sum_{k=0}^{n} P_k \, B_{k,n}(u) \qquad 0 \le u \le 1, \tag{1}$$

where $B_{k,n}(u)$ is a Bernstein polynomial

$$B_{k,n}(u) = \frac{n!}{k! \, (n-k)!} \, u^k (1-u)^{n-k}. \tag{2}$$

Bézier curves have some nice properties that are suitable for our confluent tracks. The first property is that a Bézier curve always passes its first and last control point. The second is that a Bézier curve always stays within the convex hull formed by its control points. In addition, the tangents of a Bézier curve at the endpoints are $P_1 - P_0$ and $P_n - P_{n-1}$. Thus it is easy to connect two Bézier curves while still maintaining the first order continuity: just let $P_n = P_0'$ and let the control points $P_{n-1}, P_n = P_0'$, and P_1' co-linear.

The confluent track between each node and the tree center is drawn as a Bézier curve. In our program we use cubic Bézier curves ($n = 4$ in Equation 2). Each such a curve has four control points, chosen as shown in Fig. 4.

More formally, assume we are given the following input for a biclique B_i: y_u, y_l, and y_c are the y-coordinates of the upper, lower, and tree center levels, respectively. x_i is the x-coordinate of the tree center for B_i. x_{ij}'s are the x-coordinates of nodes in biclique B_i. Let Δy be a distance parameter that controls the shape of the curve edges. When node j is in the upper level, the four control points are $P_0 = (x_{ij}, y_u)$, $P_1 = (x_{ij}, y_u + \Delta y)$, $P_2 = (x_i, y_c - \Delta y)$, and $P_3 = (x_i, y_c)$. When node j is in the lower level, the four control points are $P_0 = (x_i, y_c)$, $P_1 = (x_i, y_c + \Delta y)$, $P_2 = (x_{ij}, y_l - \Delta y)$, and $P_3 = (x_{ij}, y_l)$.

From Equation 1, it is not hard to verify that in a confluent layered drawing, two Bézier curves cross each other if and only if the corresponding straight-line

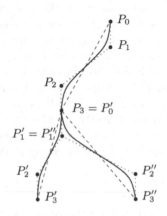

Fig. 4. Bézier curves.

edges (dashed lines in Fig. 4) of the bicliques cross each other, given that the control points are chosen as above. This should clear the doubt that appears at the end of Section 4.2.

5 Multi-depth Confluent Layered Drawings

So far we have introduced the method of confluent layered drawings: replacing subsets of edges in a two-layer graph by tree-like structures. This method can be extended to obtain drawings that display richer information. The extended drawings are called *multi-depth confluent layered drawings*.

The idea is as follows: after the biclique cover for a two-layer graph $G = (U, L, E)$ is computed, the tree center nodes are viewed as a middle layer M, and a new three layer graph $G' = (U, M, L, E')$ is constructed as in Section 4.2. The same biclique cover algorithm is then applied to G' twice, once for the subgraph induced by $U \cup M$; once for the subgraph induced by $M \cup L$. By applying this approach recursively, we get biclique covers at different depth. In the final drawing, only biclique covers at the largest depth are replaced by sets of tree-like structures. The final drawing is a multi-depth confluent layered drawing. The drawings discussed before this section are all *depth-one (confluent layered) drawings*.

In a depth-one drawing, we compute a biclique cover and lay out the biclique cover. In general, for a depth-i drawing, we need to compute $2^i - 1$ biclique covers and 2^{i-1} biclique covers are laid out.

An example drawing of depth-two is shown in Fig. 5.

Because the control points for the Bézier curves are chosen in a way such that the tangents at the endpoints of the Bézier curves are all vertical, it is guaranteed that all segments of a path are connected seamlessly and smoothly in

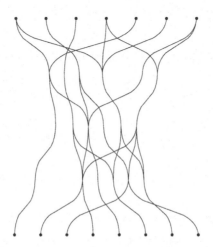

Fig. 5. Depth-two confluent drawing (on same input as the drawing of Fig. 1).

multi-depth drawings. Readers probably have already noticed some wavy edges in the drawing of Fig. 5. It is because a single edge biclique ($K_{1,1}$) is also drawn as two Bézier curves. We offer an option in our program to do a simple treatment for these single edge bicliques: draw them as a single Bézier curves instead of two. But after this special treatment is applied, the crossing property is not preserved any more. That means two curve segments could have crossing(s), even though their corresponding edges in G' don't cross each other in a straight-line drawing.

Multi-depth drawings may further reduce the number of crossings. They also show a richer structure than the depth-one drawings, which only display bicliques. For example we can observe relationships between bicliques in a depth-two confluent layered drawing. However higher depth requires more computations of biclique covers, and generates more dummy centers. The former leads to the increasing of time and space complexity, while the latter could result in a more complicate confluent drawing. We feel that drawings with depth higher than two are not very practically useful.

6   Real-World Examples

We list two example drawings of real-world graphs in Fig. 6. We implemented the algorithm of computing biclique cover using the RLF heuristic. For the center placement we implemented the barycenter method. We assume that besides the

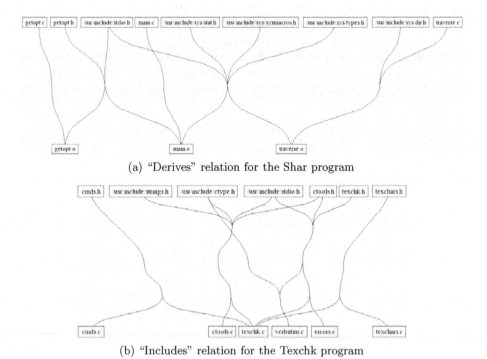

(a) "Derives" relation for the Shar program

(b) "Includes" relation for the Texchk program

Fig. 6. Confluent drawing for examples of Newbery [21].

two-layer graph, the input also includes the positions of (fixed) nodes in upper and lower levels (possibly output by other algorithms that take labels and other information into account.) The result drawing is written into a file of DOT format [17]. The neato program in the Graphviz package [7] is then used to generate the graphic file in a desired format. Fig. 6 (a) is a depth-one drawing. Fig. 6 (b) is a depth-two drawing with the special smoothing treatment applied.

7   Conclusions and Acknowledgments

In this paper we introduce a new method – confluent layered drawing, for visualizing layered graphs. It combines the layered drawing technique with the relaxed confluent drawing approach. There are still interesting open problems, e.g., how to test whether a layered graph has a crossing-free confluent layered drawing? How to minimize the crossing of the drawing among all possible biclique covers? It is also useful to investigate better ways for visualizing confluent tracks.

We would like to thank anonymous referees for their helpful comments.

References

1. W. Barth, M. Jünger, and P. Mutzel. Simple and efficient bilayer crossing counting. In M. Goodrich and S. Kobourov, editors, *Graph Drawing (Proc. GD '02)*, volume 2528 of *Lecture Notes Comput. Sci.*, pages 130–141. Springer-Verlag, 2002.
2. Brélaz. New methods to color the vertices of a graph. *Communications of the ACM*, 22(4):251–256, 1979.
3. G. Campers, O. Henkes, and J. P. Leclerq. Graph coloring heuristics: A survey, some new propositions and computational experiences on random and "Leighton's" graphs. In G. K. Rand, editor, *Operational research '87 (Buenos Aires, 1987)*, pages 917–932. North-Holland Publishing Co, 1988.
4. M. J. Carpano. Automatic display of hierarchized graphs for computer aided decision analysis. *IEEE Trans. Syst. Man Cybern.*, SMC-10(11):705–715, 1980.
5. M. Dickerson, D. Eppstein, M. T. Goodrich, and J. Y. Meng. Confluent drawing: Visualizing nonplanar diagrams in a planar way. In G. Liotta, editor, *Graph Drawing (Proc. GD '03)*, volume 2912 of *Lecture Notes Comput. Sci.*, pages 1–12. Springer-Verlag, 2003.
6. P. Eades and K. Sugiyama. How to draw a directed graph. *J. Inform. Process.*, 13:424–437, 1991.
7. J. Ellson, E. Gansner, E. Koutsofios, and S. North. Graphviz. URL: http://www.research.att.com/sw/tools/graphviz.
8. T. Eschbach, W. Günther, R. Drechsler, and B. Becker. Crossing reduction by windows optimization. In M. Goodrich and S. Kobourov, editors, *Graph Drawing (Proc. GD '02)*, volume 2528 of *Lecture Notes Comput. Sci.*, pages 285–294. Springer-Verlag, 2002.
9. P. C. Fishburn and P. L. Hammer. Bipartite dimensions and bipartite degree of graphs. *Discrete Math.*, 160:127–148, 1996.
10. M. Forster. Applying crossing reduction strategies to layered compound graphs. In M. Goodrich and S. Kobourov, editors, *Graph Drawing (Proc. GD '02)*, volume 2528 of *Lecture Notes Comput. Sci.*, pages 276–284. Springer-Verlag, 2002.

11. E. R. Gansner, E. Koutsofios, S. C. North, and K. P. Vo. A technique for drawing directed graphs. *IEEE Trans. Softw. Eng.*, 19:214–230, 1993.

12. E. R. Gansner, S. C. North, and K. P. Vo. DAG – A program that draws directed graphs. *Softw. - Pract. Exp.*, 18(11):1047–1062, 1988.

13. M. R. Garey and D. S. Johnson. *Computers and Intractability: A Guide to the Theory of NP-Completeness.* W. H. Freeman, New York, NY, 1979.

14. M. R. Garey and D. S. Johnson. Crossing number is NP-complete. *SIAM J. Algebraic Discrete Methods*, 4(3):312–316, 1983.

15. D. J. Gschwind and T. P. Murtagh. A recursive algorithm for drawing hierarchical directed graphs. Technical Report CS-89-02, Department of Computer Science, Williams College, 1989.

16. M. Jünger and P. Mutzel. 2-layer straightline crossing minimization: Performance of exact and heuristic algorithms. *J. Graph Algorithms Appl.*, 1(1):1–25, 1997.

17. E. Koutsofios and S. North. Drawing graphs with *dot*. Technical report, AT&T Bell Laboratories, Murray Hill, NJ., 1995. Available from http://www.research.bell-labs.com/dist/drawdag.

18. F. T. Leighton. A graph coloring algorithm for large scheduling problems. *Journal of Research of National Bureau of Standard*, 84:489–506, 1979.

19. E. B. Messinger. Automatic layout of large directed graphs. Technical Report 88-07-08, Department of Computer Science, University of Washington, 1988.

20. E. B. Messinger, L. A. Rowe, and R. H. Henry. A divide-and-conquer algorithm for the automatic layout of large directed graphs. *IEEE Trans. Syst. Man Cybern.*, SMC-21(1):1–12, 1991.

21. F. J. Newbery. Edge concentration: A method for clustering directed graphs. In *Proc. 2nd Internat. Workshop on Software Configuration Management*, pages 76–85, 1989.

22. F. Newbery Paulisch and W. F. Tichy. EDGE: An extendible graph editor. *Softw. - Pract. Exp.*, 20(S1):63–88, 1990. also as Technical Report 8/88, Fakultat fur Informatik, Univ. of Karlsruhe, 1988.

23. M. Newton, O. Sýkora, and I. Vrt'o. Two new heuristics for two-sided bipartite graph drawing. In M. Goodrich and S. Kobourov, editors, *Graph Drawing (Proc. GD '02)*, volume 2528 of *Lecture Notes Comput. Sci.*, pages 312–319. Springer-Verlag, 2002.

24. H. Purchase. Which aesthetic has the greatest effect on human understanding? In G. Di Battista, editor, *Graph Drawing (Proc. GD '97)*, volume 1353 of *Lecture Notes Comput. Sci.*, pages 248–261. Springer-Verlag, 1997.

25. K. Sugiyama, S. Tagawa, and M. Toda. Methods for visual understanding of hierarchical systems. *IEEE Trans. Syst. Man Cybern.*, SMC-11(2):109–125, 1981.

26. V. Waddle and A. Malhotra. An E log E line crossing algorithm for levelled graphs. In J. Kratochvíl, editor, *Graph Drawing (Proc. GD '99)*, volume 1731 of *Lecture Notes Comput. Sci.*, pages 59–71. Springer-Verlag, 1999.

Simultaneous Embedding of Planar Graphs with Few Bends*

Cesim Erten and Stephen G. Kobourov

Department of Computer Science,
University of Arizona
{cesim,kobourov}@cs.arizona.edu

Abstract. We present an $O(n)$ time algorithm for simultaneous embedding of pairs of planar graphs on the $O(n^2) \times O(n^2)$ grid, with at most three bends per edge, where n is the number of vertices. For the case when the input graphs are both trees, only one bend per edge is required. We also describe an $O(n)$ time algorithm for simultaneous embedding with fixed-edges for tree-path pairs on the $O(n) \times O(n^2)$ grid with at most one bend per tree-edge and no bends along path edges.

1 Introduction

Traditional problems in graph drawing involve the layout of a single graph. Problems in simultaneous graph drawing, involve the layout of multiple related graphs. Visualization of multiple related graphs, that is, graphs that are defined on the same set of vertices, arise in many applications. Software engineering, databases, and social network analysis, are all examples of areas where multiple relationships on the same set of objects are often studied.

Consider the case where a pair of related graphs is given and the goal is to visualize them so as to compare the two. If drawings for the two graphs are obtained independently, there would be little correspondence between the two layouts, since the viewer has no "mental map" between the two graphs. When examining a graph the user constructs a mental view of it, for example, using the positions of the vertices relative to each other. When viewing multiple graphs the user has to reconstruct this mental view after examining each graph and our goal should be to aid the user in this reconstruction while providing a readable drawing for each graph individually.

In simultaneous graph embedding, the vertices are placed in the exact same locations in all the graphs. Fixing the vertex positions in all the graphs preserves the mental map, but at the expense of readability of the individual drawings, if edges are to be drawn with straight-line segments. With this in mind, in this paper we consider the problem of drawing planar graphs on the same point-set using few bends. We describe efficient algorithms for simultaneous drawing of pairs of general planar graphs on small integer grids. We also describe better results for pairs of trees or tree-path pairs.

---

* This work is partially supported by the NSF under grant ACR-0222920 and by ITCDI under grant 003297.

J. Pach (Ed.): GD 2004, LNCS 3383, pp. 195–205, 2004.

1.1 Previous Work

The existence of simultaneous geometric embeddings for pairs of paths, cycles, and caterpillars is shown in [2]. Counter-examples for pairs of general planar graphs, pairs of outer-planar graphs, and triples of paths are also presented there. Modified force-directed algorithms are used in [1,8] to simultaneously visualize general graphs, while attempting to preserve the user's mental map and obtaining readable individual drawings.

A related notion is that of *graph thickness* [12], defined as the minimum number of planar subgraphs whose union yields the given graph. If a graph has thickness two then it can be drawn on two layers such that each layer is crossing-free and the corresponding vertices of different layers are placed in the same locations. *Geometric thickness* is a version of the thickness problem where the edges are required to be straight-line segments [6]. Thus, if two graphs have a simultaneous geometric embedding, then their union has geometric thickness at most two. Similarly, the union of any two planar graphs has graph thickness at most two. Simultaneous geometric embedding techniques are used in [7] to show that degree-four graphs have geometric thickness two.

The existence of straight-line, crossing-free drawings for planar graphs is well known [9,15,17]. It is also known that every 3-connected planar graph has a convex drawing [16]. These techniques, however, do not guarantee anything about the resolution of the drawing and thus are not well-suited for automated graph drawing. The vertex resolution problem is addressed in [5,14] where it is shown that any planar graph can be drawn with straight-lines and no crossings on a grid of size $O(n) \times O(n)$.

Simultaneous drawing of multiple graphs is also related to the problem of embedding planar graphs on a fixed set of points in the plane. Several variations of this problem have been studied. If the mapping between the vertices V and the points P is not fixed, then the graph can be drawn without crossings using two bends per edge in polynomial time [11]. However, if the mapping between V and P is fixed, then $O(n)$ bends per edge are necessary to guarantee planarity, where n is the number of vertices in the graph [13].

1.2 Our Results

Formally, the drawing $\mathcal{D}$ of a graph $G = (V, E)$ is a function that maps each vertex $u \in V$ to a distinct point $\mathcal{D}(u)$ in the plane, and each edge $(u, v) \in E$ to a simple Jordan curve $\mathcal{D}(u, v)$ with endpoints $\mathcal{D}(u)$ and $\mathcal{D}(v)$. The problem of simultaneously embedding two planar graphs G_1, G_2 is the problem of finding drawings D_1, D_2 with corresponding vertices of G_1 and G_2 mapped to the same points in the plane. The following are three variations of the simultaneous embedding problem for pairs of planar graphs:

Definition 1. Given two planar graphs $G_1 = (V, E_1)$ and $G_2 = (V, E_2)$ *simultaneous geometric embedding of G_1 and G_2* is the problem of finding plane straight-line drawings D_1 and D_2 of G_1 and G_2, respectively, such that every vertex is mapped to the same point in the plane in both D_1 and D_2.

Definition 2. Given two planar graphs $G_1 = (V, E_1)$ and $G_2 = (V, E_2)$ *simultaneous embedding of G_1 and G_2 with fixed edges* is the problem of finding plane drawings D_1 and D_2 of G_1 and G_2, respectively, such that every vertex is mapped to the same point in the plane in both D_1 and D_2 and every shared edge $e \in G_1 \cap G_2$ is represented with the same simple open Jordan curve in D_1 and D_2.

Definition 3. Given two planar graphs $G_1 = (V, E_1)$ and $G_2 = (V, E_2)$ *simultaneous embedding of G_1 and G_2* is the problem of finding plane drawings D_1 and D_2 of G_1 and G_2, respectively, such that every vertex is mapped to the same point in the plane in both D_1 and D_2.

The definitions are inclusive in the given order: simultaneous geometric embedding is a special case of simultaneous embedding with fixed edges, which is in turn a special case of simultaneous embedding.

In Section 2 we present a simple non-existence proof for simultaneous geometric embedding of a pair of graphs. Next, we present an $O(n)$ time algorithm for simultaneous embedding of pairs of planar graphs on the $O(n^2) \times O(n^2)$ grid, with at most three bends per edge, where n is the number of vertices. For the case when the input graphs are both trees, we only need one bend per edge. We also describe an $O(n)$ time algorithm for simultaneous embedding with fixed-edges for tree-path pairs on the $O(n) \times O(n^2)$ grid with at most one bend per tree-edge and no bends along the path edges. In Section 3 we briefly describe the implementation of these algorithms, show some of the resulting layouts, and conclude with several open problems.

2   Simultaneous Embedding

Simultaneous geometric embeddings are easy to find on small integer grids for pairs of paths, pairs of cycles, pairs of caterpillars, and others [2]. For pairs of general planar graphs, and even for pairs of outer-planar graphs, simultaneous geometric embeddings do not always exist. This is the main motivation for relaxing the conditions of simultaneous geometric embeddings, to just simultaneous embedding, by dropping the straight-line edge constraint. Under these weaker constraints, we can obtain simultaneous drawings with few bends per edge. Such drawings are also useful for pairs of trees, as it is not known whether simultaneous geometric embedding of pairs of trees is always possible.

2.1   Simultaneous Geometric Embedding

Here we briefly describe a simple case of a pair of planar graphs that do not admit simultaneous geometric embedding.

Theorem 1. *There exists a planar graph G and a path P such that there is no simultaneous geometric embedding of G and P.*

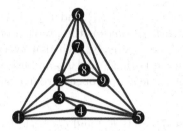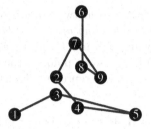

Fig. 1. Planar graph G and path P that do not allow a simultaneous geometric embedding.

Proof Sketch: Consider graph G and path P as shown in Fig. 1. Let G' be the subgraph of G induced on vertices $\{1, 2, 3, 4, 5\}$, and G'' be the subgraph of G induced on vertices $\{2, 6, 7, 8, 9\}$. Since G is 3-connected fixing the outer-face fixes an embedding for G. With the given outer-face of G, the path P contains two crossings: one involving $(2, 4)$, and the other one involving $(6, 8)$. Graph G' has six faces and unless we change the outer-face of G' such that it contains the edge $(1, 3)$ or $(3, 5)$, the edge $(2, 4)$ is involved in a crossing in the path. Similarly for G'', unless we change its outer-face such that it contains $(2, 7)$ or $(7, 9)$, the edge $(6, 8)$ is involved in a crossing in the path. However G' and G'' do not share any faces and removing both crossings depends on taking two different outer-faces, which is impossible. Thus, regardless of the choice for the outer-face of G, path P contains a crossing.   □

2.2   Relaxing the Constraints

While some classes of planar graphs allow simultaneous geometric embedding, there are other classes that do not, and still others for which it is not known whether simultaneous geometric embeddings exist. Since the latter two categories contain a large number of planar graph classes (trees, outer-planar graphs, general planar graphs), it is natural to look for simultaneous drawings with weaker constraints. One possible solution for larger classes of graphs is to relax the constraints on the edges. Instead of restricting the edges to be straight-line segments we allow each edge to be drawn as a sequence of straight-line segments. Recall that such embeddings are called *simultaneous embeddings* (rather than simultaneous geometric embeddings).

Note that it is trivial to find a simultaneous embedding of any two planar graphs, if we are willing to accept a large number of bends per edge. Given a point-set P of size n in the plane and a planar graph G with n vertices, together with a one-to-one mapping between the vertices of G and the points in P, we can find a crossing-free drawing of G on P using edges with bends [13]. This allows us to embed any number of planar graphs simultaneously. However, the resulting drawings contain $O(n)$ bends per edge. Next, we describe methods to simultaneously embed any two planar graphs so that each edge has at most three bends.

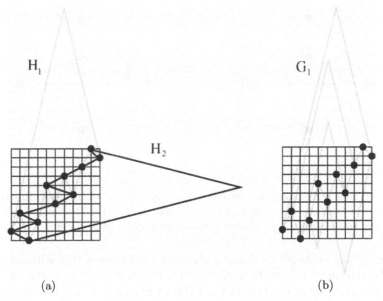

Fig. 2. (a) H_1 and H_2 drawn simultaneously. (b) Only the edges of G_1 are shown. Edges not in the hamiltonian cycle have the same slopes as the outermost edge.

2.3   Simultaneous Embedding with Few Bends

Since in this version of the problem we no longer insist on straight-line edges, the problem of simultaneously embedding two graphs boils down to finding a point-set in the plane and a mapping between the vertices of graphs and the points, with as few bends per edge as possible. The following theorem summarizes our results for pairs of general planar graphs.

Theorem 2. *Given two planar graphs G_1 and G_2 and a mapping between their vertices, we can simultaneously embed G_1 and G_2 using at most three bends per edge. The resulting drawing requires an integer grid of size $O(n^2) \times O(n^2)$ such that each vertex is placed on a grid point, and the algorithm requires $O(n)$ time, where n is the number of vertices.*

Proof Sketch: Vertex Placement: We make use of two techniques described in [2, 11]. Initially, we assume the graphs are 4-connected. We show how to remove this assumption later in the proof. First we find a hamiltonian cycle H_1 of G_1 and a hamiltonian cycle H_2 of G_2. We can do this in linear time using the algorithm of [4]. Starting at a random vertex in H_1 we traverse its vertices, assigning increasing x–coordinates to each vertex visited. Starting at a random vertex in H_2 we traverse its vertices, assigning increasing y–coordinates to each vertex visited. Not considering the final edges enclosing the cycles, this gives us an x–monotone path for H_1 and a y–monotone path for H_2; see Fig. 2(a). Since both paths are monotone the edges of the paths are crossing-free. Let δ be the largest slope of the edges on the path defined by H_1. We complete the

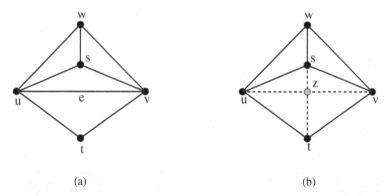

Fig. 3. (a) Removing separating triangles. (a) Edge e is part of the separating triangle (u, v, w). The two faces containing e are (u, v, s) and (u, v, t). (b) The separating triangle is removed by deleting e, introducing z and connecting it to u, v, s, and t.

drawing of the cycle H_1 by drawing the final edge between the leftmost vertex and the rightmost vertex. It is drawn with two segments such that the slope of the initial segment starting at the leftmost vertex is δ' and the slope of the second segment ending at the rightmost vertex is $-\delta'$, where δ' is slightly larger than δ. Since G_1 is hamiltonian, the cycle H_1 divides the edges into two groups: inside and outside edges (with respect to H_1). Then each of the inside edges is drawn with two line segments with slopes δ' and $-\delta'$ on the inside of H_1. Similarly, the outside edges are drawn with the same slopes on the outside of H_1. Note that some edges will overlap but postprocessing rotation can be used to remove the overlaps; see Fig. 2(b).

The edges of G_2 are handled in the same way with respect to H_2. It is easy to see that the vertex set requires grid size $n \times n$. The overall area of the drawing is larger, as the bend points lie outside the original grid. It is easy to show, however, that the entire drawing fits inside an $O(n^2) \times O(n^2)$ grid.

Making the Graphs 4-Connected: For the case when the input graphs are not 4-connected, we use techniques introduced in [11] to augment them. Given a 3-connected planar graph G we create a 4-connected planar graph by introducing new vertices. This is done by removing all the separating triangles in G. A separating triangle is a cycle of length 3 such that the removal of the vertices of the cycle disconnects G. Separating triangles of G can be easily found by the algorithm of [3]. Let $e = (u, v)$ be an edge of a separating triangle in G such that e is adjacent to the faces (u, v, s) and (u, v, t); see Fig. 3. We remove the separating triangle by inserting a dummy vertex z on e, deleting the edge e, and introducing four new edges: $(u, z), (v, z), (s, z), (t, z)$. The newly introduced vertex z is not part of any separating triangle, so each time we introduce such a vertex we decrease the number of separating triangles. Doing the same operation on all the separating triangles gives us a 4-connected planar graph.

Once G_1 and G_2 have been augmented to 4-connected graphs, we obtain the hamiltonian cycles H_1 and H_2 of G_1 and G_2. We augment the edges of H_2 with the extra vertices of G_1 and augment the edges of H_1 with the extra vertices of G_2. The placement of the hamiltonian cycles and the drawing of the remaining edges is done as before. After finishing the placement we treat the dummy vertices as bend points and ignore the edges inserted in the augmentation phase. As a result, an edge $e = (u, v)$ that got split with a dummy vertex z ends up having three bend points: one between u and z, one at the location of z, and finally one between v and z. As there are $O(n)$ dummy vertices, the bounds for the integer grid remain unchanged.

Running Time: The two non-trivial operations are finding the separating triangles and finding the hamiltonian cycles. Finding the separating triangles and making the graphs 4-connected takes linear time [3]. A Hamiltonian cycle in a 4-connected planar graphs can be found in linear time [4]. □

The corollary below follows from the above theorem by fixing the slopes of all the edges and refining the grid.

Corollary 1. *Given two planar graphs G_1 and G_2 and a mapping between their vertices, we can simultaneously embed G_1 and G_2 using at most three bends per edge on an integer grid of size $O(n^3) \times O(n^3)$, with all the vertices and bend-points at grid-points.*

Proof Sketch: Consider the original $n \times n$ grid where H_1 and H_2 are placed. Let the slope $\delta = n$, where δ and $-\delta$ are the slopes of all edge segments among edges drawn with bends. Let $e = (u, v) \in G_1$ such that u is placed to the left of v and e is drawn with a bend point p. Let x_{dist}, y_{dist} be the x-coordinate and y-coordinate distances between u and v. The x-coordinate distance between u and the point p is $(n \times x_{dist} - y_{dist})/2n$. If we place a $2n \times 2n$ grid inside each unit square of the original grid, then the x-coordinate distance between u and p is an integer. Since the slope of the segment $\overline{up}$ is n, the y-coordinate distance between u and p is also an integer, and p is on a grid point. Similar argument applies to the edges of G_2 as well. The final grid area is $O(n^3) \times O(n^3)$. □

If both input graphs are trees then it is easy to reduce the number of bends required to only one per edge. The Theorem below follows from Theorem 2 and the above corollary.

Theorem 3. *Given two trees T_1 and T_2 and a mapping between their vertices, they can be simultaneously embedded in linear time, using at most one bend per edge, on an integer grid of size $O(n^2) \times O(n^2)$ (or $O(n^3) \times O(n^3)$, if both the vertices and bend-points are on grid points).*

2.4   Simultaneous Embedding with Fixed Edges

The algorithm from the previous section simultaneously embeds two planar graphs with the corresponding vertices mapped on the same positions and thus preserves the mental map for the vertex set. There is a significant drawback with respect to preserving the mental map for the edge set. In particular, edges common to both graphs are drawn differently in the two drawings unless they happen to be on the paths defined by the hamiltonian cycles.

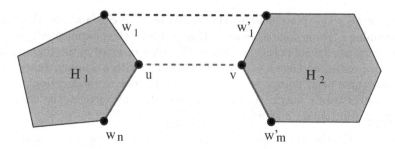

Fig. 4. Constructing the hamiltonian cycle H_T from H_1 and H_2. The common edges are shown in red.

Simultaneous embedding with fixed edges, requires that shared edges be represented the same way in both drawings. We describe an algorithm for simultaneous embedding with fixed edges for a tree and a path below.

Theorem 4. *Given a tree T, a path P, and a mapping between their vertices they can be simultaneously embedded with fixed edges in linear time, using at most one bend per edge, on an integer grid of size $O(n) \times O(n^2)$ (or $O(n^2) \times O(n^3)$), if both the vertices and bend-points are on the grid.*

Proof Sketch: The main idea is the same as that in Theorem 2, except that we ensure that the edges common to both T and P belong to the hamiltonian cycle for the tree. Then the path and the hamiltonian cycle have a simultaneous geometric embedding. The rest of the tree edges are routed as before using one bend per edge, thus yielding a simultaneous embedding with fixed edges for T and P.

Let $E_{T,P}$ be the set of edges common to both T and P. In order to obtain a hamiltonian cycle for the tree T we augment it with edges until the resulting graph T' has a hamiltonian cycle H_T that contains all edges that are in common with the path. We use a recursive divide-and-conquer procedure to construct H_T: the input to the recursive call is a subtree T and the output is the hamiltonian cycle H_T and the modified graph T'.

The base case for the recursion is a tree with just one node, $T = \{u\}$. In this case, let $H_T = (u, u)$, and $T' = T$. For all other cases, we take an edge $e = (u, v) \in E_{T,P}$ from T if such an edge exists. If not we take an arbitrary edge $e = (u, v) \in T$. Let T_1, T_2 be the two trees obtained after the removal of e from T. Assume we can construct hamiltonian cycles H_1, and H_2 of T_1 and T_2, respectively. Let T_1' and T_2' be the graphs that we get after these constructions, corresponding to T_1 and T_2, respectively. We merge the two subgraphs into the new graph $T' = T_1' \cup T_2'$ by adding e to T'.

In order to combine the hamiltonian cycles of the two subgraphs into a hamiltonian cycle for union, we need to add one more edge between the two subgraphs (as the edge e is a bridge). We add an edge between a neighbor u_{new} of u to a neighbor v_{new} of v and combine the two cycles by dropping the edges (u, u_{new}) and (v, v_{new}).

Let $H_1 = (u, w_1, w_2, \ldots, w_n, u)$ and $H_2 = (v, w_1', w_2', \ldots, w_m', v)$. If T_1' has only one vertex u we assign $u_{new} = u$, and if it has two vertices u and u' we assign $u_{new} = u'$. We do similar assignments for v_{new} if T_2' has one or two vertices. In order to find u_{new}, v_{new} for all other cases, we check the first and the last edges of the hamiltonian cycles.

Since P is a path, either $(u, w_1) \notin E_{T,P}$, or $(u, w_n) \notin E_{T,P}$ (otherwise, vertex u must have degree greater than 2 in the path). Without loss of generality, assume $(u, w_1) \notin E_{T,P}$. We assign $u_{new} = w_1$. The same holds for H_2, that is, either $(v, w_1') \notin E_{T,P}$ or $(v, w_m') \notin E_{T,P}$. Without loss of generality, assume $(v, w_1') \notin E_{T,P}$. We assign $v_{new} = w_1'$. We insert edge (u_{new}, v_{new}) in T', if $e \neq (u_{new}, v_{new})$. As a result of this insertion the new hamiltonian cycle becomes, $H_T = (u, v, w_m', w_{m-1}', \dots, w_1', w_1, w_2, \dots, w_n, u)$; see Fig. 4.

Planarity: The above recursive procedure augments the tree T to a graph T' that has a hamiltonian cycle which contains all the edges that T has in common with the path P. We still need to show that the resulting graph T' is planar. Recall the recursive procedure above and let us assume that T_1' and T_2' are planar. Then there exists a planar embedding for T_1' so that the edge (u, w_1) is on the outer-face and a planar embedding for T_2' so that the edge (v, w_1') is on the outer-face. Since all the vertices u, w_1, v, w_1' are on the outer-faces of their graphs, the inserted edges (u, v) and (w_1, w_1') do not have any crossings with the edges of T_1' and T_2'. The resulting graph T' is planar, and the resulting embedding is a planar embedding.

Running Time: We only need to show that the hamiltonian cycle construction takes linear time, since the rest of the algorithm is the same as the one described in the previous section. Note that we do not have to explicitly find planar embeddings of T_1' and T_2' at each level of the recursion. The planar embedding of the final graph T' suffices and we can find it in linear time [10]. The merging of the two hamiltonian cycles requires constant number of operations at each recursive step and thus the overall running time of the algorithm is $O(n)$.   □

3   Conclusion and Future Work

We implemented the algorithms described above using the LEDA library in C++. Fig. 5 shows the layouts obtained for a path and tree. All of the algorithms in this paper rely on the approach of augmenting planar graphs to hamiltonian planar graphs, so as to obtain simultaneous embeddings and simultaneous embeddings with fixed edges, using one or three bends. However, for simultaneous embedding with fixed edges, this technique cannot be extended from the path and tree case to pairs of trees (and hence cannot be extended to larger classes of planar graphs). We do not know of an algorithm for fixed-edge simultaneous embedding of pairs of trees. Neither do we have a counter-example. Similarly, the problem of simultaneous geometric embedding of pairs of trees is still open.

Acknowledgments

We would like to thank Michael Kaufmann and David Eppstein for discussions about these problems, and Petr Moravsky for helping with the implementation.

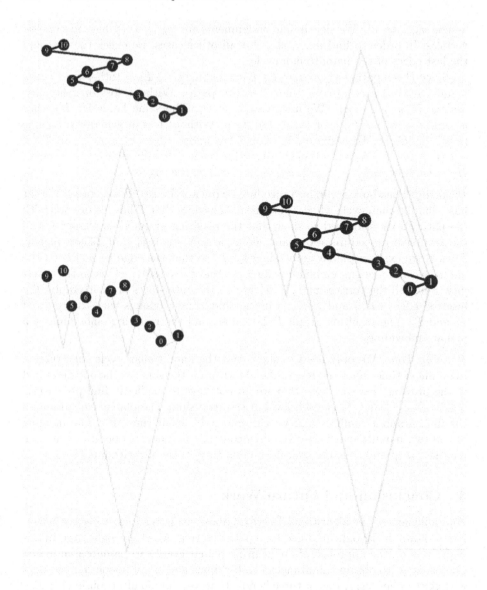

Fig. 5. A simultaneous embedding with fixed edges for a tree and a path. The path $(0, 1, \ldots, 10)$ is shown on the top left. The tree is shown on the bottom left. Note that the path and the tree share the edge $(0,1)$. The combined view of the tree and the path is shown on the right.

References

1. U. Brandes and S. R. Corman. Visual unrolling of network evolution and the analysis of dynamic discourse. In *IEEE Symposium on Information Visualization (INFOVIS '02)*, pages 145–151, 2002.

2. P. Brass, E. Cenek, C. A. Duncan, A. Efrat, C. Erten, D. Ismailescu, S. G. Kobourov, A. Lubiw, and J. S. B. Mitchell. On simultaneous graph embedding. In *8th Workshop on Algorithms and Data Structures*, pages 243–255, 2003.

3. N. Chiba and T. Nishizeki. Arboricity and subgraph listing algorithms. *SIAM J. Comput.*, 14:210–223, 1985.

4. N. Chiba and T. Nishizeki. The hamiltonian cycle problem is linear-time solvable for 4-connected planar graphs. *Journal of Algorithms*, 10(2):187–211, 1989.

5. H. de Fraysseix, J. Pach, and R. Pollack. How to draw a planar graph on a grid. *Combinatorica*, 10(1):41–51, 1990.

6. M. B. Dillencourt, D. Eppstein, and D. S. Hirschberg. Geometric thickness of complete graphs. *Journal of Graph Algorithms and Applications*, 4(3):5–17, 2000.

7. C. A. Duncan, D. Eppstein, and S. G. Kobourov. The geometric thickness of low degree graphs. In *20th Annual ACM-SIAM Symposium on Computational Geometry (SCG)*, pages 340–346, 2004.

8. C. Erten, S. G. Kobourov, A. Navabia, and V. Le. Simultaneous graph drawing: Layout algorithms and visualization schemes. In *11th Symposium on Graph Drawing (GD)*, pages 437–449, 2003.

9. I. Fáry. On straight lines representation of planar graphs. *Acta Scientiarum Mathematicarum*, 11:229–233, 1948.

10. J. Hopcroft and R. E. Tarjan. Efficient planarity testing. *Journal of the ACM*, 21(4):549–568, 1974.

11. M. Kaufmann and R. Wiese. Embedding vertices at points: Few bends suffice for planar graphs. *Journal of Graph Algorithms and Applications*, 6(1):115–129, 2002.

12. P. Mutzel, T. Odenthal, and M. Scharbrodt. The thickness of graphs: a survey. *Graphs Combin.*, 14(1):59–73, 1998.

13. J. Pach and R. Wenger. Embedding planar graphs at fixed vertex locations. *Graphs and Combinatorics*, 17:717–728, 2001.

14. W. Schnyder. Embedding planar graphs on the grid. In *Proceedings of the 1st ACM-SIAM Symposium on Discrete Algorithms (SODA)*, pages 138–148, 1990.

15. S. K. Stein. Convex maps. *Proceedings of the American Mathematical Society*, 2(3):464–466, 1951.

16. W. T. Tutte. How to draw a graph. *Proc. London Math. Society*, 13(52):743–768, 1963.

17. K. Wagner. Bemerkungen zum vierfarbenproblem. *Jahresbericht der Deutschen Mathematiker-Vereinigung*, 46:26–32, 1936.

A Fast and Simple Heuristic
for Constrained Two-Level Crossing Reduction

Michael Forster

University of Passau, 94030 Passau, Germany
forster@fmi.uni-passau.de

Abstract. The one-sided two-level crossing reduction problem is an important problem in hierarchical graph drawing. Because of its NP-hardness there are many heuristics, such as the well-known barycenter and median heuristics. We consider the constrained one-sided two-level crossing reduction problem, where the relative position of certain vertex pairs on the second level is fixed. Based on the barycenter heuristic, we present a new algorithm that runs in quadratic time and generates fewer crossings than existing simple extensions. It is significantly faster than an advanced algorithm by Schreiber [12] and Finnocchi [1, 2, 6], while it compares well in terms of crossing number. It is also easy to implement.

1 Introduction

The most common algorithm for drawing directed acyclic graphs is the algorithm of Sugiyama, Tagawa, and Toda [13]. The vertex set is partitioned into parallel horizontal levels such that all edges point downwards. For every intersection between an edge and a level line, a dummy vertex is introduced that may later become an edge bend. In a second phase, a permutation of the vertices on each level is computed that minimizes the number of edge crossings. Finally, horizontal coordinates are computed, retaining the vertex order on each level.

A small number of crossings is very important for a drawing to be understandable. Thus, the crossing reduction problem is well studied. The minimization of crossings is NP-hard [4, 8], and many heuristics exist for crossing reduction. Most of them reduce the problem to a sequence of one-sided two-level crossing reduction problems. Starting with an arbitrary permutation of the first level, a permutation of the second level is computed that induces a small number of edge crossings between the first two levels. Then the permutation of the second level is fixed and the third level is reordered. This is repeated for all levels, alternately top down and bottom up, until some termination criterion is met.

A simple and efficient heuristic for the one-sided two-level crossing reduction problem is the barycenter heuristic. For every vertex v on the second level, its barycenter value $b(v)$ is defined as the arithmetic mean of the relative positions of its neighbors $N(v)$ on the first level $b(v) = \frac{1}{|N(v)|} \sum_{v \in N(v)} \mathrm{pos}(v)$. The vertices on the second level are then sorted by their barycenter value. In practice this strategy gives good results, while keeping the running time low. An alternative is the median heuristic, which works similar but uses median values instead of

J. Pach (Ed.): GD 2004, LNCS 3383, pp. 206–216, 2004.

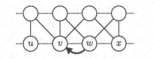

(a) The constraint is violated.   (b) The constraint is satisfied.

Fig. 1. The constrained crossing reduction problem.

the barycenter. The median heuristic can be proven [3, 5] to miss the minimum number of crossings by a factor of at most three. However, in experimental results [9, 10] it is outperformed by the barycenter heuristic.

As a variant of the crossing reduction problem we consider the constrained one-sided two-level crossing reduction problem. In addition to the permutation of the first level, some pairs of vertices on the second level have a fixed relative position. Figure 1 shows a two-level graph with one *constraint* $c = (w, v)$, visualized by the bold arrow. The constraint means that its *source vertex* w must be positioned on the left of its *target vertex* v. In Fig. 1(a), the constraint is violated, and in Fig. 1(b) it is satisfied. Obviously, constraints may increase the number of crossings, in this case from two to five.

Formally, an instance of the constrained one-sided two-level crossing reduction problem consists of a two-level graph $G = (V_1, V_2, E)$, $E \subseteq V_1 \times V_2$ with a fixed permutation of the first level V_1 and a set $C \subseteq V_2 \times V_2$ of constraints. It is our objective to find a permutation of the vertices on the second level V_2 with few edge crossings and all constraints satisfied. Clearly, this problem is NP-hard as well. A solution only exists if the *constraint graph* $G_C = (V_2, C)$ is acyclic.

While the constrained crossing reduction problem has many direct practical applications, it also appears as a subproblem in other graph drawing problems. An example is the application of the Sugiyama algorithm to graphs with vertices of arbitrary size [12] or to clustered graphs [7]. When vertices or clusters span multiple levels, constraints can be used to prevent overlap. Another application is preserving the mental map when visualizing a sequence of related graphs.

This paper is organized as follows: We survey existing approaches for the constrained two-level crossing reduction problem in the next section. In Sect. 3 we present our heuristic and prove its correctness in Sect. 4. Section 5 gives experimental results that compare our heuristic to the existing algorithms. We close with a short summary in Sect. 6.

2 Previous Work

The constrained crossing reduction problem has been considered several times. Sander [11] proposes a simple strategy to extend iterative two-level crossing reduction algorithms to handle constraints. Starting with an arbitrary admissible vertex permutation, updates are only executed if they do not violate a constraint. Together with the barycenter heuristic a modified sorting algorithm is used: The positions of two vertices are only swapped, if no constraint is violated. Waddle [14] presents a similar algorithm. After the calculation of the barycenter values it is checked for each constraint whether its target has a lower barycenter value than its source. In that case the constraint would be violated after sorting the vertices by the barycenter values. To avoid this, the barycenter value of the source vertex is changed to the barycenter value of the target vertex plus some small value. The result of both heuristics is a vertex permutation that satisfies all constraints. However, the extensions are rather restrictive and often prevent the algorithm from finding a good permutation. Accordingly, the results are significantly worse than in graphs without constraints.

Schreiber [12] and Finnocchi [1, 2, 6] have independently presented an advanced algorithm that considers constraints and crossing minimization simultaneously. Their main idea is to reduce the constrained crossing reduction problem to the weighted feedback arc set problem, which is also NP-hard [3]. First the so-called *penalty graph* is constructed. Its vertices are the vertices of the second level. For each pair (u, v) of vertices the number of crossings in the two relative orders of u and v is compared. For this, only edges incident to u or v are considered. If the number of crossings c_{uv} in the relative order $\ldots, u, \ldots, v, \ldots$ is less than the number of crossings c_{vu} in the reverse order $\ldots, v, \ldots, u, \ldots$, then an edge $e = (u, v)$ with weight $w(e) = c_{vu} - c_{uv}$ is inserted. Constraints are added as edges with infinite (or very large) weight. Figure 2 shows the penalty graph of the two-level graph in Fig. 1.

Then a heuristic for the weighted feedback arc set problem is applied to the penalty graph. It is important that the used heuristic guarantees that the edges with infinite weight are not reversed, or constraints may be violated. Finally, the vertices of the now acyclic penalty graph are sorted topologically, and the resulting permutation defines the order of the second level. If no edges had to be reversed, the number of crossings meets the obvious lower bound $c_{\min} = \sum_{u,v \in V} \min\{c_{uv}, c_{vu}\}$. Each reversed edge e increments the number of crossings by its weight. This implies that an optimal solution of the weighted feedback arc set problem is also optimal for the constrained crossing reduction problem.

Comparing the approaches of Sander [11] and Waddle [14] with those of Schreiber [12] and Finnocchi [1, 2, 6] shows a direct trade-off between quality and execution time. Schreiber presents detailed experimental results which show that the penalty graph approach generates significantly less crossings than the barycenter heuristic extensions. This is especially evident, if there are many constraints. The running times, however, are considerably higher. This is not very surprising due to the $O(|V_2|^4 + |E|^2)$ time complexity.

3   A Modified Barycenter Heuristic

The goal of our research is to build an algorithm that is as fast as the existing barycenter extensions while delivering a quality comparable to the penalty graph

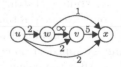

Fig. 2. The penalty graph of Fig. 1.

approach. To achieve this we use a new extension of the barycenter heuristic. We could have used the median heuristic as well, but we did not, because it is experimentally worse and in our algorithm median values are more difficult to handle.

We start by computing the barycenter values of all vertices. As long as the source of each constraint has a lower barycenter value than the target, all constraints are satisfied automatically. In the reverse case the permutation has to be corrected. In this context, we call a constraint $c = (s, t)$ *satisfied* if $b(s) < b(t)$ and *violated* otherwise.

Our algorithm is based on a simple assumption: If a constraint is violated as in Fig. 3(a), the greater barycenter value of the source vertex indicates more edges "to the right" than "to the left", $|E_3| > |E_1|$. The inverse is true for the target vertex, $|E_4| < |E_2|$. In this situation we assume that in the corrected permutation no other vertices should be positioned in between. This seems plausible, because between s and t larger subsets of adjacent edges have to be crossed than outside. Using median values it can be proven that for a vertex with only one incident edge there is always an optimal position beyond any violated constraint. This is not generally true, however, for vertices of higher degree or for the barycenter heuristic as Fig. 3(b) shows. The optimal position for vertex v is in the middle, where its edges generate 6 crossings as opposed to 8 crossings at the other two positions. Nevertheless, adopting the assumption is justified by good experimental results presented in Sect. 5.

Our heuristic, shown in Algorithm 1, partitions the vertex set V_2 into totally ordered vertex lists. Initially there is one singleton list $L(v) = \langle v \rangle$ per vertex v. In the course of the algorithm these lists are pairwise concatenated into longer lists according to violated constraints. Concatenated lists are represented by new dummy vertices and associated barycenter values. As long as there are violated constraints, each violated constraint $c = (s, t)$ is removed one by one and the lists containing s and t are concatenated in the required order. They are then treated as a cluster of vertices. This guarantees that the constraint is satisfied but prevents other vertices from being placed between s and t. Following our assumption, this does no harm. A new vertex v_c replaces s and t to represent the concatenated list $L(v_c) = L(s) \circ L(t)$. The barycenter value of v_c is computed as if all edges that are incident to a vertex in $L(v_c)$ were incident to v_c. This can be done in constant time as demonstrated in lines 8 and 9 of the algorithm. Note that this is not doable for the median value.

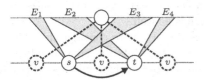 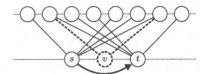

(a) Vertices with a single edge should not be positioned between the vertices of a violated constraint ($b(s) > b(t)$).

(b) In general, the optimal position for a vertex may be between the vertices of a violated constraint.

Fig. 3. The Basic Assumption of Our Algorithm.

Algorithm 1: CONSTRAINED-CROSSING-REDUCTION.

Input: A two-level graph $G = (V_1, V_2, E)$ and acyclic constraints $C \subseteq V_2 \times V_2$
Output: A permutation of V_2

begin

1   **foreach** $v \in V_2$ **do**
2        $b(v) \leftarrow \sum_{u \in N(v)} \text{pos}(u) / \deg(v)$ // *barycenter of v*
3        $L(v) \leftarrow \langle v \rangle$ // *new singleton list*

4    $V \leftarrow \{ s, t \mid (s, t) \in C \}$ // *constrained vertices*
5    $V' \leftarrow V_2 - V$ // *unconstrained vertices*

6   **while** $(s, t) \leftarrow$ FIND-VIOLATED-CONSTRAINT$(V, C) \neq \bot$ **do**
7   create new vertex v_c

8        $\deg(v_c) \leftarrow \deg(s) + \deg(t)$ // *update barycenter value*
9        $b(v_c) \leftarrow (b(s) \cdot \deg(s) + b(t) \cdot \deg(t)) / \deg(v_c)$
10      $L(v_c) \leftarrow L(s) \circ L(t)$ // *concatenate vertex lists*

11   **forall** $c \in C$ **do**
12   **if** c is incident to s or t **then**
13   make c incident to v_c instead of s or t

14      $C \leftarrow C - \{(v_c, v_c)\}$ // *remove self loops*
15      $V \leftarrow V - \{s, t\}$

16   **if** v_c has incident constraints **then** $V \leftarrow V \cup \{v_c\}$
17   **else** $V' \leftarrow V' \cup \{v_c\}$

18    $V'' \leftarrow V \cup V'$
19   sort V'' by $b()$

20    $L \leftarrow \langle \rangle$ // *concatenate vertex lists*
21   **foreach** $v \in V''$ **do**
22        $L \leftarrow L \circ L(v)$

23   **return** L

end

When no violated constraints are left, the remaining vertices and vertex lists are sorted by their barycenter value as in the standard barycenter heuristic. The concatenation of all vertex lists results in a vertex permutation that satisfies all constraints. We claim that it has few crossings as well.

For the correctness of the algorithm it is important to consider the violated constraints in the right order. In Fig. 4 the constraints are considered in the wrong order and c is processed first. This leads to a cycle in the resulting constraint graph which makes it impossible to satisfy all remaining constraints, although the original constraint graph was acyclic. If c is violated, at least one of the other constraints is also violated. Processing this constraint first leads to a correct result.

Thus, we must avoid generating constraint cycles. We use a modified topological sorting algorithm on the constraint graph. The constraints are considered

(a) Before the merge all constraints are satisfiable by the given order. Let c be violated.

(b) After merging s and t the generated constraint cycle makes it impossible to satisfy all constraints.

(c) Starting with c' leads to a correct result.

Fig. 4. Considering constraints in the wrong order.

sorted lexicographically by the topsort numbers of the target and source vertices in ascending and descending order, respectively. Using Algorithm 2 this traversal can be implemented in $O(|C|)$ time. The vertices are traversed in topological order. The incoming constraints of a vertex t are stored in an ordered list $I(t)$ that is sorted by the reverse traversal order of the source vertices. If a traversed vertex has incoming violated constraints, the topological sorting is cancelled and the first of them is returned. Note that the processing of a violated constraint can lead to newly violated constraints. Thus, the traversal must be restarted for every violated constraint.

4   Theoretical Analysis

In this section we analyse the correctness and running time of our algorithm. For the correctness we have to show that the vertex permutation computed by our algorithm satisfies all constraints. We start by analyzing Algorithm 2:

Lemma 1. *Let $c = (s, t)$ be a constraint returned by Algorithm 2. Then merging of s and t does not introduce a constraint cycle of two or more constraints.*

Proof. Assume that merging of s and t generates a cycle of at least two constraints. Because there was no cycle before, the cycle corresponds to a path p in G_C from s to t with a length of at least two. Because of the specified constraint traversal order, any constraint in p has already been considered, and thus is satisfied. This implies that $b(t) > b(s)$, and therefore contradicts the assumption.
                                                                                            □

Theorem 1. *The permutation computed by Algorithm 1 satisfies all constraints.*

Proof. Algorithm 1 maintains the invariant that the constraint graph is acyclic. Because of Lemma 1 no nontrivial cycles are introduced, and self loops are explicitly removed in line 14.

Next we analyse whether the removed self loop constraints are satisfied by the algorithm. Any such self loop c' has been generated by the lines 11–13 from

---

Algorithm 2: FIND-VIOLATED-CONSTRAINT.

Input: An acyclic constraint graph $G_C = (V, C)$ without isolated vertices
Output: A violated constraint c, or $\perp$ if none exists

begin

1     $S \leftarrow \emptyset$ *// active vertices*

2   **foreach** $v \in V$ **do**

3         $I(v) \leftarrow \langle\rangle$ *// empty list of incoming constraints*

4   **if** indeg$(v) = 0$ **then**

5             $S \leftarrow S \cup \{v\}$ *// vertices without incoming constraints*

6   **while** $S \neq \emptyset$ **do**

7   choose $v \in S$

8         $S \leftarrow S - \{v\}$

9   **foreach** $c = (s, v) \in I(v)$ in list order **do**

10   **if** $b(s) \geq b(v)$ **then**

11   **return** c

12   **foreach** outgoing constraint $c = (v, t)$ **do**

13             $I(t) \leftarrow \langle c \rangle \circ I(t)$

14   **if** $|I(t)| = $ indeg(t) **then**

15                $S \leftarrow S \cup \{t\}$

16   **return** $\perp$

end

---

a constraint between s and t. Because of the constraint $c = (s, t)$, the invariant implies that c' was not directed from t to s. Therefore, $c' = (s, t)$ is explicitly satisfied by the list concatenation in line 10.

Each remaining constraint has not been returned by Algorithm 2. Thus, the barycenter value of its source vertex is less than that of its target vertex. Then the constraint is satisfied by line 19.   □

The rest of this section analyses the running time of our algorithm. Again, we start with the analysis of Algorithm 2.

Lemma 2. *Algorithm 2 runs in $O(|C|)$ time.*

Proof. The initialization of the algorithm in lines 1–5 runs in $O(|V|)$ time. The while-loop is executed at most $|V|$ times. The nested foreach-loops are both executed at most once per constraint. The sum of these time bounds is $O(|V| + |C|)$. Because the constraint graph does not contain isolated vertices, the overall running time of the algorithm is bounded by $O(|C|)$.   □

Theorem 2. *Algorithm 1 runs in $O(|V_2| \log |V_2| + |E| + |C|^2)$ time.*

Proof. The initialization of the algorithm in lines 1–3 considers every vertex and edge once and therefore needs $O(|V_2| + |E|)$ time. The while-loop is executed at most once per constraint. It has an overall running time of $O(|C|^2)$ because the running time of one loop execution is bounded by the $O(|C|)$ running time of Algorithm 2. Finally, the sorting in line 19 needs $O(|V_2| \log |V_2|)$ time. The sum of these time bounds is $O(|V_2| \log |V_2| + |E| + |C|^2)$. All other statements of the algorithm do not increase the running time.   □

5   Experimental Analysis

To analyse the performance of our heuristic, we have implemented both our algorithm and the penalty graph approach in Java. We have tested the implementations using a total number of 37,500 random graphs: 150 graphs for every combination of the following parameters: $|V_2| \in \{50, 100, 150, 200, 250\}$, $|E|/|V_2| \in \{1, 2, 3, 4, 5, 6, 7, 8, 9, 10\}$, $|C|/|V_2| \in \{0, 0.25, 0.5, 0.75, 1.0\}$.

Figure 5 displays a direct comparison. The diagrams show, how the results vary, when one of the three parameters is changed. Because the number of crossings grows very fast in the number of edges, we do not compare absolute crossing numbers, but the number of crossings divided by the number of crossings before the crossing reduction. As expected, the penalty graph approach gives strictly better results than our heuristic. But the graphs also show that the difference is very small. For a more detailed comparison, we have also analyzed the quotient of the crossing numbers in Fig. 6. These graphs show that our algorithm is never more than 3% worse than the penalty graph approach. Mostly the difference is below 1%. Only for very sparse graphs there is a significant difference.

This is a very encouraging result, considering the running time difference of both algorithms: Figure 7 compares the running time of the algorithms. As expected, our algorithm is significantly faster than the penalty graph approach. Because of the high running time of the penalty graph approach we have not compared the algorithms on larger graphs, but our algorithm is certainly capable of processing larger graphs. For example, graphs with $|V_2| = 1000$, $|E| = 2000$, and $|C| = 500$ can be processed in less than a second, although our implementation is not highly optimized.

6   Summary

We have presented a new fast and simple heuristic for the constrained one-sided two-level crossing reduction problem. In practice, the algorithm delivers nearly the same quality as existing more complex algorithms, while its running time is significantly better. For further improvement, a traversal of the violated constraints is desired that runs faster than $O(|C|^2)$.

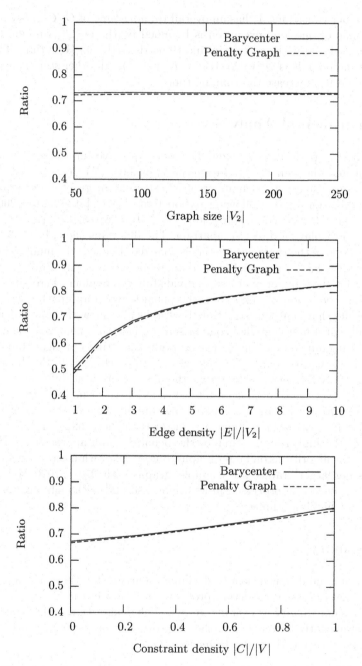

Fig. 5. The ratio of crossings before and after crossing reduction. Lesser values are better.

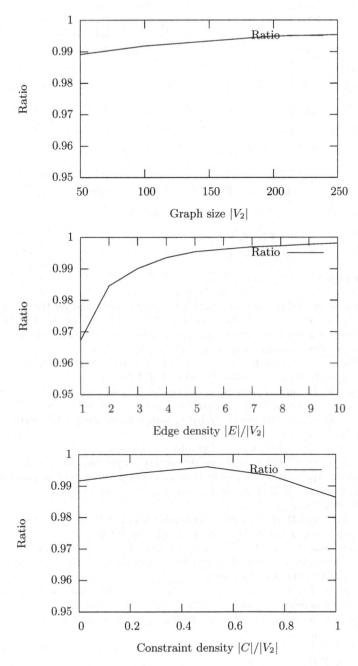

Fig. 6. Number of crossings using the penalty graph approach divided by the number of crossings using the extended barycenter approach.

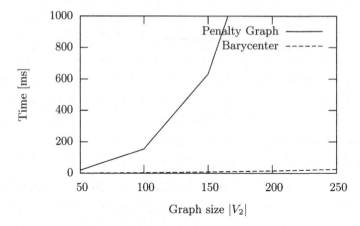

Fig. 7. Running time comparison.

References

1. C. Demetrescu and I. Finocchi. Break the "right" cycles and get the "best" drawing. In B. Moret and A. Goldberg, editors, *Proc. ALENEX'00*, pp. 171–182, 2000.
2. C. Demetrescu and I. Finocchi. Removing cycles for minimizing crossings. *ACM Journal on Experimental Algorithmics (JEA)*, 6(1), 2001.
3. G. di Battista, P. Eades, R. Tamassia, and I. G. Tollis. *Graph Drawing: Algorithms for the Visualization of Graphs*. Prentice Hall, 1999.
4. P. Eades, B. D. McKay, and N. C. Wormald. On an edge crossing problem. In *Proc. ACSC'86*, pp. 327–334. Australian National University, 1986.
5. P. Eades and N. C. Wormald. Edge crossings in drawings of bipartite graphs. *Algorithmica*, 11:379–403, 1994.
6. I. Finocchi. Layered drawings of graphs with crossing constraints. In J. Wang, editor, *Proc. COCOON'01*, volume 2108 of *LNCS*, pp. 357–368. Springer, 2001.
7. M. Forster. Applying crossing reduction strategies to layered compound graphs. In S. G. Kobourov and M. T. Goodrich, editors, *Proc. GD'02*, volume 2528 of *LNCS*, pp. 276–284. Springer, 2002.
8. M. R. Garey and D. S. Johnson. Crossing number is NP-complete. *SIAM Journal on Algebraic and Discrete Methods*, 4(3):312–316, 1983.
9. M. Jünger and P. Mutzel. Exact and heuristic algorithms for 2-layer straightline crossing minimization. In F. J. Brandenburg, editor, *Proc. GD'95*, volume 1027 of *LNCS*, pp. 337–348. Springer, 1996.
10. M. Jünger and P. Mutzel. 2-layer straightline crossing minimization: Performance of exact and heuristic algorithms. *JGAA*, 1(1):1–25, 1997.
11. G. Sander. *Visualisierungstechniken für den Compilerbau*. PhD thesis, University of Saarbrücken, 1996.
12. F. Schreiber. *Visualisierung biochemischer Reaktionsnetze*. PhD thesis, University of Passau, 2001.
13. K. Sugiyama, S. Tagawa, and M. Toda. Methods for visual understanding of hierarchical system structures. *IEEE Trans. SMC*, 11(2):109–125, 1981.
14. V. Waddle. Graph layout for displaying data structures. In J. Marks, editor, *Proc. GD'00*, volume 1984 of *LNCS*, pp. 241–252. Springer, 2001.

Contact and Intersection Representations

Hubert de Fraysseix and Patrice Ossona de Mendez

UMR 8557, CNRS, Paris, France

Abstract. A necessary and sufficient condition is given for a connected bipartite graph to be the incidence graph of a family of segments and points. We deduce that any 4-connected 3-colorable plane graph is the contact graph of a family of segments and that any 4-colored planar graph without an induced C_4 using 4 colors is the intersection graph of a family of straight line segments.

To Chantal. Her life crossed mine on a too short path.

1 Introduction

Touchings and crossings of arcs in the plane have been the subject of lively interest, giving rise to astonishingly complex problems, albeit easy to state. As an example, the Gauss problem on the characterization of crossing sequences of self-intersecting closed curves [12], which has been fully solved only recently [6, 23]. The algebraic matroidal properties used to solve this problem further led to a characterization of bipartite circle graphs [3] and then to a characterization of general circle graphs à la Whitney [4].

Intersection graphs of arcs, the so-called *string graphs*, have been independently introduced by Sinden [27], Ehrlich, Even and Tarjan [11]. Their approach appeared to be quite complex [15, 17]. The recognition problem has been proven to be NP-hard [16] and, more recently, NP-complete [21, 24].

The particular cases of intersection graphs of pseudo-segments and intersection graphs of segments [18] are of special interest, as shown by the following question by Scheinerman [25]: *Is every planar graph the intersection graph of a set of segments in the plane?*

This question is still open even for pseudo-segments, but some partial results have been obtained:

- the recognition problem of contact graphs of segments is NP-complete, even when restricted to planar graphs [14],
- bipartite planar graphs are contact graphs of a set of orthogonal segments [9, 13] (see also [1]),
- triangle-free planar graphs are necessarily contact graphs of a set of segments in three directions [2],
- 4-connected 3-colorable plane graphs are contact graphs of a set of pseudo-segments [5],
- 4-colored plane graphs without C_4-separator using 4 colors are intersection graphs of a set of pseudo-segments [5] (see Fig.1 to 2).

J. Pach (Ed.): GD 2004, LNCS 3383, pp. 217–227, 2004.

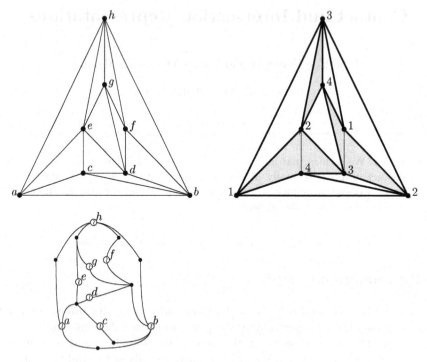

Fig. 1. Using a coloration, a graph G_0 gives rise to a bipartite plane graph.

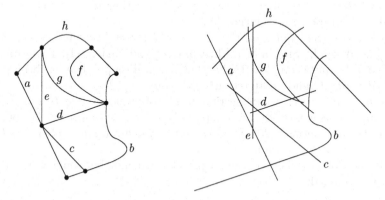

Fig. 2. From the bipartite plane graph shown in Fig 1, we obtain a contact family of pseudo-segments, which by local deformation gives rise to a representation of the graph G_0 of Fig. 1 as the intersection graph of a family of pseudo-segments.

Using the arc-stretching techniques presented in [7, 8], the last two results may be strengthened (see Fig. 3):

Theorem 1. *4-connected 3-colorable plane graphs are contact graphs of a set of segments.*

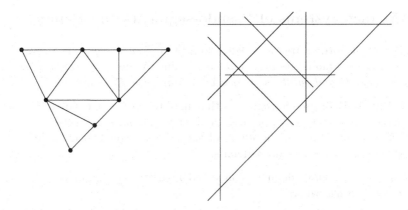

Fig. 3. Using stretching techniques, from the bipartite graph of Fig. 1, we obtain a contact family of segments. By local perturbations, this contact system gives rise to a representation of the graph G_0 of Fig 1 as the intersection graph of a family of segments (here in 4 directions).

Theorem 2. *4-colored plane graphs without C_4-separator using 4 colors are intersection graphs of a set of segments.*

We shall present a sketch of the proof of these theorems using the following characterization of incidence graphs of a family of segments, which we shall also prove:

Theorem 3. *A connected bipartite graph $G = (V_l, V_\bullet, E)$ is the incidence graph of a (one-sided) contact family of segments and points if and only if*

- *G is planar,*
- *the minimum degree of the vertices in V_l is at least 2*
- *$\forall X \subseteq V$ such that $|X \cap V_l| \geq 2$,*

$$|E(G_X)| \leq 2\,|X \cap V_l| + |X \cap V_\bullet| - 3 \tag{1}$$

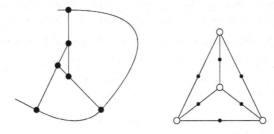

Fig. 4. Representation of K_4 by a non-stretchable contact family of pseudo-segments. The corresponding incidence graph (V_l is represented with white vertices, $V_\bullet$ with black ones), so that $|E| > 2\,|V_l| + |V_\bullet| - 3$.

2   Contact Systems of Pseudo-segments and Points

A finite set of Jordan arcs is called a family of *pseudo-segments* if every pair of arcs in the set intersects in at most one point. A *one-sided contact family of pseudo-segments and points* is defined by a couple $(\mathcal{A}, P)$, where:

- $\mathcal{A}$ is a family of pseudo-segments that may touch (on one side only at each contact point) but may not cross, and whose union is connected,
- P is a set of points in the union of the pseudo-segments, including all the extremities of the pseudo-segments.

Such a contact family defines a connected bipartite plane graph $G=(V_l,V_\bullet,E)$, its *incidence graph*, where:

- V_l corresponds to the pseudo-segment set,
- $V_\bullet$ corresponds to the point set,
- E corresponds to the set of incidences between points and pseudo-segments.

Notice that vertices in V_l have minimal degree at least 2.

Moreover, the contact family also defines an orientation of G: if $x \in V_\bullet$ corresponds to a point p on a pseudo-segment S corresponding to $y \in V_l$, $\{x,y\}$ is oriented from x to y if p is an extremity of S and from y to x, otherwise. The orientation thus obtained is such that the indegree of a vertex in V_l is exactly 2 and the indegree of a vertex in $V_\bullet$ is at most 1. We call such an orientation a $(2, \leq 1)$-*orientation*.

The following theorem is quite simple to prove (see [5]):

Theorem 4. *A bipartite graph* $G = (V_l, V_\bullet, E)$ *is the incidence graph of a (one-sided) contact family of pseudo-segments and points if and only if*

- *G is planar,*
- *G has girth at least 6,*
- *the minimum degree of the vertices in V_l is at least 2*
- $\forall X \subseteq V,$

$$|E(G_X)| \leq 2\,|X \cap V_l| + |X \cap V_\bullet| \qquad (2)$$

In general, representations by contacts of straight line segments raise important difficulties that may be collected into what we call the stretching problem:

Problem 1. When is a contact system of pseudo-segments stretchable, that is: when is it homeomorphic to a contact system of straight line segments?

This problem has been addressed in [8, 7], with the following characterization theorem:

Theorem 5. *Let $\mathcal{A}$ be a contact system of pseudo-segments. Then, the following conditions are equivalent:*

1. *$\mathcal{A}$ is stretchable,*
2. *each subsystem of $\mathcal{A}$ has at least 3 extremal points, unless it has cardinality at most one ,*
3. *$\mathcal{A}$ is extendible.*

where

- *An* extremal point *of a contact system of arcs is a point of the union of the arcs which is interior to no arc.*
- *A contact system of pseudo-segments is* extendible *if there exists an arrangement of pseudo-lines such that each pseudo-segment of the contact system is included in a corresponding pseudo-line of the family.*

Notice that the equivalence of extendibility and stretchability for contact systems of pseudo-segments is in strong contrast with the difficulty of the decidability problem concerning the stretching of arrangements of pseudo-lines (this problem is NP-hard, as proved by Mněv [19, 20]; see also Shor [26] and Richter-Gebert [22]).

3   Deficiency and $(2, \leq 1)$-Orientation

In order to make use of Theorem 5 to characterize those bipartite graphs that are representable by a contact family of segments, we first need to prove an orientation theorem. For that, we need few definitions and lemmas.

In the following, we consider a connected bipartite graph $G = (V_\wr, V_\bullet, E)$. Let $V = V_\wr \cup V_\bullet$. Given a subset $A \subseteq V$, we introduce the notation $A_\wr = A \cap V_\wr$ and $A_\bullet = A \cap V_\bullet$. By extension, if $f(x)$ is a subset of V, we employ the notation $f_\wr(x) = f(x) \cap V_\wr$ and $f_\bullet(x) = f(x) \cap V_\bullet$. We denote by $\mathcal{N}$ the *neighborhood function* defined by $\mathcal{N}(X) = X \cup \{y \in V, \exists x \in X : \{x, y\} \in E\}$. Observe that $X \subseteq \mathcal{N}(X)$.

3.1   Deficiency

Definition 1.

the deficiency ρ *of a subset* $X \subseteq V$ *is* $\rho(X) = 2\,|X_\wr| + |X_\bullet| - |E(G_X)|$

the minimal deficiency $\rho_{\min}$ *of* X *is* $\rho_{\min}(X) = \min\limits_{X \subseteq Y} \rho(Y)$

the deficiency closure Clos *of* X *is* $\mathrm{Clos}(X) = \bigcup\limits_{\substack{X \subseteq Y \\ \rho(Y) = \rho_{\min}(X)}} Y$

Lemma 1. *The function* ρ *is semimodular, that is,* $\forall X_1, X_2 \subseteq V$:

$$\rho(X_1 \cup X_2) + \rho(X_1 \cap X_2) \leq \rho(X_1) + \rho(X_2) \qquad (3)$$

Proof. This is a direct consequence of the inequality

$$|E(G_{X_1 \cup X_2})| \geq |E(G_{X_1})| + |E(G_{X_2})| \qquad \square$$

Lemma 2. *Let* $X \subseteq V$. *Then* $\rho(\mathrm{Clos}(X)) = \rho_{\min}(X)$.

Proof. Assume X_1, X_2 are subsets of V containing X such that $\rho(X_1) = \rho(X_2) = \rho_{\min}(X)$. Then, as $X \subseteq X_1 \cap X_2$, we get $\rho(X_1 \cap X_2) \geq \rho_{\min}(X)$ and, according to (3), $\rho(X_1 \cup X_2) \leq \rho_{\min}(X)$. Thus $\rho(X_1 \cup X_2) = \rho_{\min}(X)$. By induction we deduce that $\rho(\mathrm{Clos}(X)) = \rho_{\min}(X)$.   □

Lemma 3. *For any $A \subseteq V_\imath$,*

$$\mathcal{N}(A) \subseteq \mathcal{N}(\mathrm{Clos}_\imath(A)) = \mathrm{Clos}(A)$$

Proof. For any $X \in V$, we have $\rho(\mathcal{N}(X_\imath)) \leq \rho(X)$, as the addition to X of a vertex in $V_\bullet$ having at least one neighbor in X doesn't increase $\rho(X)$ and as the deletion of a vertex in $V_\bullet$ having no neighbor in X decreases $\rho(X)$ by 1. Hence $\rho(\mathcal{N}(X_\imath)) \leq \rho(X)$ and equality may only occur if $X \subseteq \mathcal{N}(X_\imath)$.

According to this property, as $\rho(\mathcal{N}(Y_\imath)) \leq \rho(Y)$ and as equality implies $Y \subseteq \mathcal{N}(Y_\imath)$, we have:

$$\mathrm{Clos}(A) = \bigcup_{\substack{A \subseteq Y \\ \rho(Y) = \rho_{\min}(A)}} Y = \bigcup_{\substack{A \subseteq Y \\ \rho(Y) = \rho_{\min}(A)}} \mathcal{N}(Y_\imath) = \mathcal{N}\Big(\bigcup_{\substack{A \subseteq Y \\ \rho(Y) = \rho_{\min}(A)}} Y_\imath \Big) = \mathcal{N}(\mathrm{Clos}_\imath(A))$$

Moreover, as $A \subseteq \mathrm{Clos}_\imath(A), \mathcal{N}(A) \subseteq \mathcal{N}(\mathrm{Clos}_\imath(A))$.   □

3.2   (2, ≤1)-Orientation

Definition 2. *A $(2, \leq 1)$-orientation $\mathcal{O}$ of a bipartite graph G is an orientation such that each vertex in $V_\imath$ has indegree exactly 2 and every vertex in $V_\bullet$ has indegree at most 1. A source of the $(2, \leq 1)$-orientation is a vertex with null indegree. Given a subset $X \subseteq V$, a vertex $x \in X$ is a relative source of X for $\mathcal{O}$ if it has a null indegree in G_X. We note $\mathrm{Source}(\mathcal{O}, X)$ the set of the relative sources of X for $\mathcal{O}$.*

The two following lemmas justify the term of deficiency for ρ.

Lemma 4 ([5]). *A connected bipartite graph G has a $(2, \leq 1)$-orientation if and only the minimal degree of vertices in $V_\imath$ is at least 2 and if*

$$\forall X \subseteq V, |E(G_X)| \leq 2 |X_\imath| + |X_\bullet| \tag{4}$$

Lemma 5. *Let G be a bipartite planar graph and a $(2, \leq 1)$-orientation of G. Let $X \subseteq V$. Then $\rho(X)$ is equal to the sum of the number of sources of G in X and of the number of arcs entering X from $V \setminus X$.*

Proof. The result is easily obtained by summing up the indegrees of the vertices in X.   □

Lemma 6. *Let $\mathcal{O}$ be a $(2, \leq 1)$-orientation of G and let $X \subseteq V$. Then there exists a $(2, \leq 1)$-orientation $\mathcal{O}'$ of G such that*

$$\mathrm{Source}(\mathcal{O}, \mathrm{Clos}(X)) \subseteq \mathrm{Source}(\mathcal{O}', \mathrm{Clos}(X)) \subseteq \mathrm{Source}(\mathcal{O}', V)$$

Proof. Let Y be the subset of $V \backslash \mathrm{Clos}(X)$ formed by the vertices y such that there exists a directed path from y to a vertex $x \in \mathrm{Clos}(X)$. We proceed iteratively, while decreasing the number of sources of G in Y.

If Y includes no source of $\mathcal{O}$, then any vertex in Y has its incoming edges incident to vertices in $Y \cup \mathrm{Clos}(X)$. Thus $\rho(\mathrm{Clos}(X) \cup Y) \leq \rho(\mathrm{Clos}(X)) = \rho_{\min}(X)$, a contradiction. Hence Y is empty, thus $\mathrm{Source}(\mathcal{O}', \mathrm{Clos}(X)) \subseteq \mathrm{Source}(\mathcal{O}', V)$.

Otherwise, let y be source of G in Y. By assumption, there exists a directed path from y to a vertex $x \in X$. According to Lemma 3, $\mathrm{Clos}(X) = \mathcal{N}(\mathrm{Clos}_{\imath}(X))$. Thus if the directed path has minimal length, $x \in \mathrm{Clos}_{\bullet}(X)$. Reorienting the directed path from x to y decreases the number of sources in Y, decreases Y and gives rise to a new $(2, \leq 1)$-orientation of G. As the reorientation may not have killed a relative source of $\mathrm{Clos}(X)$, we have $\mathrm{Source}(\mathcal{O}, \mathrm{Clos}(X)) \subseteq \mathrm{Source}(\mathcal{O}', \mathrm{Clos}(X))$. $\square$

Lemma 7. *Let $\mathcal{O}$ be a $(2, \leq 1)$-orientation of G and let $X \subseteq V$. Assume that any vertex $x \in X_{\imath}$ which has a neighbor out of X has at least two neighbors in X and that $\mathrm{Source}(\mathcal{O}, \mathrm{Clos}(X)) \subseteq \mathrm{Source}(\mathcal{O}, V)$. Then there exists a $(2, \leq 1)$-orientation $\mathcal{O}'$ of G, which differs with $\mathcal{O}$ only on $E(\mathrm{Clos}(X))$, such that:*

$$\mathrm{Source}(\mathcal{O}, \mathrm{Clos}(X)) \cap X \subseteq \mathrm{Source}(\mathcal{O}', \mathrm{Clos}(X)) \subseteq \mathrm{Source}(\mathcal{O}', V) \cap X$$

Proof. We proceed by induction on the cardinality of $\mathrm{Source}(\mathcal{O}, \mathrm{Clos}(X)) \setminus X$.

If $\mathrm{Source}(\mathcal{O}, \mathrm{Clos}(X)) \subseteq X$, we are done. Otherwise, let $s \in \mathrm{Source}(\mathcal{O}, \mathrm{Clos}(X)) \setminus X$. Let Y be the subset of the vertices $x \in \mathrm{Clos}(X)$ reachable from s by a directed path. If $Y \cap X = \emptyset$, then $\rho(\mathrm{Clos}(X) \setminus Y) < \rho(\mathrm{Clos}(X))$ although $X \subseteq \mathrm{Clos}(X) \setminus Y$, contradicting the minimality of $\rho(\mathrm{Clos}(X))$. Thus $Y \cap X \neq \emptyset$. Let $v_1 = s, \ldots, v_k$ be a minimal length directed path from s to a vertex in X. If $v_k \in X_{\imath}$, there exists an outgoing arc at v_k to a vertex $v_{k+1} \in X_{\bullet}$, as v_k has degree at least 2 in X by assumption, as the indegree of v_k is 2 and as v_k has an incoming edge from $v_{k-1} \notin X$. Reversing the orientation of the path $v_1, \ldots, v_k$ (resp. $v_1, \ldots, v_{k+1}$) if $v_k \in X_{\bullet}$ (resp. $v_k \in X_{\imath}$), we obtain a new $(2, \leq 1)$-orientation $\mathcal{O}'$ of G. As this orientation differs with $\mathcal{O}$ on $G_{\mathrm{Clos}(X)}$ only, we still have $\mathrm{Source}(\mathcal{O}', \mathrm{Clos}(X)) \subseteq \mathrm{Source}(\mathcal{O}', V)$. As we may not have killed a source in X, $\mathrm{Source}(\mathcal{O}, \mathrm{Clos}(X)) \cap X \subseteq \mathrm{Source}(\mathcal{O}', \mathrm{Clos}(X))$. Moreover, $|\mathrm{Source}(\mathcal{O}, \mathrm{Clos}(X)) \setminus X|$ decreased by one. $\square$

Definition 3. *A subset X of vertices of a connected plane graph G is a disk if any vertex of X having a neighbor out of X belongs to the outer face of G_X and $X = \mathcal{N}(X_{\imath})$.*

Theorem 6. *Assume $G = (V_{\imath}, V_{\bullet}, E)$ is a connected bipartite plane graph such that the minimum degree of vertices in $V_{\imath}$ is at least 2 and such that $\forall X \subseteq V$,*

$$|X_{\imath}| \geq 2 \Longrightarrow |E(G_X)| \leq 2|X_{\imath}| + |X_{\bullet}| - 3 \tag{5}$$

Then G has a $(2, \leq 1)$-orientation $\mathcal{O}$ such that, for any disk X with $|X_{\imath}| \geq 2$, we have

$$|\mathrm{Source}(\mathcal{O}, X) \cap \mathrm{Extr}(X)| \geq 3 \tag{6}$$

where $\mathrm{Extr}(X)$ denotes the vertex set of the outer face of G_X.

Proof. According to Lemma 4, G has a $(2, \leq 1)$-orientation $\mathcal{O}_0$.

We prove by induction over $(|V|, |V \setminus \mathrm{Extr}(V)|)$ that the required $(2, \leq 1)$-orientation $\mathcal{O}$ may be found, with the additional properties that the sources of $\mathcal{O}_0$ in $\mathrm{Extr}(V)$ are also sources of $\mathcal{O}$.

Let $A = \mathrm{Extr}(V)$. According to Lemmas 6 and 7, there exists a $(2, \leq 1)$-orientation $\mathcal{O}_1$ of G, such that

$$\mathrm{Source}(\mathcal{O}_0, V) \cap A \subseteq \mathrm{Source}(\mathcal{O}_1, \mathrm{Clos}(A)) \subseteq \mathrm{Source}(\mathcal{O}_1, V) \cap A$$

Let $B_1 = V \setminus \mathrm{Clos}(A)$, $B_2 = \mathcal{N}(B_1) \setminus B_1$ and $B_3 = \mathcal{N}(B_2) \setminus B_2$. According to Lemma 3, $B_2 \subseteq V_\bullet$ and $B_3 \subseteq V_?$. Moreover, $B_3 \subseteq \mathrm{Extr}(B_1 \cup B_2 \cup B_3)$. Let G' be the directed bipartite plane graph obtained from G as follows: First, we remove all the vertices not in $B_1 \cup B_2 \cup B_3$ and all the arcs oriented from B_2 to B_3. Finally, for every $v \in B_3$, we add two new vertices v_1 and v_2 on the outer face with arcs (v_1, v) and (v_2, v). Let C be the corresponding set of added black vertices. Then, $B_3 \cup C$ belong to the outer face of G' and the orientation of G' is a $(2, \leq 1)$-orientation having every vertex in C as a source. As $V_?(G')$ is either strictly included in $V_?(G)$ or is equal but then $|\mathrm{Extr}_?(V(G'))|$ is strictly greater than $|\mathrm{Extr}_?(V(G))|$, the induction applies. Thus there exists a $(2, \leq 1)$-orientation $\mathcal{O}_2$ of G', such that any vertex in C is a source of $\mathcal{O}_2$ and such that (6) holds for any disk X with $|X_?| \geq 2$.

Let $\mathcal{O}$ be the orientation of G induced by $\mathcal{O}_1$ on $G_{V \setminus B_1}$ and $\mathcal{O}_2$ on $G_{B_1 \cup B_2}$. By construction, $\mathcal{O}$ is a $(2, \leq 1)$-orientation of G such that $\mathrm{Source}(\mathcal{O}_0, V) \cap \mathrm{Extr}(V) \subseteq \mathrm{Source}(\mathcal{O}, V)$ and $\mathrm{Source}(\mathcal{O}, \mathrm{Clos}(A)) \subseteq \mathrm{Source}(\mathcal{O}, V) \cap \mathrm{Extr}(V)$.

Let X be a disk of G such that $|X_?| \geq 2$ and let $Y = X \cap \mathrm{Clos}(A)$.

Assume $|Y_?| \geq 2$. According to (5), we get $|E(G_Y)| \leq 2|Y_?| + |Y_\bullet| - 3$. According to Lemma 5, $|\mathrm{Source}(\mathcal{O}, Y)| \geq 3$. Moreover, as X is a disk and as $\mathrm{Clos}(A)$ has no entering arc, every relative source of Y belongs to the outer face of G_X. Thus $|\mathrm{Source}(\mathcal{O}, X) \cap \mathrm{Extr}(X)| \geq |\mathrm{Source}(\mathcal{O}, Y) \cap \mathrm{Extr}(X)| \geq 3$.

Otherwise, $X_?$ is included in the vertex set of G'. In G' we thus get

$$|\mathrm{Source}(\mathcal{O}_2, X) \cap \mathrm{Extr}(X)| \geq 3.$$

By construction of G' and $\mathcal{O}$, we deduce $|\mathrm{Source}(\mathcal{O}, X) \cap \mathrm{Extr}(X)| \geq 3$.   □

Theorem 7. *A connected bipartite graph $G = (V_?, V_\bullet, E)$ is the incidence graph of a (one-sided) contact family of segments and points if and only if*

- *G is planar,*
- *the minimum degree of the vertices in $V_?$ is at least 2*
- *$\forall X \subseteq V$ such that $|X \cap V_?| \geq 2$,*

$$|E(G_X)| \leq 2|X \cap V_?| + |X \cap V_\bullet| - 3 \tag{7}$$

Proof. According to Theorem 4, G is the contact graph of a family of pseudo-segments $\mathcal{A}$. According to Theorem 6, G has a $(2, \leq 1)$-orientation $\mathcal{O}$, such that any disk X with $|X_?| \geq 2$ has at least 3 relative sources on the outer face of G_X. Thus, according to Theorem 5, $\mathcal{A}$ is stretchable into a contact family of segments.   □

Corollary 1. *A graph $G = (V, E)$ is the intersection graph of a (one-sided) simple contact family of segments if and only if*

- *G is planar,*
- *any subgraph $H \subseteq G$ of order $n_H \geq 2$ has its size bounded by: $m_H \leq 2n_H - 3$.*

Proof. Apply Theorem 7 to the bipartite graph obtained by subdividing each edge of G exactly once.   □

In [2], it is proved that any 4-connected 3-colorable plane graphs is the contact graphs of a set of pseudo-segments. It is also proved in [2] that the assumptions of Theorem 6 hold for the incidence graph of the contact system. Thus, we get:

Corollary 2. *Any 4-connected 3-colorable plane graphs is the contact graphs of a family of segments.*

In [2], representations of planar graphs by intersection of pseudo-segments are obtained using local perturbations of contact systems of pseudo-segments. The assumptions of Theorem 6 are proved to hold for the contact system in [2]. Thus, using Theorem 7 and a perturbation argument, this theorem may be strengthened:

Corollary 3. *Any 4-colored planar graph without induced C_4 using 4 colors is the intersection graph of a family of straight line segments.*

4   Open Problems

It is not difficult to prove that any contact family of pseudo-segments is homeomorphic to a contact family of polylines composed by three segments.

Problem 2. Is any contact family of pseudo-segments homeomorphic to a contact family of polylines composed by two segments?

It is known that every planar graph is representable as the contact graph of a family of triangles[10]. Using stretching techniques, this result might extend:

Problem 3. Is any planar linear hypergraph representable as the contact hypergraph of a family of triangles?

Scheinerman's conjecture may be straightened as follows, as a self-dual statement:

Problem 4. Is any planar linear hypergraph representable as the intersection hypergraph of a family of segments?

As the coloration seems to play a central role, we may also ask:

Problem 5. Is any planar graph G representable as the intersection graph of a family of segments in $\chi(G)$ directions?

References

1. J. Czyziwicz, E. Kranakis, and J. Urrutia, *A simple proof of the representation of bipartite planar graphs as the contact graphs of orthogonal straight line segments*, Information Processing Letters (1998), no. 66, 125–126.

2. N. de Castro, F. J. Cobos, J.C. Dana, and A. Márquez, *Triangle-free planar graphs as segment intersection graphs*, Journal of Graph Algorithms and Applications **6** (2002), no. 1, 7–26.

3. H. de Fraysseix, *Local complementation and interlacement graphs*, Discrete Mathematics **33** (1981), 29–35.

4. _____, *A Characterization of Circle Graphs*, European Journal of Combinatorics **5** (1984), 223–238.

5. H. de Fraysseix and P. Ossona de Mendez, *Intersection Graphs of Jordan Arcs*, Contemporary Trends in Discrete Mathematics, DIMACS Series in Discrete Mathematics and Theoretical Computer Science, vol. 49, DIMATIA-DIMACS, 1999, Štiřin 1997 Proc., pp. 11–28.

6. _____, *On a Characterization of Gauss Codes*, Discrete and Computational Geometry **2** (1999), no. 2.

7. _____, *Barycentric systems and stretchability*, Discrete Applied Mathematics (2004), (submitted).

8. _____, *Stretching of Jordan arc contact systems*, Graph Drawing, Lecture Notes in Computer Science, vol. 2912, Springer, 2004, pp. 71–85.

9. H. de Fraysseix, P. Ossona de Mendez, and J. Pach, *Representation of planar graphs by segments*, Intuitive Geometry **63** (1991), 109–117.

10. H. de Fraysseix, P. Ossona de Mendez, and P. Rosenstiehl, *On triangle contact graphs*, Combinatorics, Probability and Computing **3** (1994), 233–246.

11. G. Ehrlich, S. Even, and R.E. Tarjan, *Intersection graphs of curves in the plane*, Journal of Combinatorial Theory **21(B)** (1986), 8–20.

12. C.F. Gauss, *Werke*, pp. 272 and 282–285, Teubner Leipzig, 1900.

13. I.B.-A. Hartman, I. Newman, and R. Ziv, *On grid intersection graphs*, Discrete Math. **87** (1991), 41–52.

14. Petr Hliněny, *Contact graphs of line segments are NP-complete*, Discrete Mathematics **235** (2001), no. 1-3, 95–106.

15. J. Kratochvíl, *String graphs I : the number of critical nonstring graphs is infinite*, J. Combin. Theory Ser. B **52** (1991), 53–66.

16. _____, *String graphs II : Recognizing string graphs is NP-hard*, J. Combin. Theory Ser. B **52** (1991b).

17. J. Kratochvíl and J. Matoušek, *String graphs requiring exponential representations*, J. Combin. Theory Ser. B **52** (1991), 1–4.

18. _____, *Intersection graphs of segments*, Journal of Combinatorial Theory **62(B)** (1994), no. 2, 289–315.

19. N.E. Mněv, *On manifolds of combinatorial types of projective configurations and convex polyhedra*, Soviet Math. Deokl. (1985), no. 32, 335–337.

20. _____, *The universality theorems on the classification problem of configuration varieties and convex polytopes varieties*, Topology and Geometry – Rohlin Seminar (Berlin) (O. Ya. Viro, ed.), Lecture Notes in Mathematics, vol. 1346, Springer-Verlag, 1988, pp. 527–544.

21. J. Pach and G. Tóth, *Which crossing number is it anyway?*, Journal of Combinatorial Theory, Ser. B **80** (2000), no. 2, 225–246.

22. J. Richter-Gebert, *Realization spaces of polytopes*, Lecture Notes in Mathematics **1643** (1996).
23. P. Rosenstiehl, *A new proof of the Gauss interlace conjecture*, Advances in Applied Mathematics **23** (1999), no. 1, 3–13.
24. M. Schaefer, E. Sedgwick, and D. Štefankovič, *Recognizing string graphs in NP*, Journal of Computer and System Sciences **67** (2003), no. 2, 365–380.
25. E.R. Scheinerman, *Intersection classes and multiple intersection parameters of graphs*, Ph.D. thesis, Princeton University, 1984.
26. P. W. Shor, *Stretchability of pseudolines is NP-hard*, DIMACS Series in Discrete Mathematics and Theoretical Computer Science **4** (1991), 531–554.
27. F.W. Sinden, *Topology of thin RC-circuits*, Bell System Tech. J. (1966), 1639–1662.
28. D. West, *Opem problems*, SIAM J. Discrete Math. Newslett. **2** (1999), no. 1, 10–12.

Dynamic Graph Drawing of Sequences of Orthogonal and Hierarchical Graphs

Carsten Görg[1], Peter Birke[1], Mathias Pohl[1], and Stephan Diehl[2]

[1] Saarland University, FR Informatik, D-66041 Saarbrücken, Germany
goerg@cs.uni-sb.de
[2] Catholic University Eichstätt, Informatik, D-85072 Eichstätt, Germany
diehl@acm.org

Abstract. In this paper we introduce two novel algorithms for drawing sequences of orthogonal and hierarchical graphs while preserving the mental map. Both algorithms can be parameterized to trade layout quality for dynamic stability. In particular, we had to develop new metrics which work upon the intermediate results of layout phases. We discuss some properties of the resulting animations by means of examples.

1   Introduction

In many applications graphs are not drawn once and for all, but change over time. In some cases all changes are even known beforehand, e.g. if we want to visualize the evolution of a social network based on an email archive, or the evolution of program structures stored in software archives. In these kinds of applications each graph can be drawn being fully aware of what graphs will follow. Unfortunately, to the best of our knowledge there exist only two algorithms that take advantage of this knowledge, namely TGRIP [6] and Foresighted Layout [8]. See Section 6 for a discussion of these and other approaches. While the former was restricted to spring embedding, the latter is actually a generic algorithm.

Recently we introduced *Foresighted Layout with Tolerance (FLT)* [7] for drawing sequences of graphs while preserving the mental map and trading layout quality for dynamic stability (tolerance). The algorithm is generic in the sense that it works with different static layout algorithms with related metrics and adjustment strategies. As an example we looked at force-directed layout. In this paper we apply FLT to orthogonal and hierarchical layout, which means that we have to develop adjustment strategies and metrics for these. We also improve FLT by introducing the importance-based backbone as a generalization of the supergraph of a sequence of graphs.

2   Improved Adjusted Foresighted Layout

In our previous work the supergraph, which is the union of all graphs in a graph sequence played a crucial role. The reason for using the supergraph was that it provided all information about the graph sequence and that its layout could be used as a sketch for all graphs in the sequence. However, the supergraph is

J. Pach (Ed.): GD 2004, LNCS 3383, pp. 228–238, 2004.

restrictive, as it induces a layout for all nodes without taking into account that they are of different relevance for the sequence.

To improve that model we now introduce the concept of a backbone of a sequence. Therefore we need a function that defines the importance of a node in the sequence $g_1, \ldots, g_n$. In the following we assume that $g_i = (V_i, E_i)$.

Definition 1 (Backbone). *Given a sequence of graphs* $g_1, \ldots, g_n$, *and a mapping* importance $: V \rightarrow \mathbb{N}$, *then* $V_B = \{v \in \bigcup_{i=1}^n V_i \mid \text{importance}(v) \geq \delta_B\}$ *and* $E_B = \{(u, v) \in \bigcup_{i=1}^n E_i \mid u, v \in V_B\}$ *define the* backbone $B = (V_B, E_B)$ *of a graph sequence* $g_1, \ldots, g_n$ *with respect to a threshold* $\delta_B \in \mathbb{N}$.

This concept of the backbone is a generalization of the concept of the supergraph: The backbone is less restrictive and is adjusted to the given graph sequence. But setting $\delta_B = 0$ will create a backbone that is equal to the supergraph.

Dependent on the choice of the importance function, the backbone represents different base models. There are several possibilities for choosing an importance function: We can define the function depending on the structure of the sequence (for example the number of occurrences of a node in the sequence: importance$(v) = |\{i \mid v \in V_i\}|$ for a graph sequence $g_1, \ldots, g_n$). If we know enough about the semantics of the graphs, we can instead choose an importance function that takes this information into account, i.e. we can use application-domain specific importance functions.

The improved algorithm for foresighted layout that uses the backbone instead of the supergraph now looks as follows:

Algorithm 1 Improved Foresighted Layout with Tolerance.

---
compute global layout L for the backbone B of $g_1, \ldots, g_n$
for $i := 1$ **to** n **do**
   $L_i := L|_{g_i}$
   $l_i := \text{adjust}(\ldots)$
end for
animate graph sequence

In this improved version the global layout does not provide initial layout information for nodes $v \in V_i - V_B$, i.e. those that are not part of the backbone. So the adjustment functions have to assign initial positions to these nodes.

3 Orthogonal Foresighted Layout with Tolerance

Brandes et al. presented in [2] an orthogonal graph drawing algorithm that produced an orthogonal layout with few bends in the Kandinsky model while preserving the general appearance of a given sketch. The angle and the bend changes can be controlled by parameters α and β. In this section, we show how to extend this approach so that it fits in our framework, i.e. it applies the backbone concept and is guided by metrics. We assume that the reader is familiar with the Kandinsky network [2, 13].

First we present the general algorithm. After computing the orthogonal layout for the given backbone and obtaining the corresponding quasi-orthogonal shape, we build a sketch S_i for each graph of the sequence and adjust it through the adjustOrth() algorithm. The sketch is a combination of the previous graph's layout and the backbone's layout restricted to the current graph. If a conflict between the layout of the backbone and the previous graph exists, we choose the one of the backbone.

Algorithm 2 Orthogonal Foresighted Layout with Tolerance

---

compute orthogonal layout L_0 for the backbone B of $g_1, \ldots, g_n$
$Q_0 := $ quasiOrthogonalShape(L_0)
for $i := 1$ **to** n **do**
    $S_i := (L_0 \oplus L_{i-1})|_{g_i}$ // $(L_i \oplus L_j)(x)$ is defined as $L_i(x)$ if $x \in dom(L_i)$ and $L_j(x)$ otherwise.
    $(Q_i, L_i) := $ adjustOrth$(S_i, g_i, L_{i-1}, Q_{i-1})$
end for
animate graph sequence

The adjustOrth() algorithm first computes the extended network of the sketch. Since the sketch was restricted to the current graph g_i, we only have to handle insertions of new nodes and edges. The insertion of a new node creates a new vertex-node in the Kandinsky network. How to insert new edges adjacent to vertex-nodes with a degree greater than 0 is presented in [2]. The insertion of a new edge adjacent to a vertex-node with a degree of 0 does not create a new face-node.

We initialize the locally (for every edge) used parameters α and β. Then we compute the quasi-orthogonal shape as described in [2]. To compare this shape with that of the previous graph, we define a new metrics for quasi-orthogonal shapes. To this end, we extend the definition of a quasi-orthogonal shape given in [2]. With $Q(f, i)$ we denote the i-th tuple of $Q(f)$, with edge(Q, f, i) the value of the edge field, with $a(Q, f, i)$ the value of the angle field, and with $b(Q, f, i)$ the value of the bend field of $Q(f, i)$. The value of the edge field of the successor tuple of $Q(f, i)$ is succEdge$(Q, f, i) = $ edge$(Q, f, (i+1) \bmod |Q(f)|)$.

Definition 2 (Quasi-orthogonal-shape metrics).
Let $\mathcal{Q}$ be the set of quasi-orthogonal shapes. The function diff$_\alpha : \mathcal{Q} \times \mathcal{Q} \to \mathcal{P}(E)$,
    $(Q_1, Q_2) \mapsto \{e = $ edge$(Q_1, f, i) \mid \exists f', j : e = $ edge$(Q_2, f', j) \ \wedge$
    succEdge$(Q_1, f, i) = $ succEdge$(Q_2, f', j) \ \wedge \ a(Q_1, f, i) \neq a(Q_2, f', j)\}$
defines the set of edges with the same successor edge, but with different angles in two quasi-orthogonal shapes. The function diff$_\beta : \mathcal{Q} \times \mathcal{Q} \to \mathcal{P}(E)$,
    $(Q_1, Q_2) \mapsto \{e = $ edge$(Q_1, f, i) \mid \forall f', j \text{ with } e = $ edge$(Q_2, f', j) :$
    $b(Q_1, f, i) \neq b(Q_2, f', j)\}$
defines the set of edges with different bends in two quasi-orthogonal shapes. Then the function Δ_α with $\Delta_\alpha(Q_1, Q_2) = |$diff$_\alpha|$ is called angle metrics and the function Δ_β with $\Delta_\beta(Q_1, Q_2) = |$diff$_\beta|$ is called bend metrics.

If the angle metrics does not fulfill the given angle threshold and there is an α that is lower than the maximal value (the maximal value $6 \cdot |V_i|$ results from the construction of the Kandinsky network and [13]), we increment the corresponding α. We deal analogously with the bend metrics and β. The construction of the modified Kandinsky network implies that incrementing β could lead also to a change of angle between two edges. If angle stability is more important than bend stability, then both β and α have to be incremented if the bend metrics does not fulfill the given bend threshold.

The last step concerns compaction. To be able to preserve the edge length of the sketch S_i, we extend the compaction algorithm from [9] by edges of prescribed length. This extension is done straightforwardly by extending the length function: let $e = (u, v)$ be an edge and (u_x, u_y) the position of u in S_i, then

$$\text{length}'(e) = \begin{cases} |u_x - v_x| + |u_y - v_y|, & \text{if } e \text{ is fixed} \\ \text{length}(e), & \text{otherwise} \end{cases}$$

An edge can be fixed if it is in the current graph as well as in the previous one, and if the values of the corresponding bend fields are equal. We compute the final layout by applying the extended compaction algorithm. If the metrics does not fulfill the given threshold we fix one more edge if there are any left.

Algorithm 3 adjustOrth($S_i, g_i, L_{i-1}, Q_{i-1}$) predecessor dependent

$N_i := $ compute extended network(S_i, g_i)
$\forall e \in E_i : \alpha_e := 0, \beta_e := 0$
repeat
    $Q_i := $ quasiOrthogonalShape(N_i, α, β)
    if $\Delta_\alpha(Q_i, Q_{i-1}) > \delta_\alpha \wedge \exists e \in \text{diff}_\alpha(Q_i, Q_{i-1}) : \alpha_e < 6 \cdot |V_i|$ **then**
        $\forall e \in \text{diff}_\alpha : \text{inc}(\alpha_e)$
    end if
    if $\Delta_\beta(Q_i, Q_{i-1}) > \delta_\beta \wedge \exists e \in \text{diff}_\beta(Q_i, Q_{i-1}) : \beta_e < 6 \cdot |V_i|$ **then**
        $\forall e \in \text{diff}_\beta : \text{inc}(\beta_e)$
    end if
until done
fixedEdges $:= \emptyset$
repeat
    $L_i = $ compact(Q_i, S_i,fixedEdges)
    if $\Delta(L_{i-1}, L_i) > \delta \wedge$ fixedEdges $\subset \{E_i \cap E_{i-1}\} - \{\text{diff}_\beta\}$ **then**
        extend fixedEdges by one edge of $\{E_i \cap E_{i-1}\} - \{\text{diff}_\beta\}$
    end if
until done
return (Q_i, L_i)

So far, we have seen how to apply orthogonal layout to the predecessor layout strategy. But it is also possible to apply it to the simultaneous layout strategy. In this case the backbone layout is used as sketch and we use global parameters α and β instead of local ones to achieve a more uniform adjustment of angles and bends over the whole sequence. The adjustOrth() algorithm first computes the quasi-orthogonal shapes for all graphs. If the condition for the angle metrics $\exists i : \Delta_\alpha(Q_{i-1}, Q_i) > \delta_\alpha \wedge \alpha < 6 \cdot |V_i|$ is not fulfilled, i.e. there is a tuple of successive shapes which do not hold the angle metrics condition and there is some space for improvement, α is increased. Analogously, β is changed depending on the bend metrics. To compute the final layouts we proceed as in the predecessor-dependent layout algorithm.

4   Hierarchical Foresighted Layout with Tolerance

The computation of a hierarchical layout of a graph following the Sugiyama approach needs several phases: First all nodes are distributed in discrete layers (the ranking phase), then the nodes of each layer are arranged, and finally the layout is computed from the layers and their arrangements. One of the problems that occur when trying to apply FLT to hierarchical layout is that there is no option for global layout adjustment such as temperature annealing in the force-directed approach. Instead, we have to divide the adjustment in standard foresighted layout into two different adjustments: an adjustment for the ranking phase and an adjustment for the rank sorting phase. However, after the ranking adjustment has been performed, we cannot apply standard metrics, as the graphs are not fully layouted. Therefore we will introduce a new kind of metrics which only concerns the rankings of two graphs.

4.1   Predecessor Dependent Layout

In this section we describe the two different adjustment steps of hierarchical foresighted layout. Starting from the input sequence, we compute the backbone first. As the nodes of the backbone are of highest importance, we try to preserve the mental map of the graph sequence by fixing these nodes to a certain rank for the entire graph sequence. A good approach is to fix the node to the median of all local rankings, which are computed in advance. So we achieve an optimal rank for at least one graph.

Definition 3 (Average ranking). *A ranking $R : V \to \mathbb{N}$ is a mapping from a node set to the set of natural numbers. Given a sequence of graphs $g_1, \ldots, g_n$ with rankings $R_1, \ldots, R_n$, the average ranking $\bar{R} : V \to \mathbb{N}$ is defined by the median of all $R_i(v)$.*

After that, we compute local rankings for each graph, with respect to the ranking of the backbone. In the second phase, we try to arrange the nodes on each rank, such that we preserve the mental map, but try to reduce the edge crossings at the same time. The general algorithm for hierarchical foresighted layout using the predecessor dependent adjustment is shown in Algorithm 4.

Rank Assignment. In this section we describe how the ranks are adjusted. We compute a new ranking by sorting g_i topologically, but all nodes of the backbone are ranked to their given backbone rank. If the metrics of the rank distance (which we describe below) between the current and the previous ranking exceeds the given threshold δ_R, we fix the rank of one more node to the rank of the previous layout. We choose a node with maximal importance from the node

---

Algorithm 4 Hierarchical Foresighted Layout with Tolerance.

compute backbone B of $g_1, \ldots, g_n$
compute average ranking R_0 of B
for $i := 1$ **to** n **do**
    $R_i^l := R_0|_{g_i}$
    $R_i := \mathsf{adjustRank}(R_i^l, R_{i-1}, g_i)$
    $(l_i, \sigma_i) := \mathsf{adjustOrder}(g_i, R_{i-1}, R_i, \sigma_{i-1}, l_{i-1})$ // with $dom(\sigma_0) = \emptyset$ and $dom(l_0) = \emptyset$
end for
animate graph sequence

Algorithm 5 $\mathsf{adjustRank}(R_i^l, R_{i-1}, g_i)$ predecessor dependent

compute R_i by sorting g_i topologically with respect to R_i^l
repeat
    if $\Delta_R(R_{i-1}, R_i) > \delta_R$ **then**
        add node $v \in \{w \mid w \in (V_i \cap V_{i-1}) - dom(R_i^l)$ and $\forall u \in (V_i \cap V_{i-1}) - dom(R_i^l) :$
        $\mathsf{importance}(w) \geq \mathsf{importance}(u)\}$ to R_i^l and let $R_i^l(v) = R_{i-1}(v)$
        compute R_i by sorting g_i topologically with respect to R_i^l
    end if
until $(V_i \cap V_{i-1}) - dom(R_i^l) = \emptyset \vee \Delta_R(R_{i-1}, R_i) \leq \delta_R$
return R_i

---

set with the following properties: the nodes are contained in the current and previous graph, but not in the backbone. Then we compute a new topological sorting. We repeat this process until the given threshold is no longer exceeded or until all nodes have fixed ranks. In the second case, we stop with a result which has minimal rank distance.

Mental Distance on Ranks. As described in our previous work, we use several metrics to check the mental distance between two layouted graphs. In the layer-assignment phase of hierarchical layout we need a metrics to check the distance between two layer-assignments, but layered graphs do not provide all necessary information for a standard metrics. The only known value is in which layer a node belongs. Therefore we introduce a new kind of metrics for the mental distance, the rank metrics.

Definition 4 (Rank metrics). *Let (g, R) be a graph g with a ranking R. Then the function Δ_R that maps $((g, R), (g', R'))$ to a positive real number is called a rank metrics. In particular, $\Delta_R((g, R,), (g', R')) = 0$ means that g and g' have a non-distinguishable ranking.*

It turns out that there is only a small degree of freedom in the choice of a reasonable rank metrics. A very general approach for such a metrics could be the distance-rank metrics.

Definition 5 (Distance-rank metrics). *Given (g, R) and (g', R'), the function Δ_D with*

$$\Delta_D((g, R), (g', R')) = \sum_{v \in V \cap V'} |R(v) - R'(v)|$$

is called distance-rank metrics.

The definition of the distance-rank metrics could be changed by using the term $(R(v) - R'(v))^2$ instead of $R(v) - R'(v)$. This change would cause the metrics to be more sensitive to nodes that jump over several ranks.

Arrangement of Layers. In this phase we try to minimize edge crossings while staying as close as possible to the predecessor arrangement of layers. Therefore we define an order in each layer.

Definition 6 (Order within ranks). *Given a ranking R of a graph $g = (V, E)$, the function $\sigma : V \to \mathbb{N}$ denotes the order within ranks, if the following property holds: $\forall v, w \in V : R(v) = R(w) \Rightarrow \sigma(v) \neq \sigma(w)$. From the function σ we derive the partial order $<_\sigma$ on nodes: $v <_\sigma w \Leftrightarrow \sigma(v) < \sigma(w)$.*

Algorithm 6 computes an initial order σ_i of nodes which fulfills the following relative orderedness conditions with respect to its predecessor (for $i > 1$):

1. $\forall v, w \in V_i \cap V_{i-1} \wedge R_i(v) = R_{i-1}(v) \wedge R_i(w) = R_{i-1}(w) :$
$v <_{\sigma_i} w \Longleftrightarrow v <_{\sigma_{i-1}} w$
2. $\forall v \in V_i \cap V_{i-1} \wedge R_i(v) \neq R_{i-1}(v) :$
$$\left| \sigma_i(v) - \frac{\sigma_{i-1}(v)}{|\{w | R_{i-1}(w) = R_{i-1}(v)\}|} \cdot |\{w | R_i(w) = R_i(v)\}| \right| \leq 1$$

The first condition states that the relative order of the nodes in the same rank in the current and predecessor graph is preserved. The second condition says that nodes which have changed their rank from the predecessor to the current layout preserve their relative layout position.

Then we compute $\hat{\sigma}_i$ by smoothly sorting the layers of g_i, where $<_{\hat{\sigma}_i}$ restricted to the j-th layer $\{v | R_i(v) = j\}$ forms a total order. As there exists no constraints for σ_1, $\hat{\sigma}_1$ is obtained by sorting the layers of g_1.

The layers of g_i can be sorted either by the barycenter heuristic or the median heuristic (see [1]). Sorting smoothly with respect to sortmax means using an arbitrary comparison-based sorting algorithm[1] where $a \leq b \cdot$ sortmax is used instead of $a \leq b$. Similarly to simulated annealing, we can use linear, logarithmic or exponential decrease of the factor sortmax.

Definition 7 (Final layout). *Given a ranking R and an order of ranks σ of graph g, then $\mathcal{L}(R, \sigma)$ is the final hierarchical layout of g.*

Computing the final layout includes all remaining phases after sorting the ranks and yields the absolute positions of all nodes and edges. Thus we can now check whether the mental map is preserved using some standard metrics. If not, we decrease sortmax and start over.

4.2   Simultaneous Layout

In this section we illustrate how to apply the simultaneous adjustment strategy to hierarchical layout. The predecessor adjustment strategy of the previous section tries to adjust a layout as much as possible with respect to its predecessor.

---

[1] E.g. bucketsort is one of the rare cases that does not belong to this class.

Algorithm 6 adjustOrder$(g_i, R_{i-1}, R_i, \sigma_{i-1}, l_{i-1})$ predecessor dependent.

> sortmax := 1
> $\sigma_i :=$ initialOrder$(\sigma_{i-1}, R_{i-1}, R_i)$
> **repeat**
> $\hat{\sigma}_i :=$ smoothSort$(g_i, \sigma_i, R_i,$ sortmax$)$
> $l_i := \mathcal{L}(R_i, \hat{\sigma}_i)$
>   dec(sortmax)
> **until** $\Delta(l_{i-1}, l_i) \leq \delta \vee$ sortmax < 0
> **return** $(l_i, \hat{\sigma}_i)$

In contrast the simultaneous adjustment strategy provides a uniform adjustment of all graphs. The main problem in applying the simultaneous adjustment strategy to hierarchical layout arises in the rank assignment phase. A possible approach in the rank phase would be to perform a topological sorting on all graphs simultaneously. But this requires that in each iteration one node in each graph is ranked and the mental distance on ranks has to be checked. If the check fails, backtracking has to be performed and the rank of the last node that was ranked has to be fixed. Indeed, this approach is not a good choice for the layer assignment of large graph sequences – in that case it is more efficient to limit the simultaneous adjustment strategy to the layer assignment phase and to use the predecessor dependent rank assignment phase.

The goal of the simultaneous arrangement of layers is to preserve the relative node order in ranks over the whole sequence. Nodes which change their ranks should preserve at least their relative position. To achieve this goal we compute a global enumeration σ^* of the nodes which is consistent throughout the entire graph sequence. Therefore we build the supergraph, layout it using a static hierarchical layout algorithm and after that we retrieve the desired enumeration by projecting the nodes on the x-axis and reading them from left to right.

A local improved enumeration σ' can be derived from σ^* by adjusting the enumeration such that nodes which have changed their rank preserve their relative position (as described in Section 4.1, second relative orderedness condition). Using σ^* and σ' we define $\sigma = (\sigma_1, \ldots, \sigma_n)$:

$$\sigma_i = \begin{cases} \sigma_1^*, \text{ if } i = 1 \\ \sigma_i^*, i > 1 \text{ and } \Delta(\mathcal{L}(R_i, \sigma_i^*), \mathcal{L}(R_{i-1}, \sigma_{i-1})) < \Delta(\mathcal{L}(R_i, \sigma_i'), \mathcal{L}(R_{i-1}, \sigma_{i-1})) \\ \sigma_i', \text{ otherwise} \end{cases}$$

In Algorithm 7, starting with this initial order, we now use the same iteration as in Algorithm 6, except that we use a global sortmax-variable.

5   Examples

In Figure 1 we show snapshots from three different animations of the same graph sequence, which consists of evolving Hesse-graphs. (Hesse-graphs represent divisibility on natural numbers: there is an edge between v and w, if w is divisible by v.) In the graphs 1 to 15 the nodes representing these numbers are inserted

Algorithm 7 adjustOrder$(((g_1, \ldots, g_n), (R_1, \ldots, R_n))$ simultaneous.

---

sortmax := 1
$\sigma^* :=$ initialGlobalOrder$((g_1, \ldots, g_n))$
$\sigma' :=$ initialLocalAdjustedOrder$(\sigma^*, (R_1, \ldots, R_n))$
$\sigma :=$ initialSimultaneousOrder$(\sigma^*, \sigma', (R_1, \ldots, R_n))$
repeat
   for $i := 1$ **to** n **do**
      $\hat{\sigma}_i :=$ smoothSort$(g_i, \sigma_i, R_i, \mathsf{sortmax})$
      $l_i := \mathcal{L}(R_i, \hat{\sigma}_i)$
   end for
   dec(sortmax)
until $\forall i : \Delta(l_{i-1}, l_i) \leq \delta \lor \mathsf{sortmax} < 0$
return $(l_1, \ldots, l_n)$

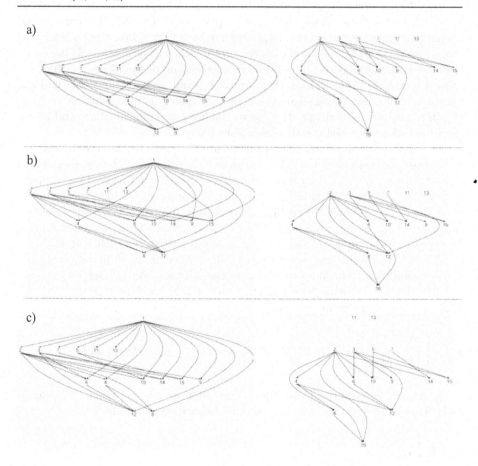

Fig. 1. Layouts of graphs 15 and 16 of evolving Hesse-graph using a) ad-hoc layout, b) FLT with small δ, $\delta_R = 0$ and c) FLT with large δ, $\delta_R = 2$.

successively. In graph 16, node 1 is deleted and node 16 is inserted. In Figure 1a) the ad-hoc approach is shown: for each graph a new layout is computed by using a static layout algorithm. The mental map is poorly preserved as all nodes change their ranks and more than half of the nodes also change their order within the ranks. In Figure 1b) the predecessor dependent layout strategy with $\delta_R = 0$ and a small δ is shown: the mental map is well preserved. No node changes its rank, and the order within the ranks is stable as well. But the local layouts are worse as there are more edge crossings. In Figure 1c) the predecessor dependent layout strategy with $\delta_R = 2$ and a large δ is shown: the left graph is equal to that produced by the ad-hoc approach. But in the next graph, all nodes contained in the backbone do not change their rank. So it is a good compromise between preserving the mental map and achieving local layout quality.

Further examples, e.g. visualization of the evolution of call graphs, are available at `http://www.cs.uni-sb.de/~diehl/ganimation`.

6 Related Work

Most work on dynamic graph drawing [4] is related to the online problem, which means that only information about the previous graphs in a sequence is used for computing a layout. This includes work on hierarchical graph drawing [12], spring embedding [3], and certain kinds of directed graphs [5]. To the best of our knowledge, the only two approaches that consider all graphs in the sequence are TGRIP and Foresighted Layout. TGRIP [6, 10] is an extension of the spring embedder GRIP for large graphs. The basic idea is very intuitive: time is modeled by springs in the third dimension. To this end each graph of the sequence is layouted in a 2D plane. Nodes representing the same vertex in subsequent graphs are connected by additional springs, but each node can only move within the 2D plane to which it belongs. In contrast to Foresighted Layout, this approach does not allow using different mental map metrics, because the metrics is built into the heuristic for minimizing the forces.

7 Conclusions

While implementing FLT for spring embedding was relatively simple, applying the approach to orthogonal and hierarchical layout turned out to require many more changes to the static layout algorithms.

Phased Algorithms. Both algorithms work in phases, and we had to introduce new metrics which work on the results of these phases instead of on the final layouts. When the mental distance of two intermediate results exceeds a given threshold, then we restrict the search space either locally, i.e. for some nodes or edges, or globally, i.e. for all nodes or edges.

Global Restrictions. For spring embedding, the global temperature was reduced to allow fewer position changes of all nodes. Similarly, for hierarchical layout the variable sortmax influences all nodes in the sorting phase.

Local Restrictions. In the ranking phase of the hierarchical layout, we fix the rank of the not yet fixed node of highest importance. Thus, all remaining nodes can still change their ranks. For orthogonal layout the metrics, in fact, also gives a hint what to restrict. As a side-effect of computing the quasi-orthogonal-shape metrics, we do get a set of edges for which we can increment the α and β parameters of one or more of these edges, i.e. restrict the number of angle and bend changes.

Future Work. The theory and implementations of FLT are now at a stage such that we can start to apply them in different domains. The effectiveness of the resulting animations is currently being studied as part of a master thesis in psychology at the Catholic University Eichstätt.

Finally, work is underway to make force-directed, orthogonal and hierarchical FLT available as web services that produce animations in the SVG format.

References

1. O. Bastert and C. Matuszewski. Layered drawings of digraphs. In *Drawing Graphs [11]*. Springer, 2001.
2. U. Brandes, M. Eiglsperger, M. Kaufmann, and D. Wagner. Sketch-Driven Orthogonal Layout. In *Proc. of Graph Drawing 2002*. Springer LNCS 2528:1-11, 2002.
3. U. Brandes and D. Wagner. A Bayesian paradigm for dynamic graph layout. In *Proc. of Graph Drawing 1997*. Springer LNCS 1353:236-247, 1997.
4. J. Branke. Dynamic graph drawing. In *Drawing Graphs [11]*. Springer, 2001.
5. R.F. Cohen, G. Di Battista, R. Tamassia, and I.G. Tollis. Dynamic graph drawings: Trees, series-parallel digraphs, and *st*-digraphs. *SIAM Journal on Computing*, 24(5), 1995.
6. C. Collberg, S. G. Kobourov, J. Nagra, J. Pitts, and K. Wampler. A system for graph-based visualization of the evolution of software. In *Proc. of ACM Symposium on Software Visualization SOFTVIS'03*, San Diego, 2003. ACM SIGGRAPH.
7. S. Diehl and C. Görg. Graphs, They are Changing – Dynamic Graph Drawing for a Sequence of Graphs. In *Proceedings of Graph Drawing 2002*. Springer LNCS 2528:23-30, 2002.
8. S. Diehl, C. Görg, and A. Kerren. Preserving the Mental Map using Foresighted Layout. In *Proceedings of Joint Eurographics – IEEE TCVG Symposium on Visualization VisSym'01*. Springer Verlag, 2001.
9. M. Eiglsperger and M. Kaufmann. Fast Compaction for Orthogonal Drawings with Vertices of Prescribed Size. In *Proceedings of Graph Drawing 2001*. Springer LNCS 2265:124-138, 2002.
10. C. Erten, P. J. Harding, S. G. Kobourov, K. Wampler, and G. Yee. GraphAEL: Graph Animations with Evolving Layouts. In *Proc. of Graph Drawing 2003*. Springer LNCS 2912:98-110, 2003.
11. M. Kaufmann and D. Wagner, editors. *Drawing Graphs – Methods and Models*, volume 2025 of *Lecture Notes in Computer Science*. Springer-Verlag, 2001.
12. S.C. North. Incremental Layout in DynaDAG. In *Proc. of Graph Drawing 1995*. Springer LNCS 1027:409-418, 1996.
13. U.Fößmeier and M. Kaufmann. Drawing high degree graphs with low bend numbers. In *Proceedings of Graph Drawing 1995*. Springer LNCS 1027:254-266, 1996.

Graph Drawing by Stress Majorization

Emden R. Gansner, Yehuda Koren, and Stephen North

AT&T Labs – Research,
Florham Park, NJ 07932
{erg,yehuda,north}@research.att.com

Abstract. One of the most popular graph drawing methods is based on achieving graph-theoretic target distances. This method was used by Kamada and Kawai [15], who formulated it as an energy optimization problem. Their energy is known in the multidimensional scaling (MDS) community as *the stress function*. In this work, we show how to draw graphs by stress majorization, adapting a technique known in the MDS community for more than two decades. It appears that majorization has advantages over the technique of Kamada and Kawai in running time and stability. We also found the majorization-based optimization being essential to a few extensions to the basic energy model. These extensions can improve layout quality and computation speed in practice.

1 Introduction

A graph is a structure $G(V=\{1,\ldots,n\}, E)$ representing a binary relation E over a set of nodes V. Visualizing graphs is a challenging problem, requiring algorithms that faithfully represent the graph's structure and the relative similarities of the nodes [4, 16]. Here we will focus on drawing undirected graphs with straight-line edges.

The most popular approach defines, sometimes implicitly, an energy, or cost function, based on some virtual physical model of the graph. Minimizing this function determines an optimal drawing. In the approach considered here, originally proposed by Kamada and Kawai [15], a nice drawing relates to good isometry. We have an ideal distance d_{ij} given for every pair of nodes i and j, modeled as a spring. Given a 2-D layout, where node i is placed at point X_i, the energy of the system is

$$\sum_{i<j} w_{ij} \left(\|X_i - X_j\| - d_{ij}\right)^2 . \tag{1}$$

We desire a layout that will minimize this function, thereby best approximating the target distances. Here, the distance d_{ij} is typically the graph-theoretical distance between nodes i and j. The normalization constant w_{ij} equals $d_{ij}^{-\alpha}$. Kamada and Kawai [15] chose $\alpha = 2$, whereas Cohen [6] also considered $\alpha = 0$ and $\alpha = 1$. Moreover, Cohen suggested setting d_{ij} to the linear-network distance to convey the clustering structure of the graph.

The function (1), with $\alpha = 0$, appeared earlier as the stress function in multidimensional scaling (MDS) [5,6,18], where it was applied to graph drawing [17]. Whereas Kamada and Kawai proposed a localized 2-D Newton-Raphson process for minimizing the stress function, researchers in the MDS field have proposed a different, more

J. Pach (Ed.): GD 2004, LNCS 3383, pp. 239–250, 2004.

global approach called *majorization*. Majorization seems to offer some distinct advantages over localized processes like Newton-Raphson or gradient descent. These include guaranteed monotonic decrease of the energy value, improved robustness against local minima and shorter running times. The main contribution of this work is the introduction of this technique in the framework of graph layout.

Three useful extensions to stress optimization require the power and flexibility of majorization optimization. The first extension, described in Section 3, deals with weighting edge lengths in a way that better utilizes the drawing area, and is especially useful for drawing real-life graphs whose degree distribution follows a power law. We have found empirically that traditional stress optimization is unstable under such a weighting, while majorization works very well. The second extension deals with sparse stress functions, where only a small fraction of all pairwise distances are considered. This is essential for reducing the time and space complexity of stress optimization, and allows in-core layout of much larger graphs. We have found that sparse stress optimization is practically impossible when using the Kamada-Kawai technique (unless one has a very good initialization). Again, with majorization, it is easy to work with sparse models.

The last extension deals with obtaining an approximate drawing of the graph by constraining the layout axes to lie within a carefully selected small vector space. Such a technique was recently introduced by Koren [14] and can be integrated into layout algorithms based on matrix algebra. Fortunately, the algebraic nature of the majorization process allows us to perform rapid subspace-restricted stress minimization. The two latter extensions are described in the full version of this work.

2   Stress Majorization

In this section, we review stress majorization as described in the MDS literature [3, 5]. We denote a d-dimensional layout by an $n \times d$ matrix X. Thus, node i is located at $X_i \in \mathbb{R}^d$ and the axes of the layout are $X^{(1)}, \ldots, X^{(d)} \in \mathbb{R}^n$. The associated stress function is

$$\text{stress}(X) = \sum_{i<j} w_{ij} \left(\|X_i - X_j\| - d_{ij} \right)^2 . \tag{2}$$

We always take $w_{ij} = d_{ij}^{-2}$, which seems to produce the best drawings in most cases. Decompose (2) to obtain

$$\text{stress}(X) = \sum_{i<j} w_{ij} d_{ij}^2 + \sum_{i<j} w_{ij} \|X_i - X_j\|^2 - 2 \sum_{i<j} \delta_{ij} \|X_i - X_j\|, \tag{3}$$

where $\delta_{ij} \stackrel{\text{def}}{=} w_{ij} d_{ij}$ for $i, j = 1, \ldots, n$.

The first term of (3), $\sum_{i<j} w_{ij} d_{ij}^2$, is a constant independent of the current layout. The second term, $\sum_{i<j} w_{ij} \|X_i - X_j\|^2$, is a quadratic sum, and can be written using the quadratic form of the weighted Laplacian L^w

$$\sum_{i<j} w_{ij} \|X_i - X_j\|^2 = \text{Tr}(X^T L^w X), \tag{4}$$

where the $n \times n$ weighted Laplacian has its ij entry, for $i, j = 1, \ldots, n$, defined as

$$L_{i,j}^w = \begin{cases} -w_{ij} & i \neq j \\ \sum_{k \neq i} w_{ik} & i = j \end{cases}.$$

The third term, $\sum_{i<j} \delta_{ij} \|X_i - X_j\|$, is more involved and we will bound it from below. We will make use of the Cauchy-Schwartz inequality

$$\|x\| \|y\| \geqslant x^T y$$

with equality when $x = y$. Consequently, given any $n \times d$ matrix Z,

$$\|X_i - X_j\| \|Z_i - Z_j\| \geqslant (X_i - X_j)^T (X_i - Z_j)$$

with equality when $X = Z$. We can now bound the third term as follows

$$\sum_{i<j} \delta_{ij} \|X_i - X_j\| \geqslant \sum_{i<j} \delta_{ij} \mathrm{inv}(\|Z_i - Z_j\|)(X_i - X_j)^T (Z_i - Z_j) \qquad (5)$$

where $\mathrm{inv}(x) = 1/x$ when $x \neq 0$ and 0 otherwise.

Inequality (5) can be written in a more convenient matrix form

$$\sum_{i<j} \delta_{ij} \|X_i - X_j\| \geqslant \mathrm{Tr}(X^T L^Z Z),$$

where the $n \times n$ matrix L^Z has its ij entry, for $i, j = 1, \ldots, n$, defined as

$$L_{i,j}^Z = \begin{cases} -\delta_{ij} \mathrm{inv}(\|Z_i - Z_j\|) & i \neq j \\ -\sum_{j \neq i} L_{i,j}^Z & i = j \end{cases}.$$

Combining all the above, we can bound the stress function using $F^Z(X)$ defined as

$$F^Z(X) = \sum_{i<j} w_{ij} d_{ij}^2 + \mathrm{Tr}(X^T L^w X) - 2\mathrm{Tr}(X^T L^Z Z). \qquad (6)$$

Thus, we have

$$\mathrm{stress}(X) \leqslant F^Z(X) \qquad (7)$$

with equality when $Z = X$.

Note that Z is a constant $n \times d$ matrix. This way we have bounded the stress with a quadratic form $F^Z(X)$. We differentiate by X and find that the minima of $F^Z(X)$ are given by solving

$$L^w X = L^Z Z.$$

Or, equivalently, for each axis we have to solve

$$L^w X^{(a)} = L^Z Z^{(a)}, \quad a = 1, \ldots, d. \qquad (8)$$

The characteristic of the minima is determined by the nature of the weighted Laplacian L^w, which is known to be positive semi-definite with a one-dimensional null space spanned by $1_n = (1, \ldots, 1) \in \mathbb{R}^n$. Hence, $F^Z(X)$ has only global minima, which are invariant under translation (addition of $\alpha \cdot 1_n$ is equivalent to translation). This makes sense, since the stress function is also invariant under translation.

Numerically, it is better to make the minimizer unique. Hence we recommend removing the translation degree-of-freedom by taking $X_1 = 0$. Therefore, we can remove the first row and column of L^w, as well as the first row of $L^Z Z$. The resulting $(n-1) \times (n-1)$ matrix is strictly diagonal dominant and hence positive definite. This is very convenient, since methods like conjugate gradient, Gauss-Seidel, and Cholesky factorization are guaranteed to work [9].

The Optimization Process

The above formulation leads to the following iterative optimization process. Given some layout $X(t)$, we want to compute a layout $X(t + 1)$ so that stress($X(t + 1)$) < stress($X(t)$). We use the function $F^{X(t)}(X)$ which satisfies $F^{X(t)}(X(t)) =$ stress($X(t)$).

We take $X(t + 1)$ as the minimizer of $F^{X(t)}(X)$ by solving

$$L^w X(t + 1)^{(a)} = L^{X(t)} X(t)^{(a)}, \quad a = 1, \ldots, d. \tag{9}$$

At this point, if $X(t + 1) = X(t)$, we terminate the process. Otherwise, we get

$$\text{stress}(X(t + 1)) \leqslant F^{X(t)}(X(t + 1)) < F^{X(t)}(X(t)) = \text{stress}(X(t)).$$

The first inequality is by (7) and the second inequality is by the uniqueness of the minimum.

In practice we terminate the process when

$$\frac{\text{stress}(X(t)) - \text{stress}(X(t + 1))}{\text{stress}(X(t))} < \epsilon, \tag{10}$$

where ϵ is the tolerance of the process. Typically, $\epsilon \sim 10^{-4}$.

To summarize, the majorization process involves iteratively solving (9). The matrix L^w is constant throughout the entire process, whereas the matrix $L^{X(t)}$ would be recomputed at each iteration.

2.1 Equation Solvers

In practice we recommend using either Cholesky factorization or conjugate gradient (CG) [9] to solve (9) (by first fixing $X_1 = 0$ as discussed above). Using Cholesky factorization implies that at a preprocessing stage we find the LL^T factorization of L^w using $n^3/3$ flops (floating point operations). Then in each iteration we solve the linear system using back substitution in time $O(n^2)$. Hence, the significant cost in Cholesky factorization is independent of the number of iterations, making it is suitable for graphs requiring many iterations of process (9).

On the other hand, CG optimization involves no preprocessing and its running time is evenly distributed among the iterations. Almost the entire solving time is devoted to performing matrix-vector multiplication. Each such multiplication takes n^2 flops. Thus, if the total number of matrix multiplications is less than about $n/3$, the CG process is expected to be faster than Cholesky factorization. Otherwise, Cholesky factorization is recommended. In practice, for most graphs we have experimented with, CG outperformed Cholesky since the total number of matrix-vector multiplications is typically less than $n/3$. Note that CG benefits by the fact that we have an initial approximate solution from the previous iteration. We observed that the overall number of iterations increases very moderately with the size of the graph. Therefore, for large graphs (over 10,000 nodes), we encountered cases where the total number of matrix-vector multiplications exceeded even n, so Cholesky factorization should do much better. In any case, all the results reported here employ CG.

2.2 Intuitive Interpretation

Let us concentrate on axis a, and denote the current coordinates by $\hat{x} = X(t)^{(a)}$. The majorization process determines the new coordinates $x = X(t+1)^{(a)}$ by solving the system of equations (9). Eliminating x_i in equation i, we rewrite the system in an equivalent form

$$x_i = \frac{\sum_{j \neq i} w_{ij} \left(x_j + d_{ij}(\hat{x}_i - \hat{x}_j)\mathrm{inv}(\|X(t)_i - X(t)_j\|) \right)}{\sum_{j \neq i} w_{ij}}. \tag{11}$$

The intuitive interpretation of this process is simple. A node j located at x_j strives to place node i (on current axis a) at $x_j + d_{ij} \frac{\hat{x}_i - \hat{x}_j}{\|X(t)_i - X(t)_j\|}$.

Based on the current placement, this is node j's best strategy to assure that node i will be at distance d_{ij} from j in the full multidimensional layout. To see this, notice that the distance between the nodes depends on all the axes. Therefore, node j's estimate of the contribution of axis a for the distance between i and j is the fraction $\alpha = |\frac{\hat{x}_i - \hat{x}_j}{\|X(t)_i - X(t)_j\|}|$. So the magnitude of displacement should be d_{ij} scaled down by α. Now, after deciding the magnitude of the 1-D displacement, the direction must be decided: should we place x_i at $x_j + \alpha d_{ij}$ or at $x_j - \alpha d_{ij}$? Again, the decision is based on the current placement, whether currently $\hat{x}_i < \hat{x}_j$ or vice versa.

This way, each node j votes for its desired placement of x_i. The final position is determined by taking the weighted average of the suggested positions. This intuition also suggests a localized optimization process, which we next describe.

2.3 Localized Optimization

Following the idea of Kamada and Kawai [15], we can fix the positions of all nodes, except some node i. Then, by the same argument given above for the full majorization process, it can be shown that the stress function is decreased by setting the position of i as follows

$$X_i^{(a)} \leftarrow \frac{\sum_{j \neq i} w_{ij} \left(X_j^{(a)} + d_{ij}(X_i^{(a)} - X_j^{(a)})\mathrm{inv}(\|X_i - X_j\|) \right)}{\sum_{j \neq i} w_{ij}}, \quad a = 1, \ldots, d. \tag{12}$$

This way we can iterate through all nodes, and in each iteration relocate all the d coordinates of node i according to (12). Each iteration is guaranteed to strictly decrease the stress until convergence. Hence, oscillations and non-convergence are impossible.

In practice, we have only used the more involved global process (9) and have no experience yet with the local version. We provide this local version here mainly because it is simple and easy to implement, requiring no equation solver[1].

2.4   Comparisons

A natural question is whether we should replace the traditional Kamada-Kawai based optimization with majorization. Based on several months of experimenting with both approaches, our definite answer is yes. We base this recommendation on several considerations.

We experimented with various example graphs. On each graph, we ran each of the two algorithms 25 times with different random initializations. At certain times during each execution, we measured the elapsed running time and the current value of the stress function, and averaged over all 25 executions. From this we obtained stress-vs.-time charts for the graphs. While it is impossible to present here all of the charts, we show a few representative ones in Figures 1-3. We can make some important observations.

Layout Quality. We observed that most of the time, the two methods eventually achieved about the same stress level. In certain cases, the Kamada-Kawai approach would yield a slightly better layout in terms of the stress value, but the difference was always small; see Figure 2. In other cases, however, the majorization approach yielded significantly better layouts as can be seen in Figure 3. Hence, probably due to its more global nature, majorization can be considered better in terms of layout quality.

Monotonicity of Convergence. A significant advantage of majorization is that iterations monotonically decrease the stress until convergence. This way, termination of the process is determined naturally by a condition like (10). However, our experience with the Kamada-Kawai approach, as implemented in Neato [7], shows that in some cases the latter process may cycle without converging, while the energy is oscillating. This requires an artificial or more convoluted termination condition.

Our experiments show that, as expected, the majorization approach was always monotonic in decreasing the stress value. The non-monotonicity problem of the Kamada-Kawai method was extremely rare (remember that we averaged over 25 executions, lessening the impact of a single bad non-monotonic execution). We did observe this non-monotonic behavior when experimenting with the Qh882 graph [1]. The result is provided in Fig. 1, which compares the average behavior of both approaches on this graph. We should note that here we weighted edges as explained in Section 3. The reader can see that after 2 seconds of running, the stress value in the Kamada-Kawai approach increases for some period. Here, this did not prevent it from converging at about the same stress level as the majorization process.

---

[1] Process (12) should not be confused with the similar Gauss-Seidel process that can be used to solve (9).

Running Time. The running time of the majorization process is consistently less than that of the Kamada-Kawai process. In all runs, it can be observed that majorization reaches the low stress level much before Kamada-Kawai.

A partial explanation is that majorization's running time is dominated by matrix operations (matrix-vector multiplication or Cholesky factorization). These operations are implemented in libraries like BLAS and LAPACK which are highly optimized on the machine instruction level for common platforms. We are using the Intel Math Kernel Library [22]; another well-known implementation is Atlas [23].

For implementations not relying on special matrix software, we found the situation to be similar to that of the stress function. Sometimes the Kamada-Kawai approach would be marginally faster; on the other hand, when the majorization process was faster, it was significantly faster. And as the size of the graphs increased, the advantage swung completely to majorization.

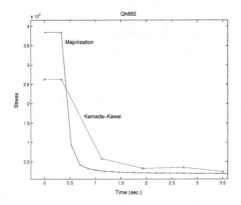

Fig. 1. Stress function vs. running time for the graph Qh882 [1] ($|V|$=882, $|E|$=1533). Here both methods reached about the same stress. Interestingly, Kamada-Kawai is not monotonic.

Before leaving this topic, we must point out that our implementation of the Kamada and Kawai process on which we based our comparisons differs slightly from the implementation originally suggested [15]. We are using the more common implementation which replaces the two nested loops with a single loop; see [2, 11]. As noted in Brandenburg, Himsolt, and Rohrer [2], this leads to a significant speed-up over the original implementation. This more efficient implementation is also the one used in Neato [7] and GraphLet [21].

3   Weighting Edge Lengths

In many real life graphs, the degree distribution decays at a much lower rate than in random graphs. Usually this distribution follows a power law and is proportional to $d^{-\lambda}$. Setting desired edge lengths to a uniform length (typically 1) inevitably makes the neighborhood of high degree nodes too dense in the layout. Consequently, we suggest weighting edges by their neighborhood size.

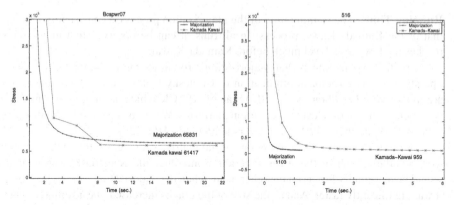

Fig. 2. Stress function vs. running time for the graphs Bcspwr07 [1] ($|V|$=882, $|E|$=1533) and 516 [19] ($|V|$=516, $|E|$=729).

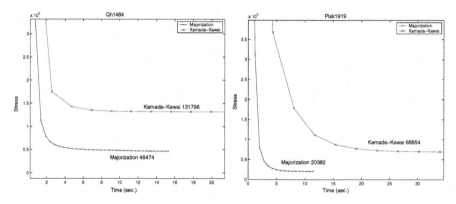

Fig. 3. Stress function vs. running time for the graphs Qh1484 [1] ($|V|$=1470, $|E|$=6420) and Plsk1919 [1] ($|V|$=1919, $|E|$=4831).

Specifically, we set the length of each edge $\langle i, j \rangle \in E$ as

$$l_{ij} = |N_i \cup N_j| - |N_i \cap N_j|, \qquad (13)$$

where $N_i = \{j | \langle i, j \rangle \in E\}$. Then, each target distance d_{ij} is the length of the shortest weighted path between i and j.

This simple change is surprisingly effective in many real life irregular graphs that have highly non-uniform degree distributions. We present here two examples. The first example is the 1138Bus graph ($|V|$=1138, $|E|$=1458) from the Matrix Market repository [1]. This graph models a network of high-voltage power distribution lines. Figure 4 shows two layouts of this graph. In one layout, edges were weighted according to (13). The other layout was made with unweighted edges. Nodes are much better dispersed in the weighted-edge-based layout. By weighting edges, more space is allocated to the dense areas, avoiding many of the edge crossings.

Another interesting example is a BGP connectivity graph representing communications between autonomous systems ($|V|$=3847, $|E|$=11539). This graph has a few nodes

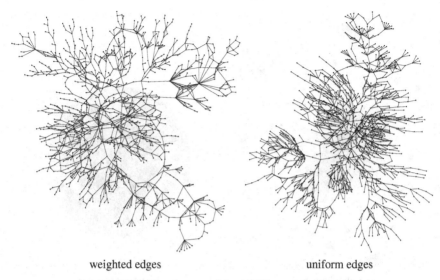

weighted edges   uniform edges

Fig. 4. Two layouts of the 1138Bus graph [1].

of high degree (e.g., one node has degree 695 and a few others are around 100), as well as 3257 nodes of degree 1. We show two layouts of this graph in Figure 5. Again, it is clear that when weighting edges, the resulting layout is much more informative. For example, in both layouts the central node is the one of degree 695. In the weighted version, its neighborhood is placed far enough from it to make it fairly visible. In the unweighted version, however, all of its neighbors are positioned densely around it, hiding its structure completely.

We have frequently found that when there are large deviations in edge lengths, as in the BGP graph, classic Kamada-Kawai optimization fails to find a nice layout. The result of Kamada-Kawai optimization on the edge-weighted BGP graph is shown in Figure 6(a). It is clearly inferior to the majorization result shown in Figure 5. We also compare the average stress-vs.-time behavior of the two methods in Figure 6(b), where it is clear the Kamada-Kawai-type optimization is pretty helpless here. Although we do not fully understand this limitation of Kamada-Kawai optimization, it seems that its local nature somehow limits its ability to deal with significantly unbalanced edge lengths.

4   Related Work

Substantial work in statistical MDS deals with the properties of the majorization process, including proofs of its convergence rate [3]. The MDS literature suggests solving equation (9) by computing $(L^w)^+$, the Moore-Penrose inverse of the singular matrix L^w. Our suggestion to set $X_1 = 0$ allows a much faster solution by Cholesky factorization.

Several studies in the graph drawing field suggest improving stress computation by multi-scale extensions [8, 10, 11], which approximate the graph by a smaller one, to quickly obtain an initial layout. We see these approaches as complementary to our

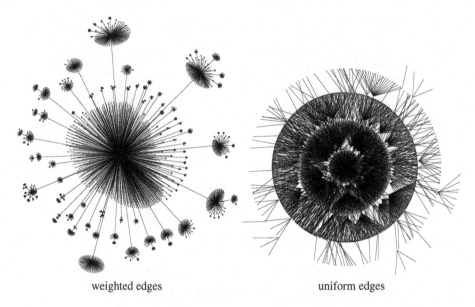

weighted edges   uniform edges

Fig. 5. Two majorization-based layouts of BGP connectivity, with a skewed degree distribution.

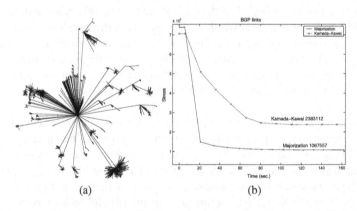

(a)   (b)

Fig. 6. (a) Layout of the edge weighted BGP connectivity graph using Kamada-Kawai optimization. **(b)** Stress-vs.-time behavior of majorization and Kamada-Kawai on weighted BGP connectivity example graph.

proposal, as one can apply majorization to optimizing the stress at each scale. In general, our recommendation is to get an initial placement either by multi-scale techniques or by subspace-restricted computation [14].

Recent work by Koren and Harel [13] describes an algorithm for monotonically decreasing the stress function in 1-D, and a heuristic extension to higher dimensions whose convergence properties are unknown. It is easy to prove that this 1-D algorithm is equivalent to 1-D majorization, although derived differently. Majorization, however, is more powerful as it can be generalized to higher dimensions. Interestingly, the optimization process of [13] is equivalent to the full, n-D Newton-Raphson process. Accordingly, we conclude that in 1-D, the majorization process is equivalent to the full, n-D Newton-Raphson process. This is unlike the Kamada-Kawai process which is based on a localized 2-D Newton-Raphson process.

5 Conclusions

Majorization, a technique developed in studies of statistical MDS, is relevant to practical graph drawing. The MDS community has studied it extensively from the standpoint of optimizing the stress function and escaping local minima. Further ideas along these lines may also prove useful in graph drawing.

The main algorithms discussed here are available in the Neato program in the Graphviz open source package [20].

References

1. R. F. Boisvert et al., "The Matrix Market: A web resource for test matrix collections", in *Quality of Numerical Software, Assessment and Enhancement*, R. F. Boisvert, ed., Chapman Hall, 1997, pp. 125–137. math.nist.gov/MatrixMarket
2. F.J. Brandenburg, M. Himsolt and C. Rohrer, "An Experimental Comparison of Force-Directed and Randomized Graph Drawing Algorithms", *Proceedings of Graph Drawing '95*, LNCS 1027, pp. 76–87, Springer Verlag, 1995.
3. J. De Leeuw, "Convergence of the Majorization Method for Multidimensional Scaling", *Journal of Classification* **5** (1988), pp. 163–180.
4. G. Di Battista, P. Eades, R. Tamassia and I. G. Tollis, *Graph Drawing: Algorithms for the Visualization of Graphs*, Prentice-Hall, 1999.
5. I. Borg and P. Groenen, *Modern Multidimensional Scaling: Theory and Applications*, Springer-Verlag, 1997.
6. J. D. Cohen, "Drawing Graphs to Convey Proximity: an Incremental Arrangement Method", *ACM Transactions on Computer-Human Interaction* **4** (1997), pp. 197–229 .
7. E. R. Gansner and S. C. North, "Improved force-directed layouts", *Proceedings of Graph Drawing '98*, LNCS 1547, pp. 364–373, Springer-Verlag, 1998.
8. P. Gajer, M. T. Goodrich and S. G. Kobourov, "A Multi-dimensional Approach to Force-Directed Layouts of Large Graphs", *Proceedings of Graph Drawing 2000*, LNCS 1984, pp. 211–221, Springer-Verlag, 2000.
9. G. H. Golub and C. F. Van Loan, *Matrix Computations*, Johns Hopkins University Press, 1996.
10. R. Hadany and D. Harel, "A Multi-Scale Method for Drawing Graphs Nicely", *Discrete Applied Mathematics* **113** (2001), pp. 3-21.
11. D. Harel and Y. Koren, "A Fast Multi-Scale Method for Drawing Large Graphs", *Journal of Graph Algorithms and Applications* **6** (2002), pp. 179–202.
12. D. Harel and Y. Koren, "Graph Drawing by High-Dimensional Embedding", *Proceedings of Graph Drawing 2002*, LNCS 2528, pp. 207–219, Springer-Verlag, 2002.
13. Y. Koren and D. Harel, "Axis-by-Axis Stress Minimization", *Proceedings of Graph Drawing 2003*, Springer-Verlag, pp. 450–459, 2003.
14. Y. Koren, "Graph Drawing by Subspace Optimization", *Proceedings 6th Joint Eurographics – IEEE TCVG Symposium Visualization (VisSym '04)*, pp. 65–74, Eurographics, 2004.
15. T. Kamada and S. Kawai, "An Algorithm for Drawing General Undirected Graphs", *Information Processing Letters* **31** (1989), pp. 7–15.

16. M. Kaufmann and D. Wagner (Eds.), *Drawing Graphs: Methods and Models*, LNCS 2025, Springer-Verlag, 2001.
17. J. Kruskal and J. Seery, "Designing network diagrams", *Proceedings First General Conference on Social Graphics* (1980), pp. 22–50.
18. J. W. Sammon, "A Nonlinear Mapping for Data Structure Analysis", *IEEE Trans. on Computers* **18** (1969), pp. 401–409.
19. C. Walshaw, "A Multilevel Algorithm for Force-Directed Graph Drawing", *Proceedings 8th Graph Drawing (GD'00)*, LNCS 1984, pp. 171–182, Springer-Verlag, 2000.
20. Graphviz. www.research.att.com/sw/tools/graphviz/
21. Graphlet. www.infosun.fmi.uni-passau.de/Graphlet/
22. Intel Math Kernel Library. www.intel.com/software/products/mkl/
23. Automatically Tuned Linear Algebra Software (ATLAS).
    math-atlas.sourceforge.net/

Computing Radial Drawings
on the Minimum Number of Circles*

Emilio Di Giacomo[1], Walter Didimo[1], Giuseppe Liotta[1], and Henk Meijer[2]

[1] Università degli Studi di Perugia, Perugia, Italy
{digiacomo,didimo,liotta}@diei.unipg.it
[2] School of Computing, Queen's University, Kingston, Ontario, Canada
henk@cs.queensu.ca

Abstract. A radial drawing is a representation of a graph in which the vertices lie on concentric circles of finite radius. In this paper we study the problem of computing radial drawings of planar graphs by using the minimum number of concentric circles. We assume that the edges are drawn as straight-line segments and that co-circular vertices can be adjacent. It is proven that the problem can be solved in polynomial time.

1 Introduction

A radial drawing is a representation of a graph in which the vertices are constrained to lie on concentric circles of finite radius. Drawing graphs radially is relevant in situations where it is important to display a graph with the constraint that some vertices are drawn "more central" than others. Examples of such applications include social networks analysis (visualization of policy networks and co-citation graphs), operating systems (visualization of filesystems), cybergeography (visualization of Web maps and communities), and bioinformatics (visualization of protein-protein interaction diagrams); see e.g. [4, 8, 9].

This paper investigates crossing-free radial drawings of planar graphs. Let G be a planar graph. A crossing-free radial drawing of G induces a partition of its vertices into levels such that vertices in the same level are co-circular in the drawing; for each level, the planarity of the drawing induces a circular ordering of the vertices in the level. Conversely, in order to construct a radial drawing of G a partition of its vertices into levels and a circular ordering within each level must be found such that vertices of the same level are drawn co-circularly and the edges can be drawn without intersecting each other.

Bachmaier et al. [1, 2] investigate the radial planarity testing problem: Given a partition of the vertices of G into levels, they want to test whether there exists a crossing-free radial drawing of G consistent with the given leveling (i.e. vertices in the same level can be drawn on the same circle and the edges can be added without crossing). In [1] it is assumed that the edges are drawn as strictly monotone curves from inner to outer circles and that no two co-circular

---

* Research supported in part by "Progetto ALINWEB: Algoritmica per Internet e per il Web", MIUR Programmi di Ricerca Scientifica di Rilevante Interesse Nazionale, and by NSERC.

J. Pach (Ed.): GD 2004, LNCS 3383, pp. 251–261, 2004.

vertices are connected by an edge. The elegant linear-time algorithm presented by Bachmaier et al. tests radial planarity by using an extension of PQ-trees, called PQR-trees. In [2] the authors extend the algorithm to the case where edges between co-circular vertices are allowed.

In this paper we study radial drawings of planar graphs from a different perspective. We assume that the partition of the vertices of G is not given and our goal is to compute a partition that minimizes the number of levels, i.e. that corresponds to a crossing-free straight-line radial drawing of G on the minimum number of circles. We call such a drawing a *minimum radial drawing* of G. In contrast with the drawing conventions adopted in [1], we assume that the edges are straight-line segments and that vertices on the same level can be adjacent. These choices are justified by different application-oriented examples of radial drawings that adopt the straight-line standard (see e.g. [12, 13]) and by the observation that allowing edges among co-circular vertices appears to be a natural approach for the reduction of the number of levels.

The contribution of the present paper is to characterize those graphs that can be drawn on a given number of concentric circles and to use this characterization to solve the above described optimization problem. More precisely:

- We show that every 2-outerplanar graph admits a crossing-free straight-line radial drawing on two circles. The proof is constructive and the radial drawing can be computed in linear time. Preliminary results on computing radial drawing of 2-outerplanar graphs appear in [6].
- We generalize this results and characterize the family of graphs that admit a crossing-free straight-line radial drawing on at most $k \geq 2$ circles. We recall that similar characterization problems for straight-line k-layered drawings are studied for the case of $k \leq 3$; see, e.g. [5]. We also recall that a planar graph admits a drawing on one circle if each edge can bend at most once [7].
- Based on the characterization above, we show that there exists a polynomial time algorithm to compute a minimum radial drawing of a planar graph. The drawing has the additional property of being "proper", i.e. an edge always connects either co-circular vertices or vertices on consecutive circles.

For reasons of space some proofs are sketched or omitted.

2   Preliminaries

A 1-*outerplanar embedded graph* (also called *1-outerplane graph*) is an embedded planar graph where all vertices are on the external face. An embedded graph is a k-*outerplanar embedded graph* (also called k-*outerplane graph*) $(k > 1)$ if the embedded graph obtained by removing all vertices of the external face is a $(k-1)$-outerplane graph. The planar embedding of a k-outerplane graph is called a k-*outerplanar embedding*. A graph is k-*outerplanar* if it admits a k-outerplanar embedding. A planar graph G has *outerplanarity* k (for an integer $k > 0$) if it is k-outerplanar and it is not j-outerplanar for $0 < j < k$. In other words, the outerplanarity of G denotes the minimum value of k for which G is k-outerplanar.

Let G be a k-outerplane graph with $k > 1$. We associate a *level* with each vertex v of G, denoted as $lev(v)$, according to the following definition: $lev(v) = 0$ if v is on the external face of G and $lev(v) = i$ ($i = 1, \ldots, k-1$) if v is on the external face after the removal of every vertex u with $lev(u) < i$. If $lev(v) = i$, we say that v is a vertex of level i. Let V_i be the set of vertices v with $lev(v) = i$. The subgraph induced by V_i is denoted by $G_i = (V_i, E_i)$. Notice that G_i is a graph of outerplanarity 1. Let $V_{i,i+1} = V_i \cup V_{i+1}$. The subgraph induced by $V_{i,i+1}$ is denoted by $G_{i,i+1} = (V_{i,i+1}, E_{i,i+1})$.

We use $C_0, C_1, \ldots, C_{k-1}$ to denote a set of k concentric circles in the plane, where the radius of C_i is greater than the radius of C_{i+1} ($i = 0, \ldots, k-2$). Let G be a planar graph and let Γ be a crossing-free straight-line drawing of G. The drawing Γ is a *radial drawing* if the vertices of G are placed on a set of concentric circles. Γ will be called a *k-radial drawing* of G if it is a radial drawing on $C_0, C_1, \ldots, C_{k-1}$. Γ is a *minimum radial drawing* if it uses the minimum number of circles. An edge (u, v) with u and v on C_i is called an *intra-level edge*. An edge (u, v) with u and v on on C_i and C_j with $i \neq j$ is called an *inter-level edge*. If all inter-level edges of a radial drawing Γ connect vertices on consecutive circles, Γ is called a *proper radial drawing*.

Let G be a k-outerplane graph. A radial drawing of G is *level-preserving* if it is a k-radial drawing and every vertex v with $lev(v) = i$ is drawn on circle C_i. A level-preserving k-radial drawing of a k-outerplane graph is proper.

3   Overview of the Approach

We study the problem of computing a radial drawing of a planar graph G on the minimum number of circles. We show that a minimum radial drawing of G can be computed in polynomial time. Namely: (a) We prove that if a graph has outerplanarity k then it admits a k-radial drawing; Also if a graph has a radial drawing on k-circles then it has outerplanarity at most k. (b) We use the above characterization and a result by Bienstock and Monma [3] to show that there exists an $O(n^5 \log n)$-time algorithm that computes a minimum radial drawing of G.

The trickiest part is to show that a graph with outerplanarity k has a k-radial drawing. We provide a linear-time algorithm that receives as input a k-outerplane graph G and computes a level-preserving k-radial drawing of G. Our approach can be summarised as follows. We start with G_0, draw the vertices in V_0 on C_0 while maintaining their circular ordering in G_0. After placing V_i on C_i we compute the radius of C_{i+1} and draw V_{i+1} on C_{i+1} without moving any vertex from V_j with $0 \leq j \leq i$. For ease of presentation, we will define *canonical k-outerplanar graphs* and show how each k-outerplane graph can be transformed into a canonical k-outerplane graph. We will also show that a k-outerplane graph has a k-radial drawing if and only if its canonical form has a k-radial drawing.

4   Canonical Graphs and Equipped $\mathcal{BC}$-Trees

Let G be a k-outerplane graph. A *mixed face* of G is a face containing vertices of two consecutive levels. G is called *inter-triangulated* if all its mixed faces are three-cycles. Assume G is inter-triangulated. Let c be a cut-vertex of G_{i+1}. Let B and B' be two blocks (i.e. biconnected components) of G_{i+1} that are consecutive when going around c in clockwise direction.

Since G is inter-triangulated, there exists at least one edge of $E_{i,i+1}$ incident on c that is encountered between B and B' when going around c in the clockwise direction. Such an edge of $E_{i,i+1}$ is called a *separating edge* because it separates blocks B and B' around c. G is said to be *canonical* if it is inter-triangulated and for any i ($i = 0, \ldots, k - 2$) and for any two clockwise consecutive blocks B, B' of G_{i+1} around a cut-vertex, there is exactly one separating edge.

Every connected k-outerplane graph can be made canonical as stated by the following lemma.

Lemma 1. *Let G be a connected k-outerplane graph with n vertices. There exists an $O(n)$-time algorithm that computes an augmented graph G' such that: (1) G' is k-outerplane, (2) G' is canonical, and (3) the levels of the vertices of G are preserved in G'.*

We now introduce equipped BC-trees. Let G be a k-outerplane graph with $k > 1$. We extend the block cut-vertex tree data structure [11] to identify specific subgraphs of G. Because of Lemma 1, we can (and will) restrict our attention to canonical graphs. Let K be a connected component of G_i with $i > 1$. An *equipped BC-tree* T of K is an embedded rooted tree such that (for an illustration see Figures 1(a), 1(b), 1(c)):

- T has three types of nodes: (a) A *B-node* for each block B_K of K, referred to as the *B-node of B_K*. (b) A *C-node* for each cut-vertex c of K, referred to as the *C-node of c*. (c) A *D-node* for each separating edge e of G that is incident on a cut-vertex of K, referred to as the *D-node of e*.
- If K is biconnected, T consist of a single B-node. If K is not biconnected, we choose an arbitrary C-node as the root of T.
- The edges of T are of two types: (a) Edges connecting a C-node of a cut-vertex c to a B-node of a block that contains c. (b) Edges connecting a C-node of a cut-vertex c to a D-node of a separating edge incident on c.
- The planar embedding of T reflects the embedding of G: if e is a separating edge incident on a cut-vertex c and e is between blocks B and B' in clockwise ordering around c, then the D-node of e is between the B-nodes of B and B' in clockwise ordering around the C-node of c.

For example cut vertex 1 is chosen as the root of the tree shown in Figure 1(c). Separating edge $(2, b)$ in Figure 1(b) separates blocks A and D. Correspondingly, the D-node of $(2, b)$ appears between the B-nodes of A and D in the circular clockwise ordering around the C-node of cut-vertex 2 in the equipped BC-tree of Figure 1(c).

Let μ be the C-node of a cut-vertex c of K, if μ is not the root of T, the parent of μ is a B-node and the leftmost child and the rightmost child of μ are D-nodes. If μ is the root of T we arbitrarily choose the leftmost child of μ as a D-node; as a consequence the rightmost child of the root is a B-node. See for example Figure 1(c) where the rightmost child of the root is the B-node of B, while the leftmost child is the D-node of $(1, e)$.

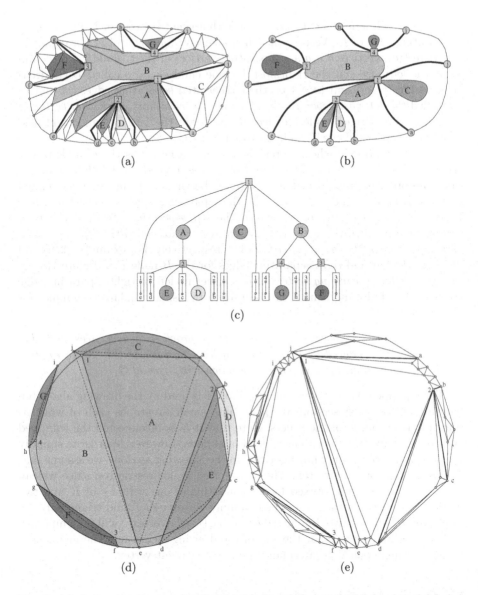

(a)

(b)

(c)

(d)

(e)

Fig. 1. (a) A 2-outerplane graph G. The blocks of G_1 are highlighted and labeled with capital letters. The cut-vertices of G_1 are numbered squares, and their separating edges are bold. (b) A schematic representation of the structure of G. The skeleton is highlighted with thick edges. (c) An equipped BC-tree of G rooted at the C-node of cut-vertex 1. (d) A drawing of the skeleton of G. The labels of the regions reflect those of the corresponding blocks. (e) A level-preserving 2-radial drawing of G.

Let ν be a C-node of a cut-vertex c of K that is not the root of T. Let (c, v) be a separating edge of K. Vertex v is called a *separating vertex* of ν. For example, in Figure 1(c), vertices b, c, d are separating vertices of the C-node 2. Let ν_l and ν_r be the leftmost and the rightmost D-node children (also called D-children) of ν, and let (c, v_l) and (c, v_r) be their associated separating edges; v_l and v_r are called the *leftmost separating vertex* and the *rightmost separating vertex* of ν, respectively. For example, vertices d and b are the leftmost and rightmost separating vertices of the C-node 2 in Figure 1(c).

Let μ be a B-node whose parent is a node ν where ν is the C-node of cut vertex c. Let ν_r, ν_l be the D-children of ν that precede and follow μ in the clockwise ordering around ν. Let $e_l = (c, v_l)$ be the separating edge associated with the D-node ν_l and $e_r = (c, v_r)$ be the separating edge associated with the D-node ν_r. Edges e_l and e_r are called the *left separating edge* and the *right separating edge* of μ, respectively. Also, v_l and v_r are called the *left separating vertex* and the *right separating vertex* of μ, respectively. For example, $(2, d)$ and $(2, c)$ are the left and right separating edges for the B-node E in Figure 1(c). A *separating edge* (*separating vertex*) of μ is either its left or right separating edge (vertex). The following lemma can be proven by using standard techniques for BC-trees [10].

Lemma 2. *Let G be a canonical 2-outerplane graph with n vertices such that the subgraph G_1 induced by the vertices of level 1 is connected. There exists a $O(n)$-time algorithm that computes an equipped BC-tree of G_1.*

The equipped block-cut-vertex tree T of K is used by the drawing algorithm described in the next section to split G into small subgraphs each of which is drawn independently of the others. Note that if K is biconnected the equipped BC-tree of K has only one node, which is a B-node. However, in order to simplify the description of the drawing algorithm in the following sections, we assume that K has at least one cut-vertex. The assumption is not restrictive, since we can always find a triangular mixed face f consisting of one vertex c of K and two vertices u, w of G_0, attach to c a new dummy vertex v in f, and triangulate the face again by adding two dummy edges $(v, u), (v, w)$. Vertex v will be temporary added to V_1 and removed at the end of the drawing algorithm. The augmented graph is still canonical 2-outerplane and c is now a cut-vertex.

5 Radial Drawings of 2-Outerplanar Graphs

Let G be a 2-outerplane graph. In this section we show how to compute a 2-radial drawing of G. This result will be a basic building block for the drawing techniques and the characterization of Section 6.

By Lemma 1 we can assume that G is canonical. Also, from the observation at the end of the previous section, we may assume that each connected component of G_1 has at least one cut-vertex. Let K be a connected component of G_1. The subgraph of G consisting of the separating edges of K is called the *skeleton* of K and is denoted as $skel(K)$. The *skeleton* of G is the union of all $skel(K)$, for every connected component K of G_1. We denote it by $skel(G)$. For example in Figure 1(a) and 1(b) the bold edges highlight the skeleton of the graph.

In order to use the algorithm as the basic tool to compute a k-radial drawing of a graph with outerplanarity k (see Section 3), we assume that it receives as input a drawing Γ_0 of G_0 on a circle C_0 and that it computes a drawing Γ of G without changing Γ_0, i.e. $\Gamma_0 \subset \Gamma$. We do not put any restrictions on the drawing Γ_0; the only hypothesis is that it preserves the planar embedding of G_0. The algorithm consists of four main steps:

1. **Choice of C_1:** The radius of circle C_1 is determined.
2. **Drawing the Skeleton of G:** For each connected component K of G_1, the drawing of $skel(K)$ is computed. Figure 1(d) shows a drawing of the skeleton of the graph of Figure 1(a).
3. **Associating Blocks with Regions:** Let Σ_K be the drawing of $skel(K)$. Σ_K induces a set of connected regions in the plane; each region is bounded by C_0 and by two separating edges. Each block is associated with a region and it will be drawn inside its region. Let B_K be a block of K, let T be an equipped BC-tree of K, and let μ be the B-node of B_K in T. Let (c, v_l) and (c, v_r) be the left and right separating edges of μ, respectively; let $\overline{cv_l}$, $\overline{cv_r}$ be the segments representing the separating edges of μ in Σ_K. Block B_K is associated with the region bounded by $\overline{cv_l}$, $\overline{cv_r}$ and by the arc of C_0 from v_r to v_l in clockwise direction. For example, let Σ_K be the drawing of Figure 1(d) and consider block D of Figure 1(b). Block D is associated with the region bounded in Figure 1(d) by segments $\overline{2c}$, $\overline{2b}$, and by the arc of C_0 from b to c in clockwise direction.
4. **Drawing the Blocks of Each Connected Component:** The drawing of G is computed by defining the coordinates of the vertices of G_1 that are not cut-vertices. To do that, the algorithm draws each connected component K of G_1 independently. For each block B_K of K it draws B_K inside its corresponding region, as defined in the previous step. Figure 1(e) shows a level-preserving 2-radial drawing of the graph of Figure 1(a).

5.1   Choice of C_1

Let K be a connected component of G_1. The radius r_1 of C_1 depends on the drawing of G_0. Radius r_1 is chosen so that when a drawing of $skel(G)$ is computed the region associated with each block contains an arc of C_1. This will be useful when drawing the vertices of the blocks inside their associated regions.

Let $K_0, K_1, \ldots, K_h$ be the connected components of G_1 and let T_j be the equipped BC-tree of K_j $(j = 0, \ldots, h)$. For each B-node of T_j with separating vertices v_l, v_r, compute the distance between the point representing v_l and the point representing v_r in Γ_0. Let δ_j be the minimum of these distances over all B-nodes of T_j and let $\delta = \min\{\delta_j : j = 0, \ldots, h\}$. We define the radius of C_1 to be such that C_1 intersects the chords of C_0 with length δ. Computing the radius of C_1 can be performed in a time that is linear in the number of blocks of G, and therefore linear in the number of vertices of G, since the graph is planar.

5.2 Drawing the Skeleton of G

In this section we assume that the root of the equipped BC-tree T of K has at least three B-children. The case when the root has only two B-children can be handled similarly. For reasons of space we do not discuss this case.

The algorithm computes a drawing Σ_K of $skel(K)$ such that Σ_K is inside the polygon of Γ_0 representing the face of G_0 that contains K in G. The drawing algorithm computes Σ_K so that for each block B_K, the separating edges that define the region associated with B_K form a convex angle inside this region (the angle is called *corner of the region* in the following). This invariant will be used in the next steps of the algorithm and is important to prove the planarity of the computed drawing of G. The algorithm performs a top-down left-to-right visit of T. When a C-node ν is visited, the associated cut-vertex c is drawn on C_1 together with all its incident separating edges. To better describe this algorithm we need to introduce some more terminology.

Let Γ be a 2-radial drawing of G on two circles C_0 and C_1 and let p be a point on C_1. A *free arc* of p is a maximal arc of C_1 having p as one end-point and containing neither vertices of Γ nor crossings between an edge of Γ and C_1. Point p has always two free arcs, one moving from p clockwise (the *left free arc of p*) and the other moving from p counterclockwise (the *right free arc of p*). Given any circle C, and two points a and b on C, the arc of C traversed when moving from a to b clockwise will be denoted as $< a, b >$. Points a and b will be called the *first point* and the *last point* of the arc, respectively. Each point of the arc distinct from a and b will be referred to as an internal point of $< a, b >$. Finally, let q be a point outside C. A point p of C is *visible* from q if the segment $\overline{pq}$ does not cross C. The set of points of C that are visible from q is an arc called the *visible region of q on C*. Note that, the first and the last points of the visible region of q on C are the intersection points between C and the straight lines through q tangent to C. The algorithm distinguishes among two cases:

- *Node ν Is the Root of T.* Let $\mu_1, \ldots, \mu_h$ be the B-children of ν and let $\mathcal{P}$ be the polygon defined by their separating vertices. From the choice of C_1, every side of $\mathcal{P}$ crosses C_1 in two distinct points. This implies that $\mathcal{P}$ contains a set of arcs of C_1. Draw c as a point of one of these arcs. See for example the cut-vertex 1 in Figure 1(d).

- *Node ν Is Not the Root of T.* Let u_l and u_r be the leftmost and the rightmost separating vertices of ν, respectively. By the choice of C_1 (Subsection 5.1), segment $\overline{u_l u_r}$ crosses circle C_1 in two distinct points; let p be the intersection point that is closer to u_r, and denote by γ the intersection of the left free arc of p and the visible region of u_r. Draw c as a point of γ. It can be proven that any point in γ guarantees that the corners of the regions of the B-children of ν are convex. However, in order to correctly complete the drawing of the blocks of K without changing the drawing of the skeleton (Subsection 5.4), the algorithm may need to make γ smaller for some cases. Details about how to reduce γ are omitted.

Lemma 3. *Let G be a 2-outerplane graph with n vertices, let $G_0 \subset G$ be the subgraph of level 0, and let skel(G) be the skeleton of G. Let Γ_0 be an embedding preserving 1-radial drawing of G_0. There exists an $O(n)$-time algorithm that computes an embedding preserving 2-radial drawing Γ of $G_0 \cup skel(G)$ such that $\Gamma_0 \subset \Gamma$. Also, Γ is a level-preserving drawing.*

5.3   Associating Blocks with Regions

Since G is canonical, the interior of each face f of G_0 contains exactly one connected component K of G_1. K is drawn inside the region f of the plane and is denoted by $\mathcal{R}_f$. Note that $\mathcal{R}_f$ is defined by Γ_0. Let T be the BC-tree of K; $\mathcal{R}_f$ is recursively subdivided into connected sub-regions $\mathcal{R}_\mu$, one for each B-node μ of T. As explained in Subsection 5.4, block μ is drawn inside $\mathcal{R}_\mu$. We formally define the regions induced by Σ_K in the following.

For each B-node μ of T let (c, v_l) and (c, v_r) be the separating edges of μ. Denote by $\mathcal{R}_\mu^*$ the region of the plane delimited by the segments $\overline{cv_l}$, $\overline{cv_r}$, and by the arc $< v_r, v_l >$ of C_0. The drawing technique of Subsection 5.4 will draw the blocks of the subtree of T rooted at μ inside $\mathcal{R}_\mu^*$. The region $\mathcal{R}_\mu$ containing the single block associated with μ is as follows: (a) If μ is a leaf, let $\mathcal{R}_\mu = \mathcal{R}_\mu^*$. (b) If μ is an internal node with grandchildren $\mu_1, \ldots, \mu_h$, let $\mathcal{R}_\mu = \mathcal{R}_\mu^* \backslash (\mathcal{R}_{\mu_1}^* \cup \cdots \cup \mathcal{R}_{\mu_h}^*)$.

5.4   Drawing the Blocks

Let B_K be a block of a connected component K of G_1, and let μ be the B-node of T representing B_K. As in Section 5.2, we assume in the following that the root of T has at least three B-children.

A vertex of G_1 is an *internal joint vertex* if it is adjacent to at least two vertices of G_0. A vertex u of G_0 is an *external joint vertex* if it is adjacent to at least two vertices of G_1. The algorithm that draws B_K inside its associated region $\mathcal{R}_\mu$ distinguishes among two cases:

Case 1: μ Is a Leaf of T. Let c be the cut-vertex associated with the parent of μ and let $c = a_0, a_1, \ldots, a_t$ be the internal joint vertices of B_K in the clockwise order they appear on the external face of B_K. Since G is inter-triangulated then the internal joint vertices a_l and a_{l+1} ($l = 0, \ldots, t-1$) are adjacent to an external joint vertex, which is denoted by u_{l+1}. Also a_t and a_0 are adjacent to an external joint vertex, which is denoted as u_{t+1}. Since G is canonical, edges (c, u_1) and (c, u_{t+1}) are the separating edges of μ.

The algorithm first places the internal joint vertices $a_0, \ldots, a_t$ in this order. At step l ($l = 1, \ldots, t$) vertex a_l and its incident edges are added to the drawing. Each vertex a_l is placed on C_1 as follows. If edge (u_l, a_{l-1}) crosses C_1 then let p be its crossing, otherwise let p be coincident with a_{l-1}. Vertex a_l is drawn as an internal point of the intersection between the right free arc of p and the visible region of u_l.

Once all the internal joint vertices are placed, the algorithm draws the remaining vertices of B_K and their incident edges. More precisely, let $v_1, \ldots, v_{r_l}$

be the vertices that are between vertices a_l and a_{l+1} ($l = 0, \ldots, t - 1$) in clockwise ordering on the external face of B_K. All these vertices are adjacent to the external joint vertex u_{l+1}. If edge (a_l, u_{l+1}) crosses C_1 then let q_1 be its crossing, otherwise let q_1 be coincident with a_l. Analogously, if edge (a_{l+1}, u_{l+1}) crosses C_1 then let q_2 be its crossing, otherwise let q_2 be coincident with a_{l+1}. Vertices $v_1, \ldots, v_{r_l}$ are drawn in this order as points of the arc $< q_1, q_2 >$.

Case 2: μ Is an Internal Node of T. This case can be handled with techniques similar to those used for the previous case. We omit the decription of them for reasons of space.

Theorem 1. *Let G be a a 2-outerplane graph with n vertices. G admits a level-preserving 2-radial drawing that preserves the embedding of G. Also there exists an $O(n)$-time algorithm that computes such a drawing.*

6   Minimum Radial Drawings of Planar Graphs

In this section we first characterize the family of graphs that admit a radial drawing on at most k-concentric circles and then use the characterization to solve in polynomial time the problem of computing a minimum radial drawing of a planar graph.

Theorem 2. *Let G be a graph with outerplanarity k and n vertices. Then G admits a proper k-radial drawing. Also, there exists an $O(n)$-time algorithm that computes such a drawing.*

Proof. Since G has outerplanarity k then it has a k-outerplanar embedding. We show how to compute a level-preserving k-radial drawing Γ of G that preserves this embedding. This implies that Γ is proper. An algorithm to compute Γ is based on first drawing the subgraph induced by the vertices of level 0 on a circle C_0 and then by adding at each step the vertices of level i on a circle C_i ($i = 1, \cdots k - 1$). At Step i the subgraph $G_{i-1,i}$ is drawn by using the algorithm described in Section 5. Since $G = \bigcup_{i=0}^{k-2} G_{i,i+1}$ the computed drawing is a radial drawing of G. The fact that no two edges cross is a consequence of Theorem 1. It follows that the above described algorithm computes a level-preserving k-radial drawing of G. As for the time complexity, it follows from Theorem 1 that the computation of drawing $G_{i-1,i}$ requires $O(n_i)$ time where n_i is the number of vertices in $G_{i-1,i}$. Therefore the overall time complexity is $O(n)$. □

Lemma 4. *If a graph G admits a k-radial drawing then it has outerplanarity at most k.*

Proof. A k-radial drawing Γ of G defines an embedding of G. All vertices on the outerface of this embedding are drawn on C_0. Removal of all vertices on C_0 results in a $(k - 1)$-radial drawing. So we can use induction to show that Γ is a k-outerplane graph. It follows that the outerplanarity of G is at most k. □

Theorem 3. *Let G be a planar graph. G admits a radial drawing on at most k-circles if and only if the outerplanarity of G is at most k.*

Proof. Assume that G has a k-radial drawing. Then by Lemma 4 G has outer-planarity at most k. Conversely, if G has outerplanarity $j \leq k$, by Theorem 2 it admits a j-radial drawing with $j \leq k$. ☐

Theorem 4. *Let G be a planar graph with n vertices. There exists an $O(n^5 \log n)$-time algorithm that computes a radial drawing of G on the minimum number of concentric circles. Furthermore the computed drawing is proper.*

Proof. Bienstock and Monma [3] describe an algorithm to compute the outerplanarity k of G and to determine a k-outerplanar embedding of G. This algorithm takes $O(n^5 \log n)$ time. The result in [3] together with Theorem 3 imply that k is the minimum number of circles for which there exists a radial drawing of G. The fact that such a drawing is proper is a consequence of Theorem 2. Again by Theorem 2 it follows that the time complexity of the whole algorithm is dominated by the technique in [3]. ☐

References

1. C. Bachmaier, F. Brandenburg, and M. Forster. Radial level planarity testing and embedding in linear time. In *Proc. GD'03*, volume 2912 of *LNCS*, pages 393–405, 2003.
2. C. Bachmaier, F. Brandenburg, and M. Forster. Track planarity testing and embedding. In *Proc. SOFSEM'04*, volume 2, pages 3–17, 2004.
3. D. Bienstock and C. L. Monma. On the complexity of embedding planar graphs to minimize certain distance measures. *Algorithmica*, 5(1):93–109, 1990.
4. S. Bornholdt and H. Schuster, editors. *Handbook of Graphs and Networks: From the Genome to the Internet.* Wiley-VCH, 2003.
5. S. Cornelsen, T. Schank, and D. Wagner. Drawing graphs on two and three lines. In *Proc. GD'02*, volume 2528 of *LNCS*, pages 31–41, 2002.
6. E. Di Giacomo and W. Didimo. Straight-line drawings of 2-outerplanar graphs on two curves. In *Proc. GD'03*, volume 2912 of *LNCS*, pages 419–424, 2003.
7. E. Di Giacomo, W. Didimo, G. Liotta, and S. K. Wismath. Curve-constrained drawings of planar graphs. *Comp. Geometry: Theory and Appl.* to appear.
8. M. Dodge and R. Kitchin. *Atlas of Cyberspace.* Addison Wesley, 2001.
9. S. N. Dorogstev and J. F. F. Mendes. *Evolution of Networks, From Biological Nets to the Internet and WWW.* Oxford University Press, 2003.
10. F. Harary. *Graph Theory.* Addison-Wesley, 1972.
11. F. Harary and G. Prins. The block-cutpoint-tree of a graph. *Publ. Math Debrecen*, 13:103–107, 1966.
12. M. Jünger and P. Mutzel, editors. *Graph Drawing Software.* Springer-Verlag, 2003.
13. K. Sugiyama. *Graph Drawing and Applications for Software and Knowldege Engineers.* World Scientific, 2002.

Hamiltonian-with-Handles Graphs and the k-Spine Drawability Problem[*]

Emilio Di Giacomo[1], Walter Didimo[1],
Giuseppe Liotta[1], and Matthew Suderman[2]

[1] Università degli Studi di Perugia, Perugia, Italy
{digiacomo,didimo,liotta}@diei.unipg.it
[2] School of Computer Science, McGill University, Montreal, Canada
msuder@cs.mcgill.ca

Abstract. A planar graph G is k-spine drawable, $k \geq 0$, if there exists a planar drawing of G in which each vertex of G lies on one of k horizontal lines, and each edge of G is drawn as a polyline consisting of at most two line segments. In this paper we: (i) Introduce the notion of hamiltonian-with-handles graphs and show that a planar graph is 2-spine drawable if and only if it is hamiltonian-with-handles. (ii) Give examples of planar graphs that are/are not 2-spine drawable and present linear-time drawing techniques for those that are 2-spine drawable. (iii) Prove that deciding whether or not a planar graph is 2-spine drawable is $\mathcal{NP}$-Complete. (iv) Extend the study to k-spine drawings for $k > 2$, provide examples of non-drawable planar graphs, and show that the k-drawability problem remains $\mathcal{NP}$-Complete for each fixed $k > 2$.

1 Introduction

Many graph drawing applications require that the vertices of the graph be placed on some set of horizontal lines. Such drawings have applications in visualization, DNA mapping, and VLSI layout [10, 8]. A common aesthetic requirement is that it be easy to locate the end-vertices of each edge. One way to achieve this is by representing edges as polylines composed of a small number of line segments, and by placing the vertices so that polylines from different edges cross a minimum number of times, if at all. Hence, we have the k-spine drawability problem: Given a planar graph G and an integer $k \geq 0$, is there a planar drawing of G such that the vertices of G lie on k horizontal lines called *spines* and each edge is drawn as a polyline consisting of at most two line segments? For $k \geq 0$, we say that a graph is k-spine drawable, or has a k-spine planar drawing, if it is a yes-instance to the k-spine drawability problem.

The k-spine drawability problem for $k = 1$ is a classic topic in the graph drawing and computational geometry literature, where 1-spine drawings are commonly called 2-page book embeddings or 2-stack layouts. Bernhart and

---
[*] The authors would like to thank Sue Whitesides for the useful discussion about the topic of this paper. Research supported in part by "Progetto ALINWEB: Algoritmica per Internet e per il Web", MIUR Programmi di Ricerca Scientifica di Rilevante Interesse Nazionale.

Kainen [1] show that a planar graph has a 2-page book embedding if and only if it is sub-hamiltonian, which implies that the 1-spine drawability testing problem is in general $\mathcal{NP}$-hard. Meaningful subclasses of planar graphs that admit 2-page book embeddings (i.e. they are 1-spine drawable) are described in the literature (see, e.g. [1, 3]).

The k-spine drawability problem for $k \geq 1$ has also been widely investigated in the case that the edges cannot bend, i.e they are straight-line segments. There are several papers devoted to this problem, both under the assumption that no two vertices on the same spine can be adjacent (see, e.g. [5, 7]) and under the assumption that there can be intra-spine edges (see, e.g. [4, 6, 9]). In particular, Cornelsen, Shank, and Wagner [4] characterize the family of graphs that admit a straight-line 2-spine drawing with intra-spine edges. They show that the graphs in this family are a proper subset of outerplanar graphs and describe a linear time test algorithm.

The present paper studies k-spine drawings for $k \geq 2$. It is assumed that edges can bend at most once and that two edges on the same spine can be adjacent. We are interested in testing whether or not a graph G admits a k-spine drawing, and, if so, computing such a drawing. The main results in this paper are as follows:

- We introduce and study the notion of hamiltonian-with-handles planar graphs. We show that a planar graph admits a 2-spine drawing if and only if it is sub-hamiltonian-with-handles.
- We study the relationship between hamiltonian-with-handles graphs and planar graphs. Namely, we show that there exist planar graphs that are not sub-hamiltonian-with-handles; consequently, they do not admit a 2-spine drawing. We also prove that every 2-outerplanar graph G is sub-hamiltonian-with-handles and that an embedding-preserving 2-spine drawing of G can be computed from a 2-outerplanar embedding in linear time.
- Motivated by these results, we study the problem of deciding whether or not a planar graph admits a 2-spine drawing. We show that this problem is $\mathcal{NP}$-Complete.
- We extend the investigation to $k > 2$ spines and prove that in this case not all planar graphs are k-spine drawable. We show that the problem of testing k-spine drawability remains $\mathcal{NP}$-Complete for any fixed integer $k > 2$.

For reason of space, some proofs are sketched or omitted.

2   Preliminaries

A k-*spine planar drawing* of G ($k \geq 1$) is a planar drawing of G in which the vertices of G are drawn as points on one of k horizontal straight lines (called *spines*), and the edges of G are drawn as polylines consisting of at most two segments (i.e. each edge is drawn with at most one bend). If G admits a k-spine planar drawing, then G is said to be k-*spine drawable*.

Let Γ be a k-spine planar drawing of G. A *jumping segment* to vertex v is a straight-line segment $\overline{pv}$ contained in an edge incident on v in Γ such that p and

v lie on different spines. We say that p is its first endpoint and v is its second endpoint. A *jumping sequence* J from a vertex v to a vertex w is a sequence $f_0, f_1, \ldots, f_h$ of jumping segments in Γ such that:

1. The first endpoint of f_0 is on the same spine as v, coinciding with v or to the right of v;
2. The second endpoint of f_h is on the same spine as w coinciding with w or to the left of w;
3. If f_i and f_{i+1} are consecutive segments in J, and p is the second endpoint of f_i and q is the first endpoint of f_{i+1}, then p and q lie on the same spine and p is to the left of q.

The *landing segments* of J are the horizontal line segments between the second endpoint of each f_i and the first endpoint of its successor in J, along with the horizontal segment between v and the first endpoint of f_0 and the horizontal segment between the second endpoint of f_h and w. Thus, the *landing sequence* $L_{v,w}(J)$ from v to w of the jumping sequence J is the sequence of landing segments of J whose order corresponds to the order of the segments in J. The *jumping vertex sequence* $V_{v,w}(J)$ of jumping sequence J from vertex v to vertex w is the sequence of vertices that lie on the landing segments of $L_{v,w}(J)$. The order of the vertices corresponds to the order that their segments appear in $L_{v,w}(J)$, and then to their left-to-right order in Γ. Whenever the jumping vertex sequence $V_{v,w}(J)$ is a simple path with $prev(w) = \emptyset$ and $next(w) = \emptyset$, we call it a *cutting path* of G in Γ. Similarly, if $V_{v,w}(J)$ can be augmented by edge addition while maintaining planarity to be a simple path with $prev(w) = \emptyset$ and $next(w) = \emptyset$, then we call it an *augmenting cutting path* of G in Γ.

Cutting paths will be essential to our characterization of 2-spine drawable graphs later. Very roughly, a cutting path splits the graph into two subgraphs that are each 1-spine drawable. The following lemma can be proved.

Lemma 1. *For each 2-spine planar drawing Γ of a planar graph G, there exists an augmenting cutting path of G in Γ.*

3   Hamiltonian-with-Handles Graphs

In this section we characterize the class of 2-spine drawable graphs. First, we require a few additional definitions.

Let G be an embedded planar graph. A *base path* of G is a simple path Π of G such that the first and the last end-vertices of Π are on the external face of G. Let Π be a base path and let η be a simple path of G such that no vertex of η is a vertex of Π. Path η is a *handle* of Π if for each end-vertex of η there exists an edge e, called a *bridge*, connecting the end-vertex to Π. The end-vertex of e in Π is called an *anchor vertex* of η. Its other end-vertex is called an *extreme vertex* of η. The subpath of Π between the anchor vertices of η is called the *co-handle* of η and is denoted $\widehat{\eta}$. The subgraph of G composed of the cycle C_η formed by η, its bridge edges and $\widehat{\eta}$, along with any edges and vertices inside C_η is called the *handle graph* of η and is denoted G_η.

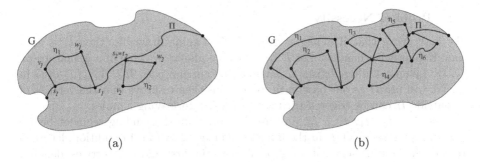

(a)   (b)

Fig. 1. (a) Illustration of handles along a path Π. η_1 is a non-dangling left handle and η_2 is a dangling right handle. Edges (s_1, v_1) and (t_1, w_1) are the bridges of η_1, and edges (s_2, v_2) and (t_2, w_2) are the bridges of η_2. Vertices s_1 and t_1 are the anchor vertices of η_2, and vertex $s_2 = t_2$ is the anchor vertex of η_2. Vertices v_1 and w_1 are the extreme vertices of η_1, and vertices v_2 and w_2 are the extreme vertices of η_2. (b) Some examples of interleaving handles.

If the two anchor vertices of a handle coincide, then the handle is called a *dangling handle*. If we walk along path Π from one end to the other, then every edge of G that is not in Π is either on the left-hand side of Π or on the right-hand side. Handles on the left-hand side are called *left handles*, and handles on the right-hand side are called *right handles*. Figure 1(a) illustrates these definitions.

Let η_1 and η_2 be two handles, and let s_1 and t_1 be the anchor vertices of η_1 such that s_1 is encountered before t_1 when walking along Π. Similarly, let s_2 and t_2 be the anchor vertices of η_2 such that s_2 is encountered before t_2 when walking along Π. Handles η_1 and η_2 are said to be *interleaving* if one of the following two cases holds:

- G_{η_1} and G_{η_2} share more than one vertex or share a vertex that is not an anchor for η_1 or η_2 (see, for example, handles η_1 and η_2 or handles η_5 and η_6 in Figure 1(b)); or
- η_1 is a left dangling handle, η_2 is a right dangling handle, and $s_1 = s_2 = t_1 = t_2$ (see, for example, handles η_3 and η_4 in Figure 1(b)).

A planar graph G is *hamiltonian-with-handles* if either G has at most two vertices or, for some planar embedding of G, the vertices of G can be covered by a cycle C and a set of paths $\eta_1, \eta_2, \ldots \eta_p$ such that: (i) C is a simple cycle, (ii) C is the union of a base path Π and an edge, and (iii) $\eta_1, \eta_2, \ldots \eta_p$ are non-interleaving handles of Π. G is *sub-hamiltonian-with-handles* if it can be augmented by adding edges in such a way that the resulting augmented graph is still planar and hamiltonian-with-handles.

4   Characterizing 2-Spine Drawable Graphs

In this section we prove the following characterization:

Theorem 1. *A planar graph G is 2-spine drawable if and only if it is sub-hamiltonian-with-handles.*

4.1   Proof of Necessity

We first prove that if a planar graph G is 2-spine drawable, then G is sub-hamiltonian-with-handles. Let Γ be a 2-spine planar drawing of a planar graph G. By Lemma 1, there exists an augmenting cutting path $\Pi = V_{v,w}(J)$ of G in Γ. We will use Π as our base path. It remains then to prove that the vertices of G outside Π can be covered with a set of non-interleaving handles.

Let $J = f_0, f_1, \ldots, f_h$, and use λ_i to denote the landing segment before each jumping segment f_i in the landing sequence $L_{v,w}(J)$. In addition, let λ_{h+1} denote the landing segment after f_h. We call the first and last vertices, denoted v_i and w_i, of each λ_i its *corner vertices*. We use π_i $(i = 0, \ldots, h + 1)$ to denote the subpath of Π consisting of:

- the vertex immediately preceding v_i, if it exists;
- all the vertices in λ_i; and,
- the vertex immediately following w_i, if it exists.

We call each π_i a *pocket*. Each pocket has an associated portion of a spine called its *pocket lead*:

- Pocket lead $\widehat{\pi_0}$ is the portion of spine that is before λ_1;
- Pocket lead $\widehat{\pi_i}$ $(i = 1, \ldots, h)$ is the portion of spine that is between λ_{i-1} and λ_{i+1}; and,
- Pocket lead $\widehat{\pi_{h+1}}$ is the portion of spine that is after λ_h.

A maximal sequence of consecutive vertices in a pocket lead is called *candidate handle*.

Lemma 2. *Let Γ be a 2-spine planar drawing of a planar graph G, and let Π be a cutting path of G in Γ. Let π_i be a pocket of Π and let $\widehat{\pi_i}$ be the pocket lead of π_i $(0 \le i \le h + 1)$. Let η be a candidate handle in $\widehat{\pi_i}$, and let v_η and w_η be the first vertex and the last vertex of η, respectively. Then, there exist two vertices $s_\eta, t_\eta \in \pi_i$ such that either there exist edges (v_η, s_η) and (w_η, t_η) in Γ or these edges can be added to Γ while maintaining the planarity of Γ. Furthermore, vertex s_η is on the spine that does not contain the vertices of η.*

Lemma 2 shows that G can be augmented by edge addition so that the resulting augmented graph can be covered by the cutting path Π plus a set of handles of Π. In order to prove that G is sub-hamiltonian-with-handles we need to prove that these handles are pairwise non-interleaving.

Lemma 3. *Let Γ be a 2-spine planar drawing of a planar graph G, let Π be the cutting path of G in Γ, and let $\eta_1, \eta_2, \ldots, \eta_p$ be a set of candidate handles of G in Γ. Then, Γ can be augmented so that $\eta_1, \eta_2, \ldots, \eta_p$ are pairwise non-interleaving handles.*

Proof. By Lemma 2, Γ can be augmented so that each η_j is a handle, and, if η_j is in pocket lead $\widehat{\pi_i}$, then its anchors s_j and t_j belong to π_i. We now prove that each pair of handles is non-interleaving. Without loss of generality, we consider the pair η_1 and η_2. By way of contradiction, assume that η_1 and η_2 are interleaving. According to the definition there are two cases.

- G_{η_1} and G_{η_2} share more than one vertex or share a vertex that is not an anchor for η_1 or η_2.

  By definition, η_1 and η_2 are disjoint so, by Lemma 2, the vertices that G_{η_1} and G_{η_2} share are also shared by the pockets corresponding to η_1 and η_2. We first consider the case where η_1 and η_2 belong to different pockets. Two pockets share vertices only if they are consecutive so we assume, without loss of generality, that η_1 belongs to pocket π_i and η_2 belongs to the next pocket π_{i+1}. In that case, the pockets share two vertices, w_i and v_{i+1}, which are consecutive on path Π. Thus, η_1 belongs to the same spine as v_{i+1} and is left of v_{i+1}. On the opposite spine, η_2 is to the right of w_i. By Lemma 2, s_2 does not belong to the spine of η_2 so s_2 appears after v_{i+1} in path Π, or coincides with v_{i+1}. Vertex t_2 appears after s_2 in Π or coincide with s_2. Hence G_{η_1} and G_{η_2} can share at most an anchor vertex. Therefore, η_1 and η_2 must belong to the same pocket.

  Since η_1 and η_2 belong to the same pocket, we assume, without loss of generality, that η_1 is to the left of η_2 on some spine. Let w_{η_1} be the last vertex of η_1 and let v_{η_2} be the first vertex of η_2. The two handles are interleaving only if the subpaths s_1 to t_1 of Π and s_2 to t_2 of Π share an edge. This implies that t_1 is to the right of s_2. By definition, $next(w_{\eta_1})$ is a crossing c_1 and $prev(v_{\eta_2})$ is also a crossing c_2 to the right of c_1. In addition, an edge incident on t_1 contains the segment $\overline{c_1 t_1}$ and another edge incident on s_2 contains the segment $\overline{c_2 s_2}$. Since c_1 is left of c_2 and s_2 is left of t_1, we have an edge crossing so η_1 and η_2 do not interleave.

- η_1 is a left dangling handle, η_2 is a right dangling handle, and $s_1 = s_2 = t_1 = t_2$. Since η_1 is a left dangling handle and η_2 is a right dangling handle then they are on different spines. By Lemma 2 also s_1 and s_2 are on different spines, but this is impossible since they coincide. $\square$

Together, Lemmas 2 and 3 prove the necessary condition of our characterization:

Lemma 4 (Necessary Condition). *If a graph G is 2-spine drawable, then G is sub-hamiltonian-with-handles.*

4.2   Proof of Sufficiency

To prove the sufficiency of the characterization of Theorem 1, we describe an algorithm that constructs a 2-spine planar drawing of any graph that is sub-hamiltonian-with-handles. For reasons of space only an outline of the algorithm is given.

Suppose that G is sub-hamiltonian-with-handles for some planar embedding and base path Π. Thus, Π divides G into two subgraphs, one to the left of Π and the other to the right of Π. Very roughly, the algorithm first draws the base path on the two spines so that it is possible to draw the subgraph that is to the left of Π, above the drawing of Π, and the subgraph that is to the right of Π, below the drawing of Π (see also Figure 2). The algorithm performs the following steps:

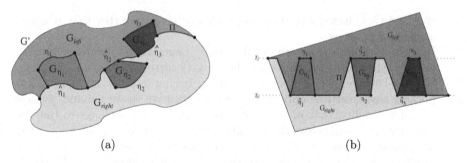

(a)   (b)

Fig. 2. Illustration of the drawing algorithm. (a) The graph G', obtained after the removal of the dangling handles, can be decomposed into two graphs G_{left} and G_{right} plus one handle graph for each handle. (b) The drawing technique assigns vertices to each spine so that the left handles can be drawn on spine T_1 and the right handles can be drawn on spine T_0. G_{left} is drawn completely above Π, and G_{right} is drawn completely below Π. The drawings of G_{left} and G_{right} share only vertices of Π.

Drawing the Vertices of Π: The algorithm starts by drawing the vertices of Π in G on the two spines. Each vertex is assigned to one of the two spines so that each co-handle of a left handle is on the lower spine and each co-handle of a right handle is on the higher spine. A position on the spine, i.e. an x-coordinate, is also assigned to each vertex of Π.

Removing the Dangling Handles: In order to simplify the algorithm, the dangling handles are removed and replaced with a set of new edges. The resulting graph G' then has only non-dangling handles but may have multiple edges. The removed handles are re-inserted back into the graph in the last step of the algorithm.

Drawing the Vertices of the Non-dangling Handles: The vertices of G' that are not in Π (i.e. the vertices of the non-dangling handles of G) are assigned an x-coordinate and a spine.

Drawing the Edges of G_{left} and of G_{right}: Recall that Π divides G' into two subgraphs, one to the left and the other to the right. We roughly define G_{left} to be the subgraph induced by the edges to the left of Π minus any handle graph edges. We similarly roughly define G_{right} to the be the subgraph induced by the edges to the right of Π minus any handle graph edges. Thus, the algorithm draws the edges of G_{left} and G_{right} separately, using the same technique for each, and then merges the two drawings together.

Drawing the Edges of the Handle Graph: After the edges of G_{left} and G_{right} are drawn, the edges of each handle graph are added to the drawing.

Re-inserting the Dangling Handles: Finally, the dangling handles are re-inserted into the drawing after removing the edges that were inserted earlier to replace the handle.

Lemma 5 (Sufficient Condition). *If a planar graph G is sub-hamiltonian-with-handles, then G is 2-spine drawable.*

Together, Lemmas 4 and 5 prove Theorem 1.

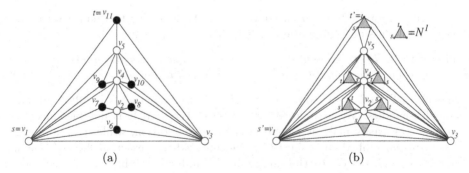

Fig. 3. (a) Maximal planar graph N; (b) The graph N^2 for the proofs of Theorem 2.

5   2-Spine Drawability Testing

The characterization result of Theorem 1 naturally raises two related questions: (i) Is every planar graph 2-spine drawable? (ii) How hard is it to decide whether or not a planar graph is 2-spine drawable? In this section we address both questions.

Theorem 2. *There exists a planar graph that is not 2-spine drawable.*

Sketch of Proof: Let N^1 be the maximal planar graph of Figure 3(a). Graph N^1 is non-hamiltonian [2] and therefore not 1-spine drawable [1]. Let H_5 be the subgraph of N^1 obtained by removing the vertices of degree three (the black vertices) from N^1. Given the embedding of H_5 in Figure 3(a), let N^2 be the maximal planar graph obtained by inserting a copy of N^1 into each face of H_5 and then triangulating the result (see Figure 3(b)).

We prove that graph N^2 is not 2-spine drawable. To this aim we consider a weaker version of the necessary condition in Theorem 1: if maximal planar graph G is 2-spine drawable, then G contains a simple cycle C such that $G \setminus C$ is 1-spine drawable. If G is 2-spine drawable, then, by Lemma 1, there exists an augmenting cutting path Π for a 2-spine planar drawing Γ of G. The endvertices of Π are on the external face of G, so, since G is maximal, they are adjacent. Therefore, Π plus the edge connecting its end-vertices form a simple cycle C. Since no edge of G crosses Π in Γ, if we remove C from the drawing of G, we are left with a set of subgraphs of G that are drawn on one spine and are therefore 1-spine drawable.

We now prove that N^2 is not 2-spine drawable. Suppose, by way of contradiction, that N^2 is 2-spine drawable. By the above necessary condition, there exists a simple cycle in N^2 such that $N^2 \setminus C$ is 1-spine drawable. Since N^1 is not 1-spine drawable, then C must contain at least one vertex from each copy of N^1. In the embedding of N^2 in Figure 3(b), each copy of N_1 is inside a different face of H_5. Thus, given any two vertices v_1 and v_2 from different copies of N_1, there must be a vertex of H_5 between v_1 and v_2 in C. Since there are six copies of N^1 and five vertices in H_5, then all the vertices of H_5 are in C. Thus, C contains at least one vertex from each copy of N^1 and all the vertices of H_5; however, this implies that there exists a hamiltonian circuit in N^1, a contradiction.   □

While Theorem 2 gives a negative result, the following theorem describes a meaningful class of 2-spine drawable graphs.

Theorem 3. *Every embedded 2-outerplanar graph is 2-spine drawable and a 2-spine planar drawing of G can be computed in linear time.*

Sketch of Proof: By Theorem 1 it is sufficient to prove that G is sub-hamiltonian-with-handles. We assume that G is biconnected. If it is not biconnected, then we can easily make it biconnected by edge addition, while maintaining a 2-outerplanar embedding. Since G is biconnected, the external face of G is a simple cycle C. Let G_0 be the subgraph of G induced by the vertices of C. We choose our base path Π to be C minus an edge. Each internal vertex, that is, each vertex that is not on the external face, is either adjacent to a vertex of the external face or can be made adjacent to a vertex of the external face by adding an edge. Each internal vertex v is a handle of length one and the edge connecting v to a vertex of the external face is its bridge. As for the time complexity, we remark that finding C and the handles takes linear time, and that the drawing procedure described in Section 4.2 requires linear time if C and the handles are given.   □

Based on the above theorem, one can ask whether embedded 2-outerplanar graphs can be drawn on less than two spines. We observe that the graph of Figure 3(b) is 2-outerplanar and that, as observed in the proof of Theorem 2, it is not 1-spine drawable.

Motivated by the results in Theorems 2 and 3, we investigate the complexity of deciding whether a planar graph is 2-spine drawable. The next theorem states that this problem is $\mathcal{NP}$-complete. In fact, we prove that the problem is $\mathcal{NP}$-complete when restricted to embedded maximal planar graphs and embedding-preserving 2-spine planar drawings. The original problem and this restricted version are polynomially equivalent because maximal planar graphs have a linear number of planar embeddings that can be efficiently computed.

The reduction is from HC-EMP: given an embedded maximal planar graph G, determine whether or not G is external hamiltonian, i.e. G has a hamiltonian circuit with an edge on the external face. Wigderson [11] has proved that HC-MP (the hamiltonian circuit problem for maximal planar graphs) is $\mathcal{NP}$-Complete. These two problems are polynomially equivalent, once again because each maximal planar graph has a linear number of embeddings. The proof of the next theorem is omitted for reasons of space.

Theorem 4. *The problem of determining whether or not a planar graph is 2-spine drawable is $\mathcal{NP}$-complete.*

6 k-Spine Drawability Testing

We extend the study of the 2-spine drawability to the case of the k-spine drawability. The following results can be proved by inductively generalizing the the proofs for the 2-spine drawing results.

Theorem 5. *For each fixed integer $k > 2$, there exists a planar graph that is not k-spine drawable.*

Sketch of Proof: The proof of this theorem is an extension of the proof of Theorem 2 and is based on a necessary condition for a planar graph to be k-spine drawable: if planar graph G is k-spine drawable, then G contains a simple cycle C such that $G \setminus C$ is $(k-1)$-spine drawable. We inductively describe a sequence of maximal planar graphs N^k that are not k-spine drawable for $k \geq 1$: (i) N^1 is the graph of Figure 3(a); (ii) N^k, for $k \geq 2$, is obtained from H_5 by inserting a copy of N^{k-1} into each face of H_5 (assuming the embedding of H_5 in Figure 3(a)) and then triangulating. We prove that N^k is not k-spine drawable by induction on k. N^1 is not 1-spine drawable since it is not hamiltonian. Assume that N^{k-1} is not $(k-1)$-spine drawable and, suppose, by way of contradiction, that N^k is k-spine drawable. By the necessary condition above, there exists a simple cycle C of N^k such that $N^k \setminus C$ is $(k-1)$-spine drawable. Since N^{k-1} is not $(k-1)$-spine drawable, then C must contain at least one vertex from each copy of N^{k-1}. In the planar embedding of N^k, each copy of N^{k-1} is inside a different face of H_5. Thus, given any two vertices v_1 and v_2 from different copies of N^{k-1}, there must be a vertex of H_5 between v_1 and v_2 in C. Since there are six copies of N^{k-1} and five vertices in H_5, then all the vertices of H_5 are in C. Thus, C contains at least one vertex from each copy of N^{k-1} and all the vertices of H_5. This implies that there exists a hamiltonian circuit in N^1 which is impossible. □

The proof of $\mathcal{NP}$-Completeness for 2-spine drawability testing can be extended to k-spine drawability for $k > 2$.

Theorem 6. *For each fixed integer $k > 2$, the problem of determining whether or not a planar graph is k-spine drawable is $\mathcal{NP}$-Complete.*

References

1. F. Bernhart and P. Kainen. The book thickness of a graph. *Journal Combinatorial Theory, Series B*, 27(3):320–331, 1979.
2. I. Cahit and M. Ozel. The characterization of all maximal planar graphs, Manuscript. http://www.emu.edu.tr/~cahit/prprnt.html, 2003.
3. F. R. K. Chung, F. T. Leighton, and A. Rosenberg. Embedding graphs in books: A layout problem with applications to VLSI design. *SIAM Journal on Algebraic and Discrete Methods*, 8:33–58, 1987.
4. S. Cornelsen, T. Schank, and D. Wagner. Drawing graphs on two and three lines. In *Graph Drawing (Proc. GD'02)*, volume 2528 of *LNCS*, pages 31–41. Springer-Verlag, 2002.
5. P. Eades, B. D. McKay, and N. C. Wormald. On an edge crossing problem. In *Proc. of the 9th Australian Computer Science Conference, ACSC9*, pages 327–334, 1986.
6. S. Felsner, G. Liotta, and S. K. Wismath. Straight-line drawings on restricted integer grids in two and three dimensions. *Journal of Graph Algorithms and Appl.*, 7(4):363–398, 2003.

7. U. Fößmeier and M. Kaufmann. Nice drawings for planar bipartite graphs. In *3rd Italian Conference on Algorithms and Complexity (Proc. CIAC '97)*, volume 1203 of *Lecture Notes in Computer Science*, pages 122–134. Springer-Verlag, 1997.
8. T. Lengauer. *Combinatorial Algorithms for Integrated Circuit Layout*. Wiley, 1990.
9. M. Suderman. Pathwidth and layered drawings of trees. *Int. Journal of Comp. Geometry*, 14(3):203–225, 2004.
10. M. S. Waterman and J. R. Griggs. Interval graphs and maps of DNA. *Bulletin of Mathematical Biology*, 48(2):189–195, 1986.
11. A. Wigderson. The complexity of the hamiltonian circuit problem for maximal planar graphs. Technical Report 298, Princeton University, EECS Department, 1982.

Distributed Graph Layout for Sensor Networks

Craig Gotsman[1] and Yehuda Koren[2]

[1] Department of Computer Science,
Harvard University, Cambridge, MA 02138
gotsman@eecs.harvard.edu
[2] AT&T Labs – Research,
Florham Park, NJ 07932
yehuda@research.att.com

Abstract. Sensor network applications frequently require that the sensors know their physical locations in some global coordinate system. This is usually achieved by equipping each sensor with a location measurement device, such as GPS. However, low-end systems or indoor systems, which cannot use GPS, must locate themselves based only on crude information available locally, such as inter-sensor distances. We show how a collection of sensors, capable *only* of measuring distances to close neighbors, can compute their locations in a purely distributed manner, i.e. where each sensor communicates only with its neighbors. This can be viewed as a distributed graph drawing algorithm. We experimentally show that our algorithm consistently produces good results under a variety of simulated real-world conditions, and is relatively robust to the presence of noise in the distance measurements.

1 Introduction

Sensor networks are a collection of (usually miniature) devices, each with limited computing and (wireless) communication capabilities, distributed over a physical area. The network collects data from its environment and should be able to integrate it and answer queries related to this data. Sensor networks are becoming more and more attractive in environmental, military and ecological applications (see [12] for a survey of this topic).

The advent of sensor networks has presented a number of research challenges to the networking and distributed computation communities. Since each sensor can typically communicate only with a small number of other sensors, information generated at one sensor can reach another sensor only by routing it thru the network, whose connectivity is described by a graph. This requires ad-hoc routing algorithms, especially if the sensors are dynamic. Traditional routing algorithms relied only on the connectivity graph of the network, but with the introduction of so-called *location-aware* sensors, namely, those who also know what their physical location is, e.g. by being equipped with a GPS receiver, this information can be used to perform more efficient *geographic* routing. See [10] for a survey of these routing techniques.

Beyond routing applications, location-aware sensors are important for information dissemination protocols and query processing. Location awareness is achieved primarily by equipping the sensors with GPS receivers. These, however, may be too expensive, too large, or too power-intense for the desired application. In indoor environments, GPS

J. Pach (Ed.): GD 2004, LNCS 3383, pp. 273–284, 2004.

does not work at all (due to the lack of line-of-sight to the satellites), so alternative solutions must be employed. Luckily, sensors are usually capable of other, more primitive, geometric measurements, which can aid in this process. An example of such a geometric measurement is the distance to neighboring sensors. This is achieved either by Received Signal Strength Indicator (RSSI) or Time of Arrival (ToA) techniques. An important question is then whether it is possible to design a distributed protocol by which each sensor can use this local information to (iteratively) compute its location in some global coordinate system.

This paper solves the following sensor layout problem: *Given a set of sensors distributed in the plane, and a mechanism by which a sensor can estimate its distance to a few nearby sensors, determine the coordinates of every sensor via local sensor-to-sensor communication.* These coordinates are called a *layout* of the sensor network.

As stated, this problem is not well-defined, because it typically will not have a unique solution. A unique solution would mean that the system is *rigid*, in the sense that the location of any individual sensor cannot be changed without changing at least one of the known distances. When all $\binom{n}{2}$ inter-sensor distances are known, the solution is indeed unique, and is traditionally solved using the Classical Multidimensional Scaling (MDS) technique [1]. When only a subset of the distances are known, more sophisticated techniques must be used.

When multiple solutions exist, the main phenomenon observed in the solutions is that of *foldovers*, where entire pieces of the graph fold over on top of others, without violating any of the distance constraints. The main challenge is to generate a solution which is fold-free. Obviously the result will have translation, orientation and reflection degrees of freedom, but either these are not important, or can be resolved by assigning some known coordinates to three sensors.

In real-world sensor networks, noise is inevitable. This manifests in the inter-sensor noise measurements being inaccurate. Beyond the obvious complication of the distances possibly no longer being symmetric, thus violating the very essence of the term "distance", there may no longer even exist a solution realizing the measured edge lengths. The best that can be hoped for, in this case, is a layout whose coordinates are, up to some acceptable tolerance, close to the true coordinates of the sensors.

In order to be easily and reliably implemented on a sensor network, the solution to the layout problem should be fully distributed (decentralized). This means that each sensor should compute based on information available *only* at that sensor and its immediate neighbors. The class of neighbors is typically characterized by a probabilistic variant of the disk graph model: Any sensor within distance R_1 is reachable, any sensor beyond distance R_2 is not reachable, and any sensor at a distance between R_1 and R_2 is reachable with probability p. Of course, information from one sensor may eventually propagate thru the network to any other sensor, but this should not be done explicitly.

2   Related Work

The problem of reconstructing a geometric graph given its edge-lengths has received some attention in the discrete geometry and computational geometry communities, where it is relevant for molecule construction and protein folding applications. Deciding whether a given graph equipped with edge lengths is rigid in 2D – i.e. admits a unique layout realizing the given edge lengths – is possible in polynomial time for the dense class of graphs known as *generic* graphs [7]. However, computing such a layout is in general NP-hard [14]. This does not change even if a layout is known to exist (as in our case).

The problem of distributed layout of a sensor network has received considerable attention in the sensor network community. A recent work of Priyantha *et al* [11] classifies these into *anchor-based* vs. *anchor-free* algorithms and *incremental* vs. *concurrent* algorithms. Anchor-based algorithms rely on the fact that a subset of the sensors are already aware of their locations, and the locations of the others are computed based on those. In practice a large number of anchor sensors are required for the resulting location errors to be acceptable. Incremental algorithms start with a small core of sensors that are assigned coordinates. Other sensors are repeatedly added to this set by local trigonometric calculations. These algorithms accumulate errors and cannot escape local minima once they are entered. Concurrent algorithms work in parallel on all sensors. They are better able to avoid local minima and avoid error accumulation. Priyantha *et al* [11] review a number of published algorithms and their classifications. All of them, however, are not fully distributed.

The algorithm we describe in this paper is most similar in spirit to the so-called Anchor-Free Localization (AFL) algorithm proposed by Priyantha *et al* [11]. The AFL algorithm operates in two stages. In the first stage a heuristic is applied to try generate a well-spread fold-free graph layout which "looks similar" to the desired layout. The second stage applies a "stress-minimization" optimization procedure to correct and balance local distance errors, converging to the final result. The heuristic used in the first stage involves the election of five reference sensors. Four of these sensors are well-distributed on the periphery of the network, and serve as north, east, south and west poles. A fifth reference sensor is chosen at the center. Coordinates are then assigned to all nodes, using these five sensors, reflecting their assumed positions. Unfortunately, this process does not lend itself easily to distribution. The second stage of the AFL algorithm attempts to minimize the partial stress energy using a gradient descent technique. At each sensor, the coordinates are updated by moving an infinitesimal distance in the direction of the spring force operating on the sensor. This is a fully distributed protocol. It, however, involves a heuristic choice of the infinitesimal step, and can be quite slow.

Our algorithm also involves two stages with similar objectives. The first aims to generate a fold-free layout. This is done based on a distributed Laplacian eigenvector computation which typically spreads the sensors well. The second stage uses the result of the first stage as an initial layout for an iterative stress-minimization algorithm. As opposed to AFL, it is not based on gradient descent, rather on a more effective *majorization* technique.

Once again we emphasize that the main challenge is to design algorithms which are *fully* distributed. This is a major concern in sensor network applications, and there is an increasing interest in designing such solutions. These turn out sometimes to be quite non-trivial. Probably the simplest example is a distributed algorithm to compute the sum (or average) of values distributed across the network; see [13].

3   The Problem

We are given a graph $G(V = \{1, \ldots, n\}, E)$, and for each edge $\langle i, j \rangle \in E$ – its Euclidean "length" l_{ij}. Denote a 2D layout of the graph by $x, y \in \mathbb{R}^n$, where the coordinates of vertex i are (x_i, y_i). Denote $d_{ij} = \sqrt{(x_i - x_j)^2 + (y_i - y_j)^2}$.

In the non-noisy version of the problem, we know that there exists a layout of the sensors that realizes the given edge lengths (i.e. $d_{ij} = l_{ij}$). Our goal is then to reproduce this layout. This layout is usually not unique. For example consider a $2n \times 2n$ square grid, where each internal sensor is connected to its four immediate neighbors with an edge of length one. We can realize all lengths using the degenerate 1D layout where half of the sensors are placed on 0 and the other half is placed on 1.

Fortunately, there is additional information that we may exploit to eliminate spurious solutions to the layout problem – we know that the graph is a full description of the close sensors. Consequently, the distance between each two nonadjacent sensors should be greater than some constant r, which is larger than the longest edge. This can further constrain the search space and eliminate most undesired solutions. Formally, we may pose our problem as follows:

Layout Problem. Given a graph $G(\{1, \ldots, n\}, E)$, and for each edge $\langle i, j \rangle \in E$ – its length l_{ij}, find an **optimal layout** $(p_1, \ldots, p_n)$ ($p_i \in \mathbb{R}^d$ is the location of sensor i), which satisfies for all $i \neq j$:

$$\begin{cases} \|p_i - p_j\| = l_{ij} \text{ if } \langle i, j \rangle \in E \\ \|p_i - p_j\| > R \text{ if } \langle i, j \rangle \notin E \end{cases}$$

Where $R = \max_{\langle i,j \rangle \in E} l_{ij}$. For the rest of this paper we assume $d = 2$.

It seems that an optimal layout is unique (up to translation, rotation and reflection) in many practical situations. For example, it overcomes the problem in the $2n \times 2n$ grid example described above. An optimal layout is similar to that generated by common force-directed graph drawing algorithms that place adjacent nodes closely while separating nonadjacent nodes. Therefore, we may exploit some known graph drawing techniques. For example, separating nonadjacent sensors can be achieved by solving an electric-spring system with repulsive forces between these sensors [2, 3]. Another possibility is to somehow estimate the distances l_{ij} between nonadjacent sensors (e.g., as the graph-theoretic distance) and then to minimize the *full stress energy*: $\sum_{i<j} \frac{(d_{ij} - l_{ij})^2}{l_{ij}^2}$ using an MDS-type technique; see [8].

However, since we aim at a distributed algorithm which should minimize communication between the sensors, dealing with repulsive forces or long-range target distances is not practical, as this will involve excessive inter-sensor interaction, which is very expensive in this scenario. To avoid this, we propose an algorithm which is based only on direct information sharing between adjacent sensors, avoiding all communication between nonadjacent sensors or any centralized supervision. Note that such a restriction rules out all common algorithms for general graph drawing problem; we are not aware of any layout algorithm that satisfies it.

In the real-life noisy version of the problem, the measured distances l_{ij} are contaminated by noise: $l_{ij} = d_{ij} + \epsilon_{ij}$. This means that there might not even exist a solution to the optimal layout problem. In this case we would like to minimize the difference between the true location of the sensors and those computed by the algorithm.

4   Smart Initialization and Eigen-projection

A useful energy function which is minimized by the desired layout is the *localized stress energy*:

$$\text{Stress}(x, y) = \sum_{\langle i,j \rangle \in E} (d_{ij(x,y)} - l_{ij})^2 \tag{1}$$

Note that this energy is not normalized, as opposed to the full stress energy. This non-convex energy function may have many local minima, which an optimizer may get stuck in. However, since in the non-noisy case, we are guaranteed the existence of a layout where $d_{ij} = l_{ij}$, namely $\text{Stress}(x, y)$ achieves the global minimum of zero, it is reasonable to hope that if we start with the optimization process at a "smart" initial layout, the process will converge to this global minimum. To construct such an initial layout, we exploit the fact that nonadjacent sensors should be placed further apart. This means that we seek a layout that spreads the sensors well. We first deal with the one-dimensional case. We will design an energy function which is minimized by such a layout, and can be optimized in a strictly distributed fashion. The function is defined as follows:

$$E(x) = \frac{\sum_{\langle i,j \rangle \in E} w_{ij} \|x_i - x_j\|^2}{\sum_{i<j} \|x_i - x_j\|^2} \tag{2}$$

Here, w_{ij} is some measure for the similarity of the adjacent sensors i and j. It should be derived from l_{ij}, e.g., $w_{ij} = 1/(l_{ij} + \alpha)$ or $w_{ij} = \exp(-\alpha l_{ij})$, $\alpha \geqslant 0$; in our experiments we used $w_{ij} = \exp(-l_{ij})$. Minimizing $E(x)$ is useful since it tries to locate adjacent sensors close to each other while separating nonadjacent sensors. It can also be solved fairly easily. Denote by D the diagonal matrix whose i'th diagonal entry is the sum of the i'th row of W: $D_{ii} = \sum_{j:\langle i,j \rangle \in E} w_{ij}$. The global minimum of $E(x)$ is the eigenvector of the related weighted Laplacian matrix $L^w = D - W$ associated with the smallest positive eigenvalue; see [6,9]. In practice, it is better to work with the closely related eigenvectors of the transition matrix $D^{-1}W$, which have some advantages over the eigenvectors of L^w; see [9]. Note that the top eigenvalue of $D^{-1}W$ is $\lambda_1 = 1$, associated with the constant eigenvector $v_1 = 1_n = (1, 1, \ldots, 1)$, so the desired solution is actually the *second* eigenvector v_2.

The vector v_2 can be computed in a distributed manner by iteratively averaging the value at each sensor with the values of its neighbors:

$$x_i \leftarrow a \left(x_i + \frac{\sum_{\langle i,j \rangle \in E} w_{ij} x_j}{\sum_{\langle i,j \rangle \in E} w_{ij}} \right) \tag{3}$$

Readers familiar with numerical linear algebra will recognize this process as power iteration of the matrix $I + D^{-1}W$. Power iteration usually converges to the eigenvector of the iterated matrix corresponding to the eigenvalue with highest absolute value. However, here we initialize the process by a vector y which is D-orthogonal to v_1, namely $y^T D v_1 = 0$, using a distributed method that will be described shortly. Hence, the process will converge to v_2 – the next highest eigenvector of $I + D^{-1}W$; see [9]. D-orthogonality, rather than simple orthogonality, is required because $D^{-1}W$ is not symmetric. The constant $a > 0$ controls the growth of $\|x\|$; in our implementation we used $a = 0.51$.

4.1   Two Dimensional Layout

We now turn our attention to the two-dimensional layout problem. $E(x)$ is defined also in higher dimensions (where x is short for (x, y)), and a "smart" initial 2D layout is achieved by taking the x coordinate to be v_2 – the second eigenvector of $D^{-1}W$, and the y coordinate to be v_3 – the third eigenvector of $D^{-1}W$. Unfortunately, the power iteration (3) will not detect v_3, as it is dominated by v_2, unless we start the process (3) with a vector D-orthogonal to $x = v_2$.

Constrained by the distributed computation requirement, it is not easy to initialize the process with a vector D-orthogonal to v_2. We resort to the following lemma:

Lemma 1. *Given two vectors x and y and matrices D and A, the vector Ay is D-orthogonal to x if $A^T Dx = 0$.*

Proof. Since $A^T Dx = 0$, then $y^T A^T Dx = 0$. Equivalently $(Ay)^T Dx = 0$ and the lemma follows.   □

Therefore, it suffices to construct a "local matrix" A such that $A^T Dx = 0$. By "local" we mean that $A_{i,j} \neq 0$ only if $\langle i, j \rangle \in E$. This will enable a distributed computation. In our case when D is diagonal, a suitable matrix is the following:

$$A_{i,j} = \begin{cases} -x_j/D_{ii} & \langle i, j \rangle \in E \\ 0 & \langle i, j \rangle \notin E, i \neq j \\ -\sum_k A_{i,k} & i = j \end{cases} \qquad i, j = 1, \ldots, n$$

It is easy to verify that $A^T Dx = 0$.

To summarize, to obtain $y = v_3$, we pick some random vector u, and initialize y with Au. Note that the computation of Au involves only local operations, and can be easily distributed. Then, we run the power iteration (3) on the vector y.

While the initial vector is D-orthogonal to v_2, it is not necessarily D-orthogonal to $v_1 = 1_n$. Hence, after many iterations, the result will be $y = \alpha v_1 + \epsilon v_3$, for some very small ϵ. While the process ultimately converges to what seems to be an essentially useless vector, its values near the limit is what is interesting. Since v_1 is the constant vector – 1_n, these values are essentially a scaled version of v_3 displaced by some fixed value (α) and they still retain the crucial information we need.

However when the numerical precision is low and the ratio α/ϵ is too high we might lose the v_3 component. Fortunately, we can work around this by *translating and scaling y* during the power iteration. Specifically, every βn iterations (we use $\beta = 1/2$) compute $\min_i y_i$ and $\max_i y_i$. A distributed computation is straightforward and can be completed with number of iterations bounded by the diameter of the graph (at most $n - 1$). Then, linearly transform y by setting

$$y_i \leftarrow \frac{y_i - \min_i y_i}{\max_i y_i - \min_i y_i} - \frac{1}{2}, \qquad i = 1, \ldots, n \tag{4}$$

After this, $\min_i y_i = -0.5$ and $\max_i y_i = 0.5$. Since translation is equivalent to addition of γv_1 and scaling cannot change direction, we can still express y as $\hat{\alpha} v_1 + \hat{\epsilon} v_3$.

Now assume, without loss of generality, that $\max_i v_3 - \min_i v_3 = 1$, and recall that $v_1 = (1, 1, \ldots, 1)$. The D-orthogonality of v_3 to 1_n implies: $\max_i v_3 > 0$ and

$\min_i v_3 < 0$. In turn, $\min_i y_i = -0.5$ and $\max_i y_i = 0.5$ imply that $|\hat{\alpha}| < 0.5$. Moreover, since all the variability of y is due to its v_3 component, we get $\hat{\epsilon} = 1$. Therefore, (4) guarantees that the magnitude of the v_3 component is larger than that of the v_1 component, avoiding potential numerical problems.

4.2 Balancing the Axes

Obviously, the process described in Section 4.1 can yield x and y coordinates at very different scales. Usually, we require that $||x|| = ||y||$, but this is difficult to achieve in a distributed manner. An easier alternative that is more suitable for a distributed computation is a balanced aspect ratio, i.e.: $\max_i x_i - \min_i x_i = \max_i y_i - \min_i y_i$.

Since the computation of the y-coordinates already achieved $\max_i y_i - \min_i y_i = 1$, it remains to ensure that the x coordinates have the same property. We achieve this by performing: $x_i \leftarrow x_i / (\max_i x_i - \min_i x_i)$, $\quad i = 1, \ldots, n$.

5 Optimizing the Localized Stress Energy

At this point we have reasonable initial locations for both the x- and y-coordinates, and are ready to apply a more accurate 2D optimization process for minimizing the localized stress energy (1). A candidate could be simple gradient descent, which is easily distributed, as in [11]. Each sensor would update its x-coordinates as follows:

$$x_i(t+1) = x_i(t) + \delta \sum_{j:\langle i,j \rangle \in E} \frac{(x_j(t) - x_i(t))}{d_{ij}(t)} (d_{ij}(t) - l_{ij}), \qquad (5)$$

where $d_{ij}(t) = \sqrt{(x_i(t) - x_j(t))^2 + (y_i(t) - y_j(t))^2}$. The y-coordinates are handled similarly. This involves a scalar quantity δ whose optimal value is difficult to estimate. Usually a conservative value is used, but this slows down the convergence significantly.

A more severe problem of this gradient descent approach is its sensitivity to the scale of the initial layout. Obviously the minimum of $E(x)$ is scale-invariant, since $E(cx) = E(x)$ for $c \neq 0$. However, the minimum of $\text{Stress}(x)$ is certainly not scale-invariant as we are given concrete target edge lengths. Therefore before applying gradient descent we have to scale the minimum of $E(x)$ appropriately.

Fortunately, we can avoid the scale problem by using a different approach called *majorization*. Besides being insensitive to the original scale, it is usually more robust and avoids having to fix a δ for the step size. For a detailed description of this technique, we refer the interested reader to multidimensional scaling textbooks, e.g., [1]. Here we provide just a brief description.

Using the Cauchy-Schwartz inequality we can bound the localized 2D stress of a layout (x, y) by another expression of (x, y) and (a, b), as follows:

$$\text{Stress}(x, y) \leqslant x^T L x + y^T L y + x^T L^{(a,b)} a + y^T L^{(a,b)} b + c, \quad x, y, a, b \in \mathbb{R}^n, \quad (6)$$

with equality when $x = a$ and $y = b$. The constant c is independent of x, y, a, b. L is the graph's $n \times n$ Laplacian matrix (also independent of x, y, a, b) defined as:

$$L_{i,j} = \begin{cases} -1 & \langle i,j \rangle \in E \\ 0 & \langle i,j \rangle \notin E \\ -\sum_{j \neq i} L_{i,j} & i = j \end{cases} \quad i,j = 1,\dots,n$$

The weighted Laplacian $n \times n$ matrix $L^{(a,b)}$ is defined as:

$$L_{i,j}^{(a,b)} = \begin{cases} -l_{ij} \cdot \mathrm{inv}\left(\sqrt{(a_i - a_j)^2 + (b_i - b_j)^2}\right) & \langle i,j \rangle \in E \\ 0 & \langle i,j \rangle \notin E \\ -\sum_{j \neq i} L_{i,j}^{(a,b)} & i = j \end{cases} \quad i,j = 1,\dots,n$$

where we define $\mathrm{inv}(x) = 1/x$ for $x \neq 0$ and $\mathrm{inv}(x) = 0$ otherwise.

Given a layout a, b, we can find another layout (x, y) which minimizes the r.h.s. $x^T Lx + y^T Ly + x^T L^{(a,b)}a + y^T L^{(a,b)}b + c$ by solving the linear equations:

$$Lx = L^{(a,b)}a$$
$$Ly = L^{(a,b)}b$$

Using inequality (6) we are guaranteed that the stress of the layout has decreased when going from (a, b) to (x, y), i.e., $\mathrm{Stress}(x, y) \leqslant \mathrm{Stress}(a, b)$. This induces an iterative process for minimizing the localized stress. At each iteration, we compute a new layout $(x(t + 1), y(t + 1))$ by solving the following linear system:

$$L \cdot x(t + 1) = L^{(x(t),y(t))} \cdot x(t)$$
$$L \cdot y(t + 1) = L^{(x(t),y(t))} \cdot y(t) \tag{7}$$

Without loss of generality we can fix the location of one of the sensors (utilizing the translation degree of freedom of the localized stress) and obtain a strictly diagonally dominant matrix. Therefore, we can safely use Jacobi iteration [4] for solving (7), which is easily performed in a distributed manner as follows.

Assume we are given a layout $(x(t), y(t))$ and want to compute a better layout $(x(t + 1), y(t + 1))$ by a single iteration of (7). Then we iteratively perform for each $i = 1,\dots,n$:

$$x_i \leftarrow \frac{1}{deg_i} \sum_{j:\langle i,j \rangle \in E} (x_j + l_{ij}(x_i(t) - x_j(t)) \, \mathrm{inv}(d_{ij}(t)))$$
$$y_i \leftarrow \frac{1}{deg_i} \sum_{j:\langle i,j \rangle \in E} (y_j + l_{ij}(y_i(t) - y_j(t)) \, \mathrm{inv}(d_{ij}(t))) \tag{8}$$

Note that $x(t)$, $y(t)$ and $d_{ij}(t)$ are *constants* in this process which converges to $(x(t+1), y(t+1))$. Interestingly, when deriving $(x(t + 1), y(t + 1))$ only the angles between sensors in $(x(t), y(t))$ are used. Therefore, this process is independent of the scale of the current layout.

It is possible to simplify the 2D majorization process somewhat. When the iterative process (8) converges the layout scale issue is resolved. Hence, instead of continuing with another application of (7) to obtain a newer layout, it is possible to resort to a faster local process (which, in contrast, *is* scale-dependent). In this process each sensor uses a local version of the energy where all other sensors are fixed. By the same majorization argument the localized stress decreases when applying the following iterative process:

$$x_i \leftarrow \frac{1}{deg_i} \sum_{j:\langle i,j\rangle \in E} (x_j + l_{ij}(x_i - x_j)\text{inv}(d_{ij}))$$

$$y_i \leftarrow \frac{1}{deg_i} \sum_{j:\langle i,j\rangle \in E} (y_j + l_{ij}(y_i - y_j)\text{inv}(d_{ij}))$$

(9)

Here, as usual $d_{ij} = \sqrt{(x_i - x_j)^2 + (y_i - y_j)^2}$. This process is similar to (8), except that x_i, x_j and d_{ij} are no longer constants. We have used this in our implementation, and it seems to accelerate the convergence. Note that this is quite close to the gradient descent (5) when using $\delta = 1/deg_i$, a different stepsize per sensor.

6   Experimental Results

We have implemented our algorithm and the AFL algorithm [11], and compared their performance on a variety of inputs. In the first experiment, we constructed a family of random graphs containing 1000 sensors distributed uniformly in a 10×10 square. Each two sensors are connected if they are in range R, where we used $R = 0.5, 0.6, 0.7, 0.8, 0.9, 1$. If the graph is disconnected, the largest connected component was taken. We measure the sensitivity of the algorithms to noise controlled by the fractional range measurement error parameter σ. The distances fed as input to the algorithms are the true distances l_{ij}, to which uniformly distributed random noise in the range $[-\sigma l_{ij}, +\sigma l_{ij}]$ is added; $\sigma = 0, 0.05, 0.1, 0.25, 0.5$. Consequently, each graph in this family is characterized by the values of R and σ. For each pair (R, σ) we generated 250 corresponding random graphs. Some properties of these graphs are given in [5].

It seems that the key to successful results is a good initial layout from which the stress minimization routine can start. To compare the performance of our algorithm to that of the AFL algorithm and a more naive method, we ran three different initialization methods on each input followed by the same stress minimization algorithm: (1) Stress majorization with random initialization (RND). (2) Stress majorization with AFL initialization (AFL). (3) Stress majorization with eigen-projection initialization (EIGEN). For each method the quality of the final solution is measured by its Average Relative Deviation (ARD), which measures the accuracy of all resulting pairwise distances:

$$ARD = \frac{2}{n(n-1)} \sum_{i<j} \frac{|d_{ij} - l_{ij}|}{\min(l_{ij}, d_{ij})}$$

Note that here we sum over *all* distances between sensors, not just the short range distances, as reflected by the edges of the graph. The results are summarized in Table 1, where each cell shows the average ARD of RND/AFL/EIGEN for 250 different graphs characterized by the same (R, σ) pair. For all graphs, EIGEN and AFL outperformed RND by a significant margin. Also, consistently, EIGEN outperformed AFL by a small margin. As expected, the algorithm performance improves as the graphs become denser, revealing more information about the underlying geometry. The sparser graphs contain nodes of degree smaller than 3, which are inherently non-rigid thereby preventing accurate recovery. We can also see that optimization is quite robust in the presence of noise and performance deteriorates only moderately as σ grows. In Figure 1 we show typical results of EIGEN, before and after stress minimization. For comparison, we also provide the original layout and the AFL initialization for the same graph.

In another experiment, we worked with 350 sensors distributed uniformly on a ring, with external radius 5 and internal radius 4. Again, the graphs are characterized by the range and noise parameters (R, σ), and for each such a pair we generated 250 corresponding random graphs. Here we worked with a different range of R, producing average degrees similar to those of the previous experiment; see [5]. Note that we avoided working with $R \leqslant 0.6$ as for these values the largest connected component broke the ring topology with high probability, making recovery impossible. We ran RND, AFL and EIGEN on these graphs, the results summarized in Table 2. The topology of the ring

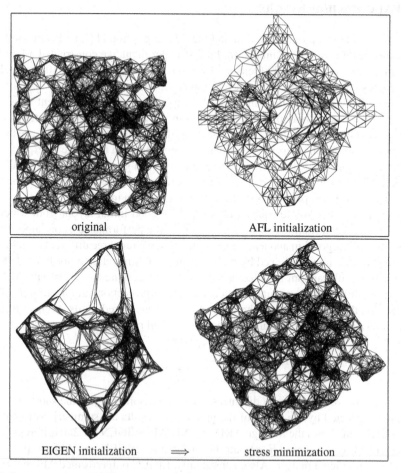

Fig. 1. Reconstructing a 1000-sensor proximity graph using EIGEN; here $R = 0.8, \sigma = 0$. Original layout and alternative AFL initialization are also shown.

Table 1. Average relative deviation (ARD) of square-based proximity graphs with varying (R, σ) generated by RND / AFL / EIGEN. Each result is averaged over 250 graphs.

	$\sigma = 0$			$\sigma = 0.05$			$\sigma = 0.1$			$\sigma = 0.25$			$\sigma = 0.5$		
	RND	AFL	EIGEN	RND	AFL	EIGEN	RND	AFL	EIGEN	RND	AFL	EIGEN	RND	AFL	EIGEN
$R = 0.5$	12.6	0.099	0.079	12.6	0.10	0.079	12.4	0.10	0.092	12.3	0.12	0.091	11.6	0.26	0.22
$R = 0.6$	11.2	0.026	0.0093	11.0	0.028	0.013	10.8	0.031	0.019	11.0	0.046	0.031	10.4	0.12	0.10
$R = 0.7$	9.70	0.013	0.0031	9.79	0.015	0.0048	9.77	0.017	0.0076	9.71	0.026	0.018	9.53	0.060	0.050
$R = 0.8$	8.51	0.0086	0.0016	8.52	0.0097	0.0033	8.42	0.012	0.0059	8.58	0.020	0.014	8.49	0.041	0.034
$R = 0.9$	7.29	0.0064	0.0011	7.37	0.0082	0.0028	7.28	0.011	0.0051	7.37	0.017	0.013	7.50	0.033	0.028
$R = 1.0$	6.31	0.0054	0.0008	6.40	0.0068	0.0025	6.51	0.0079	0.0047	6.33	0.016	0.012	6.52	0.030	0.026

Table 2. Average relative deviation (ARD) of disk-based proximity graphs with varying (R, σ) constructed using RND / AFL / EIGEN. Each result is averaged over 250 graphs.

	$\sigma = 0$			$\sigma = 0.05$			$\sigma = 0.1$			$\sigma = 0.25$			$\sigma = 0.5$		
	RND	AFL	EIGEN	RND	AFL	EIGEN	RND	AFL	EIGEN	RND	AFL	EIGEN	RND	AFL	EIGEN
$R = 0.7$	4.96	0.34	0.14	5.16	0.26	0.13	4.94	0.26	0.13	4.66	0.33	0.15	4.88	0.39	0.21
$R = 0.8$	7.69	0.19	0.091	7.53	0.23	0.091	7.54	0.020	0.090	7.81	0.19	0.10	7.41	0.29	0.16
$R = 0.9$	7.52	0.14	0.064	7.35	0.16	0.065	7.56	0.14	0.065	7.27	0.18	0.080	7.14	0.22	0.13
$R = 1.0$	6.61	0.10	0.041	6.62	0.11	0.045	6.41	0.11	0.046	6.54	0.13	0.055	6.40	0.15	0.091
$R = 1.1$	5.77	0.10	0.029	5.72	0.098	0.031	5.69	0.10	0.035	5.62	0.12	0.044	5.69	0.14	0.070
$R = 1.2$	4.97	0.11	0.021	4.98	0.11	0.021	4.88	0.11	0.026	5.08	0.13	0.032	4.97	0.16	0.058

is different than that of the square, and resulted in a lower quality results. However, all the observations from the square-based experiment still hold here. Note, that in a ring there is no natural central node. Therefore, the AFL initialization that identifies one node as the center is less appropriate here. A surprising finding is that the performance of AFL seems to deteriorate when increasing R from 1.1 to 1.2, instead of improving, as would be expected. We observed this also with other types of graphs we experimented with. We believe that this is due to the fact that the first phase of AFL models the network as an unweighted graph. Thus, as the variance of the true edge lengths becomes larger, this model is less accurate.

7   Conclusion

We have presented an algorithm to generate sensor network layouts in a fold-free manner based on noisy measurements of short-range inter-sensor distances. This algorithm is fully distributed (decentralized), and relies on no explicit communication other than that between immediate neighbors. The fully distributed nature of the algorithm is crucial for a practical implementation which avoids excessive communication. To the best of our knowledge, this is the first fully distributed algorithm for graph drawing. Beyond this important feature, judging from our experiments, our algorithm seems to be superior to the state-of-the-art in the sensor network literature. We discuss several extensions of the basic algorithm in [5].

References

1. I. Borg and P. Groenen, *Modern Multidimensional Scaling: Theory and Applications*, Springer-Verlag, 1997.
2. P. Eades, "A Heuristic for Graph Drawing", *Congressus Numerantium* **42** (1984), 149–160.

3. T.M.G. Fruchterman and E. Reingold, "Graph Drawing by Force-Directed Placement", *Software-Practice and Experience* **21** (1991), 1129–1164.
4. G.H. Golub and C.F. Van Loan, *Matrix Computations*, Johns Hopkins University Press, 1996.
5. C. Gotsman and Y. Koren, "Distributed Graph Layout for Sensor Networks", Harvard University Computer Science TR #20-04, 2004.
6. K. M. Hall, "An r-dimensional Quadratic Placement Algorithm", *Management Science* **17** (1970), 219–229.
7. B. Hendrickson, "Conditions for Unique Graph Realizations", *SIAM J. Comput.*, **21** (1992), 6–84.
8. T. Kamada and S. Kawai, "An Algorithm for Drawing General Undirected Graphs", *Information Processing Letters* **31** (1989), 7–15.
9. Y. Koren, "On Spectral Graph Drawing", *Proc. 9th Inter. Computing and Combinatorics Conference (COCOON'03)*, LNCS 2697, Springer-Verlag, pp. 496–508, 2003.
10. M. Mauve, J. Widmer and H. Hartenstein. "A Survey on Position-Based Routing in Mobile Ad-Hoc Networks", *IEEE Network*, **15**(2001), 30–39.
11. N.B. Priyantha, H. Balakrishnan, E. Demaine and S. Teller, "Anchor-Free Distributed Localization in Sensor Networks", *Proc. 1st Inter. Conf. on Embedded Networked Sensor Systems (SenSys 2003)*, 2003, pp. 340–341. Also *TR #892, MIT LCS*, 2003.
12. M. Tubaishat, S. Madria. "Sensor Networks : An Overview", *IEEE Potentials*, **22** (2003), 20–23.
13. L. Xiao, S. Boyd. "Fast Linear Iterations for Distributed Averaging", *Systems and Control Letters*, **53** (2004), 65–78.
14. Y. Yemini, "Some Theoretical Aspects of Location-Location Problems", *Proc. 20th Annu. IEEE Sympos. Found. Comput. Sci.*, 1979, pp. 1-8.

Drawing Large Graphs with a Potential-Field-Based Multilevel Algorithm

Extended Abstract

Stefan Hachul and Michael Jünger

Universität zu Köln, Institut für Informatik,
Pohligstraße 1, 50969 Köln, Germany
{hachul,mjuenger}@informatik.uni-koeln.de

Abstract. Force-directed graph drawing algorithms are widely used for drawing general graphs. However, these methods do not guarantee a sub-quadratic running time in general. We present a new force-directed method that is based on a combination of an efficient multilevel scheme and a strategy for approximating the repulsive forces in the system by rapidly evaluating potential fields. Given a graph $G = (V, E)$, the asymptotic worst case running time of this method is $O(|V| \log |V| + |E|)$ with linear memory requirements. In practice, the algorithm generates nice drawings of graphs containing 100000 nodes in less than 5 minutes. Furthermore, it clearly visualizes even the structures of those graphs that turned out to be challenging for some other methods.

1 Introduction

Given a graph $G = (V, E)$, force-directed graph drawing methods generate drawings of G in the plane in which each edge is represented by a straight line connecting its two adjacent nodes. The computation of the drawings is based on associating G with a physical model. Then, an iterative algorithm tries to find a placement of the nodes so that the total energy of the physical system is minimal. Such algorithms are quite popular, since they are easy to implement and often generate nice drawings of general graphs. In practice, classical force-directed algorithms like [5, 12, 6, 4] are not well suited for drawing large graphs containing many thousands of vertices, since their worst case running time is at least quadratic. Significantly accelerated force-directed algorithms have been developed by [15, 14, 7, 9, 17]. These algorithms generate nice drawings of a big range of large graphs in reasonable time. Some of these methods guarantee a sub-quadratic running time in special cases or under certain assumptions but not in general. Others are not sub-quadratic in any case. Besides force-directed algorithms other very fast methods for drawing large graphs have been invented by Harel and Koren [10] and Koren et al. [13] that do not use a physical force model.

In Section 2 we sketch the most important parts of a new force-directed graph drawing algorithm that guarantees a sub-quadratic worst case running time. An

J. Pach (Ed.): GD 2004, LNCS 3383, pp. 285–295, 2004.

excerpt of the experimental results is given in Section 3. For space restrictions, we can neither describe every basic component of this algorithm in detail, nor compare our method with the existing ones in a satisfactory way. This will be presented in the full version of this paper.

2   The Fast Multipole Multilevel Method (FM^3)

We describe the most important parts of the new method that is a combination of an efficient multilevel technique with an $O(|V| \log |V|)$ approximation algorithm to obtain the repulsive forces between all pairs of nodes/particles. Other important parts like a preprocessing step that enables the algorithm to draw graphs with nodes of different sizes and a part that is designed to handle disconnected graphs are not described here. Therefore, we simply assume that the given graph G is a connected weighted graph. The edge weight of each edge represents its individual desired edge length.

2.1   The Force Model

First, we must choose a force model. This is done by identifying the nodes with charged particles that repel each other and by identifying edges with springs, like in most classical force-directed methods. If in $\mathbb{R}^2$ two charges u, v are placed at a distance d from each other, the repulsive forces between u and v are proportional to $1/d$. Our choice of the spring forces is not strictly related to physical reality. We found that choosing the spring force of an edge e to be proportional to $\log(d/\textit{desired\_edgelength}(e)) \cdot d^2$ gives very good results in practice.

2.2   The Multilevel Strategy

Since in classical force-directed algorithms many iterations are needed to transform an initial (random) drawing of a large graph into the final drawing, one might hope to reduce the constant factor of force-directed algorithms by using a multilevel strategy. Multilevel strategies have been introduced into force-directed graph drawing by [7, 9, 17] and share the following basic ideas: Given $G = (V, E) =: G_0$ they create a series of Graphs $G_1, \ldots, G_k$ with decreasing sizes. Then, the smallest graph G_k at *level* k is drawn using (a variation of) a classical force-directed *(single-level)* algorithm. This drawing is used to get an initial layout of the next larger graph G_{k-1} that is drawn afterwards. This process is repeated until the original graph G_0 is drawn.

   Unlike previous approaches, we want to design a multilevel algorithm that has provably the same asymptotic running time as the single-level algorithm that is used to draw all graphs G_i with $i = 0, \ldots, k$.

   The idea of our multilevel step is as follows: First, we partition the node set of G into disjoint subsets. The induced connected subgraphs are called *solar systems*. A solar system S consists of one central *sun* node (*s*-node). Each of its neighbors is called *planet* node (*p*-node) and is also contained in S. The rest of the nodes in a solar system are called *moon* nodes (*m*-nodes), and each *m*-node is required to have graph-theoretic distance 2 to its associated *s*-node in G. Each *m*-node is assigned to its nearest neighboring *p*-node in S. This *p*-node is relabeled *planet with moon* node (*pm*-node), indicating that at least one *m*-node is assigned to it. Thus, the subgraph of G that is induced by a solar system has diameter at most 4.

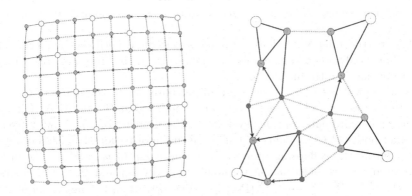

Fig. 1. (left) Drawing of $G = G_0$. (right) Drawing of G_1.

Figure 1(left) shows an example of a grid graph that is partitioned into 17 solar systems. The sun, planet, and moon nodes are represented by the white big, grey medium, and black small circles, respectively. The solid edges represent *intra* solar system edges, whereas the edges connecting nodes of two different solar systems (*inter* solar system edges) are dashed edges. The edges that connect an *m*-node and its assigned *pm*-node are drawn as directed edges, indicating that the *m*-node is assigned to this planet node.

We sketch a linear time method for constructing a solar system partitioning of a graph G that works in three steps: First, we create the sun nodes. We store a candidate set V' that is a copy of V and randomly select a first sun node s_1 from V'. Then, s_1 and all nodes that have a graph-theoretic distance at most 2 from s_1 in G are deleted from V'. We iteratively select the next sun nodes in the same way, until V' is empty and $Suns = s_1, \ldots, s_l$ is the list of all sun nodes. Second, for each $s_i \in Suns$ all its neighbors are labeled as planet nodes. Finally, there might be some nodes in V that are neither labeled as planet nodes, nor as sun nodes. These nodes are the moon nodes, and we assign each moon node to the planet node that is its nearest neighbor in G.

Given a solar system partition of the node set of $G = G_0$, we construct a smaller graph G_1 by collapsing (shrinking) the node set of each solar system into one single node and deleting parallel edges (see Figure 1(right)). The smaller graph should reflect the attributes of the bigger graph as much as possible. Therefore, we initialize the desired edge length of an edge $e_1 = (u_1, v_1)$ in G_1 as follows: Suppose that *p*-node u_0 belongs to the solar system S_0 with sun node s_0 in G_0 and *p*-node v_0 belongs to the solar system T_0 with sun node

t_0 in G_0. Let us also assume that the edge $e_0 = (u_0, v_0)$ is the unique inter solar system edge connecting S_0 and T_0. Furthermore, we assume that nodes u_1 and v_1 in G_1 are obtained by collapsing S_0 and T_0. Then, we set the desired edge length of e_1 to $desired\_edgelength((s_0, u_0)) + desired\_edgelength(e_0) + desired\_edgelength((v_0, t_0))$. For later use, we denote the corresponding path P_0 and its length p_0. If more than one inter solar system edge in G_0 connects nodes of S_0 with nodes of T_0, we just take the average of the previously calculated desired edge lengths. The case that u_0 and/or v_0 is a moon node is treated similarly.

This partitioning and collapsing process is iterated until the smallest graph G_k contains only a constant number of nodes. Then, this graph is drawn by an algorithm that is introduced later.

Going upwards to G_{k-1}, we assign initial positions to the nodes of G_{k-1} in two steps: First, we place each sun node s of G_{k-1} at the position of its ancestor (that represents its solar system) in the drawing of G_k. Now, we place the other nodes of G_{k-1}. This is done by using information that has been generated during the collapsing process: Given u_0, v_0, s_0, t_0, p_0, and P_0 like in the example above, we place u_0 on the line connecting s_0 and t_0 at the position $\text{Pos}(s_0) + \frac{desire\_edgelength((s_0, u_0))}{p_0}(\text{Pos}(t_0) - \text{Pos}(s_0))$. If u_0 belongs to more than one such path P_0, we take the barycenter of all these positions. The case that u_0 is a moon node is treated similarly.

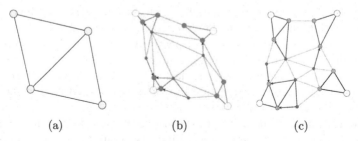

Fig. 2. (a) Drawing of G_2. (b) Initial placement of G_1. (c) Drawing of G_1.

Figure 2 demonstrates this procedure. Figure 2(a) is a drawing of the multi-level graph G_2 of Figure 1(left). Figure 2(b) is the initial position of the drawing of G_1 that is obtained from the drawing of G_2. Figure 2(c) shows G_1 that is drawn with a new force-directed single-level algorithm.

The total running time of the multilevel strategy is $t_{\text{mult}}(|V|, |E|) = \sum_{i=0}^{k-1} t_{\text{create}}(|V_i|, |E_i|) + \sum_{i=0}^{k-1} t_{\text{init\_pl}}(|V_i|, |E_i|) + \sum_{i=0}^{k} t_{\text{single}}(|V_i|, |E_i|)$. Here, $t_{\text{create}}(|V_i|, |E_i|)$ denotes the time that is needed to create the multilevel graph G_{i+1} from G_i. $t_{\text{init\_pl}}(|V_i|, |E_i|))$ denotes the time that is needed to get an initial placement of the nodes of the multilevel graph G_i from the drawing of G_{i+1}. $t_{\text{single}}(|V_i|, |E_i|)$ is the time that the chosen single-level algorithm needs to draw G_i.

Since every node of G_i belongs to a solar system, and a solar system contains at least two nodes, G_{i+1} contains at most $|V_i|/2$ nodes. Therefore, $k \leq \log |V|$.

Let us assume that $|E_{i+1}| \leq |E_i|/2$ for all $i = 0, \ldots, k-1$. Since both $t_{create}(|V_i|, |E_i|)$ and $t_{\text{init\_pl}}((|V_i|, |E_i|))$ are linear in $|V_i| + |E_i|$ we get $\sum_{i=0}^{k-1} t_{create}$ $(|V_i|, |E_i|) + \sum_{i=0}^{k-1} t_{\text{init\_pl}}(|V_i|, |E_i|) = O(|V| + |E|)$. Furthermore, we get the following estimation on t_{single}:

$$\sum_{i=0}^{k} t_{\text{single}}(|V_i|, |E_i|) \leq \sum_{i=0}^{k} t_{\text{single}}\left(\frac{|V|}{2^i}, \frac{|E|}{2^i}\right) \leq t_{\text{single}}(|V|, |E|) \sum_{i=0}^{k} \frac{1}{2^i} \leq$$

$2\, t_{\text{single}}(|V|, |E|)$. The second inequality is true for sufficiently large values of $|V|$ and $|E|$, since $t_{\text{single}}(|V|, |E|) = \Omega(|V| + |E|)$. Therefore, $t_{\text{mult}}(|V|, |E|)$ and $t_{\text{single}}(|V|, |E|)$ have the same asymptotic running time.

Certainly, it cannot be guaranteed that the number of edges decreases by a factor $\frac{1}{2}$ as well. This might result in an additional factor $k = \log |V|$ on the parts of the algorithm that touch edges. However, it can be shown by an analogous argumentation that $t_{\text{mult}}(|V|, |E|)$ and $t_{\text{single}}(|V|, |E|)$ have the same asymptotic running time if $|E_{i+1}| \leq |E_i|/d$ for all $i = 0, \ldots, k-1$ and a fixed divisor $d > 1$. Therefore, it is sufficient to stop the multilevel process, whenever the algorithm has generated more than a constant number of graphs G_i that do not satisfy the inequality $|E_{i+1}| \leq |E_i|/d$ for some small $1 < d \leq 2$.

2.3   The Force Calculation Step

In order to save running time, the multilevel algorithms [7, 9, 17] use the grid-variant method of [6] or variations of [12] that are comparatively inaccurate approximative variations of the original single-level algorithms [6, 12].

Unlike this, the single-level algorithm that is used in FM^3 follows the basic strategy of [15, 14] by approximating the repulsive forces between all pairs of distinct nodes/particles with high accuracy and calculating the forces induced by the edges/springs exactly. Then, these forces are added, and the nodes are moved in the direction of the resulting forces. This process is repeated a constant number of iterations. (In practice, we let the constant decrease from 300 iterations for G_k to 30 iterations for G_0, although convergence is reached even faster for many tested graphs.)

In the following, we concentrate on the calculation of the repulsive forces. Greengard [8] has invented an N-body simulation method that is based on the evaluation of the field of the potential energy of $N := |V|$ particles. This is done by evaluating multipole expansions using a hierarchical data structure called quadtree. However, Aluru et al. [1] have shown that the running time of his method depends on the particle distribution and cannot be bounded in the number of particles. They also have proven that the running time of the popular center of mass approximation method of Barnes and Hut [3] that is used in the graph drawing methods [15, 14] cannot be bounded in the number of particles. Based on the techniques and analytical tools of Greengard [8], Aluru et al. [1] have presented an $O(N \log N)$ approximative multipole algorithm that is distribution independent.

Based on the work of Greengard [8] and Aluru et al. [1], we have developed a new $O(N \log N)$ multipole method that – in practice – is faster than Aluru et al.'s [1] method. It works in two steps. Given a distribution of N particles in the plane, first a special quadtree data structure is constructed. Then, each node of the quadtree is assigned information that is used for approximating the potential energy of the system. In particular, a constant number of coefficients of a so called *multipole expansion* (to be introduced later) are associated with each tree node and are used to obtain the repulsive forces.

Construction of the Reduced Quadtree. Suppose that N particles are distributed on a square D and we fix a *leaf capacity* $c \geq 1$. (In practice we choose $c = 25$.) Furthermore, suppose one recursively subdivides D into four squares of equal size, until each square contains at most c particles. This process can be represented by an ordered rooted tree of maximum child degree four (with the root representing D) that is called *quadtree*. The particles are stored in the leaves of the quadtree. A *degenerate path* $P = (v_1, \ldots, v_p)$ in a quadtree is a path in which v_1 and v_p have at least 2 nonempty children and $v_2, \ldots, v_{p-1}$ each have exactly one nonempty child. A *reduced quadtree* T can be obtained from a quadtree by shrinking degenerate paths $P = (v_1, \ldots, v_p)$ to edges (v_1, v_p). Figure 3 shows an example.

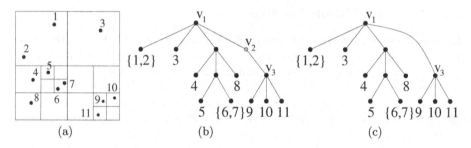

(a)   (b)   (c)

Fig. 3. (a) A distribution of $N = 11$ particles in the plane. (b) The quadtree with leaf capacity $c = 2$ associated with (a). $P = (v_1, v_2, v_3)$ is a degenerate path in the quadtree. (c) The reduced quadtree with leaf capacity $c = 2$ associated with (a).

A reduced quadtree has only $O(N)$ nodes independently of the distribution of the particles. This allows the development of a linear time method (excluding the time needed for constructing the reduced quadtree) for approximating the repulsive forces, using this structure.

Aluru et al. [1] present an $O(N \log N)$ method that constructs a reduced quadtree with $c = 1$. As can be shown by a reduction from sorting, it is neither possible to construct a quadtree nor a reduced quadtree for arbitrary distributions of the particles in $o(N \log N)$ time.

We have developed a new $O(N \log N)$ method that is omitted here for brevity, since it quite technical. Instead, we will explain another new tree construction method that is conceptionally simpler and in practice faster. But (motivated by the assumptions in [15]) it restricts the possible particle distributions: We force the particles to be placed on a large square grid with a resolution that is polynomial in N. This can be realized by rounding the x, y coordinates of each particle to integers in the range $[0, \mathcal{P}(N)]$, where $\mathcal{P}(N)$ is any whole-numbered polynomial in N of maximum degree l, and by treating pathological cases in which particles have same coordinates efficiently. In practice, it is sufficient to set $\mathcal{P}(N) = d \cdot N^2$, with a big constant d, say 1000. This bounds the depth of the reduced quadtree to $O(\log(\mathcal{P}(N))) = O(l \cdot \log N) = O(\log N)$.

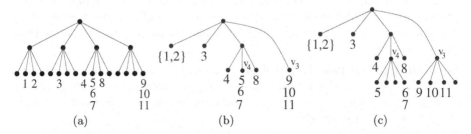

$$(a) \qquad\qquad\qquad (b) \qquad\qquad\qquad (c)$$

Fig. 4. Building up the reduced quadtree T with leaf capacity $c = 2$ and $N = 11$ particles for the distribution of Figure 3(a). (a) First step: Building up the complete subtree T^1. (b) Second step: Thinning out T^1. (c) Recursion: Building up the complete quadtrees $T^2(v_3)$ and $T^2(v_4)$.

First, we build up a complete truncated subtree T^1 with depth max$\{1, \lfloor \log N/c \rfloor\}$. Then, all particles are assigned to the leaves of T^1. Since T^1 contains $O(N)$ nodes and its structure is predefined, this step can be performed in linear time. Figure 4(a) shows an example of T^1 that corresponds to the distribution of Figure 3(a). In the next step, we thin out T^1. This is done by traversing the tree bottom up and thereby calculating for each internal tree node the number of particles that are contained in the square region that it represents. This also needs time linear in N. Then, we traverse the subtree T^1 top down, delete all nodes that do not contain particles and shrink degenerate path to edges. If (during this process) we visit an internal node v that is the root of a subtree containing at most c particles, this subtree is deleted, and all the particles that were stored in the deleted subtree are assigned to v. Figure 4(b) shows the thinned out subtree T^1.

If none of the leaves of T^1 contains more than c particles, the procedure ends and $T^1 = T$ has been constructed in linear time. Otherwise, we repeat the previous steps recursively. For example, the nodes v_3 and v_4 in Figure 4(b) both contain $3 > c$ particles. Therefore, we build up complete subtrees $T^2(v_3)$ rooted at v_3 and $T^2(v_4)$ rooted at v_4. Both subtrees have depth max$\{1, \lfloor \log 3/c \rfloor\} = 1$. Now, the particles $5, 6, 7$ are assigned to the leaves of $T^2(v_4)$ and the particles $9, 10, 11$ are assigned to the leaves of $T^2(v_3)$ (see Figure 4(c)). After thinning out $T^2(v_3)$ and $T^2(v_4)$ the desired tree T (see Figure 3(c)) is created.

What is the total running time of this approach? Building up T^1 needs $O(N)$ time. If T^1 is not the reduced quadtree, we build up subtrees $T^2(v_1), \ldots, T^2(v_k)$ for all leaves $v_1, \ldots, v_k$ of T^1 that contain more than c particles. This needs $O(N)$ time in total, since the sum of the tree nodes contained in all $T^2(v_i)$ is at most $O(N)$. Then, we possibly have to build up subtrees rooted at the leaves of the T^2 trees and so forth. Since for each $j \geq 1$ the sum of the tree nodes of all T^j is bounded above by $O(N)$, the total running time is $O(|\text{recursion\_levels}| \cdot N)$. Therefore, the running time is bounded by $O(N \log N)$. The construction of the tree needs linear time whenever $|\text{recursion\_levels}|$ is a constant.

Evaluating Multipole Expansions. Unlike the construction of the tree, the calculation of the forces – using the tree data structure – is quite complex. Therefore, we only sketch the basic ideas. The most essential part is the following theorem of Greengard [8]. First, we identify each point $p = (x, y) \in \mathbb{R}^2$ with a point $z = x + iy \in \mathbb{C}$.

Theorem 1 (Multipole Expansion) *Suppose that m charges of strengths q_i, $\{i = 1, \ldots, m\}$ are located within a circle of radius r around the center z_0. Then, for any $z \in \mathbb{C}$ with $|z - z_0| > r$, the potential Energy $\mathcal{E}(z)$ induced by the m charged particles is given by:*

$$\mathcal{E}(z) = Q \log(z - z_0) + \sum_{k=1}^{\infty} \frac{a_k}{(z - z_0)^k} \text{ with } Q = \sum_{i=1}^{m} q_i \text{ and } a_k = \sum_{i=1}^{m} \frac{-q_i(z_i - z_0)^k}{k}$$

The corresponding force is $\mathcal{F}(z) = (\text{Re}(\mathcal{E}'(z)), -\text{Im}(\mathcal{E}'(z)))$. Based on this theorem, the idea is to develop the infinite series only up to a constant index p. In practice, choosing $p = 4$ has turned out to be sufficient to keep the error of the approximation less than 10^{-2}. The resulting truncated Laurent series is called *p-term multipole expansion*. Estimations of the error and several other fundamental theorems for working with such series can be found in [8].

We demonstrate the use of this theorem for speeding up force-calculation algorithms on an example: Suppose that m particles are located within a circle C_0 of radius r with center z_0. Suppose that another m particles are located within a circle C_1 of radius r with center z_1, and let $|z_0 - z_1| > 3r$ (see Figure 5).

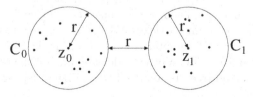

Fig. 5. An example distribution, showing the use of the Multipole Expansion Theorem.

Computing the repulsive forces acting on each particle in C_0 due to all particles in C_1 naively would need $\Theta(m^2)$ time. Now, suppose that we first compute the coefficients of a p-term multipole expansion of the potential due to the particles in C_1. This needs $\Theta(pm)$ time. Evaluating the resulting p-term multipole expansion for all particles within C_0 needs also $\Theta(pm)$ time. Therefore, we obtain an accurate approximation of the potential energy of all particles placed in C_0 due to the particles placed in C_1 in $\Theta(m)$ time. Since we are interested in the forces rather than the energy, we first calculate the derivative of the p-term multipole expansion before evaluating it for each particle in C_0. Since the multipole expansion is a simple Laurent series, the calculation of the derivative of the p-term multipole expansion needs only $O(p)$ additional time.

Now we sketch the idea how this theorem is used for calculating the forces: First, the p-term multipole expansions of the particles in the leaves of the reduced quadtree are calculated. Then, the reduced quadtree is traversed bottom up, and thereby p-term multipole expansions of the interior nodes are obtained. Afterwards, the tree is traversed top down, and suitable coefficients of p-term multipole expansions are used to calculate coefficients of special power series that are called *p-term local expansions*. Finally, these expansions are evaluated to obtain the repulsive forces. All these operations together take time linear in the number of particles.

To get a better impression how this algorithm really works, we refer the interested reader to [8, 1]. Our method for evaluating the multipole expansions is an extension of the method of Aluru [1] et al. for the general case in which $c \geq 1$.

It is important to note that our multipole method remains $O(N \log N)$ – even if we allow arbitrary particle positions during the computation – if we use our other tree construction method or the tree construction method of Aluru et al. [1].

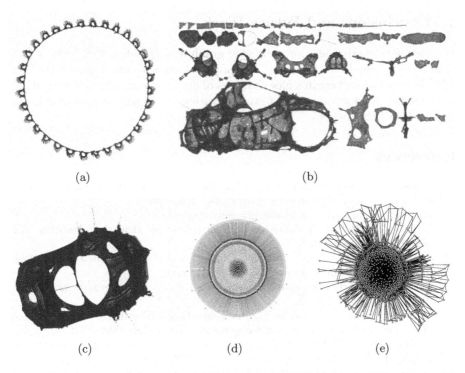

(a)

(b)

(c)   (d)   (e)

Fig. 6. (a) *finan512*: $|V| = 74752$, $|E| = 261120$, CPU-time $= 158.2$ seconds. (b) *fe_body*: $|V| = 44775$, $|E| = 163734$, CPU-time $= 96.5$ seconds. (c) *bcsstk31*: $|V| = 35588$, $|E| = 572914$, edge density $= 16.1$, CPU-time $= 83.6$ seconds. (d) *dg_1087*: $|V| = 7602$, $|E| = 7601$, maximum degree $= 6566$, CPU-time $= 18.1$ seconds. (e) *ug_380*: $|V| = 1104$, $|E| = 3231$, maximum degree $= 856$, CPU-time $= 2.1$ seconds.

3   Remarks on the Experimental Results

The method FM^3 has been implemented in C++ within the framework of *AGD* [11]. We tested our method on a 2.8 GHz PC running Linux. The tested graphs are the graphs contained in the graph partitioning archive of C. Walshaw [16] with up to 200000 nodes and the biggest graphs from the AT&T graph collection [2]. Furthermore, we generated artificial graphs containing up to 100000 nodes. For example, these graphs include grid graphs, sierpinski graphs, random disconnected graphs, graphs that contain many biconnected components, graphs with a very high edge density, and graphs that contain nodes with a very high degree. The tested graphs containing less than 1000, 10000, and 100000 nodes have been drawn in less than 2, 24, and 263 seconds, respectively. Figure 6 shows example drawings that are generated by FM^3. Our practical experiments indicate that the combination of our multilevel strategy with a highly accurate force approximation algorithm increases the quality of the generated drawings.

4   Conclusions and Future Work

We have developed a new force-directed graph drawing algorithm (FM^3) that runs in $O(|V|\log|V| + |E|)$ time. The practical experiments demonstrate that FM^3 is very fast and creates nice drawings of even those graphs that turned out to be challenging for some other tested algorithms. This will be presented in the full version of this paper.

References

1. S. Aluru et al. Distribution-Independent Hierarchical Algorithms for the N-body Problem. *Journal of Supercomputing*, 12:303–323, 1998.
2. The AT&T graph collection: www.graphdrawing.org
3. J. Barnes and P. Hut. A hierarchical $\mathcal{O}(N\log N)$ force-calculation algorithm. *Nature*, 324(4):446–449, 1986.
4. R. Davidson and D. Harel. Drawing Graphs Nicely Using Simulated Annealing. *ACM Transaction on Graphics*, 15(4):301–331, 1996.
5. P. Eades. A heuristic for graph drawing. *Congressus Numerantium*, 42:149–160, 1984.
6. T. M. J. Fruchterman and E. M. Reingold. Graph Drawing by Force-directed Placement. *Software–Practice and Experience*, 21(11):1129–1164, 1991.
7. P. Gajer et al. A Multi-dimensional Approach to Force-Directed Layouts of Large Graphs. In J. Marks, editor, *Graph Drawing 2000*, volume 1984 of *Lecture Notes in Computer Science*, pages 211–221. Springer-Verlag, 2001.
8. L. F. Greengard. *The Rapid Evaluation of Potential Fields in Particle Systems*. ACM distinguished dissertations. The MIT Press, Cambridge, Massachusetts, 1988.
9. D. Harel and Y. Koren. A Fast Multi-scale Method for Drawing Large Graphs. In J. Marks, editor, *Graph Drawing 2000*, volume 1984 of *Lecture Notes in Computer Science*, pages 183–196. Springer-Verlag, 2001.

10. D. Harel and Y. Koren. Graph Drawing by High-Dimensional Embedding. In M. T. Goodrich and S. G. Kobourov, editors, *Graph Drawing 2002*, volume 2528 of *Lecture Notes in Computer Science*, pages 207–219. Springer-Verlag, 2002.

11. M. Jünger et al. *Graph Drawing Software*, volume XII of *Mathematics and Visualization*, chapter AGD - A Library of Algorithms for Graph Drawing, pages 149–169. Springer-Verlag, 2004.

12. T. Kamada and S. Kawai. An Algorithm for Drawing General Undirected Graphs. *Information Processing Letters*, 31:7–15, 1989.

13. Y. Koren et al. Drawing Huge Graphs by Algebraic Multigrid Optimization. *Multiscale Modeling and Simulation*, 1(4):645–673, 2003.

14. A. Quigley and P. Eades. FADE: Graph Drawing, Clustering, and Visual Abstraction. In J. Marks, editor, *Graph Drawing 2000*, volume 1984 of *Lecture Notes in Computer Science*, pages 197–210. Springer-Verlag, 2001.

15. D. Tunkelang. JIGGLE: Java Interactive Graph Layout Environment. In S. H. Whitesides, editor, *Graph Drawing 1998*, volume 1547 of *Lecture Notes in Computer Science*, pages 413–422. Springer-Verlag, 1998.

16. C. Walshaw's graph collection: `www.gre.ac.uk/~c.walshaw/partition`

17. C. Walshaw. A Multilevel Algorithm for Force-Directed Graph Drawing. In J. Marks, editor, *Graph Drawing 2000*, volume 1984 of *Lecture Notes in Computer Science*, pages 171–182. Springer-Verlag, 2001.

Building Blocks of Upward Planar Digraphs

Patrick Healy and Karol Lynch

CSIS Department, University of Limerick, Limerick, Ireland
{patrick.healy,karol.lynch}@ul.ie

Abstract. We show that a digraph is upward planar if and only if its biconnected components have certain properties.

1 Introduction

A drawing of a digraph is *planar* if no edges cross and *upward* if all edges are monotonically increasing in the vertical direction. A digraph is *upward planar* (UP) if it admits a drawing that is *both* upward and planar. The upward planarity of digraphs has been much studied and the area is surveyed by Di Battista et al. [1]. In this paper we show that a digraph is UP if and only if its blocks have UP drawings satisfying the conditions specified in Theorems 6–8. Following preliminary definitions in Section 2, lemmas concerned with sufficient conditions for upward planarity appear in Section 3, lemmas concerned with necessary conditions for upward planarity appear in Section 4, our main results appear in Section 5, and we conclude in Section 6.

2 Preliminaries

Let G be a digraph. We denote the *node set* of G by $V(G)$ and the *edge set* of G by $E(G)$. The *union* of two digraphs G and F, denoted $G \cup F$, is the digraph with node set $V(G \cup F) = V(G) \cup V(F)$ and edge set $E(G \cup F) = E(G) \cup E(F)$. A *null digraph* is a digraph whose node and edge sets are empty. A *strongly embedded digraph* G_φ is an equivalence class of planar drawings of a digraph G which belong to the same embedded digraph and which have the same outer face. Such a choice φ for an *embedding* and an outer face is called a *strong embedding* of G. A node v of G_ϕ is *bimodal* if the outgoing (or incoming) edges incident on v appear consecutively around v. If all the nodes of G_ϕ are bimodal then G_ϕ is *bimodal*. An *upward planar straight line* (UPSL) drawing of a digraph is an UP drawing in which each edge is represented by a straight line segment. Let G_ϕ be an embedded digraph with a node v. We use $\phi(v)$ to denote the clockwise ordering of the edges incident on v in G_ϕ. If e is an edge incident on v we use $\phi^e(v)$ to denote the linear sequence of edges with first edge e that is consistent with the circular sequence of edges $\phi(v)$. The *angles* of G_ϕ are ordered triples $\langle a, v, b \rangle$, where a and b are edges and v is a node incident on both a and b, such that either a directly precedes b in $\phi(v)$ or v is a node of degree 1. An angle $\langle a, v, b \rangle$ of G_ϕ is *incident* on the node v. An angle $\langle a, v, b \rangle$ is an *S-angle* (resp.,

J. Pach (Ed.): GD 2004, LNCS 3383, pp. 296–306, 2004.

T-angle) if both a and b *leave* (resp., *enter*) v; and an I-angle if one of the edges a, b leaves v and the other enters v. Let f be a face of G_ϕ. We use the term *facial boundary* of f to refer to the circular sequence of nodes and edges W defined by traversing the boundary of f such that W contains a subsequence $\langle x, y, z \rangle$, where x and z are edges and y is a node if and only if $\langle x, y, z \rangle$ is an angle in G_ϕ. A face f is said to *contain* an edge, angle, or node x if x is a sub-sequence of the facial boundary of f. The angles of G_ϕ are mapped to geometric angles in an UPSL drawing Γ of G_ϕ. Let $\langle a, v, b \rangle$ be an angle of Γ. If $a \neq b$ the size of the corresponding geometric angle of $\langle a, v, b \rangle$ in Γ equals the number of radians one has to rotate a in the clockwise direction around v in order to reach b. If $a = b$ the size of the corresponding geometric angle of $\langle a, v, b \rangle$ is 2π. An angle of Γ is *large* (resp., *small*) if its corresponding geometric angle is greater (resp., smaller) than π. If f is a face we use $S_a(f)$ (resp., $L_a(f)$) to represent the number of S-angles (resp., *(large S-angles + large T-angles)*) contained by f. If v is a node we use $L_a(v)$ to represent the number of *(large S-angles + large T-angles)* incident on v. A *block* of a digraph G is a maximal connected subdigraph B of G such that no node of B is a cut-vertex of B. Let G be a digraph with a node v. A *component of G with respect to v* is formed from a connected component H of $G \setminus v$ by adding to H the node v and all edges between v and H. Let C be a component of G with respect to v. We use $G \setminus C$ to denote the digraph derived from G by deleting all the nodes in $C \setminus v$ from G. C is an *S-component*, *T-component*, or *I-component* of G with respect to (w.r.t.) v if v is a source node, sink node or internal node respectively in C. We use $S(v)$, $T(v)$, and $I(v)$ to refer to the subdigraph of G consisting of the union of all S-components, T-components, and I-components of G with respect to v, respectively. Definition 1 attempts to formalise the intuitive operation of adding two strongly embedded digraphs by identifying a node from each.

Definition 1. *Let G_{φ_G} and F_{φ_F} be strongly embedded digraphs such that $V(G) \cap V(F) = \{u\}$, the outer face E_G of G_{φ_G} contains the angle $\langle g_1, u, g_2 \rangle$ and the face X_F of F_{φ_F} contains the angle $\langle f_1, u, f_2 \rangle$. The result of adding G_{φ_G} and F_{φ_F} by inserting $\langle g_1, u, g_2 \rangle$ within $\langle f_1, u, f_2 \rangle$ is the strongly embedded digraph H_φ where $H = G \cup F$, $\varphi(v) = \varphi_G(v), \forall \in V(G) \setminus u, \varphi(v) = \varphi_F(v), \forall v \in V(F) \setminus u$, $\varphi(u) = \langle \varphi_G^{g_2}(u), \varphi_F^{f_2}(u) \rangle$, and the outer face is the face of H_φ whose facial boundary contains the facial boundary of the outer face of F_{φ_F} as a (not necessarily proper) subsequence.*

Property 1. The set of faces of H_φ contains all the faces of G_{φ_G} except E_G, all the faces of F_{φ_F} except the face, X_F, and the "new" face X_H whose facial boundary is formed by concatenating the facial boundaries of E_G and X_F. Thus the facial boundary of X_H is $\langle u, g_2, ..., g_1, u, f_2, ..., f_1 \rangle$, where the list of nodes and edges from g_2 to g_1 concatenated with u is the facial boundary of E_G and the list of nodes and edges from f_2 to f_1 concatenated with u is the facial boundary of X_F. Therefore X_H contains all the angles contained by E_G except $\langle g_1, u, g_2 \rangle$, all the angles contained by X_F except $\langle f_1, u, f_2 \rangle$, and the new angles $\langle f_1, u, g_2 \rangle$ and $\langle g_1, u, f_2 \rangle$.

The following four preliminary lemmas have been proved by us previously [4].

Lemma 1. *If G and F are acyclic digraphs such that $V(G) \cap V(F) = \{u\}$, then $G \cup F$ is acyclic.*

Lemma 2. *Let $H_i, i = 1, \ldots, c$ be the components of a digraph H with respect to a node u. Let $\mathcal{H}$ be a drawing of H and let $\mathcal{H}_i$ be the sub-drawing induced on $H_i, i = 1, \ldots, c$. $\mathcal{H}$ is UP only if at least $c - 1$ of the sub-drawings $\mathcal{H}_i$ for $i = 1$ to c are UP drawings whose outer face contains u.*

Lemma 3. *Let H be a planar digraph with a node u such that there are exactly two components of H w.r.t. u, which we label G and F. Let $\mathcal{H}$ be a planar drawing of H and let $\mathcal{G}$ (resp., $\mathcal{F}$) be the sub-drawing induced on G (resp., F). Then all of $\mathcal{G}$ (resp., $\mathcal{F}$) lies in a single face of $\mathcal{F}$ (resp., $\mathcal{G}$) and at least one of $\mathcal{G}$ or $\mathcal{F}$ lies in the outer face of the other.*

Lemma 4. *Let H be a bimodal planar digraph with an internal node u such that there are exactly two components, G and F, of H with respect to u. Let $\mathcal{H}$ be a bimodal planar drawing of H and let $\mathcal{G}$ (resp., $\mathcal{F}$) be the subdrawing induced on G (resp., F). If G and F are I-components then all of $\mathcal{G}$ (resp., $\mathcal{F}$) lies in a single face of $\mathcal{F}$ (resp., $\mathcal{G}$) which contains an I-angle incident on u and at least one of $\mathcal{G}$ or $\mathcal{F}$ lies in the outer face of the other.*

We now present some previously published properties of UP digraphs. Lemma 5 and Theorem 1 are proved by Bertolazzi et al. [2]. Let G_φ be a strongly embedded digraph. Consider an *assignment* M that maps each source or sink v of G_φ to a face $M(v)$ of G_φ which contains v. Such an assignment M is *consistent* if the number of nodes assigned to the outer face h of G_φ equals $S_a(h) + 1$ and the number of nodes assigned to each internal face f of G_φ equals $S_a(f) - 1$. For each face z in G_φ we use $M^{-1}(z)$ to denote the set of nodes assigned to z by M.

Theorem 1 (Bertolazzi et al.). *A strongly embedded digraph G_φ is UP if and only if it is acyclic, bimodal and admits a consistent assignment of sources and sinks to its faces.*

Lemma 5 (Bertolazzi et al.). *The following properties hold for any UPSL drawing Γ of a digraph G. $L_a(f) = S_a(f) + 1$ if f is the outer face of Γ and $L_a(f) = S_a(f) - 1$ if f is an internal face of Γ. Also $L_a(v) = 0$ if v is an internal node of G and $L_a(v) = 1$ if v is a source or sink node of G.*

3 Sufficient Conditions for Upward Planarity

The same technique is used to prove each of the Lemmas 6–9. We present the proof of Lemma 6 here; proofs of Lemmas 7–9 can be found in [4]. Although the proofs of these lemmas are quite detailed the high level strategy is not. Given two UP strongly embedded digraphs G_{φ_G} and F_{φ_F} with certain properties we simply show that the strongly embedded digraph H_φ that results from adding G_{φ_G} and F_{φ_F} by inserting a certain angle from G_{φ_G} within a certain angle from the F_{φ_F} is acyclic, bimodal and has a consistent assignment of sinks and sources to its faces. It then follows from Theorem 1 that H_φ is UP.

Lemma 6. *Let G and F be UP digraphs such that $V(G) \cap V(F) = \{u\}$ and u is a source (resp., sink) node in both G and F. Let $H = G \cup F$. The following three statements hold. **1.** H is UP if at least one of G or F has an UP drawing whose outer face contains u. **2.** H has an UP drawing whose outer face contains u if G and F both have an UP drawing whose outer face contains u. **3.** H has an UPSL drawing whose outer face contains a large angle incident on u if G and F both have UPSL drawings whose outer face contains a large angle incident on u.*

Proof. Assume that u is a source node (the case when u is a sink node follows by symmetry) and that G has an UPSL drawing $\mathcal{G}$ whose outer face E_G contains the angle $\langle g_1, u, g_2 \rangle$. Let $\mathcal{F}$ be an UPSL drawing of F and let X_F be the face of $\mathcal{F}$ that contains $\langle f_1, u, f_2 \rangle$, the large angle incident on u in $\mathcal{F}$. Thus G (resp., F) has a strong embedding φ_G (resp., φ_F) such that G_{φ_G} (resp., F_{φ_F}) contains $\mathcal{G}$ (resp., $\mathcal{F}$) and has a consistent assignment, M_G (resp., M_F) of sinks and sources to faces, such that M_G (resp., M_F) assigns u to a face X_G (resp., the face X_F) of G_{φ_G} (resp., F_{φ_F}) (Theorem 1). Let H_φ be the strongly embedded digraph that results from adding G_{φ_G} and F_{φ_F} by inserting $\langle g_1, u, g_2 \rangle$ within $\langle f_1, u, f_2 \rangle$. We now show that H_φ is acyclic, bimodal and has a consistent assignment. That H is acyclic follows directly from Lemma 1. It follows from Definition 1 that $\varphi(v) = \varphi_G(v), \forall v \in V(G) \setminus u$ and that $\varphi(v) = \varphi_F(v), \forall v \in V(F) \setminus u$. But G_{φ_G} and F_{φ_F} are bimodal. Thus all nodes in $V(H) \setminus u$ are bimodal. As u is a source node it is also bimodal. Thus all nodes in H_φ are bimodal. The set of faces of H_φ consists of all the faces of G_{φ_G} except E_G, all the faces of F_{φ_F} except X_F, and the "new" face X_H (Property 1). We now show that H_φ has a consistent assignment M of sinks and sources to faces. M is defined as follows: **P1.** For each source or sink v of $V(H) \cap V(G)$, if $M_G(v) \neq E_G$ then $M(v) = M_G(v)$; otherwise $M(v) = X_H$. **P2.** For each source or sink v of $V(H) \cap V(F) \setminus u$, if $M_F(v) \neq X_f$ then $M(v) = M_F(v)$; otherwise $M(v) = X_H$. We now show that the assignment M is consistent. The number of S-angles in each face $z \neq X_H$ of H_φ is equal to the number of S-angles in its corresponding face in either G_{φ_G} or F_{φ_F}. It follows from **P1** that $|M_G^{-1}(z)| = |M^{-1}(z)|$ for each face $z \neq X_H$ of H_φ that is a face in both G_{φ_G} and H_φ. It follows from **P2** that $|M_F^{-1}(z)| = |M^{-1}(z)|$ for each face $z \neq X_H$ of H_φ that is a face in both F_{φ_F} and H_φ (because u was assigned to X_F by M_F). Thus from the consistency of M_G and M_F that the assignment M is consistent for each face $z \neq X_H$ of H_φ. We now consider the face X_H. All the S-angles contained by E_G (resp., X_f) except $\langle g_1, u, g_2 \rangle$ (resp., $\langle f_1, u, f_2 \rangle$) are also contained by X_H (Property 1). X_H also contains two "new" S-angles $\langle f_1, u, g_2 \rangle$ and $\langle g_1, u, f_2 \rangle$ (Property 1). Therefore $S_a(X_H) = S_a(E_G) - 1 + S_a(X_F) - 1 + 2 = S_a(E_G) + S_a(X_F)$. It follows from the consistency of M_G that the number of nodes in $V(G) \cap V(H)$ that are assigned to X_H by M equals $S_a(E_G) + 1$ (from **P1**). We separate the cases when X_F is an internal face of F_{φ_F} and X_F is the outer face of F_{φ_F}. **Case 1.** Assume X_F is an internal face of F_{φ_F}. Thus X_H is an internal face of H_φ. It follows from the consistency of M_F that the number of nodes in $V(F) \cap V(H) \setminus u$ that are assigned to X_H by M equals $S_a(X_F) - 2$ (from **P2**). Therefore $|M^{-1}(X_H)| = S_a(E_G) + 1 + S_a(X_F) - 2 = S_a(X_F) + S_a(E_G) - 1$. So M is consistent for X_H. **Case 2.** Assume X_F is the outer face of F_{φ_F}. Thus X_H

is the outer face of H_φ. It follows from the consistency of M_F that the number of nodes in $V(F) \cap V(H) \setminus u$ that are assigned to X_H by M equals $S_a(X_F)$ (from **P2**). So M is consistent for X_H. Thus M is a consistent assignment of nodes to the faces of H_φ for case 1 and for case 2. But H_φ is also acyclic and bimodal. Therefore H_φ is UP (Theorem 1). Thus H is UP. So Statement 1 is true. If the outer face of F_{φ_F} contains u then the outer face of H_{φ_H} also contains u (Definition 1). Thus Statement 2 holds. If $M_G(E_G) = u$ and X_F is the outer face of F_{φ_F} then X_H is the outer face of H_φ (Definition 1) and $M(u) = X_H$ (from **P1**). Therefore M assigns u to the outer face of H_φ. Thus H has an UPSL drawing whose outer face contains the large angle incident on u and so Statement 3 holds.

Lemma 7. *Let G and F be UP digraphs such that $V(G) \cap V(F) = \{u\}$ and u is an internal node in both G and F. Let $H = G \cup F$. Then the following four statements hold. **1.** H is UP if at least one of G or F has an UP drawing whose outer face contains an I-angle incident on u. **2.** H has an UP drawing whose outer face contains an I-angle incident on u if G and F both have an UP drawing whose outer face contains an I-angle incident on u. **3.** H has an UP drawing whose outer face contains u, but does not contain an edge which enters (resp., leaves) u if at least one of G or F has an UP drawing whose outer face contains u, but does not contain an edge which enters (resp., leaves) u and the other has an UP drawing whose outer face contains an I-angle incident on u. **4.** H has an UP drawing whose outer face contains u if at least one of G or F has an UP drawing whose outer face contains u and the other has an UP drawing whose outer face contains an I-angle incident on u.*

Lemma 8. *Let G and F be UP digraphs such that $V(G) \cap V(F) = \{u\}$ and u is a source node in G and a sink node in F. If $H = G \cup F$ then H is UP if at least one of G or F has an UPSL drawing whose outer face contains a large angle incident on u and H has an UP drawing whose outer face contains an I-angle incident on u if G and F both have an UPSL drawing whose outer face contains a large angle incident on u.*

Lemma 9. *Let G and F be UP digraphs such that $V(G) \cap V(F) = \{u\}$, u is a source (resp., sink) node in G, and u is an internal node in F. If $H = G \cup F$ then H is UP if G has an UPSL drawing whose outer face contains the large angle incident on u or F has an UP drawing whose outer face contains an edge which leaves (resp., enters) u. H has an UP drawing whose outer face contains an edge which enters (resp., leaves) u if G has an UPSL drawing whose outer face contains a large angle incident on u and F has an UP drawing whose outer face contains an edge which enters (resp., leaves) u.*

Lemmas 10 and 11 follow by induction from Lemma 6, whilst Lemma 12 follows by induction from Lemma 7.

Lemma 10. *Let G be a digraph with a source or sink node u. G has an UP drawing whose outer face contains u if every component of G w.r.t. u has an UP drawing whose outer face contains u.*

Lemma 11. *Let G be a digraph with a source (resp., sink) node u. G has an UPSL drawing whose outer face contains the large angle incident on u if every component of G w.r.t. u has an UPSL drawing whose outer face contains the large angle incident on u.*

Lemma 12. *Let G be a digraph with an internal node u such that all components of G w.r.t. u are I-components. G has an UP drawing whose outer face contains an I-angle incident on u if every component of G w.r.t. u has an UP drawing whose outer face contains an I-angle incident on u.*

Lemma 13. *Let G be a digraph with a node u such that all components of G w.r.t. u are I-components. G has an UP drawing whose outer face contains u, but no edge which enters (resp., leaves) u if some component G_x of G w.r.t. u has an UP drawing whose outer face contains u, but does not contain an edge which enters (resp., leaves) u and all other components of G w.r.t. u have an UP drawing whose outer face contains an I-angle incident on u.*

Proof. Let $G_{\overline{x}}$ be the union of all components of G w.r.t. u which are distinct from G_x. $G_{\overline{x}}$ has an UP drawing whose outer face contains an I-angle incident on u (Lemma 12). Therefore $G = G_x \cup G_{\overline{x}}$ has an UP drawing outer face contains u, but no edge which enters (resp., leaves) u (Lemma 7).

Lemma 14. *Let G be a digraph with a node u such that all components of G w.r.t. u are I-components. G has an UP drawing whose outer face contains u if all components of G with respect to u have an UP drawing whose outer face contains u and at most one component of G w.r.t. u does not have an UP drawing whose outer face contains an I-angle incident on u.*

Proof. If all components of G w.r.t. u have an UP drawing whose outer face contains an I-angle incident on u then G has an UP drawing whose outer face contains an I-angle incident on u (Lemma 12). Suppose some component G_y of G w.r.t. u has no UP drawing whose outer face contains an I-angle incident on u, but does have an UP drawing whose external face contains u. Let $G_{\overline{y}}$ be the union of the all components of G with respect to u which are distinct from G_y. If all the components of $G_{\overline{y}}$ w.r.t. u have an UP drawing whose external face contains an I-angle incident on u then $G_{\overline{y}}$ has an UP drawing whose outer face contains an I-angle incident on u (Lemma 12). It follows from Lemma 7 that $G = G_y \cup G_{\overline{y}}$ has an UP drawing whose external face contains u.

4   Necessary Conditions for Upward Planarity

Lemma 15. *Let G and F be DAGs such that $V(G) \cap V(F) = \{u\}$ and u is a source node in G and a sink node in F. If $H = G \cup F$ then H is UP only if at least one of G or F has an UPSL drawing whose outer face contains a large angle incident on u and the other is UP.*

Proof. If neither F nor G admit an UPSL drawing whose external contains a large angle incident on u then some components F' of F and G' of G w.r.t. u do not admit an UPSL drawing whose external face contains a large angle incident on u (Lemma 11). Let $H' = F' \cup G'$, let $\mathcal{H}'$ be an UPSL drawing of H' and let $\mathcal{G}'$ and $\mathcal{F}'$ be the sub-drawings induced on G' and F', respectively by $\mathcal{H}'$. At least one of $\mathcal{G}'$ or $\mathcal{F}'$ lies in the outer face of the other (Lemma 3). As u is an internal node in H' it follows from Lemma 5 that both $\mathcal{G}'$ and $\mathcal{F}'$ must lie in the face of the other that contains the large angle incident on u. This is a contradiction.

Lemma 16. *Let $H = G \cup F$ where G is an UP digraph with a source node u, F is an UP digraph with a node u such that all components of F w.r.t. u are I-components, and $V(G) \cap V(F) = \{u\}$. H is UP only if G has an UPSL drawing whose outer face contains a large angle incident on u or F has an UP drawing whose outer face contains an edge which leaves u.*

Proof. (outline) As in the proof of Lemma 15 we show that if both G and F do not have drawings with the stated properties then some component G' of G w.r.t. u does not admit an UPSL drawing whose outer face contains a large angle incident on u, and some component F' of F w.r.t. does admit an UP drawing whose external face contains an edge which leaves u. We then use Lemmas 5 and 3 to show that $F' \cup G'$ is not upward planar.

Lemma 17. *Let $G_i, i = 1, \ldots, c$ be the components of a digraph G with respect to a node u such that all components of G with respect to u are I-components. Let $\mathcal{G}$ be a drawing of G and let $\mathcal{G}_i$ be the sub-drawing induced on G_i, for $i = 1, \ldots, c$. Then $\mathcal{G}$ is an UP drawing whose outer face contains an I-angle incident on u only if each $\mathcal{G}_i$ for $i = 1, \ldots, c$ is an UP drawing whose external face contains an I-angle incident on u.*

Proof. Let a be any integer such that $1 \leq a \leq c$. Suppose that $\mathcal{G}_a$ is not an UP drawing whose outer face contains an I-angle incident on u. If $\mathcal{G}_a = \mathcal{G}$ then it is trivially true that $\mathcal{G}$ is not an UP drawing whose outer face contains an I-angle incident on u. Otherwise let b be any integer such that $(1 \leq b \leq c) \wedge (b \neq a)$ and let $\mathcal{G}_a \cup \mathcal{G}_b$ be the sub-drawing of $\mathcal{G}$ induced on $G_a \cup G_b$. Because the outer face of $\mathcal{G}_a$ does not contain an I-angle incident on u it follows from Lemma 4 that $\mathcal{G}_a \cup \mathcal{G}_b$ is bimodal planar only if all of $\mathcal{G}_b$ is drawn within an internal face of $\mathcal{G}_a$. Therefore $\mathcal{G}_a \cup \mathcal{G}_b$ is UP only if the outer face of $\mathcal{G}_a$ is also the outer face of $\mathcal{G}_a \cup \mathcal{G}_b$. Therefore $\mathcal{G}$ is UP only if the outer face of $\mathcal{G}_a$ is also the outer face of $\mathcal{G}$. Therefore $\mathcal{G}$ is an UP drawing whose outer face contains an I-angle incident on u only if each $\mathcal{G}_i$ for $i = 1, \ldots, c$ is an UP drawing whose outer face contains an I-angle incident on u.

Lemma 18. *Let $G_i, i = 1, \ldots, c$ be the components of a digraph G with respect to a node u. Let $\mathcal{G}$ be a drawing of G and let $\mathcal{G}_i$ be the sub-drawing induced on $G_i, i = 1, \ldots, c$. $\mathcal{G}$ is an UP drawing whose outer face contains u only if each $\mathcal{G}_i$ for $i = 1$ to c is an UP drawing whose outer face contains u.*

Proof. Let x be any integer such that $1 \leq x \leq c$. Suppose that $\mathcal{G}_x$ is an UP drawing whose outer face $\mathcal{E}_x$ does not contain u. Therefore u is a point in the interior of some closed curve $\mathcal{E}'_x$ that is a sub-drawing of $\mathcal{E}_x$. But $\mathcal{E}'_x$ is a sub-drawing of $\mathcal{G}$. Therefore u is not contained by the outer face of $\mathcal{G}$. Therefore $\mathcal{G}$

is an UP drawing whose outer face contains u only if each $\mathcal{G}_i$, $i = 1, \ldots, c$, is an UP drawing whose outer face contains u.

Proofs of the next two lemmas appear previously [4].

Lemma 19. *Let $G_i, i = 1, \ldots, c$ be the components of a digraph G with respect to a source (resp., sink) node u. Let $\mathcal{G}$ be a drawing of G and let $\mathcal{G}_i$ be the sub-drawing induced on G_i, for $i = 1, \ldots, c$. Then $\mathcal{G}$ is an UPSL drawing whose outer face contains a large angle incident on u only if each $\mathcal{G}_i$, $i = 1, \ldots, c$, is an UPSL drawing whose outer face contains a large angle incident on u.*

Lemma 20. *Let $G_i, i = 1, \ldots, c$ be the components of a digraph G w.r.t. a node u such that all components of G with respect to u are I-components. Let $\mathcal{G}$ be a drawing of G and let $\mathcal{G}_i$ be the sub-drawing induced on G_i, for $i = 1, \ldots, c$. Then $\mathcal{G}$ is an UP drawing whose outer face contains u, but does not contain an edge which enters (resp., leaves) u only if exactly one of the drawings $\mathcal{G}_i, i = 1, \ldots, c$ is an UP drawing whose outer face contains u, but does not contain an edge which enters (resp., leaves) u and exactly $c - 1$ of the drawings $\mathcal{G}_i, i = 1, \ldots, c$ are UP drawings whose outer face contains an I-angle incident on u.*

5   Main Results

Theorem 2. *Let G be a digraph with a source or sink node u. G is UP if and only if all components of G w.r.t. u are UP and at most one component of G w.r.t. u does not have an UP drawing whose outer face contains u.*

Proof. **Suf.** If all components of G w.r.t. u have an UP drawing whose outer face contains u then G is UP (Lemma 10). Suppose some UP component X of G w.r.t. u does not have an UP drawing whose outer face contains u, but that all components of $G \setminus X$ w.r.t. u do have an UP drawing whose outer face contains u. Therefore $G \setminus X$ has an UP drawing whose outer face contains u (Lemma 10). Thus $G = X \cup G \setminus X$ is UP (Lemma 6). **Nec.** Clearly if some component Y of G w.r.t. u is not UP then G is not UP because Y is a subdigraph of G. Also if two components Y_1 and Y_2 of G with respect to u do not have an UP drawing whose outer face contains u then it follows from Lemma 2 that G is not UP.

Theorem 3. *Let G be a digraph with a internal node u such that all components of G w.r.t. u are either S-components or T-components. G is UP if and only if all components of G with respect to u are UP, at most one component of G w.r.t. u does not have an UP drawing whose outer face contains u and at least one of $S(u)$ or $T(u)$ has an UP drawing whose outer face contains a large angle incident on u.*

Proof. *(outline)* The sufficiency follows from Theorem 2 and Lemma 8 whilst the necessity follows from Lemma 2 and Lemma 15.

Theorem 4. *Let G be a digraph with a node u such that all components of G w.r.t. u are I-components. G is UP if and only if all components of G w.r.t. u are UP and at most one component of G w.r.t. u does not have an UP drawing whose outer face contains an I-angle incident on u.*

Proof. **Suf.** Let X be a component of G w.r.t. u such that all components of $G \backslash X$ w.r.t. u admit an UP drawing whose outer face contains an I-angle incident on u. Therefore $G \backslash X$ has an UP drawing whose outer face contains an I-angle incident on u (Lemma 12). Therefore $G = X \cup G \backslash X$ is UP if X is UP (Lemma 7). **Nec.** Suppose that two components G^1 and G^2 of G w.r.t. u have no UP drawing whose outer face contains an I-angle incident on u. Let $H = G_1 \cup G_2$. Let $\mathcal{H}$ be a planar drawing of H and let $\mathcal{G}^1$ (resp., $\mathcal{G}^2$) be the sub-drawing induced on G^1 (resp., G^2). It follows from Lemma 4 that $\mathcal{H}$ is bimodal planar only if $\mathcal{G}^1$ (resp., $\mathcal{G}^2$) lies in a face of $\mathcal{G}^1$ (resp., $\mathcal{G}^2$) which contains an I-angle incident on u and at least one of G^1 or G^2 lies in the external face of the other. Therefore H has no UP drawing. But H is a subgraph of G.

A proof of Theorem 5 appears previously [4].

Theorem 5. *Let G be a digraph with an internal node u such that there are no T-components (resp., S-components) of G w.r.t. u. G is UP if and only if all components of G w.r.t. u are UP, at most one I-component of G w.r.t. u does not have an UP drawing whose outer face contains an I-angle incident on u, and at least one of the following two statements is true:*

1. *All S-components (resp., T-components) of G w.r.t. u have an UPSL drawing whose outer face contains the large angle incident on u.*
2. *All I-components of G w.r.t. u have an UP drawing whose outer face contains an edge which leaves (resp., enters) u and at most one S-component (resp., T-component) of G with respect to u does not have an UP drawing whose outer face contains u.*

Theorem 6. *Let G be a digraph and let u be any node in G. G is UP if and only if all components of G w.r.t. u are UP, at most one I-component of G w.r.t. u does not have an UP drawing whose outer face contains an I-angle incident on u, at most one S-component or T-component of G w.r.t. u does not have an UP drawing whose outer face contains u, and at least one of the following three statements is true:*

1. *There are no S-components or T-components of G w.r.t. u which do not admit an UP drawing whose outer face contains a large angle incident on u.*
2. *There are no S-components of G w.r.t. u which do not admit an UP drawing whose outer face contains a large angle incident on u and there are no I-components of G w.r.t. u which do not admit an UP drawing whose outer face contains an edge which enters u.*
3. *There are no T-components of G w.r.t. u which do not admit an UP drawing whose outer face contains a large angle incident on u and there are no I-components of G w.r.t. u which do not admit an UP drawing whose outer face contains an edge which leaves u.*

Proof. **Suf.** Assume that all components of G w.r.t. u are UP and that at most one I-component of G w.r.t. u does not have an UP drawing whose outer face contains an I-angle incident on u. Thus $I(u)$ is UP (Theorem 4). Also assume that at most one S-component or T-component does not have an UP drawing whose outer face contains u. Thus $S(u)$ and $T(u)$ are both UP (Theorem 2). We consider three cases. **Case 1.** Assume Statement 1 is true. Therefore $S(u)$ and $T(u)$ both have an UPSL drawing whose outer face contains a large angle incident on u (Lemma 11). Thus $S(u) \cup T(u)$ has an UP drawing whose outer face contains an I-angle incident on u (Lemma 8). But $I(u)$ is UP. Therefore $G = I(u) \cup (S(u) \cup T(u))$ is UP (Lemma 7). **Case 2.** Assume Statement 2 is true. Thus $S(u)$ has an UP drawing whose outer face contains a large angle incident on u (Lemma 11) and $I(u)$ has an UP drawing whose external face contains an edge which enters u (Lemmas 14 and 20). It follows from Lemma 9 that $S(u) \cup I(u)$ has an UP drawing whose outer face contains an edge which enters u. But $T(u)$ is UP. Thus $G = (S(u) \cup I(u)) \cup T(u)$ is UP (Lemma 9). **Case 3.** Assume Statement 3 is true. It follows by symmetry from the sufficiency of case 2 that G is UP. **Nec.** Every component of G w.r.t. u is a subgraph of G. Therefore G is UP only if all components of G with respect to u are UP. It follows from Theorem 4 that $I(u)$ is UP only if at most one I-component of G with respect to u does not have an UP drawing whose outer face contains an I-angle incident on u. It follows from Lemma 2 that $S(u) \cup T(u)$ is UP only if at most one S-component or T-component of G w.r.t. u does not have an UP drawing whose outer face contains u. We now prove the necessity of at least one of Statements 1, 2 or 3 of Theorem 6 being true. We will need the following four statements. Statement $\mathcal{A}$ (resp., Statement $\mathcal{B}$) is that all components of $S(u)$ (resp., $T(u)$) w.r.t. u have an UPSL drawing whose outer face contains a large angle incident on u. Statement $\mathcal{C}$ (resp., Statement $\mathcal{D}$) is that all components of $I(u)$ w.r.t. u have an UP drawing whose outer face contains an edge which leaves (resp., enters) u. $S(u) \cup T(u)$ is UP only if at least one of Statement $\mathcal{A}$ or Statement $\mathcal{B}$ is true (Theorem 3). $I(u) \cup S(u)$ is UP only if at least one of Statement $\mathcal{A}$ or Statement $\mathcal{C}$ is true (Theorem 5). $I(u) \cup T(u)$ is UP only if at least one of Statement $\mathcal{B}$ or Statement $\mathcal{D}$ is true (Theorem 5). Therefore $G = I(u) \cup S(u) \cup T(u)$ is UP only if Statements $\mathcal{A}$ and $\mathcal{B}$ are both true, and/or Statements $\mathcal{A}$ and $\mathcal{D}$ are both true, and/or Statements $\mathcal{B}$ and $\mathcal{C}$. But (Statement $\mathcal{A} \wedge$ Statement $\mathcal{B}$) is equivalent to Statement 1 of Theorem 6; (Statement $\mathcal{A} \wedge$ Statement $\mathcal{D}$) is equivalent to Statement 2 of Theorem 6; (Statement $\mathcal{B} \wedge$ Statement $\mathcal{C}$) is equivalent to Statement 3 of Theorem 6. Thus G is UP only if at least one of Statements 1, 2 or 3 of Theorem 6 is true.

Proofs of the following two theorems are similar to that of Theorem 6.

Theorem 7. *Let G be a digraph with a node u, such that C is an S-component (resp., T-component) of G w.r.t. u that contains a node w. G has an UP drawing whose outer face contains a certain type of angle incident on w if and only if C has an UP drawing whose outer face contains the same type of angle incident on w, there are no S-components (resp., T-components) of $G \setminus C$ w.r.t. u which do not admit an UP drawing whose outer face contains u, there are no T-components (resp., S-components) of $G \setminus C$ with respect to u which do not admit an UP drawing whose outer face contains a large angle incident on u, there are no I-components of $G \setminus C$ w.r.t. u which do not admit an UP drawing whose outer face contains an edge which leaves (resp., enters) u and at most one*

I-component of $G \setminus C$ w.r.t. u does not have an UP drawing whose outer face contains an I-angle incident on u.

Theorem 8. *Let G be a digraph with a node u, such that C is an I-component of G w.r.t. u that contains a node w. G has an UP drawing whose outer face contains a certain type of angle incident on w if and only if C admits an UP drawing whose outer face contains the same type of angle incident on w, there are no S-components or T-components of $G \setminus C$ with respect to u which do not admit an UP drawing whose outer face contains a large angle incident on u, and all I-components of $G \setminus C$ w.r.t. u have an UP drawing whose external face contains an I-angle incident on u.*

6 Conclusions

Bertolazzi et al.'s algorithm for testing an embedded digraph for upward planarity [2] can be modified to detect for the properties listed in Theorems 6–8. A divide and conquer approach to upward planarity testing follows that involves recursively splitting a digraph at its cut-vertices and testing the individual blocks for the given properties. Bertolazzi et al. also describe a branch-and-bound algorithm that tests the upward planarity of biconnected digraphs [3]. An interesting question is could their algorithm also be tailored to detect the properties listed in Theorems 6–8.

References

1. G. Di Battista, P. Eades, R. Tamassia, and I. G. Tollis. *Graph Drawing: Algorithms for the Visualization of Graphs.* Prentice-Hall, 1999.
2. P. Bertolazzi, G. Di Battista, G. Liotta, and C. Mannino. Upward drawings of triconnected digraphs. *Algorithmica*, 6(12):476–497, 1994.
3. P. Bertolazzi, G. Di Battista, and W. Didimo. Quasi-upward planarity. *Algorithmica*, 32:474–506, 2002.
4. P. Healy and W. K. Lynch. Investigations into upward planar digraphs. Technical Report TR-04-02, Dept. of CSIS, University of Limerick, 2004.
   http://www.csis.ul.ie/Research/TechRpts.htm

A Linear Time Algorithm for Constructing Maximally Symmetric Straight-Line Drawings of Planar Graphs*

Seok-Hee Hong and Peter Eades

National ICT Australia; School of Information Technologies,
University of Sydney, Australia
{shhong,peter}@it.usyd.edu.au

Abstract. This paper presents a linear time algorithm for constructing maximally symmetric straight-line drawings of *biconnected* and *one-connected* planar graphs.

1 Introduction

Symmetry is one of the most important aesthetic criteria that clearly reveals the structure and properties of a graph. Symmetric drawings of a graph G are clearly related to the automorphisms of G, and algorithms for constructing symmetric drawings have two steps:

1. Find the *geometric automorphisms* [3], and
2. Draw the graph displaying these automorphisms as symmetries.

This paper presents a linear time algorithm for constructing maximally symmetric straight-line drawings of *biconnected* and *one-connected* planar graphs. The first polynomial time algorithm which runs in quadratic time has appeared [4]. Here we present a linear time algorithm. A linear time algorithm for *triconnected* planar graphs [6] and *disconnected* graphs are dealt with in [9]. The following theorem summarizes our main result.

Theorem 1. *There is a linear time algorithm that constructs maximally symmetric planar drawings of biconnected and one-connected planar graphs, with straight line edges.*

In the next section, we review necessary background. In Section 3 and Section 4, we present a linear time algorithm for finding maximum number of symmetries (*planar automorphisms*) of biconnected and one-connected planar graphs. In Section 5, we describe the symmetric drawing algorithms.

---

* This research was partially supported by a grant from the Australian Research Council. For the full version of this paper, see [7, 8]. National ICT Australia is funded by the Australian Government's Backing Australia's Ability initiative, in part through the Australian Research Council.

2   Symmetries and Planar Automorphisms

An *automorphism* of a graph is a permutation of the vertex set that preserves adjacency. Symmetry in graph drawing is closely related to automorphisms of graphs: a symmetry of a graph drawing induces an automorphism of the graph. In this case, we say that the drawing *displays* the automorphism, and the automorphism is *geometric* [3]. *Note that not every automorphism is geometric.*

An automorphism α of a graph G is a *planar automorphism* if there is a *planar* drawing D of G which displays α. Note that not every geometric automorphism is planar. Further, the product of two planar automorphisms is not necessarily planar (because they may be displayed by different drawings). An automorphism group A of a graph is a *planar automorphism group* if there is a *single* planar drawing of the graph that displays every element of A [6]. The central problem of this paper is to find a planar automorphism group of maximum size.

Planar Automorphism Problem
Input: A planar graph G.
Output: A maximum size planar automorphism group A of G.

Previous research on the Planar Automorphism Problem has concentrated on *subclasses* of planar graphs [5, 11]. Our aim is to give a linear time algorithm for planar graphs in general. We use connectivity to divide the problem into cases.

1. Triconnected graphs: a linear time algorithm is presented in [6].
2. Biconnected graphs: this is the topic of this paper.
3. One-connected graphs: this is the topic of this paper.
4. Disconnected graphs: a linear time algorithm is presented in [9].

Note that each case relies on the result of the previous case. The triconnected case was solved in [6]; the algorithm finds a plane embedding of G that has maximum size planar automorphism group. Generators of the planar automorphism group of G with given plane embedding are then derived. They also give a linear time drawing algorithm.

3   The Biconnected Case

If the input graph G is biconnected, then we break it into *triconnected components* in a way that is suitable for the task. The overall algorithm is composed of three steps.

Algorithm Biconnected_Planar
1. Construct the SPQR-tree T_1 of G, and root T_1 at its center.
2. *Reduction*: For each level i of T_1 (from the lowest level to the root level)
   (a) For each leaf node on level i, compute labels.
   (b) For each leaf node on level i, label the corresponding *virtual edge* of the parent node with the labels.
   (c) Remove the leaf nodes on level i.
3. Compute a maximum size planar automorphism group at the labeled center.

We briefly describe each step of the algorithm. The first step is to construct the SPQR-tree for the input biconnected planar graph. The SPQR-tree represents a decomposition of a biconnected planar graph into triconnected components. There are four types of nodes in the SPQR-tree T_1 and each node v in T_1 is associated with a graph which is called as the *skeleton* of v (*skeleton(v)*). The node types and their skeletons are:

1. Q-node: The skeleton consists of two vertices which are connected by two multiple edges.
2. S-node: The skeleton is a simple cycle with at least 3 vertices.
3. P-node: The skeleton consists of two vertices connected by at least 3 edges.
4. R-node: The skeleton is a triconnected graph with at least 4 vertices.

In fact, we use slightly different version of the SPQR-tree. We use the SPQR-tree without Q nodes. The SPQR-tree is unique for each biconnected planar graph. Let v be a node in T_1 and u a parent node of v. The graph *skeleton(u)* has one common *virtual edge* with *skeleton(v)*, which is called as a *virtual edge* of v. For details, see [2].

The second step is the *reduction*. The reduction process takes the SPQR-tree of a biconnected graph, rooted at the center, based on the following lemma.

Lemma 1. *[1] The center of the SPQR-tree is fixed by a planar automorphism group of a biconnected planar graph.*

The reduction process proceeds the SPQR-tree from the leaf nodes to the center level by level, computing labels. The labels are a pair of integers, and boolean values that capture some information of the planar automorphisms of the leaf nodes. First it computes the labels for the leaf nodes. Then it labels the corresponding virtual edge in the parent node and delete each leaf node. The reduction process stops when it reaches the root.

The reduction process clearly does not decrease the planar automorphism group of the original graph. This is not enough; we need to also ensure that the planar automorphism group is not increased by reduction. This is the role of the labels. As a leaf v is deleted, the algorithm labels the virtual edge e of v in *skeleton(u)* where u is a parent of v. Roughly speaking, the labels encode information about the deleted leaf to ensure that every planar automorphism of the labeled reduced graph extends to a planar automorphism of the original graph.

The last step is to compute a maximum size planar automorphism group at the center using the information encoded on the labels.

3.1   The Labels and Labeling Algorithms

Standard Labels. Let v be an internal node of T_1. We say that a virtual edge e of *skeleton(v)* is a *parent (child) virtual edge* if e corresponds to a virtual edge of u which is a parent (child) node of v. We define a *parent separation pair* $s = (s_1, s_2)$ of v as the two endpoints of a parent virtual edge e.

When we compute the labels of v, we need to delete the parent virtual edge e from *skeleton(v)*. We denote the resulting graph by *skeleton⁻(v)*. The union of the graphs $skeleton^-(u)$ for all descendants u of v, including v itself, is denoted by $G^+(v)$.

Suppose that nodes $v_1, v_2, \ldots v_k$ of the SPQR-tree T_1 are deleted at one iteration of the reduction process. These nodes correspond to virtual edges $e_1, e_2, \ldots, e_k$ in the level above the current level. For each v, we need to compute the following *standard* labels.

1. isomorphism code: a pair $Iso(v)$ of integers.
2. axial codes:
   (a) $A_{swap}(v)$: a boolean label indicating whether $G^+(v)$ has an axial symmetry that *swaps* the parent separation pair.
   (b) $A_{fix}(v)$: a boolean label indicating whether $G^+(v)$ has an axial symmetry that *fixes* the parent separation pair.
3. rotation code: a boolean label $Rot(v)$ indicating whether $G^+(v)$ has a rotational symmetry of 180 degrees that *swaps* the parent separation pair.

Note that we need these labels when the virtual edge is fixed by a planar automorphism of the parent node. Further we need to define *special labels*, which are motivated by the special case below and plays important role to give a linear time algorithm.

Special Case. The aim of labeling is to encode information about planar automorphisms of the $skeleton^-(v)$ of the non-root node v of the SPQR tree in the parent virtual edges. In this way we can find planar automorphisms of the whole graph by finding planar automorphisms of the labeled $skeleton(c)$ of the root node c. This strategy has a simple flaw: the edges may not model the topological properties of the $skeleton^-(v)$ correctly. In particular, while $skeleton^-(v)$ of a child node v can enclose $skeleton(u)$ of a parent node u (see Figure 1(a)), the child virtual edge e in the $skeleton(u)$ of parent node u (see Figure 1(b)) cannot enclose $skeleton(u)$, since it is purely a one-dimensional curve. An embedding in which $skeleton(u)$ is on an inside face of $G^+(v)$, where u is the parent of v, is called an *enclosing composition*.

Figure 2(a) shows an example of a drawing constructed by an enclosing composition; this shows the maximum number of symmetries of a graph, two rotational and two axial symmetries. Figure 2(b) shows the SPQR tree of the graph. Here $skeleton(c)$ of the root node c of the SPQR tree is enclosed by the $skeleton(v)$ of its child node v. The enclosing child node may be in turn enclosed by one of its children, again fixed setwise. Note that the graph can be

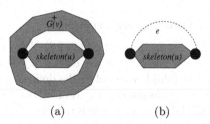

(a)   (b)

Fig. 1. An enclosing composition: v is a child of u.

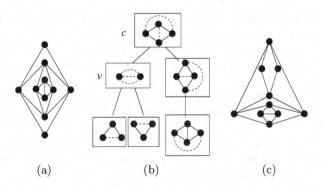

Fig. 2. The special (enclosing) case.

drawn without enclosing nodes, based on $skeleton(c)$, as in Figure 2(c); but this displays less symmetry, only one axial symmetry.

Special Labels. To take care of special cases such as that illustrated in Figure 2, we define special labels. The most important property of this special case is that the whole planar automorphism group fixes two faces of $G^+(v)$: the outside face, and the internal face containing the separation pair. Thus we define the following labels.

1. *special* axial codes:
   (a) $A^*_{swap}(v)$: a boolean label indicating whether $G^+(v)$ has an axial symmetry that *swaps* the parent separation pair and with the parent separation pair on an inside face.
   (b) $A^*_{fix}(v)$: a boolean label indicating whether $G^+(v)$ has an axial symmetry that *fixes* the parent separation pair and with the parent separation pair on an inside face.
2. *special* rotation code: a boolean label $Rot^*(v)$ indicating whether $G^+(v)$ has a rotational symmetry that swaps the parent separation pair and with the parent separation pair on an inside face.

In fact, it is enough to require that the planar automorphism fixes two faces: one incident to the parent separation pair, and the other not incident to the parent separation pair, as described in the following lemma.

Lemma 2. *Suppose that D is a drawing of a planar graph G and u and w are vertices of G that share a face in D. Suppose that D displays an axial planar automorphism ϕ that fixes $\{u, w\}$. Then ϕ fixes at least one face not incident to u and w if and only if there is a drawing D' of G that displays ϕ, with u and w not on the outside face.*

Lemma 2 implies, for example, that $A^*_{swap}(v) = true$ if and only if $G^+(v)$ has an axial planar automorphism that swaps the parent separation pair, fixes a face incident to the separation pair, and fixes at least one other face.

Computation of an Isomorphism Code. The isomorphism code $Iso(v)$ consists of a pair of integers, because the $skeleton(v)$ has an orientation with respect to the parent separation pair. The isomorphism code can be computed in linear time using a planar graph isomorphism algorithm [10]. For details, see [7].

Computation of Axial Codes. An axial symmetry either *swaps* the parent separation pair or *fixes* the parent separation pair. First we describe an algorithm for $A_{swap}(v)$.

Note that the axial symmetry should *respect* the isomorphism code of the child virtual edge. Further, if the axial symmetry fixes a child virtual edge, then we need to test its label. Further, from Lemma 2, $A^*_{swap}(v)$ is *true* if and only if $A_{swap}(v)$ is *true* and the axial symmetry of $skeleton^-(v)$ can be extended to an axial symmetry of $G^+(v)$ that fixes more than one face. Thus, we can compute both $A_{swap}(v)$ and $A^*_{swap}(v)$ as follows.

Algorithm Compute_Axial_Code_Swap
1. Test whether $skeleton^-(v)$ has an axial symmetry α which swaps the parent separation pair and respects the isomorphism codes of child virtual edges.
2. If α exists, then
   (a) For each child virtual edge e_j that is fixed by α, check the followings:
      i. if α fixes the endpoints of e_j, then $A_{fix}(v_j) = true$.
      ii. if α swaps the endpoints of e_j, then $A_{swap}(v_j) = true$.
   (b) If one of these properties fails,
      then $A_{swap}(v) := false$; else $A_{swap}(v) := true$.
   else $A_{swap}(v) := false$.
3. If $A_{swap}(v) = false$ then $A^*_{swap}(v) := false$;
   else if either
   - α fixes more than one face, or
   - α swaps the endpoints of a child virtual edge f such that $A^*_{swap}(f) = true$, or
   - α fixes the endpoints of a child virtual edge f such that $A^*_{fix}(f) = true$,
   then $A^*_{swap}(v) := true$; else $A^*_{swap}(v) := false$.

Algorithms for computing $A_{fix}(v)$ and $Rot(v)$ are similar to the algorithm for computing $A_{swap}(v)$. Note that the labeling algorithm correctly computes labels and runs in linear time. For proofs, see [7].

When v is a P-node, we use similar algorithms to the case of *parallel* compositions in series parallel digraphs [5]. When v is an S-node, we use similar algorithms to the case of *series* compositions in series parallel digraphs [5]. For details, see [7].

3.2 Computing a Maximum Size Planar Automorphism Group at the Center

The center of the SPQR-tree may be a node or an edge. If the center is a node c, then we can further divide into three cases by its type. If c is a R-node, then we use the triconnected case [6] to compute a maximum size planar automorphism group. If c is a P-node, then we use a similar algorithm to the case of a *parallel* composition in series parallel digraphs (see [5, 7]). If c is an S-node, then we can use the algorithm for outerplanar graphs [11].

However, there may exist some other node v which is fixed by a planar automorphism group as in the special case. Thus to find a maximum size planar automorphism group at the center c, we consider both cases and then find the maximum. Here we need an algorithm to find a child node which gives the best result for the enclosing composition. This can be done in linear time by using *generators*. For details, see [7].

If the center is an edge, then we find the maximum among three cases: parallel composition, reduction composition and enclosing composition. Parallel composition means that we construct a drawing with two labeled edges such as a parallel composition in series parallel digraphs. Reduction composition means that we compute labels of one node u and then delete u by labeling the corresponding virtual edge e of the other node v. Then we compute a planar automorphism group at v using the center node case. Enclosing composition means that we construct a drawing such as the special case. Each of these cases can be computed in linear time, see [7].

4   The One-Connected Case

The algorithm for computing a maximum size planar automorphism group of one-connected planar graph has a similar flavor to the biconnected case: we use reduction approach. We also use algorithm **Biconnected_Planar** as a subroutine. The reduction process is similar to the biconnected case. In this case we take the *block-cut vertex tree* (BC-tree) and then compute labels at each leaf node (block or cut vertex). However, the labels are different.

4.1   The Labels and Labeling Algorithms

We need two types of labels: isomorphism code and axial code. However, these are further divided into the case of a cut vertex or a block. Let B represent a block and C represent a cut vertex.

1. isomorphism code : an integer $Iso_B(v)$ (or $Iso_C(v)$).
2. axial code : an integer $A_B(v)$ (or $A_C(v)$) indicating whether B (or C) has an axial symmetry which fixes the parent node.

Note that we need these labels when the block or cut vertex is fixed by a planar automorphism of the parent node.

Computation of an Isomorphism Code of a Block. Suppose that $B_1, B_2,$ $\dots, B_m$ are the blocks on the lowest level and $p_1, p_2, \dots, p_m$ are the parent cut vertices for the blocks. We compute isomorphism code $Iso_B(v_i)$ for each B_i using a planar graph isomorphism algorithm which takes linear time [10]. Note that the isomorphism should respect the isomorphism code of the child cut vertex. We now describe the algorithm.

Algorithm Compute_Iso_B
for each $B_i, i = 1, 2, \ldots, m$,
    if there is an isomorphism α between B_i and B_j such that
    (a) $\alpha(p_i) = p_j$, and
    (b) for each cut vertex c_k of B_i,
        i. $\alpha(c_k)$ is a cut vertex, and
        ii. $Iso_C(c_k) = Iso_C(\alpha(c_k))$.
    then assign isomorphism codes such that $Iso_B(v_i) = Iso_B(v_j)$.

Computation of an Axial Code of a Block. The label $A_B(v)$ represents whether the block B has an axial symmetry which fixes the parent cut vertex p. Let $c_1, c_2, \ldots, c_k$ be the child cut vertices of B. In fact, the algorithm computes a ternary value for $A_B(v)$. The interpretation of $A_B(v)$ is:

1. $A_B(v) = 1$ if $G^+(v)$ has an axial symmetry that fixes p and one face incident to p.
2. $A_B(v) = 2$ if $G^+(v)$ has an axial symmetry which fixes p and two faces incident to p.
3. $A_B(v) = 3$ if $G^+(v)$ has no axial symmetry that fixes p and a face incident to p.

First we find an axial symmetry α of B which fixes the parent cut vertex using Biconnected_Planar. Then we check whether each fixed child cut vertex c_j preserves the axial symmetry. For this purpose, we need some information about the axial code $A_C(v)$. The interpretation of values of $A_C(v)$ is:

1. $A_C(v) = 0$ if $G^+(v)$ has an axial symmetry which does not fix any $G^+(v_i)$ for any i.
2. $A_C(v) = 1$ if
    (a) every axial symmetry of $G^+(v)$ fixes at least one $G^+(v_i)$, and
    (b) there is an axial symmetry α such that if α fixes $G^+(v_i)$ then $A_B(v_i) < 3$, and there is at most one j such that α fixes $G^+(v_j)$ and $A_B(v_j) = 1$.
3. $A_C(v) = 2$ if
    (a) every axial symmetry of $G^+(v)$ fixes at least two $G^+(v_i)$ for which $A_B(v_i) = 1$, and
    (b) there is an axial symmetry α such that if α fixes $G^+(v_i)$ then $A_B(v_i) < 3$, and there are at most two indices j such that α fixes $G^+(v_j)$ and $A_B(v_j) = 1$.
4. $A_C(v) = 3$ otherwise.

Finally we assign the value, depending on the fixed faces which are adjacent to p. We now state the algorithm.

Algorithm Compute_Axial_B
Apply Biconnected_Planar [7] to the labelled graph B; if B has an axial planar automorphism α such that
(a) α fixes p and respects the isomorphism partition labels on B;
(b) For each child cut vertex c_j fixed by α, the number of faces incident to c_j and fixed by α is at least as large as $A_C(c_j)$.
then $A_B(v) :=$ the number of faces fixed by α and incident to p.
else $A_B(v) := 3$.

Computation of an Isomorphism Code of a Cut Vertex. Suppose that $c_1, c_2, \ldots, c_k$ are the cut vertices on the lowest level. We compute $Iso_C(c_i)$ for each c_i, $i = 1, 2, \ldots, k$, which represents an isomorphism code of c_i. More specifically, $Iso_C(c_i) = Iso_C(c_j)$ if and only if the subgraph which is rooted at c_i is isomorphic to the subgraph which is rooted at c_j. We now state the algorithm.

> **Algorithm Compute_Iso_C**
> 1. For each c_i:
>   (a) Let $B_{i1}, B_{i2}, \ldots, B_{im}$ be the child blocks of c_i.
>   (b) $s(c_i) := (Iso_B(B_{i1}), Iso_B(B_{i2}), \ldots, Iso_B(B_{im}))$.
>   (c) Sort $s(c_i)$.
> 2. Let Q be the list of $s(c_i)$, $i = 1, 2, \ldots, k$.
> 3. Sort Q lexicographically.
> 4. For each c_i, compute $Iso_C(c_i)$ as follows: Assign the integer 1 to c_i whose $s(c_i)$ is the first distinct tuple of the sorted sequence Q. Assign the integer 2 to c_j whose $s(c_j)$ is the second distinct tuple, and so on.

Computation of an Axial Code of a Cut Vertex. The label $A_C(v)$ represents whether a cut vertex c has an axial symmetry which fixes the parent block. Let B_p be the parent block of v and $B_1, B_2, \ldots, B_m$ be the child blocks of v. Suppose that α is an axial symmetry which fixes B_p. We use $A_C(v)$ to decide whether c preserves α of B_p. More specifically, this indicates that whether $B_1, B_2, \ldots, B_m$ can be attached to c, preserving α.

To compute $A_C(v)$, we use $A_B(B_j)$. The label $A_B(B_j)$ indicates that whether B_j has an axial symmetry which fixes c. Further, it indicates that whether there is a fixed face adjacent to c. We now state the algorithm.

> **Algorithm Compute_Axial_C**
> 1. Partition $B_1, B_2, \ldots, B_m$ into isomorphism classes P_ℓ using $Iso_B(B_j)$, and compute the size s_ℓ of each isomorphism class P_ℓ.
> 2. If all s_ℓ are even, then $A_C(v) := 0$; exit.
> 3. If there is an s_ℓ that is odd and $A_B(v_j) = 3$ for each $j \in P_\ell$ then $A_C(v) := 3$; exit.
> 4. Let f be the number of odd s_ℓ such that if $j \in P_\ell$ then $A_B(v_j) = 1$.
>   (a) If $f > 2$ then $A_C(v) := 3$.
>   (b) If $f = 2$ then $A_C(v) := 2$.
>   (c) If $f = 1$ then $A_C(v) := 1$.

Note that all the labeling algorithms correctly compute labels and run in linear time, see [8].

4.2 Computing a Maximum Size Planar Automorphism Group at the Center

We can compute a maximum size planar automorphism group of the whole graph by computing a maximum size planar automorphism group at the labeled center, based on the following lemma.

Lemma 3. *The center of the BC-tree is fixed by a planar automorphism group of a one-connected planar graph.*

The algorithm can be divided into two cases, since the center of the BC tree may be a block or a cut vertex. Roughly speaking, if the center is a block B, then we use algorithm `Biconnected_Planar` in Section 3. If the center is a cut vertex c, then we use a similar method that was used in the case of trees [11].

However, the algorithm is not as simple as this, mainly because there are some special cases. Namely, there may exist a fixed cut vertex when the center is a block, and there may exist a fixed block when the center is a cut vertex. To illustrate the special cases, see the graph Figure 3, with its BC tree. The center of the BC tree is a block B with 5 cut vertices. The symmetries of the drawing of this block fix the cut vertex c with 4 children in the BC tree. Maximizing symmetry for the whole graph is not merely a matter of the reduction process plus maximizing symmetry of B; we must also arrange the children of the fixed cut vertex in a symmetrical way. Essentially this means merging the symmetries of $G^+(c)$ with the symmetries of $G^+(B)$. Similar case can happen when the center is a cut vertex. Thus to compute the maximum, we need to consider these special cases. Again, we need an algorithm to find the child vertex of the center which can gives the maximum. However, this can be done in linear time, see [8].

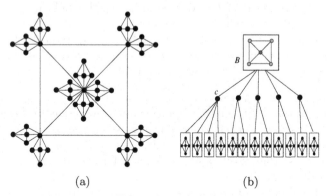

(a)   (b)

Fig. 3. Special case.

5   The Drawing Algorithms

The algorithms presented in the preceding sections take a biconnected planar graph and one-connected planar graph as input and has two outputs: a planar automorphism group of maximum size, and an embedding of the graph. In this section, we show how to use this information to construct a straight line symmetric drawing of the graph.

Given an embedding of a biconnected planar graph, we use "augmentation": we increase the connectivity by adding new edges and new vertices to make it triconnected, while preserving the planar automorphism group. Then we can apply the algorithm for constructing symmetric drawings of triconnected planar graphs with straight-line edges to construct a symmetric drawing [6]. The algorithm runs in linear time [6]. Finally we delete the added edges and vertices.

Given an embedding of a one-connected planar graph, we use "attachment": first, we draw each block using the algorithm for constructing symmetric drawings of biconnected planar graphs with straight-line edges to construct a symmetric drawing. Then we attach each drawing at each cut vertex, preserving planarity. Clearly, this algorithm runs in linear time.

References

1. L. Babai, Automorphism Groups, Isomorphism, and Reconstruction, Chapter 27 of *Handbook of Combinatorics* Volume 2 (ed. Graham, Groetschel and Lovasz), Elsevier Science, 1995.
2. G. Di Battista and R. Tamassia, On-Line Planarity Testing, *SIAM Journal on Computing* 25(5), pp. 956-997, 1996.
3. P. Eades and X. Lin, Spring Algorithms and Symmetries, *Theoretical Computer Science* 240, pp. 379-405, 2000.
4. S. Hong, P. Eades and S. Lee, An Algorithm for Finding Geometric Automorphisms in Planar Graphs, *Algorithms and Computation,* Proc. of ISAAC 98, Lecture Notes in Computer Science 1533, pp. 277-286, Springer Verlag, 1998.
5. S. Hong, P. Eades and S. Lee, Drawing Series Parallel Digraphs Symmetrically, *Computational Geometry: Theory and Applicatons* 17(3-4), pp. 165-188, 2000.
6. S. Hong, B. McKay and P. Eades, Symmetric Drawings of Triconnected Planar Graphs, *Proc. of SODA 2002*, pp. 356-365, 2002.
7. S. Hong and P. Eades, Drawing Planar Graphs Symmetrically II: Biconnected Graphs, Technical Report CS-IVG-2001-01, Basser Department of Computer Science, The University of Sydney, 2001, *Submitted.*
8. S. Hong and P. Eades, Drawing Planar Graphs Symmetrically III: One-connected Graphs, Technical Report CS-IVG-2001-02, Basser Department of Computer Science, The University of Sydney, 2001, *Submitted.*
9. S. Hong and P. Eades, Symmetric Layout of Disconnected Graphs, *Algorithms and Computation,* Proc. of ISAAC 2003, Lecture Notes in Computer Science 2906, pp. 405-414, Springer Verlag, 2003.
10. J. E. Hopcroft and J. K. Wong, Linear Time Algorithm for Isomorphism of Planar Graphs, *Proc. of the Sixth Annual ACM Symposium on Theory of Computing,* pp. 172-184, 1974.
11. J. Manning, *Geometric Symmetry in Graphs*, Ph.D. Thesis, Purdue Univ., 1990.

Train Tracks and Confluent Drawings

Peter Hui[1], Marcus Schaefer[1], and Daniel Štefankovič[2]

[1] Department of Computer Science, DePaul University, Chicago, Illinois 60604
phui@students.depaul.edu, mschaefer@cti.depaul.edu
[2] Department of Computer Science, University of Chicago, Chicago, Illinois 60637
stefanko@cs.uchicago.edu

Abstract. Confluent graphs capture the connection properties of train tracks, offering a very natural generalization of planar graphs, and – as the example of railroad maps shows – are an important tool in graph visualization. In this paper we continue the study of confluent graphs, introducing strongly confluent graphs and tree-confluent graphs. We show that strongly confluent graphs can be recognized in **NP** (the complexity of recognizing confluent graphs remains open). We also give a natural elimination ordering characterization of tree-confluent graphs which shows that they form a subclass of the chordal bipartite graphs, and can be recognized in polynomial time.

1 Introduction

The area of graph drawing deals with the visualization of graphs, where the visualization meets certain aesthetic or technical constraints [1]. Typically, the goal of the drawing of a graph is to minimize some parameter such as the crossing number, or, for grid drawings, the area, the number of times an edge bends, or the total length of the edges. Among these parameters, the crossing number has probably drawn the most attention. A crossing number of zero corresponds to planarity, for which linear time algorithms are known, but, in general, determining the crossing number of a graph is **NP**-complete [4], making it a hard parameter to minimize. Recently, Dickerson, Eppstein, Goodrich, and Meng [2] suggested an extension of the notion of planarity called confluency that, while allowing crossings, hides them in the drawing. At the core is an idea similar to the train tracks of Thurston [6]: we allow edges in the drawing to merge, like train tracks, into a single track. The merging device is called a switch. Figure 1 shows how to draw complete graphs and complete bipartite graphs confluently.

Dickerson, Eppstein, Goodrich, and Meng [2] identified several classes and families of graphs which are confluent, including interval graphs and cographs. They also gave examples for graphs which are not confluent (their smallest example is obtained from the Petersen graph by removing a single vertex), and a heuristic algorithm to recognize whether a graph is confluent or not. Interestingly, they did not study the complexity of the recognition problem.

The main contribution of this paper is to show that a natural strengthening of confluency in graphs can be recognized in **NP**. In Section 2 we define the

J. Pach (Ed.): GD 2004, LNCS 3383, pp. 318–328, 2004.

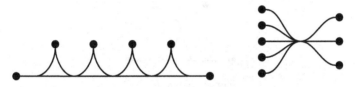

Fig. 1. How to draw K_6 and $K_{5,3}$ confluently.

notions of confluency and strong confluency. Their relationship is investigated in Section 3 by studying their underlying train tracks. Section 4 shows that strong confluency can be recognized in **NP** by giving a polynomial upper bound on the number of switches necessary to represent a graph. We think it is not unlikely that the problem will turn out to be **NP**-complete.

If confluency does turn out to be **NP**-hard, it will be of interest to identify large, and natural, subclasses which can be recognized efficiently. One immediate way of obtaining interesting classes of confluent graphs is by taking well-known graph classes whose definition depends on planarity, and replace planarity with confluency. In that manner we obtain outer-confluent graphs (see Section 6), and tree-confluent graphs, whose confluent drawings are treelike. In Section 5 we will show that the tree-confluent graphs can be recognized efficiently with the help of an elimination order characterization.

2   Train Tracks and Confluent Drawings

Definition 1. *A* curve *is a continuous mapping of* $[0, 1]$ *into the Euclidean plane; we often identify a curve and its image. A curve is* smooth, *if it is differentiable (intuitively, it cannot make sharp turns). A smooth curve which does not self-intersect is called* locally monotone *[2].*

Definition 2. *A* train track drawing *with vertices* V *and switches* S *is a subset* D *of the Euclidean plane such that (i)* $V \cap S = \emptyset$, *(ii) there is a injective mapping of* $V \,\dot\cup\, S$ *into* D *(we identify a point in* $V \,\dot\cup\, S$ *with its image), (iii) any curve in* D *not containing a switch must be smooth, and (iv) any two overlapping curves in* D *must have a common tangent at any point of overlap; that is, they have to join smoothly.*

A curve in a train track drawing which shares exactly its two endpoints with $V \,\dot\cup\, S$ *is called a* branch.

Based on this notion of a train track drawing, we derive two graph drawing concepts.

Definition 3. *We call a graph* $G = (V, E)$ confluent, *if there is a train-track drawing* D *on* V *such that* $uv \in E$ *if and only if there is a locally monotone curve in* D *with endpoints* u *and* v *that does not contain any other points of* V. *In this case we call* D *a* confluent drawing *of* G.

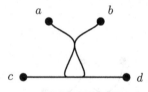

Fig. 2. K_4 or $K_4 - e$?

For an example, consider the train track drawing in Figure 2. We can easily trace locally monotone curves connecting all pairs of vertices – with the exception of a and b. There is a smooth curve connecting a to b, but it is not locally monotone, since it has to self-intersect. Hence, the train track drawing in Figure 2 is a confluent drawing of a $K_4 - e$.

When tracing a train track drawing visually, the requirement to avoid self-intersections seems to force a reader to backtrack to determine whether two points are connected. Removing this requirement leads to the following notion.

Definition 4. *We call a graph $G = (V, E)$ strongly confluent, if there is a train track drawing D on V such that $uv \in E$ if and only if there is a smooth curve in D with endpoints u and v that does not contain any other points of V. In this case we call D a strongly confluent drawing of G.*

Using this new definition, we would say that the train track drawing in Figure 2 is a strongly confluent drawing of a K_4.

Remark 1. The notion of confluency was introduced by Dickerson, Eppstein, Goodrich, and Meng in [2]; at a first glance it might seem that confluency is a stronger requirement than strong confluency. The opposite, however, is true; every strongly confluent graph is confluent (as we will see in Corollary 1), and there is a confluent graph which is not strongly confluent.

By definition, any point of D at which several curves combine is a switch. A switch has two sides, each with an arbitrary number of incoming curves. Every such switch can be replaced by a series of simple switches, where a *simple switch* is a switch in which two curves merge into a single curve. For example, the drawing of K_6 in Figure 1 uses simple switches, whereas the drawing of $K_{5,3}$ in the same figure uses a single switch which is not simple. Figure 3 shows how to draw $K_{5,3}$ using only simple switches.

For the rest of the paper we will use switch synonymously with simple switch, unless explicitly stated otherwise.

3   Train Tracks

We want to capture the combinatorial structure of a train track drawing D in graph-theoretic terms, abstracting from the particular embedding. To this end, we call a vertex-labelled graph $H = (V \,\dot\cup\, S, F, o)$ a *train track* if the following holds:

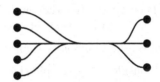

Fig. 3. How to draw $K_{5,3}$ using only simple switches.

(i) The vertex set of H consists of two types of vertices V and S, we call *vertices* and *switches* of H,

(ii) switches have degree 3,

(iii) $o(s) \in V \cup S$ is one of the three neighbors of s (for every switch $s \in S$).

Remark 2. We will often consider H as a directed graph, for example to specify in which direction the undirected edge uv is traversed. In that case we will write (u, v) or (v, u), and we will tacitly consider the graph as symmetric; that is, for every directed edge, the reverse edge also belongs to the graph.

We think of $o(s)$ as determining the orientation of the switch s: if we enter the switch s coming from $o(s)$, it forks into two branches. A curve in a train track drawing now corresponds to a walk in the train track which *respects the orientation of the switches* in the sense that for every part (u, s, v) of the walk s is a switch, and $o(s)$ is either u or v (and u and v are different from each other). We call such a walk *acceptable*. We can now rephrase our notions of confluency and strong confluency in terms of train tracks.

Lemma 1. *A graph $G = (V, E)$ is confluent if there is a planar train track H such that $uv \in E$ if and only if there is an acceptable path from u to v in H. The graph is strongly confluent if there is a planar train track H such that $uv \in E$ if and only if there is an acceptable walk from u to v in H.*

Proof. Consider a train track drawing D with vertices V and switches S. Construct a train track H as follows: $V \cup S$ will make up the vertex set of H. Include an edge (u, v) in F if in D there is a curve from u to v which does not cross through any vertices or switches. We assumed that switches are simple, hence there are three branches leaving s. Let $o(s)$ be the endpoint (other than s) of the edge corresponding to the branch that extends the other two branches smoothly. Then H is a planar train track in which every acceptable walk corresponds to a smooth curve in D, and every acceptable path to a locally monotone curve. ∎

Remark 3. Given a train track, the graph it represents can be found in polynomial time. In the case of strong confluency this is obvious, for confluency the problem can be reduced to a matching problem [3].

Theorem 1. *If $G = (V, E)$ is strongly confluent, then it is represented by a planar train track $H = (V \cup S, F, o)$ such that $ab \in E$ if and only if there is an acceptable path P_{ab} in H from a to b.*

We omit the proof of Theorem 1. Together with Lemma 1 it immediately implies a relationship between the two notions of confluency we introduced.

Corollary 1. *Any strongly confluent graph is confluent.*

The inclusion is strict, consider for example the graph drawn in Figure 4. By adding some more vertices and edges on the outside, we can force that all the switches occur within the circle. The resulting graph is confluent, but not strongly confluent (since vertex 6 will always be connected to vertex 4 in a strongly confluent drawing).

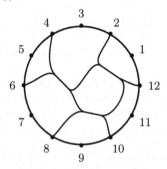

Fig. 4. Construction for graph which is confluent but not strongly confluent.

4   Strong Confluency in NP

Lemma 2. *If $G = (V, E)$ is confluent, then there is a train track H representing G, and acceptable paths P_e for every edge $e \in E$ such that the following condition holds:*

> *If P is a longest path contained in both some P_e and some P_f, then P is a single edge.*

Proof. We need a measure of overlap between two paths P_e and P_f. To this end, we introduce the numbers

$$o_{ef} := \sum_{P \text{ maximal subpath of } P_e \cap P_f} |P|^2.$$

With this we can establish the following claim:

> Suppose H is chosen to minimize the number o_{ef}. In that case, if P is a path contained in both P_e and P_f, then P is a single edge.

If the conclusion of the claim is false, there is a path (u, x, v) belonging to both P_e and P_f. Let y be the endpoint of the third edge incident to x; without loss of generality, we can assume that $o(x) = u$. Since P_e and P_f are paths, the edge xy cannot belong to either of them. Modify H as follows: remove edges ux and xv and add two new vertices u', v' and edges uu', $u'v'$, $v'v$, $u'x$ and xv'; set $o(u') = u$, $o(v') = v$, and $o(x) = u'$. Modify P_e and P_f such that one of them uses

(u, u', x', v', v) and the other (u, u', v', v). This will split the maximal common subpath of $P_e \cap P_f$ containing (u, x, v) into two parts. Since $(i + j)^2 > i^2 + j^2$ for $i, j \geq 1$, this strictly reduces o_{ef} showing that H did not minimize it. This establishes the claim.

In the modification made to H in the proof of the claim, we can route any other P_g through (u, u', v', v) if it used (u, x, y) or through (u, u', x, y) if it used (u, x, v); in either case the length of another P_g path will be increased by at most one.

More importantly, if any maximal subpath of P_g and P_h is an edge, then the modification to H will not change that: if P_g and P_h were affected by the modification and shared a single edge on the vertices u, x, v, and y, it must have been ux, and one of P_g and P_h must have used xv and the other xy; hence, after the modification they will only share uu'.

Let $e_1 f_1, e_2 f_2, \ldots, e_k f_k$ be an ordering of all pairs of distinct edges of G. The above observation immediately implies that if we choose H so as to minimize (in lexicographic ordering) the vector

$$(o_{e_1 f_1}, o_{e_2 f_2}, \ldots, o_{e_k f_k})$$

then any paths P_e and P_f intersect in isolated edges. ■

For the rest of this section we will concentrate on strongly confluent graphs. Because of Corollary 1, we can still apply Lemma 2 in that case, concluding that overlap between a P_e and a P_f consists of non-adjacent edges. Moreover, these overlaps between P_e and P_f correspond to crossings in a planar drawing of H. That is, if we have the path (u_e, s, t, v_e), part of P_e and (u_f, s, t, v_f), part of P_f then u_e and v_e cannot be on the same side of st in the planar drawing of H, since otherwise we could have reduced o_{ef} by having two separate paths (u_e, v_e) and (u_f, v_f) as shown in Figure 5.

There is one scenario which would prohibit the application of the move shown in Figure 5, namely if there was a third P_g making use of the edge st. This, however, is not possible, since one pair from P_e, P_f, P_g would share a path of length ≥ 2.(Note that this operation would not be valid if the representation was just confluent, since lifting the path could introduce new connections between vertices not possible before.)

Our goal is to show that we can assume the number of switches in H to be polynomial in the number of vertices. To this end we equip the train track with an edge labelling that contains connectivity information.

Given a train track $H = (V \mathrel{\dot{\cup}} S, F, o)$ for $G = (V, E)$ we define a labelling of the directed edges of H as follows:

$$\ell(u, v) = \{a \in V : \text{ there is an acceptable walk from } a \text{ to } v \text{ in } H$$
$$\text{passing through } (u, v)\}.$$

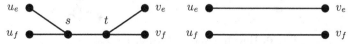

Fig. 5. Lifting a path (in a strongly confluent representation).

From the definition it follows that ℓ is the minimal labelling fulfilling

(i) $a \in \ell(a, u)$ for any edge $(a, u) \in F \cap (V, V \cup S)$,

(ii) $N_G(a) = \bigcup_{(u,a) \in F} \ell(u, a)$ for any $a \in V$, where $N_G(a) = \{b : ab \in E\}$ is the neighborhood of a in G,

(iii) for any switch $s \in S$ and its neighbors $u = o(s)$, v, w:

$$\ell(u, s) \subseteq \ell(s, v) \cap \ell(s, w), \text{ and}$$
$$\ell(s, u) \supseteq \ell(s, v) \cup \ell(w, s).$$

By the results proved so far we know that if $G = (V, E)$ is strongly confluent, then there is $H = (V \cup S, F, o)$ such that $ab \in E$ if and only if there is an acceptable path P_{ab} from a to b in H.

Lemma 3. *If $G = (V, E)$ is strongly confluent, then it is represented by a train track H with $O(|V|)^6$ vertices and switches.*

Let $uv \in E$. Consider a path P_{uv} in H whose inner vertices are all switches, and the function $\ell(\vec{e})$ as we move the directed edge e along P_{uv} from u to v. This yields a monotone function, namely, if e occurs before f on P_{uv} and both edges are directed towards v, then $\ell(\vec{e}) \subseteq \ell(\vec{f})$. Therefore, $\ell(\vec{e})$ can take on at most $|V| + 1$ different values along P_{uv}. Similarly, if we move an edge $\overleftarrow{e}$ directed towards u along P_{uv} from u to v, the corresponding label sets are monotonously decreasing, and, hence, also take at most $|V| + 1$ different values along P_{uv}. Consequently, the expression $(\ell(\vec{e}), \ell(\overleftarrow{e}))$ can change value less than $2(|V| + 1)$ times as we travel along P_{uv} from u to v.

For each $uv \in E$ we color those switches at which $(\ell(\vec{e}), \ell(\overleftarrow{e}))$ changes red, and the remaining switches blue. Note that at most $(2|V| + 1)|E|$ switches are colored red. We call the maximal segments of P_{uv} which do not contain red switches blue segments. There are at most $4(|V| + 1)|E|$ blue segments. We will show that there is a drawing such that any two blue segments intersect at most once. Hence there is a drawing with at most $(2|V| + 1)|E| + 2(4(|V| + 1)|E|)^2 = O(|V|^6)$ switches.

Consider two edges e and f in H that are adjacent crossings of a blue segment P with other blue segments. There are two possible scenarios depending on whether the crossings are parallel or not (as earlier, the sharp angles represent the forking part of a switch, and thus define o). Figure 6 shows how the order of two parallel crossings can be swapped.

We can use a similar move for non-parallel crossings, as shown in Figure 7.

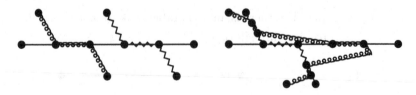

Fig. 6. How to swap two parallel crossings.

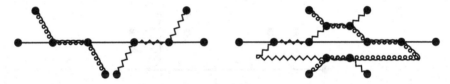

Fig. 7. How to swap two non-parallel crossings.

Note that in both cases we can extend the labelling of H to the newly introduced edges so that the graph represented by H remains the same: we simply label the new edges with $(\ell(\overrightarrow{e}), \ell(\overleftarrow{e}))$.

Suppose that two blue segments P and R cross more than once. Let e_1, e_2 be crossings of P and R such that there are no crossings of P and R between e_1 and e_2 on R. There may be crossings of P and R between e_1 and e_2 on P. Label them by i if after cutting R between e_1 and e_2 they would be in the same component of R as e_i. There is a pair of neighboring crossings f_1, f_2 labeled by $1, 2$, respectively. Using the swap moves on edges intersecting R we can make e_1, e_2 adjacent on R and then shortcut P eliminating half of the intersections created by the swap moves. Similarly using the swap moves on edges intersecting P we can make f_1, f_2 adjacent on P and then shortcut R eliminating half of the intersection created by the swap moves. In one of the cases we decrease the total number of intersections while preserving the property that any two paths intersect in paths of length 1. Hence there is a train track in which any two blue segments intersect at most once.

Corollary 2. *Strong confluency can be tested in* **NP**.

Proof. Corollary 2 shows that if G is strongly confluent, then there is a train track representing G of size polynomial in $|G|$. In **NP** we can guess any such train track, and verify that it represents G. ∎

5 Tree-Confluent Graphs

We call a train track drawing D *tree-like*, if it does not contain a closed curve (the curve would not have to be smooth or locally monotone). For example, Figure 8 shows a tree-like train track drawing. On the other hand, Figure 1 shows a train track drawing representing K_6 which is not tree-like. We call a confluent graph which can be represented by a tree-like train track drawing *tree-confluent* (the strong case being the same). We will see later that all tree-confluent graphs are bipartite, so there is no tree-like train track drawing representing K_6.

In graph theoretic terms, the underlying abstract train track of a tree-confluent graph has to be a tree. A graph is tree-confluent if and only if it is represented by a planar train track which is a tree. We now give a characterization of tree confluent graphs in terms of a vertex elimination ordering. This characterization leads to a fast recognition algorithm.

Theorem 2. *A graph is tree confluent if and only if repeatedly removing (i) vertices of degree 1, and (ii) vertices u such that there is another vertex v with $N(u) = N(v) \neq \emptyset$, leads to the trivial graph (containing only a single vertex).*

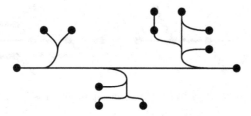

Fig. 8. A tree-like train track drawing.

Proof. First observe that if G is tree confluent then it will still be tree confluent after the removal of vertices of type (i) or (ii). Furthermore, if G is not tree confluent, it cannot become tree confluent by removing a vertex of type (i) or (ii): if $G - \{v\}$ were tree confluent and v has degree 1 in G, then it has degree 1 in the underlying train track, hence G is tree confluent; similarly if $G - \{v\}$ is tree confluent, and G contains another vertex u with $N(u) = N(v)$, then we can replace u in the train track for $G - \{v\}$ by a switch that branches to u and v, showing that G is tree confluent (note that G does not contain the edge uv, since $N(u) = N(v)$).

Since the trivial graph is tree confluent, this observation implies that any graph which can be reduced to the trivial graph by removing vertices of type (i) and (ii) is tree confluent.

Furthermore, for the other direction, the observation shows that the order of removal is irrelevant, and it is sufficient to show that if G is tree confluent, there is some order $\mathcal{E}$ in which to remove vertices of type (i) and (ii) such that we end up with the trivial graph.

Suppose $G = (V, E)$ is tree confluent; then there is a planar train track $H = (V \,\dot{\cup}\, S, F, o)$ which represents G. Consider F as a set of directed edges. We define a function p from F to $\mathbb{N}$ as follows:

$$p(u, v) = |\{w \in V : \text{there is a walk from } v \text{ to } w \text{ that does not use } u\}|.$$

Figure 9 shows an example of p.

Using p we define a second function r from S to $\mathbb{N}$. Let the neighbors of s be u, v, w, then

$$r(s) = \min\{p(s, u) + p(s, v), p(s, u) + p(s, w), p(s, v) + p(s, w)\}.$$

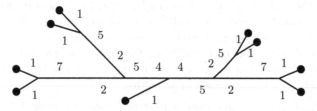

Fig. 9. A train track. Every edge is labelled with the number of vertices contained in the portion of the train track to which the edge points.

We begin the construction of $\mathcal{E}$ by repeatedly removing vertices of degree 1 in G until there are no such vertices left. If G has not turned into the trivial graph at that point, it has to contain at least one switch (otherwise it would be a tree on its vertices and would have to contain a vertex of degree 1).

Choose the switch s with minimal $r(s)$. We claim that $r(s) = 2$. First note that $r(s) \geq 2$, because p is always at least 1. Now suppose that $r(s) > 2$. Then s must be adjacent to at least one edge su such that $p(s, u) \geq 2$, since otherwise $r(s)$ would equal 2. The portion of the train track to which su leads cannot contain any switches, for if it contained a switch t, then $r(t)$ would be strictly less than $r(s)$ violating the minimality of $r(s)$. Therefore, the vertices in the portion of H to which (s, u) leads must form a tree. However, the leaves of this tree would have degree 1, and we already eliminated all those vertices.

Let s be a switch with $r(s) = 2$. Then there are u and v such that $p(s, u) + p(s, v) = 2$, and therefore $p(s, u) = p(s, v) = 1$. Hence, u and v have to be vertices of G. Since neither of them can have degree 1 in G, the switch s must be oriented to fork into u and v which implies that $N(u) = N(v)$. Hence we can continue the construction of $\mathcal{E}$ by selecting u, for example.

We continue in this fashion, eliminating vertices of degree 1 as long as possible, and then identifying leaf vertices of switches s with $r(s) = 2$. We showed that the only reason such a switch would not exist is that the graph G has turned into the trivial graph, which is what we had to show. ∎

The elimination characterization of Theorem 2 leads to a randomized linear time algorithm for recognizing tree-confluent graphs.

In [5], Golumbic and Goss introduced the now well-known class of graphs known as the *chordal bipartite graphs*, which are those bipartite graphs in which every cycle of length at least 6 contains a chord (that is, no cycle of length at least 6 is induced). Removing a vertex of degree 1 or a vertex u such that there is another vertex v for which $N(u) = N(v)$ from a graph does not change the property of a graph being chordal bipartite. This observation gives us the following lemma.

Lemma 4. *Every tree confluent graph is chordal bipartite.*

The reverse is not true as witnessed by a C_6 with a single chord.

6   Open Problems

While we have shown that strong confluency can be recognized in **NP**, we currently have no such result for confluency. Although the two notions are very similar, their combinatorial nature seems to be quite different. At this point we cannot even rule out the possibility that a confluent graph needs an exponential number of switches to be realized (although that would not necessarily affect membership in **NP**, as witnessed by the example of string graphs [7]).

Identifying large classes of confluent graphs remains a challenging task. We suggest the notion of outer-confluency (confluent graphs that can be drawn in a disk with all the vertices on the boundary of the disk). As in the case of confluency there are examples of graphs that are outer-confluent but not strongly outer-confluent (see Figure 4). Dickerson, Eppstein, Goodrich, and Meng [2] showed – in effect – that all cographs are outer-confluent (even strongly outer-confluent), thereby also showing that outer-confluency is a strict superclass of tree-confluency. It does not seem unlikely that outer-confluent graphs can be recognized in polynomial time.

We seem to have a good understanding of tree-confluent graphs; the main missing piece is a deterministic linear time recognition algorithm.

References

1. Giuseppe Di Battista, Peter Eades, Roberto Tamassia, and Ioannis G. Tollis. *Graph Drawing: Algorithms for the Visualization of Graphs*. Prentice Hall, 1999.
2. Matthew T. Dickerson, David Eppstein, Michael T. Goodrich, and Jeremy Yu Meng. Confluent drawings: visualizing non-planar diagrams in a planar way. In *Proc. 11th Int. Symp. Graph Drawing (GD 2003)*, Lecture Notes in Computer Science. Springer-Verlag, 2003.
3. David Eppstein, 2004. personal communication.
4. Michael R. Garey and David S. Johnson. Crossing number is NP-complete. *SIAM Journal on Algebraic and Discrete Methods*, 4(3):312–316, 1983.
5. M. Golombic and C. F. Goss. Perfect elimination and chordal bipartite graphs. *J. Graph Theory*, 2:155–163, 1978.
6. R. C. Penner and J. L. Harer. *Combinatorics of train tracks*, volume 125 of *Annals of mathematics studies*. Princeton University Press, 1992.
7. Marcus Schaefer, Eric Sedgwick, and Daniel Štefankovič. Recognizing string graphs in NP. *J. Comput. System Sci.*, 67(2):365–380, 2003. Special issue on STOC2002 (Montreal, QC).

The Three Dimensional Logic Engine

Matthew Kitching[1,*] and Sue Whitesides[2,**]

[1] Dept. of Comp. Sci., U. of Toronto, Canada M5S 3G4
kitching@cs.toronto.edu
[2] School of Comp. Sci., McGill U., Montreal, Canada H3A 2A7
sue@cs.mcgill.ca

Abstract. We consider the following graph embedding question: given a graph G, is it possible to map its vertices to points in 3D such that G is isomorphic to the mutual nearest neighbor graph of the set P of points to which the vertices are mapped? We show that this problem is NP-hard. We do this by extending the "logic engine" method to three dimensions by using building blocks inpired by the structure of diamond and by constructions of A.G. Bell and B. Fuller.

1 Introduction

Proximity graphs are an important and well studied area of computer science and find applications, for example, in architecture, pattern recognition, and geography. Proximity graphs are defined to capture some kind of spatial relationship between pairs of points on the plane or in space. Two points, regarded as vertices of a graph, are joined by an edge if, and only if, the points satisfy some given proximity criterion. Examples of proximity graphs include mutual nearest neighbour graphs, Gabriel graphs, and the Delauney triangulation. For a review of proximity graphs, see [11, 16].

Given an abstract combinatorial graph whose vertices are labeled, and a proximity criterion, the *recognition problem* is to determine whether there is some set P of points, typically required to lie in 2D or 3D, such that the graph is isomorphic to the proximity graph on P defined by the given proximity criterion. The *realization problem* is to produce such a set P if one exists. This paper proves that the problem of recognizing mutual nearest neighbour graphs in 3D is NP-hard, an open problem in graph drawing (see [4]). Our proof builds a 3D version of a "logic engine". The building blocks we designed are based on the structure of diamond; they may prove useful in extending the logic engine approach to obtain complexity results for other 3D layout and proximity problems studied previously in two dimensions (e.g., [3, 4, 6, 7, 12–14]).

The rest of this section contains background material. Section 2 extends the logic engine approach to 3D, using the *octet truss* of Buckminster Fuller, and gives our NP-hardness result. Section 3 concludes.

* Research undertaken while in the M.Sc. program in the School of CS at McGill U.
** Partially supported by NSERC and FCAR.

Preliminaries. A *mutual nearest neighbour graph* of a set P of points is a proximity graph for which each pair x,y of vertices arising from points x,y is connected by an edge if, and only if, point x is a nearest neighbor of y, and point y is a nearest neighbor of x. The set P of points typically lies in 2D or 3D. Note that in 2D, points x,y determine an edge if the interior of the union of the two discs centred at x and y, and having radius equal to their separation distance, is empty of points other than x and y.

A simple example of a 3D mutual nearest neighbor graph is given by the combinatorial structure of the vertices and edges of the regular tetrahedron. The four vertices of the regular tetrahedron of unit edge length are positioned at points in space so that each point is unit distance from each of the other three. Thus each pair of points gives rise to an edge in the mutual nearest neighbor graph of the points, which is thus K_4, the complete graph on four vertices.

As we will later see, if we start with a combinatorial graph K_4 whose vertices are labeled and ask whether it can be realized in 3D as the mutual nearest neighbor graph of some set P of four points, we find that there are exactly two realizations, up to translation, rotation, and scaling, and that these two realizations are mirror images of each other.

Similarly, the vertex-edge incidence structure of the regular octahedron can be thought of as a 3D mutual nearest neighbour graph, and the combinatorial graph has exactly two realizations, up to translation, rotation, and scaling.

In the early 1900's, Alexander Graham Bell used rigid tetrahedra and octahedra to construct kites, an unsuccessful flying machine, and a tall tower [1]. The structures that Bell assembled are mutual nearest neighbour graphs, as we will prove. These structures were later rediscovered by Buckminster Fuller [2], who patented the *octet truss* and used it extensively.

Mutual Nearest Neighbour Graph Recognition (MNNGR). Given an undirected graph G, is G realizable as a mutual nearest neighbour graph?

For 2D, MNNGR was proved NP-Hard by Eades and Whitesides [9], by a reduction from Not-All-Equal-3-Satisfiability (NAE3SAT) to MNNGR via a method they called the "logic engine" approach, reviewed below. Recall that NAE3SAT is NP-complete and that an instance consists of m clauses each containing three distinct literals, and that a satisfying assignment must contain at least one true and at least one false literal in each clause ([10]). It may be assumed that no clause contains both a literal and its complement.

The Logic Engine Approach. The logic engine is a virtual mechanical device that encodes instances of NAE3SAT. The device can be positioned a certain way in the plane if and only if the instance of NAE3SAT that it encodes can be satisfied. The idea for obtaining hardness results for proximity graph recognition problems is to design a graph whose only possible realizations imitate the correctly positioned mechanical device.

The (m,n) logic engine contains a rigid "frame" and a "shaft". To the shaft are attached a series of "armatures" $A_j, 1 \leq j \leq n$, one for each literal x_j in the instance of NAE3SAT. Each armature can rotate about the shaft independently of the others, although the position of each armature along the shaft is fixed.

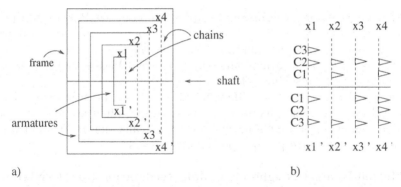

a)   b)

Fig. 1. a) Schematic for a (3,4) logic engine, and b) Encoding for NAE-3SAT instance $c_1 = \{x_1, x'_2, x_3\}$, $c_2 = \{x'_1, x'_2, x'_3\}$, $c_3 = \{x_2, x_3, x_4\}$.

Each armature A_j in turn has two "chains" attached to it: a_j from one end of the armature to the shaft, and a'_j from the other end of the armature to the shaft. The length of a chain equals the distance between the shaft and the ends of its armature. Each chain has at least m links, which are numbered 1,2,...,m, outward from the shaft; the first m links correspond to clauses with the same indices, respectively.

The a_j chain of armature A_j represents the uncomplemented literal x_j, and its ith link represents the possible occurrence of x_j in clause c_i; similarly for link i of chain a'_j and the possible occurence of the complemented literal x'_j in c_i. A "flag" attached to link i of armature chain a_j indicates that x_j does NOT occur in clause c_i; similarly, a flag on link i of chain a'_j indicates the non-occurence of x'_j in c_i. See Figure 1 for a (3,4) logic engine encoding of a NAE-3SAT instance.

The entire structure can move in the following ways: each armature can lie in one of two positions, either with a_i above the shaft, or with a'_i above the shaft; and each flag can face to the right, or can be flipped to face the left.

Although each flag is free to rotate, flags that lie on the same row and lie on adjacent armatures must not face one another. If they do face each other, the flags *collide*. Similarly, any flag in armature A_n collides with the frame if it faces outward; any flag in armature A_1 collides with the frame if it faces inward. We later use the following lemma.

Lemma 1. *(from [9]) A given instance of NAE3SAT has a satisfying solution if, and only if, there exists a collision-free configuration for the logic engine.*

2   Mutual Nearest Neighbor Graphs in Three Dimensions

Here we prove the NP-hardness of MNNGR in 3D, settling a problem from [4]. While the result was anticipated in [9], this is the first concrete proof.

3D Nearest Neighbour Rule: Vertex v_i is a nearest neighbour of v_j if, and only if, the open sphere of radius $d(v_i, v_j)$ around v_j contains only v_j.

3D Mutual Nearest Neighbour Graph: Suppose that P is a set of points in 3D. Then a 3DMNNG, defined by a set P of points, is an undirected graph G with a vertex v_i for every point $p_i \in P$. For every pair of vertices $v_i, v_j \in G$, there is an edge between them if, and only if, v_i is a 3D nearest neighbour of v_j, and v_j a 3D nearest neighbour of v_i.

A graph G is *realizable* as a **3DMNNG** if for some point set P in 3D, the 3DMNNG on P is isomorphic to G. In that case we often use "G" to denote both the combinatorial graph and its geometric realization in 3D, and we use the same labels for corresponding points and vertices.

3D Mutual Nearest Neighbour Graph Recognition (3DMNNGR):
Given an undirected connected graph G, is G realizable as a 3DMNNG?

We will use the logic engine paradigm to transform NAE3SAT to 3DMNNGR in polynomial time, thus showing that 3DMNNGR is NP-hard.

Lemma 2. (straight-forward from Lemma 2 of [9]) *Suppose that G is a connected 3DMNNG. Then all edge segments of G have the same length.*

From now on, we assume all edges in a connected 3DMNNG have *unit* length.

Lemma 3. (by Lemma 2) *Suppose that H is an induced connected subgraph of a combinatorial graph G. Then any 3DMNNG realization of G includes a 3DMNNG realization of H.*

Two realizations of a labeled graph are "the same" if, following possible translation, rotation, and scaling, vertices with the same label coincide. Thus, all mirror images of a given labeled structure are the same in this sense. However, mirror images are not, in general, the same as their initial labeled structure. Taking the mirror image of a vertex-labeled tetrahedron turns it inside out; the mirror image cannot be superimposed on the original, with labels matching, by translation and rotation.

In the next lemma, and throughout the paper, we let h = $\sqrt{6}/3$, which is the distance from the base of a tetrahedron to the top, if all edges are unit length.

Lemma 4. (straight-forward) *The labeled graph K_4 has exactly two realization as a 3DMNNG, both of which are regular tetrahedra.*

Once we prove a labeled combinatorial graph has exactly two realizations as a 3DMNNG, we use the term "graph" to refer to either realization.

Lemma 5. (straight-forward) *Let H be a labeled graph isomorphic to the combinatorial structure of Figure 2c). Then H has exactly two realizations as a 3DMNNG, namely a regular octahedron and its mirror image.*

Lemma 6. *Let H be the labeled combinatorial graph (called an **octet truss**) arising from the geometric structure in Figure 3a). H can be realized as a 3DMNNG in exactly two ways: points a,b,c,d,e,f must lie on a single **base plane**, while points u,v,w must lie on a parallel **lid plane** at a distance h from the base plane.*

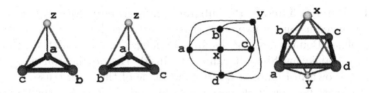

Fig. 2. a) and b) Two realizations of labeled K_4; c) and d) Labeled graph H and one of its realizations as an octahedron.

Proof. By Lemma 2, all edges must have unit length. By Lemma 5, the points b,c,e,u,v,e construct a regular octahedron. When the plane of b,c,e is viewed from above, then the plane of u,v,w is parallel to it, and may lie on either side of it. (In Figure 3a, the plane of u,v,w is shown closer to the viewer than the plane of b,c,e). Now add the points a,d,f to the graph. These points become members of tetrahedra. Consider each of a,d,f in turn. Although by Lemma 4, a labeled tetrahedron has two realizations, one of these would place the point (a,d or f) inside the octahedron and violate a distance constraint: no vertex in a connected MNNG can lie distance less than one from another vertex. Hence the remaining points lie in the plane of b,c,e as shown.   □

A **base plane** is defined by the six coplanar vertices of an octet truss (in Figure 3a, the plane of a,b,c,d,e,f). The remaining three vertices of the truss define the **lid plane** (in Figure 3a, the plane of v,u,w). A **base vertex** is any one of the six vertices on the base of an octet truss. A **lid vertex** is any one of the three vertices on the lid of an octet truss. The plane parallel to the base plane, at distance 4h on the same side of the base plane as the lid plane, is the **mid-plane**. The plane parallel to the base plane, at distance 8h on the same side of the base plane as the lid plane, is the **sky plane**.

Lemma 7. *Let H be the labeled combinatorial graph arising from the geometric structure in Figure 3b) (called the **hexagonal octet truss**). H can be realized in exactly two ways. Furthermore, u,v,w and x are coplanar, and the remaining vertices are coplanar, and these planes are parallel.*

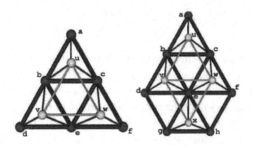

Fig. 3. a) Octet truss and b) Extended octet truss.

Proof. By Lemmas 3 and 6, the subgraph induced by vertices a,b,c,d,e,f,u,v,w must be realized as one of two octet trusses, with the lid plane on either side of the base plane. Point x lies unit distance from each of v,e,w, with which it forms a regular tetrahedron. Since it cannot lie inside the octahedron b,c,e,u,v,w, it must lie as shown in Figure 3b), in the lid plane of the octet truss.

Point g must lie on a circle perpendicular to d,e, centred at the mid-point of d,e. Likewise, g must lie on a circle centred at the mid-point of x,e. The circles intersect in two points, so g must lie either as shown in 3b), or at the position occupied by vertex v, which is clearly not allowed. Similarly for point f.   □

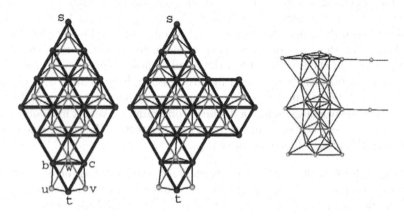

Fig. 4. a) Link graph b) Flagged link graph c) Tower with mid-wire and sky-wire.

The **link graph**, the **flagged link graph** and the **k-tower** are the labeled graphs having the combinatorial structures shown in Figures 4 a), b), and c), respectively. All three graphs can be realized as 3DMNNG's. The next lemma proves that each graph has exactly two realizations, namely, the realizations shown in the figure, together with their mirror images. We also define the *direction* of a link graph realization, or flagged link graph realization, to be the vector of the directed line segment from the s to the t vertices of the realization. The line defined by the points s and t is called the **s-t axis**.

Lemma 8. *The link graph, flagged link graph, and k-tower each have only two realizations.*

Proof. The combinatorial structure of the k-tower has the form $H_1 \cup ... \cup H_k$, where H_i is an octet truss, and for i odd, $H_i \cap H_{i+1}$ consists of three shared lid vertices, and for i even, $H_i \cap H_{i+1}$ consists of six shared base vertices. By Lemma 6, each H_i has two realizations. However, once a realization is chosen for an octet truss, say H_1, the remaining realizations are determined. Thus a k-tower has exactly two realizations, which are mirror images of each other.

The link graph and flagged link graph also have two realizations. By Lemma 7, the position of each vertex, with the exception of u, v, and t, is determined once a realization of the octet truss containing s is determined, since all vertices are connected by extended octet trusses. To place u,v,t, note that points b,w,c,u,v,t form a regular octahedron for which there are exactly two realizations by Lemma 5. However, one of these realizations violates a distance constraint with respect to the rest of the graph, so the octahedron is as shown in Figure 5.   □

The **k-flagged link graph** is a sequence of k link graphs or flagged link graphs joined together as shown in Figure 5. Each link graph in the k-flagged link graph is called a **block**. Since the 3DMNNG realization of a flagged link graph has exactly two realizations, which are mirror images of each other, the Euclidean distance from vertex s to vertex t is a constant, which we denote by $d_{flagged\_link}$.

Lemma 9. (from the definition of $d_{flagged\_link}$) *If the 3DMNNG realization of a k-flagged link graph spans a distance of $k \cdot d_{flagged\_link}$, then the s-t axis of all link graphs and flagged link graphs must coincide.*

Such a realization is a **taut realization** of a k-flagged link graph.

Lemma 10. *In a taut realization of a k-flagged link graph, the base planes of all the flagged link graph realizations must be the same, and similarly, the lid planes must be the same.*

Proof. To show that all the base planes form a common plane, consider two joining link graphs X and Y (see Figure 5). Note that t_{x1},t_{x2},t_{y0},t form a regular tetrahedron. Fix block X in space, thereby determining the position of points t,t_{x1},t_{x2}, and therefore the position of t_{y0}. This prevents the rotation of block Y about the s-t axis. The midpoint of t_{y1} and t_{y2} must lie on the s-t axis at a known position. Points t_{y1} and t_{y2} must lie on a circle centred at the midpoint of

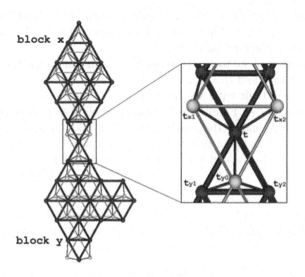

Fig. 5. 2-flagged link graph.

t_{y1}, t_{y2}, and also, on a circle centred at midpoint of t, t_{y0}, which is also a known position. These two distinct circles intersect in two points, namely the positions occupied by t_{y1} and t_{y2} in Figure 5. These positions lie in the base plane of block X; hence blocks X and Y have the same base plane. Since the position of t_{y0} is determined by block X, the lid plane of block Y is thus the same as the lid plane of block X.   □

Lemma 11. *Any taut realization of a k-flagged link graph has exactly 2^k realizations as a 3DMNNG.*

Proof. As seen in Lemma 10, a taut realization of a k-flagged link graph must have a common base plane and a common lid plane. However, any block can have two realizations, which are mirror images of each other. To see this, consider a realization of a block, and take its mirror image with respect to the plane containing the s-t axis and perpendicular to the base plane. This second realization has the same s-t axis, base plane, and lid plane as the first.   □

Note that for blocks that are flagged link graphs, the two mirror images point in opposite directions. We will use the taking of mirror images to imitate rotations in the virtual logic engine.

Now, we build an (m,n) 3D logic graph, to imitate a logic engine, by constructing the following components, as seen in Figure 6. A **frame** consists of a series of octet trusses, together with two 8-towers (the locations of which are shown in gray in Figure 6). Attached to one side of the tower is a **mid-wire** at height 4h, and a **high wire** at height 8h (see Figure 4c). Each wire consists of a path of vertices and runs between both frame towers. By a similar argument to that in the proof of Lemma 9, we can ensure that all vertices of the mid-wire are colinear by forcing them to span a set distance to the other tower. The same is true for the sky wire.

Each of these wires intersects the towers of each of the armatures in a path of three vertices. As we will later prove, this ensures that the frame and all the armatures share a common base plane.

We define the π plane to be the plane perpendicular to the base plane, containing the mid-wire and the high wire.

An armature is built with a series of octet truss components forming three sides of a rectangle (see Figure 6). The last side consists of two m-flagged link graphs: a_i from one end of the armature to the π plane, and a_i' from the other end of the armature to the π plane. The armature also has a tower of height 8h intersecting the mid-wire and the high-wire in paths of three vertices lying on the π plane.

Lemma 12. *The frame component has two realizations, both of which have a single base, lid, mid, and sky plane. The armature component has a single base, lid, mid, and sky plane. The base planes of all armatures and the frame are coplanar. The same is true for the lid, mid, and sky planes.*

Proof. The frame is built with overlapping hexagonal octet trusses in such a way that all trusses must share a common base plane and lid plane. The towers of the

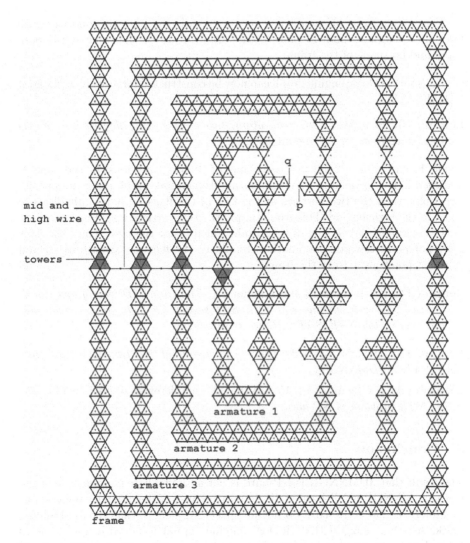

Fig. 6. A 2D projection of the entire logic engine. Towers project to the gray regions, and the mid-wire and sky-wire project to a common line. Edges between lid vertices are not shown, and edges between lid vertices and base vertices are not shown. There is a distance violation between vertices p and q; however a valid realization results when flag p is flipped to the right.

frame intersect the rest of the frame in octet trusses, and this also determines the orientation of the towers. Similarly, within each armature the base, lid, mid, and sky planes are the same for all but the k-flagged link graph. Since the distance between the extreme vertices of the k-flagged link graph is determined by the rest of the armature to be $k \cdot d_{flagged\_link}$, Lemma 9 applies to the k-flagged link graph. The k-flagged link graph is connected to the rest of the armature with the same connection seen between flagged link graphs.

The intersections of each armature tower with the mid-wire and sky wire force the base plane and lid plane of the armature to be the same as the base plane and lid plane of the frame.   □

An (m,n) 3D logic engine graph can now be constructed for any given instance of NAE3SAT.

Lemma 13. *In a 3DMNNG realization, two neighbouring flagged link graphs may not have flags facing one another.*

Proof. From Lemma 12, all the base planes of all armatures must be the same. If a flagged linked graph and its neighbouring flagged link graph point in opposite directions, then the two vertices at the tips of the flags must be unit distance apart in the logic engine. This would imply the two vertices would be connected in the combinatorial version of the graph, which is not the case. Thus, the flags must not point towards one another. Similarly, the flags on armatures A_1 and A_n must point away from the frame.   □

Lemma 14. (by Lemma 13 and Lemma 1) *The (m,n) 3D logic graph can be customized to encode instances of NAE3SAT so that the logic graph is realizable if, and only if, the NAE3SAT instance is satisfiable.*

Lemma 15. (straight-forward) *There is a polynomial time transformation from NAE3SAT to 3DMNNG.*

Theorem 1. (by Lemmas 14, 15 and the NP-completeness of NAE3SAT). *The 3DMNNGR problem is NP-hard.*

3   Conclusion

The result that 3DMNNGR is NP-hard is not an immediate consequence of the fact that MNNGR is NP-hard in 2D, in part because of the difficulty pointed out by Fuller: regular tetrahedra do not fill space. We believe that the 3D building blocks seen here suggest that the logic engine approach indeed is applicable to three dimensional problems in graph drawing.

References

1. A.G. Bell, "Tetrahedral principle in kite structure", *National Geographic Magazine*, 14(6), June 1903, pp. 219-251.
2. R. Buckminster Fuller. Inventions, the Patented Works of R. Buckminster Fuller. St. Martin's Press, 1983.
3. P. Bose, W. Lenhart, and G. Liotta, "Characterizing proximity trees", *Algorithmica*, 16, 1996, pp. 83-110, 1996.
4. F.J. Brandenburg, D. Eppstein, M. T. Goodrich, S. G. Kobourov, G. Liotta, and P. Mutzel, "Selected open problems in graph drawing", *Proc. 11th Int. Symp. Graph Drawing, 2003*, Springer-Verlag LNCS vol. 2912, pp. 515-539.

5. G. Di Battista, P. Eades, R. Tamassia, I.G. Tollis. Graph Drawing. Prentice Hall, 1999, Chapter 11.2.
6. M.B. Dillencourt, "Realizability of Delaunay triangulations", *Informa. Process. Lett.*, 33(6), Feb. 1990, pp. 283-287.
7. M.B. Dillencourt and W.D. Smith, "Graph-theoretical conditions for inscribability and Delaunay realizability", *Proc. 6th Canad. Conf. Comput. Geom.*, 1994, pp. 287-292.
8. P. Eades and S. Whitesides, "The logic engine and the realization problem for nearest neighbour graphs", *Theoretical Computer Science*, 169, 1996, pp. 23-37.
9. P. Eades and S. Whitesides, "The realization problem for Euclidean minimum spanning trees is NP-hard", *Algorithmica*, 16, 1996, pp. 60-82.
10. M. Garey and D. Johnson. Computers and Intractability: a Guide to the Theory of NP-Completeness. W.H. Freeman, 1979.
11. J.W. Jaromczyk and G.T. Toussaint, "Relative neighborhood graphs and their relatives", *Proc. IEEE*, 80(9), 1992, pp. 1502-1517.
12. W. Lenhart and G. Liotta, "The drawability problem for minimum weight triangulations", *Theoret. Comp. Sci.* vol. 27, 2002, pp. 261-286.
13. G. Liotta and G. Di Battista, "Computing proximity drawings of trees in the 3-dimensional space", *Proc. 4th Workshop Algorithms and Data Structures WADS 1995*, Springer-Verlag LNCS vol. 955, pp. 239-250.
14. G. Liotta, A. Lubiw, H. Meijer, and S.H. Whitesides, "The rectangle of influence drawability problem", *Comput. Geom. Theory Appl.*, 10(1):1-22, 1998.
15. G. Liotta and H. Meijer, "Drawing of trees", *Computational Geometry: Theory and Applications*, 24(3), 2003, pp. 147-178.
16. G. Toussaint, "A graph-theoretical primal sketch", in Computational Morphology. North-Holland, 1988, pp. 229-260.

Long Alternating Paths in Bicolored Point Sets

Jan Kynčl[1], János Pach[2,*], and Géza Tóth[3,**]

[1] Department of Applied Mathematics and
Institute for Theoretical Computer Science,
Charles University, Prague, Czech Republic
`jankyncl@centrum.cz`
[2] City College, CUNY and Courant Institute of Mathematical Sciences,
New York University, New York, NY, USA
`pach@cims.nyu.edu`
[3] Rényi Institute, Hungarian Academy of Sciences, Budapest, Hungary
`geza@renyi.hu`

Abstract. Given n red and n blue points in convex position in the plane, we show that there exists a noncrossing alternating path of length $n + c\sqrt{\frac{n}{\log n}}$. We disprove a conjecture of Erdős by constructing an example without any such path of length greater than $\frac{4}{3}n + c'\sqrt{n}$.

1   Introduction

It is a basic problem in geometric graph theory to decide which graphs can be drawn on a given point set with noncrossing straight-line edges. For instance, it is known that every outerplanar graph (i.e., triangulated cycle) G of n vertices can be drawn on any set of n points in general position in the plane [GMPP91]. Moreover, if G is a rooted tree, one can find such an embedding even if the image of the root is specified [IPTT94,T96]. An unsolved problem of this kind is to find the size of the smallest "universal" set in the plane, on which one can draw every planar graph of n vertices with noncrossing straight-line edges [dFPP90,CK89].

We obtain many interesting new questions by considering *colored* point sets; see [KK04] for a survey. It is a well known mathematics contest problem to prove that between any set R of n *red* and any set B of n *blue* points in general position in the plane there is a noncrossing matching, i.e., a one-to-one correspondence between their elements so that the segments connecting the corresponding point pairs are pairwise disjoint. Moreover, if R and B are separated by a line, one can also find an *alternating Hamilton path*, i.e., a noncrossing polygonal path passing through every element of $R \cup B$ such that any two consecutive vertices have opposite colors [AGH97]. If we do not assume that R and B are separated, then the last statement is known to be false for $n \geq 8$, even if $R \cup B$ is in *convex position*, i.e., its elements form the vertex set of a convex $2n$-gon. The following problem was communicated to the second named author by Erdős around 1989.

* János Pach has been supported by NSF Grant CCR-00-98246, by PSC-CUNY Research Award 65392-0034, and by OTKA T-032452.
** Géza Tóth has been supported by OTKA-T-038397.

Problem. *Determine or estimate the largest number $\ell = \ell(n)$ such that, for every set of n red and n blue points on a circle, there exists a noncrossing alternating path consisting of ℓ vertices.*

Of course, the condition that the points are on a circle is equivalent to the assumption that they are in convex position.

Erdős and others conjectured that the asymptotically extremal configuration was the following. Suppose n is divisible by four. Cut the circle into four intervals, and place in them $\frac{n}{2}$ red, $\frac{n}{4}$ blue, $\frac{n}{2}$ red, and $\frac{3}{4}n$ blue points, respectively. It is easy to see that in this construction the number of vertices in the longest noncrossing alternating path is $\frac{3}{2}n + 2$. That is, we have $\ell(n) \leq \frac{3}{2}n + 2$. The main aim of this note is to disprove Erdős's conjecture by exhibiting a better construction in Section 2. A similar construction was found independently and at about the same time by Abellanas et al. [AGHT03].

From the other direction, it is easy to argue that $\ell(n) \geq n$. Indeed, divide the circle into two arcs, each containing n points. At least half of the points belonging to the first arc are of the same color, say, red. Then the second arc must contain the same number of blue points. Enumerate the red (resp. blue) points of the first (resp. second) arc in clockwise (resp. counterclockwise) order. Starting with the first red point on the first arc, and connecting each point with the next available element of opposite color on the other arc, we obtain a noncrossing alternating path of length $2\lceil\frac{n}{2}\rceil \geq n$. In Section 3, we improve this bound by a term that tends to infinity. Our results can be summarized as follows.

Theorem 1. *There exist constants $c, c' > 0$ such that*

$$n + c\sqrt{\frac{n}{\log n}} < \ell(n) < \frac{4}{3}n + c'\sqrt{n}.$$

It is an annoying feature of this problem that it is not clear whether the assumption that the points are in convex position plays any significant role. In particular, the above argument for finding an alternating path of length n easily generalizes to arbitrary 2-colored sets, on the other hand, our proof for the lower bound in Theorem 1 relies heavily on the fact that the points are in convex position. We conjecture that the upper bound in Theorem 1 is asymptotically tight, that is,

$$|\ell(n) - \frac{4}{3}n| = o(n).$$

See also our Conjecture at the end of the paper.

The problem of covering a set of n red and n blue points with *several* noncrossing alternating paths was discussed by Kaneko, Kano, and Suzuki [KKS04]. Alternating Hamiltonian *cycles* with at most $n - 1$ crossings were found by Kaneko, Kano, and Yoshimoto [KKY00]. Their result cannot be improved. Many other interesting questions about partitioning the plane into a given number of convex pieces, each containing roughly or exactly the same number of red and blue points, were studied in [BKS00,BM01,S02]. Analogous questions can be asked when we color by red and blue all the $\binom{n}{2}$ *segments* between n points in general position in the plane. Furthermore, instead of long alternating paths, we may be interested in finding long *monochromatic* ones [KPTV98]. Merino et al. [MSU05] studied alternating paths in k-colored point sets for $k \geq 3$.

2   Upper Bound

Consider a '2-equicolored' set C of $2n$ points in convex position in the plane. That is, let half of the elements of C be colored red and half of them blue. An *uninterrupted run* (or, in short, *run*) is a maximal set of consecutive points of C that have the same color. The *length* of a run is the number of its elements. We say that C is a *k-configuration* if it consists of k red and k blue runs.

A set of pairwise disjoint segments, each of which connects two points of different colors, is called a *matching*. The *size* of a matching is defined as the total number of points participating in it, that is, twice the number of segments. A matching is said to be *separated* if there is a straight-line that intersects the interior of each of its segments.

Lemma 2.1. *Let C be a k-configuration for some $k > 0$, which has a noncrossing alternating path of length l. Then C has a separated matching whose size is at least $l - 4k - 1$.*

Proof. Suppose without loss of generality that all elements of C lie on a circle. Consider a noncrossing alternating path p of length l. Fix a chord c of the circle that crosses the first and last segments along p, but does not pass through any point of C. Let M_1 denote the matching consisting of all odd-numbered segments of p. Clearly, the size of M_1 is at least $l - 1$. Let $M_2 \subseteq M_1$ be the set of all segments in M_1 that cross c. By definition, M_2 is a separated matching.

To establish the lemma, it is enough to show that the number of elements of M_1 that do not cross c is at most $2k$. Let us call these segments *outer* segments. For each pair of consecutive points of C that have different colors, pick a point between them on the circle. Any two consecutive runs of C are separated by at least one such point, so the number of points we selected is precisely $2k$. Every outer segment s divides the circle into two (closed) arcs. One of them, $I(s)$, contains both endpoints of c; let the other one be denoted by $J(s)$. Since s connects two points of different colors, $J(s)$ must contain at least one of the selected points. On the other hand, both endpoints of the alternating path p belong to $I(s)$, so $J(s)$ cannot contain the endpoints of any outer segment other than the endpoints of s. Thus, for any two outer segments, s and s', $J(s)$ and $J(s')$ are disjoint, so the selected points lying in the corresponding arcs $J(s)$ and $J(s')$, resp., are different. Hence, the number of outer segments cannot exceed the total number of selected points, which is $2k$. See Fig. 1.   □

Represent any k-configuration with runs $S_1, S_2, \ldots, S_{2k}$ by the sequence $(|S_1|, |S_2|, \ldots, |S_{2k}|)$. We assume that the odd-numbered runs are red and the even-numbered are blue.

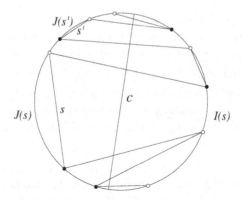

Fig. 1. $J(s)$ and $J(s')$ are disjoint.

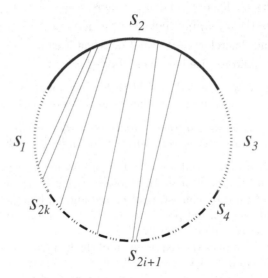

Fig. 2. The k-configuration $(k\frac{n}{3k-2}, (2k-1)\frac{n}{3k-2}, k\frac{n}{3k-2}, \frac{n}{3k-2}, \ldots, \frac{n}{3k-2})$.

Lemma 2.2. *Let $k \geq 2$ and assume that n is divisible by $3k-2$. Then the size of any separated matching in the k-configuration $(k\frac{n}{3k-2}, (2k-1)\frac{n}{3k-2}, k\frac{n}{3k-2}, \frac{n}{3k-2},$
$\ldots, \frac{n}{3k-2})$ is at most $2n\frac{2k-1}{3k-2}$.*

Proof. Let $S_1, S_2, \ldots, S_{2k}$ denote the consecutive runs of the k-configuration $(k\frac{n}{3k-2}, (2k-1)\frac{n}{3k-2}, k\frac{n}{3k-2}, \frac{n}{3k-2}, \ldots, \frac{n}{3k-2})$ (see Fig. 2). Let M be a separated matching. We distinguish five cases, according to the set of runs that are connected to S_2 by at least one edge in M.

Case 1: No edge of M has an endpoint in S_2.
Then M uses at most $n - n\frac{2k-1}{3k-2}$ blue points, so its size is at most $2n - 2n\frac{2k-1}{3k-2} = 2n\frac{k-1}{3k-2} < 2n\frac{2k-1}{3k-2}$.

Case 2: No edge of M runs between S_2 and $S_1 \cup S_3$.
Now the points in S_2 can be connected only to the elements of $S_5, S_7, \ldots, S_{2k-1}$. Hence, at most $n\frac{k-2}{3k-2}$ points of S_2 are matched and at least $n\frac{k+1}{3k-2}$ are missed by M. The size of M is at most $2n - 2n\frac{k+1}{3k-2} < 2n\frac{2k-2}{3k-2}$.

Case 3: S_2 is connected by an edge of M to both S_1 and S_3.
Since M is a separated matching, the blue points in $S_4, S_6, \ldots, S_{2k}$ are not matched, so M uses at most $n\frac{2k-1}{3k-2}$ blue points. Thus, the size of M is at most $2n\frac{2k-1}{3k-2}$.

Case 4: S_2 is connected by an edge of M to S_1, but not to S_3.
Suppose that the size of M exceeds $2n\frac{2k-1}{3k-2}$. Then M matches more than $n\frac{k}{3k-2}$ points of S_2, so at least one edge of M must connect S_2 to a red run different from S_1. Let $i > 1$ denote the smallest integer such that there is an edge of M between S_2 and S_{2i+1}. Then M matches at most $n\frac{k+(k-i)}{3k-2} = n\frac{2k-i}{3k-2}$ blue points from S_2 and misses the $n\frac{k-i}{3k-2}$ blue points in $S_{2i+2}, S_{2i+4}, \ldots, S_{2k}$, because M is a separated matching. Therefore, it does not match at least $n\frac{(2k-1-(2k-i))+(k-i)}{3k-2} = n\frac{k-1}{3k-2}$ blue points, and its size is at most $2n\frac{2k-1}{3k-2}$.

Case 5: S_2 is connected by an edge of M to S_3, but not to S_1.
By symmetry, the same argument applies as in the previous case.   □

Lemma 2.3. *For any positive integers k and $n \geq k$, there exists a k-configuration of $2n$ points with no alternating path longer than $2n\frac{2k-1}{3k-2} + 16k$.*

Proof. The statement is trivial for $k = 1$, and also for $n \leq 8k$. Suppose that $k \geq 2$, and $n > 8k$. Let $n_0 \leq n$ be the largest integer divisible by $3k - 2$. Let C_0 denote a k-configuration consisting of n_0 red and n_0 blue points, considered in Lemma 2.2. Add $n - n_0$ red points to S_1 and $n - n_0$ blue points to S_2, and denote the resulting k-configuration by C.

We claim that C satisfies the requirement of the lemma. Let p be an alternating path of length $l(p)$ in C. By Lemma 2.1, there is a separated matching M_1 in C, whose size is $l(M_1) \geq l(p) - 4k - 1$. Remove from M_1 the $2n - 2n_0$ points that were added later and all edges in M_1 incident to them. We obtain a separated matching M_0 of C_0 of size $l(M_0) \geq l(M_1) - 4(n - n_0) \geq l(M_1) - 4(3k - 2) = l(M_1) - 12k + 8 \geq l(p) - 16k$. By Lemma 2.2, we have $l(M_0) \leq 2n_0\frac{2k-1}{3k-2}$, so that $l(p) \leq 2n\frac{2k-1}{3k-2} + 16k$.   □

The upper bound in Theorem 1 immediately follows from Lemma 2.3. For any n, set $k = \lfloor\sqrt{n}\rfloor$. Applying Lemma 2.3, we obtain a configuration of n red and n blue points in which the length of any noncrossing alternating path is at most $\frac{2k-1}{3k-2} \cdot 2n + 16k < \frac{4n}{3} + 20\sqrt{n}$, as required.

3   Lower Bound

As before, let C be the vertex set of a convex $2n$-gon, with n red and n blue elements. Suppose without loss of generality that the elements of C lie on a circle. A set of consecutive vertices of C (of not necessarily the same color) is said to be an *interval*. The *length* of an interval is its cardinality.

Assume that we can find a separated matching M of size $2l$, all of whose segments are crossed by a chord c. Then we can easily construct a noncrossing alternating path of length $2l$. To see this, enumerate the segments $s_1, s_2, \ldots, s_l$ of M according to the order of their intersection points with c. Let r_i and b_i be the red and blue endpoints of s_i, respectively. Then $p = (r_1 b_1, b_1 r_2, r_2 b_2, \ldots, b_{l-1} r_l, r_l b_l)$ is a noncrossing alternating path of length $2l$.

Therefore, it is sufficient to establish a lower bound on the size of a separated matching in a k-configuration of $2n$ points. We divide the proof into two steps. Lemmas 3.1 and 3.2 provide reasonably good bounds when k is small and when k is large, respectively. Their combination implies the general lower bound in Theorem 1.

Lemma 3.1. *Let k, m, n be positive integers such that $k = 2^m$ divides n. Then every k-configuration C of $2n$ points contains a separated matching of size at least $n \left(1 + \frac{1}{k(m+1)} \right)$.*

Proof. Let S be the run of length at least $\frac{n}{k}$. Let I_0 denote a monochromatic interval in S, whose length is precisely $\frac{n}{k}$. For $1 \leq i \leq m+1$, let I_i be an interval of length $2^{i-1} \frac{n}{k}$ such that $I_0, I_1, \ldots, I_{m+1}$ are consecutive in the clockwise direction (see Fig. 3). These intervals form a partition of the underlying set C consisting of all $2n$ vertices. Assume without loss of generality that all elements of I_0 are blue. □

Suppose for contradiction that there is no separated matching whose size is at least $n \left(1 + \frac{1}{k(m+1)} \right)$.

Claim. *For every $0 \leq i \leq m+1$, the interval $J_i = I_0 \cup I_1 \cup \ldots \cup I_i$ has at least $\left(2^{i-1} \frac{1}{k} + \frac{1}{2k} - \frac{i}{2k(m+1)} \right) n$ blue points. Moreover, strict inequality holds if $i > 0$.*

Note that Lemma 3.1 immediately follows from the Claim. Indeed, for $i = m+1$, we obtain that there are more than $\left(2^m \frac{1}{k} + \frac{1}{2k} - \frac{m+1}{2k(m+1)} \right) n = n$ blue points on the circle, which is a contradiction.

Proof of Claim. We proceed by induction on i. For $i = 0$, the statement obviously holds. Assume that for some $i \in \{0, 1, 2, \ldots, m\}$, there are at least $\left(2^{i-1} \frac{1}{k} + \frac{1}{2k} - \frac{i}{2k(m+1)} \right) n$ blue points in the interval J_i. We show that there are more than $\left(2^{i-1} \frac{1}{k} - \frac{1}{2k(m+1)} \right) n$ blue points in I_{i+1}.

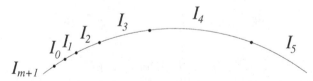

Fig. 3. $I_0, I_1, \ldots, I_{m+1}$ form a partition of the vertices.

Suppose this is not the case. Then there are at least $l = \left(2^{i-1}\frac{1}{k} + \frac{1}{2k(m+1)}\right)n$ red points in I_{i+1}. Since we have $2^{i-1}\frac{1}{k} + \frac{1}{2k} - \frac{i}{2k(m+1)} \geq 2^{i-1}\frac{1}{k} + \frac{1}{2k(m+1)}$, the number of blue points in J_i is at least l. Thus, there is a separated matching of size $2l$ between the blue points of J_i and the red points of I_{i+1}. Let M be the most 'economical' such matching of size $2l$. That is, if $b_1, b_2, \ldots, b_s$ $(s \geq l)$ denote the blue points of J_i listed in counterclockwise order and $r_1, r_2, \ldots, r_t$ $(t \geq l)$ denote the red points of I_{i+1} listed in clockwise order, then let M consist of the segments $b_1 r_1, b_2 r_2, \ldots, b_l r_l$.

Let K denote the interval between b_l and r_l, oriented clockwise. All blue points from K not matched by M lie in I_{i+1}, so their number is at most $u_b = \left(2^{i-1}\frac{1}{k} - \frac{1}{2k(m+1)}\right)n$. All red points of K not matched by M lie in J_i, so their number is at most

$$u_r = \left(2^i\frac{1}{k} - \left(2^{i-1}\frac{1}{k} + \frac{1}{2k} - \frac{i}{2k(m+1)}\right)\right)n$$

$$= \left(2^{i-1}\frac{1}{k} - \frac{1}{2k} + \frac{i}{2k(m+1)}\right)n.$$

Using the fact that $u_b - u_r = \left(\frac{1}{2k} - \frac{i+1}{2k(m+1)}\right)n \geq 0$, we obtain that $u_b \geq u_r$.

Let L denote the complement of K in the set C of all points. Clearly, L has at least $n_0 = n - l - u_b = \left(1 - 2^i\frac{1}{k}\right)n$ points of each color. Divide L into two intervals L_1, L_2, each of length at least n_0. Obviously, at least one of the following two conditions is satisfied:

(a) There are at least $\frac{n_0}{2}$ blue points in L_1 and at least $\frac{n_0}{2}$ red points in L_2.
(b) There are at least $\frac{n_0}{2}$ red points in L_1 and at least $\frac{n_0}{2}$ blue points in L_2.

We can assume without loss of generality that (a) is true. Then there exists a separated matching M' of size at least n_0 between intervals L_1 and L_2. The union of M and M' is also a separated matching. (One endpoint of the corresponding chord lies between the intervals J_i and I_{i+1}, and the other between L_1 and L_2.) The size of $M \cup M'$ is at least

$$2l + n_0 = \left(2^{i-1}\frac{1}{k} + \frac{1}{2k(m+1)}\right)2n + \left(1 - 2^i\frac{1}{k}\right)n = \left(1 + \frac{1}{k(m+1)}\right)n,$$

which is a contradiction. Hence, our assumption was wrong: there are more than $\left(2^{i-1}\frac{1}{k} - \frac{1}{2k(m+1)}\right)n$ blue points in I_{i+1}.

Consequently, the number of blue points in the interval $J_{i+1} = I_0 \cup I_1 \cup \ldots \cup I_{i+1}$ is larger than

$$\left(2^{i-1}\frac{1}{k} + \frac{1}{2k} - \frac{i}{2k(m+1)}\right)n + \left(2^{i-1}\frac{1}{k} - \frac{1}{2k(m+1)}\right)n$$

$$= \left(2^i\frac{1}{k} + \frac{1}{2k} - \frac{i+1}{2k(m+1)}\right)n,$$

completing the induction step, the proof of the Claim, and hence the proof of Lemma 3.1.

Lemma 3.2. *For $n \geq k \geq 1$, every k-configuration of $2n$ points admits an alternating path whose length is at least $n + k - 1$.*

Proof. Let $v_1, v_2, \ldots, v_{2n}$ be the vertices of a k-configuration, in clockwise direction. Assume that v_1 is red. For any $1 \leq i < 2n$, if v_i is red and v_{i+1} is blue, then v_i and v_{i+1} are called special vertices. There are $2k$ special and $2n - 2k$ non-special vertices. Moreover, exactly half of the special and exactly half of the non-special vertices are red, and blue. Let m be the smallest number such that there are $n - k$ non-special vertices in the set $\{v_i \mid 1 \leq i \leq m\}$. Assume that $t \geq \lceil \frac{n-k}{2} \rceil$ of them are red. (The other case can be settled analogously.) Denote those red points by $u_1, u_2, \ldots, u_t$, in clockwise direction. Then the set $\{v_i \mid m+1 \leq i \leq 2n\}$ also contains $n - k$ non-special vertices, and t of them are blue. Denote those blue points by $w_1, w_2, \ldots, w_t$, in counterclockwise direction.

The vertices $u_1, u_2, \ldots, u_t$ divide the set $\{v_i \mid 1 \leq i \leq m\}$ into $t+1$ intervals of consecutive vertices, denote them by $I_0, I_1, \ldots, I_t$, in clockwise direction. For $0 \leq 1 \leq t$, if I_i contains some special vertices, denote them by $u_{i,1}, u_{i,2}, \ldots u_{i,\alpha_i}$, in clockwise direction. Since $u_1, u_2, \ldots, u_t$ are non-special, $u_{i,j}$ is red if j is odd, and blue if j is even.

Similarly, the vertices $w_1, w_2, \ldots, w_t$ divide the set $\{v_i \mid m + 1 \leq i \leq 2n\}$ into $t + 1$ intervals of consecutive vertices, denote them by $J_0, J_1, \ldots, J_t$, in counterclockwise direction. For $0 \leq 1 \leq t$, if J_i contains some special vertices, denote them by $w_{i,1}, w_{i,2}, \ldots w_{i,\beta_i}$, also in counterclockwise direction. Now $w_{i,j}$ is blue if j is odd, and red if j is even.

Finally, consider the following path:

$$u_{0,1}, u_{0,2}, \ldots u_{0,\alpha_0}, u_1, \ w_{0,1}, w_{0,2}, \ldots w_{0,\beta_0}, w_1, \ u_{1,1}, u_{1,2}, \ldots u_{1,\alpha_1}, u_2,$$

$$w_{1,1}, w_{1,2}, \ldots w_{1,\beta_1}, w_2, \ldots, u_{t,1}, u_{t,2}, \ldots u_{t,(\alpha_t - 1)}, \ w_{t,1}, w_{t,2}, \ldots w_{t,\beta_t}.$$

It is a noncrossing, alternating path of length $2t + 2k - 1 \geq n - k + 2k - 1 = n + k - 1$. This concludes the proof of the lemma. $\qquad\square$

Now we are ready to prove the lower bound in Theorem 1. Suppose that we have a k-configuration of $2n$ points. We distinguish two cases. If $k \geq \frac{1}{10}\sqrt{\frac{n}{\log n}}$, then, by Lemma 3.2, there exists an alternating path of length at least $n + \frac{1}{10}\sqrt{\frac{n}{\log n}} - 1$.

We are left with the case $k < \frac{1}{10}\sqrt{\frac{n}{\log n}}$. Let m be the least positive integer such that $k \leq 2^m$. Then $m < 1 + \log k$. Let $n' = 2^m \lfloor \frac{n}{2^m} \rfloor \geq n - 2^m$ and choose any subconfiguration C' of n' red and n' blue points from C. C' is a k_0-configuration for some $k_0 \leq k$. So it has a run of length at least $\frac{n'}{k_0} \geq \frac{n'}{k} \geq \frac{n'}{2^m}$. Now, according to Lemma 3.1, C has a separated matching (and also an alternating path) whose size is at least

$$n' + \frac{n'}{2^m(m + 1)} \geq n - 2k + \frac{n}{2^{m+1}(m + 1)} \geq$$

$$n - 2k + \frac{n}{4k(\log k + 2)} \geq n - 2k + \frac{n}{2k \log n} \geq$$

$$n - \frac{1}{5}\sqrt{\frac{n}{\log n}} + \frac{5n}{\log n}\sqrt{\frac{\log n}{n}} \geq n + 4\sqrt{\frac{n}{\log n}}.$$

This completes the proof of the lower bound in Theorem 1.

Conjecture. *For any fixed k and large n, every k-configuration of $2n$ points admits a separated matching of size at least $2n\frac{2k-1}{3k-2} + o(n)$.*

Acknowledgement

We are very grateful to the anonymous referee for his useful comments.

References

[AGH97] M. Abellanas, J. Garcia, G. Hernández, M. Noy, and P. Ramos: Bipartite embeddings of trees in the plane, in: *Graph Drawing* (S. North, ed.), *Lecture Notes in Computer Science* **1190**, Springer-Verlag, Berlin, 1997, 1–10. Also in: *Discrete Appl. Math.* **93** (1999), 141–148.

[AGHT03] M. Abellanas, A. Garcia, F. Hurtado, and J. Tejel: Caminos alternantes, in: *X Encuentros de Geometría Computacional* (in Spanish), Sevilla, 2003, 7–12.

[BM01] I. Bárány and J. Matoušek: Simultaneous partitions of measures by k-fans, *Discrete Comput. Geom.* **25** (2001), 317–334.

[BKS00] S. Bespamyatnikh, D. Kirkpatrick, and J. Snoeyink: Generalizing ham sandwich cuts to equitable subdivisions, *Discrete Comput. Geom.* **24** (2000), 605–622.

[dFPP90] H. de Fraysseix, J. Pach, and R. Pollack: How to draw a planar graph on a grid, *Combinatorica* **10** (1990), 41–51.

[CK89] M. Chrobak and H. Karloff: A lower bound on the size of universal sets for planar graphs, *SIGACT News* **20**/4 (1989), 63–86.

[GMPP91] P. Gritzmann, B. Mohar, J. Pach, and R. Pollack: Embedding a planar triangulation with vertices at specified points (solution to Problem E3341), *Amer. Math. Monthly* **98** (1991), 165–166.

[IPTT94] Y. Ikeba, M. Perles, A. Tamura, and S. Tokunaga: The rooted tree embedding problem into points on the plane, *Discrete Comput. Geom.* **11** (1994), 51–63.

[KK04] A. Kaneko and M. Kano: Discrete geometry on red and blue poins in the plane – a survey, in: *Discrete and Computational Geometry* (B. Aronov et al., eds.), Springer-Verlag, Berlin, 2004, 551–570.

[KKS04] A. Kaneko, M. Kano, and K. Suzuki: Path coverings of two sets of points in the plane, in: *Towards a Theory of Geometric Graphs* (J. Pach, ed.), *Contemporary Mathematics* **342** (2004), 99–111.

[KKY00] A. Kaneko, M. Kano, and K. Yoshimoto: Alternating Hamiltonian cycles with minimum number of crossings in the plane, *Internat. J. Comput. Geom. Appl.* **10** (2000), 73–78.

[KPTV98] Gy. Károlyi, J. Pach, G. Tóth, and P. Valtr: Ramsey-type results for geometric graphs II, *Discrete Comput. Geom.* **20** (1998), 375–388.

[MSU05] C. Merino, G. Salazar, and J. Urrutia: On the length of longest alternating paths for multicolored point sets in convex position, manuscript.

[S02] T. Sakai: Balanced convex partitions of measures in R^2, *Graphs and Combinatorics* **18** (2002), 169–192.

[T96] S. Tokunaga: On a straight-line embedding problem of graphs, *Discrete Math.* **150** (1996), 371–378.

Intersection Reverse Sequences
and Geometric Applications

Adam Marcus[1,*] and Gábor Tardos[2,**]

[1] Department of Mathematics, Georgia Institute of Technology,
Atlanta, GA 30332-0160
adam@math.gatech.edu
[2] Alfréd Rényi Institute of Mathematics,
Pf.127, H-1364 Budapest, Hungary
tardos@renyi.hu

Abstract. Pinchasi and Radoičić [1] used the following observation to bound the number of edges in a topological graph with no self-intersecting cycles of length 4: if we make a list of the neighbors for every vertex in such a graph and order these lists cyclically according to the connecting edge, then the common elements in any two lists have reversed cyclic order. Building on their work we give an estimate on the size of the lists having this property. As a consequence we get that a topological graph on n vertices not containing a self-intersecting C_4 has $O(n^{3/2} \log n)$ edges. Our result also implies that n pseudo-circles in the plane can be cut into $O(n^{3/2} \log n)$ pseudo-segments, which in turn implies bounds on point-curve incidences and on the complexity of a level of an arrangement of curves.

1   Introduction

In this paper we consider *cyclicly ordered sequences* of distinct symbols from a finite alphabet. We say that two such sequences are *intersection reverse* if the common elements appear in reversed cyclic order in the two sequences.

A *topological graph* is a graph without loops or multiple edges drawn in the plane (vertices correspond to distinct points, edges correspond to Jordan curves connecting the corresponding vertices). We assume no edge passes through a vertex other than its endpoints and every two edges have a finite number of common interior points and they properly cross at each of these points. For a vertex v of a topological graph G let $L_G(v)$ be the list of its neighbors ordered cyclically counterclockwise according to the initial segment of the connecting edge.

Pinchasi and Radoičić [1] noticed the following simple fact:

Fact 1. *If the lists $L_G(u)$ and $L_G(v)$ are not intersection reverse for all distinct vertices u and v of the topological graph G, then G contains a self-crossing cycle*

* Research was done while a Visiting Researcher at The Alfréd Rényi Institute, Budapest and was supported by the Fulbright Program in Hungary.
** Partially supported by the Hungarian National Research Fund OTKA T029255.

of length 4. Moreover, u and v are opposite vertices of a cycle of length 4 in G that has two edges that cross an odd number of times.

For the proof one only has to consider drawings of the complete bipartite graph $K_{2,3}$ (see details in [1]). Pinchasi and Radoičić used Fact 1 to bound the number of edges of a topological graph not containing a self-crossing C_4. They showed that such a graph on n vertices has $O(n^{8/5})$ edges. Following in their footsteps, we use the same property to improve their bound to $O(n^{3/2} \log n)$. This bound is tight apart from the logarithmic factor as one has (abstract) simple graphs on n vertices with $\Omega(n^{3/2})$ edges containing no C_4-subgraph. Our main technical result is the following:

Theorem 1. *Let us be given m cyclicly ordered lists of d element subsets of a set of n symbols. If the lists are pairwise intersection reverse, then*

$$d = O\left(\sqrt{n}\log n + \frac{n}{\sqrt{m}}\right) .$$

We give the proof of this theorem in Section 2. In Section 3 we prove several consequences, among them the following bound:

Theorem 2. *If an n-vertex topological graph does not contain a self-crossing C_4 it has $O(n^{3/2} \log n)$ edges. The same holds if every pair of edges in every C_4 subgraph crosses an even number of times.*

The most important consequence of Theorem 1 deals with collections of *pseudo-circles*: simple closed Jordan curves, any two of which intersect at most twice, with proper crossings at each intersection. The result readily generalizes to unbounded open curves such as *pseudo-parabolas*, the graphs of continuous real functions defined on the entire real line such that any two intersect at most twice and they properly cross at these intersections.

Tamaki and Tokuyama [2] were the first to consider the problem of *cutting* pseudo-parabolas into *pseudo-segments*, i.e., subdividing the original curves into segments such that any two segments intersect at most once. Such a separation turns out to be quite useful since pseudo-segments are much easier to work with than pseudo-parabolas and pseudo-circles, as will be seen in Section 3.

Tamaki and Tokuyama [2] proved that n pseudo-parabolas can be cut into $O(n^{5/3})$ pseudo-segments. This was extended to x-monotone pseudo-circles by Aronov and Sharir [3] and by Agarwal, et al. [4]. It was also improved for certain collections of curves that admit a three parameter algebraic parameterization to $n^{3/2} \log^{\alpha^{O(1)}(n)}(n)$, where α is the inverse Ackermann function.

Previously, the best bound for arbitrary pseudo-parabolas and x-monotone pseudo-circles was $O(n^{8/5})$ [4], which uses the result of Pinchasi and Radoičić on topological graphs without self-crossing C_4. With our improvement of the latter result, we can prove that n pseudo-parabolas can be cut into $O(n^{3/2} \log n)$ pseudo-segments. This substantially improves the previous bounds for arbitrary collections and is still slightly better than results on families with algebraic parameterization; we replace a little-over-polylog factor with a single log factor. In doing so, we are able to simplify the results in [4, 1, 2], as well as generalize them to the cases when the pseudo-parabolas and pseudo-circles are not necessarily x-monotone.

In Section 3 we show the above result, as well as its applications to point-curve incidences and the level complexities of curve arrangements. See [4, 3, 5, 6, 2] for more details and applications.

Finally in Section 4 we discuss a few related problems that are still open. All logarithms in this paper are binary.

2   Intersection Reverse Sequences

In this section we prove our main technical result, Theorem 1. Much of the proof follows the argument of Pinchasi and Radoičić [1]. We start with an overview of their techniques.

Pinchasi and Radoičić break the cyclicly ordered lists into linearly ordered blocks. They consider pairs of blocks from separate lists and pairs of symbols contained in both blocks. They distinguish between *same pairs* and *different pairs* according to whether the two symbols appear in the same or in different order. They observe that any pair of symbols that appears in many blocks must produce almost as many same pairs as different pairs. On the other hand the intersection reverse property forces two cyclicly ordered lists – unless most of their intersection is concentrated into a single pair of blocks – to contribute many more different than same pairs. Exceptional pairs of cyclicly ordered lists are treated separately with techniques from extremal graph theory. They optimize in their choice for the length of the blocks.

We follow almost the same path, but instead of optimizing for block length we consider many block lengths (an exponential sequence) simultaneously. For two intersection reverse lists no block length yields significantly more same pairs than different pairs, but we will show that some block length actually gives many more different pairs than same pairs. As a consequence we do not have to bound "exceptional pairs" of lists separately.

Definition. We will use the term *sequence* to denote a linearly ordered list of distinct symbols and the term *cyclic sequence* to denote a cyclicly ordered list of distinct symbols. Clearly, if one breaks up a cyclic sequence into blocks, then the blocks are (linearly ordered) sequences. For a sequence or cyclic sequence A we write $\overline{A}$ for the set of symbols in A. We define intersection reverse for sequences just as for cyclic sequences: we say that the sequences A and B are *intersection reverse* if they induce inverse linear orders on $\overline{A} \cap \overline{B}$. If two sequences are not intersection reverse, we call them *singular*. Note that if two sequences A and B have $|\overline{A} \cap \overline{B}| \leq 1$, then the sequences are trivially intersection reverse. The same holds for cyclic sequences A and B if $|\overline{A} \cap \overline{B}| \leq 2$.

For a sequence B and symbols $a \neq b$ we define

$$f(B, a, b) = \begin{cases} 0 & \text{if } a \notin \overline{B} \text{ or } b \notin \overline{B}, \\ 1 & \text{if } a \text{ precedes } b \text{ in } B, \\ -1 & \text{if } b \text{ precedes } a \text{ in } B. \end{cases}$$

For two sequences B and B' we let $f(B, B', a, b) = f(B, a, b)f(B', a, b)$. Notice that $f(B, B', a, b) = 1$ for same pairs and $f(B, B', a, b) = -1$ for different pairs, and that $\sum f(B, B', a, b)$ corresponds to the difference between the number of same pairs and different pairs.

The next lemma is taken from [1]. The summation $\sum_{a \neq b}$ here and later in this section is taken for all ordered pair of distinct symbols a and b.

Lemma 3. *Let the cyclic sequences A and A' consist of the (linearly ordered) blocks $B_1, \ldots, B_k$ and $B'_1, \ldots, B'_{k'}$, respectively. If A and A' are intersection reverse, then at most one of the pairs B_i, B'_j is singular. For this singular pair we have*

$$\sum_{a \neq b} f(B_i, B'_j, a, b) \leq |\overline{B_i} \cap \overline{B'_j}| .$$

For the rest of this section, let $A^1, \ldots, A^m$ be the m cyclic sequences in the theorem, $p^{ij} = |\overline{A^i} \cap \overline{A^j}|$, and $p = \sum_{i \neq j} p^{ij}$. The following bounds follow directly from the inequality between the arithmetic and quadratic means assuming $dm > 2n$ (otherwise the statement of Theorem 1 is immediate).

Lemma 4. *We have $p \geq \frac{d^2 m^2}{2n}$ and $\sum_{i \neq j}(p^{ij})^2 \geq \frac{p^2}{m^2}$.*

We split each A^i into two almost equal size consecutive blocks A_0^i and A_1^i. In general, for a 0–1 sequence s we split the block A_s^i of A^i into two almost equal halves (differing in size by at most 1): A_{s0}^i and A_{s1}^i. The cyclic order of A^i linearly orders all of these blocks. Let $k = \lceil \log d \rceil < \log n + 1$. Clearly, any 0–1 sequence s of length k satisfies $|\overline{A_s^i}| \leq 1$.

For $1 \leq i \leq m$ and $1 \leq j \leq m$ we let

$$S^{ij} = \sum_{a \neq b} \sum_{l=1}^{k} \sum_{|s|=|t|=l} w_l f(A_s^i, A_t^j, a, b) ,$$

where the summation is taken over all pairs of symbols $a \neq b$, and 0–1 sequences s and t of size l. The weights w_l in the formula are positive and we set them later. Our goal is to contrast a lower bound on $\sum_{i \neq j} S^{ij}$ (or rather on the partial sum for fixed symbols $a \neq b$) with upper bounds on the individual S^{ij}. The pair (i, j) (just as the pair (a, b) above) is considered ordered resulting in double (quadruple, really) counting.

The lower bound is straightforward:

Lemma 5. $\sum_{i \neq j} S^{ij} \geq -md^2 \sum_{l=1}^{k} \frac{w_l}{2^l}$

Proof. Notice that for fixed a, b, and length $|s| = |t|$ we get a perfect square when summing over all i and j. In particular,

$$\sum_{i=1}^{m} \sum_{j=1}^{m} S^{ij} = \sum_{l=1}^{k} w_l \sum_{a \neq b} \left(\sum_{i=1}^{m} \sum_{|s|=l} f(A_s^i, a, b) \right)^2 \geq 0 .$$

We can bound the S^{ii} terms separately as they are merely the weighted sum of the number of pairs in the blocks:

$$\sum_{i \neq j} S^{ij} = \sum_{i=1}^{m} \sum_{j=1}^{m} S^{ij} - \sum_{i=1}^{m} S^{ii} \geq 0 - md^2 \sum_{l=1}^{k} \frac{w_l}{2^l} \qquad \square$$

The upper bound, however, requires more effort.

Lemma 6. *For* $i \neq j$ *we have*

$$S^{ij} \leq p^{ij} \sum_{l=1}^{k} w_l - \frac{(p^{ij})^2}{4 \sum_{l=1}^{k} \frac{1}{w_l}}.$$

Proof. We fix the indices $i \neq j$ and consider the following quantities:

$$r_{st} = |\overline{A_s^i} \cap \overline{A_t^j}| \text{ and}$$
$$Q_{st} = \sum_{a \neq b} f(A_s^i, A_t^j, a, b)$$

where s and t are 0–1 sequences of equal length.

For a fixed $1 \leq l \leq k$ the blocks A_s^i with $|s| = l$ form a subdivision of A^i, while the blocks A_t^j with $|t| = l$ form a subdivision of A^j. We use Lemma 3: there is at most one singular pair (A_s^i, A_t^j) for every length $|s| = |t| = l$. For these singular pairs we have $Q_{st} \leq r_{st}$, while for the intersection reverse ones straightforward calculation gives $Q_{st} = r_{st} - r_{st}^2$. Note that for $|s| = |t| = k$ the corresponding pairs are trivially intersection reverse.

For a 0–1 sequence s of length $|s| > 1$ let s' denote the sequence obtained from s by deleting its last digit; hence $A_{s'}^i$ contains A_s^i. We call a pair (s, t) of equal length 0–1 sequences a *leader* if (A_s^i, A_t^j) is intersection reverse and either $|s| = |t| = 1$ or the pair $(A_{s'}^i, A_{t'}^j)$ is singular. Clearly, there are at most 4 leaders in any given length. Furthermore, any symbol $a \in \overline{A^i} \cap \overline{A^j}$ is contained in $\overline{A_s^i} \cap \overline{A_t^j}$ for exactly one leader pair (s, t): the longest intersection reverse pair of blocks containing them (recall that we only consider pairs of blocks with equal length subscripts). Thus we have $\sum_{(s,t) \in L} r_{st} = p^{ij}$ for the set L of leader pairs.

We use $Q_{st} = r_{st} - r_{st}^2$ for leader pairs (s, t) only. For all other pairs, intersection reverse or singular, we use $Q_{st} \leq r_{st}$:

$$S^{ij} \leq \sum_{l=1}^{k} w_l \sum_{|s|=|t|=l} r_{st} - \sum_{(s,t) \in L} w_{|s|} r_{st}^2 = p^{ij} \sum_{l=1}^{k} w_l - \sum_{(s,t) \in L} w_{|s|} r_{st}^2$$

since $\sum_{|s|=|t|=l} r_{st} = p^{ij}$ for any fixed l. The Cauchy-Schwarz inequality gives

$$\left(\sum_{(s,t) \in L} w_{|s|} r_{st}^2 \right) \left(\sum_{(s,t) \in L} \frac{1}{w_{|s|}} \right) \geq \left(\sum_{(s,t) \in L} r_{st} \right)^2 = (p^{ij})^2 .$$

Here $\sum_{(s,t)\in L}(1/w_{|s|}) \le 4\sum_{l=1}^{k}(1/w_l)$, so we conclude that

$$S^{ij} \le p^{ij} \sum_{l=1}^{k} w_l - \frac{(p^{ij})^2}{4\sum_{l=1}^{k} \frac{1}{w_l}}$$

as claimed.   □

Comparing the two estimates in the previous lemmas gives the theorem.

Proof (of Theorem 1). Using Lemmas 5, 6, and 4 (respectively) we obtain

$$-md^2 \sum_{l=1}^{k} \frac{w_l}{2^l} \le \sum_{i\ne j} S^{ij} \le \sum_{i\ne j} p^{ij} \sum_{l=1}^{k} w_l - \frac{\sum_{i\ne j}(p^{ij})^2}{4\sum_{l=1}^{k} \frac{1}{w_l}}$$

$$\le p \sum_{l=1}^{k} w_l - \frac{p^2}{4m^2 \sum_{l=1}^{k} \frac{1}{w_l}}.$$

Using the fact that $p \ge d^2 m^2/(2n)$ from Lemma 4 we get

$$d \le 4\sqrt{n} \sqrt{\sum_{l=1}^{k} w_l} \sqrt{\sum_{l=1}^{k} \frac{1}{w_l}}$$

or

$$d \le \frac{6n}{\sqrt{m}} \sqrt{\sum_{l=1}^{k} \frac{w_l}{2^l}} \sqrt{\sum_{l=1}^{k} \frac{1}{w_l}}.$$

We choose the weights w_l now. Equal weights ($w_l = 1$) yield $d = O(\sqrt{n}\log n + n\sqrt{\log n}/\sqrt{m})$, but we can improve on this bound by choosing

$$w_l = \frac{1}{1 + \frac{k}{2^{l/2}}}.$$

In this case $\sum_{l=1}^{k} w_l \le k$, $\sum_{l=1}^{k}(1/w_l) \le 4k$, and $\sum_{l=1}^{k}(w_l/2^l) \le 3/k$. Thus we either have $d \le 8k\sqrt{n}$ or $d \le 21n/\sqrt{m}$ and the theorem follows.   □

3   Consequences

In this section we present several geometric applications of Theorem 1.

3.1   Self-intersecting Cycles of Length 4

Any bound for the $n = m$ case of Theorem 1 carries over to the number of edges of a topological graph not containing self-intersecting C_4 by [1]. Using the following corollary, however, the proof is even simpler:

Corollary 7. *Let us be given m cyclic sequences over an n-element set of symbols. If the cyclic sequences are pairwise intersection reverse, then the sum of their sizes is $O\left(m\sqrt{n}\log n + n\sqrt{m}\right)$.*

The simple deduction from Theorem 1 is omitted.

The proof of Theorem 2 is now straightforward. The statements are direct consequences of Corollary 7 using Fact 1. Notice that the sum of the sizes of the lists of neighbors (the sum of the degrees) is twice the number of edges.

3.2   Cutting Number

Tamaki and Tokuyama [2] considered the *cutting number* of a collection of curves. This is the number of cuts needed to obtain a collection of shorter curves, each pair of which intersects at most once.

The restriction of the next corollary to so called *x-monotone* pseudo-circles can be derived from Theorem 2 using a combination of techniques in the papers [4, 2]. A simple and direct deduction from Corollary 7 that does not require any additional assumption on the pseudo-circles will appear in the final version of this paper. Recall that this result slightly improves the best previous bound for (*x*-monotone) pseudo-circles with a three parameter algebraic representation as defined in [4] (such as circles or axis-aligned parabolas) and substantially improves the previous bounds for pseudo-circles lacking such representation. For the definition of pseudo-circles see Section 1.

Corollary 8. *An arrangement of n pseudo-circles can be cut at $O(n^{3/2}\log n)$ points such that the resulting curves form a system of pseudo-segments.*

Corollary 8 naturally generalizes to collections of open Jordan curves including, for example, pseudo-parabolas. We omit the straightforward deduction. We call a collection of simple closed and open Jordan curves a *generalized pseudo-circle collection* if both ends of every open curve are at infinity, any two curves have at most two points of intersection, and the curves cross properly at each intersection.

Corollary 9. *A generalized pseudo-circle collection C of n curves can be cut at $O(n^{3/2}\log n)$ points such that the resulting curve segments form a system of pseudo-segments.*

3.3   Levels

Corollary 9 also has many consequences in the study of *levels* in arrangements of curves. Tamaki and Tokuyama [2] first showed the usefulness of cutting numbers in this area, and progress has been made by Chan [5, 6].

Definition. Let C be the set of the graphs of the real functions $f_1, f_2, \ldots, f_n$. We assume that each f_i is continuous and defined everywhere on the real line, and that any pair of curves in C intersects a finite number of times. We define the kth *level* of C to be the closure of the locus of points (x, y) on the curves in C with $|\{i : f_i(x) \leq y\}| = k$. The kth level consists of portions of the curves in C, delimited by intersections between these curves. We will call the total number of curve segments in a level its *complexity*.

Chan [6] derives an upper bound on the complexity of a given level using the number of cuts needed to turn a collection of pseudo-parabolas into pseudosegments. Our Corollary 9 improves his analysis. We omit the details.

Corollary 10. *Let C be a collection of n pseudo-parabolas. The maximum complexity of any level of C is $O(n^{3/2} \log^2 n)$.*

The above corollary represents a substantial improvement over the previous bound in [6] for an arbitrary collection of pseudo-parabolas. For a collection possessing a three parameter algebraic representation (as defined in [4]) the improvement is marginal, replacing a little-over-polylog term by $\log^2 n$. These improvements carry over to levels of arrangements of algebraic curves of degree higher than two by the technique of *bootstrapping*, as developed in [6]. We do not state the slightly improved bounds here.

3.4   Incidences and Faces

Aronov and Sharir [3] also used cutting numbers in their analysis of incidences between curves and points in the plane.

Definition. Let C be a set of curves and P a set of points in the plane. We define $I(C, P)$ to be the number of *incidences* between C and P, that is the number or pairs $(c, p) \in C \times P$ such that curve c contains point p. We also define $K(C, P)$ to be the sum of the complexities of the faces of the arrangement C containing points in P (assuming now that they are not on the curves). The *complexity* of a face is defined to be the number of curve segments that comprise the face.

The results in [3] relate the values of $K(C, P)$ and $I(C, P)$ to the cutting numbers discussed above. Applying their results and Corollary 9, we get

Corollary 11. *If C is a collection of n generalized pseudo-circles and P a set of m points, then*

1. $I(C, P) = O(m^{2/3} n^{2/3} + m + n^{3/2} \log n)$
2. $K(C, P) = O(m^{2/3} n^{2/3} + m + n^{3/2} \log^3 n)$

4   Open Problems

The results in this paper raise a number of interesting questions. Theorem 2 is tight except possibly for the logarithmic factor as graphs with n vertices and $\Omega(n^{3/2})$ edges are known which do not contain any C_4. This also implies that the $n = m$ special case of Theorem 1 and Corollary 7 are almost tight. Nevertheless, it would be interesting to know if the logarithmic factor is needed.

Problem 1. Is the logarithmic factor needed in Theorem 2?

The geometric consequences use Theorem 1 in the $n = m$ special case, but it is interesting to give bounds in the asymmetric cases as well. We define $R(n, m)$ to be the maximum total length of m pairwise intersection reverse cyclic sequences over an alphabet of size n. With this notation Corollary 7 gives $R(n,m) = O(m\sqrt{n}\log n + n\sqrt{m})$. We collect here a few simple lower and upper bounds for $R(n,m)$.

A trivial consequence of the property that a collection of cyclic sequences are pairwise intersection reverse is that no three symbols appear together in three cyclic sequences. By the Kővári–T. Sós–Turán Theorem [7], we have that $R(n,m) = O(nm^{2/3} + m)$ and $R(n,m) = O(n^{2/3}m + n)$. The first bound supersedes the bound in Corollary 7 if $m \geq n^{3/2}$. The second bound supersedes the bound in Corollary 7 if $m < n^{2/3}$. So for these extremely large or small values of m Corollary 7 is not tight.

The simplest constructions of intersection reverse cyclic sequences are constructions for collections of *subsets* intersecting each other in at most two elements. No matter how we order these subsets the resulting collection of cyclic sequences is pairwise intersection reverse. A simple construction for such subsets is any collection of circles in the plane over a finite field where -1 is not a square. Taking all points of the plane and a subset of the circles gives $R(n,m) = \Omega(m\sqrt{n})$ for $m \leq n^{3/2}$. Taking all circles and a subset of the points gives $R(n,m) = \Omega(nm^{2/3})$ for $m \geq n^{3/2}$. A collection of singleton sets gives the trivial bound $R(n,m) \geq m$, which is better than the previous bounds for $m > n^3$. Pairwise disjoint sets provide the other trivial $R(n,m) \geq n$ bound, which is better than the other bounds for $m \leq \sqrt{n}$.

The solid lines in the logarithmic scale diagram in Figure 1 shows the lower and upper bounds mentioned above. These bounds determine $R(n,m)$ up to a constant factor for $m \geq n^{3/2}$ and $m \leq n^{1/3}$ and up to a logarithmic factor for $n \leq m \leq n^{3/2}$. In any construction proving better lower bounds then the ones above, a typical pair of cyclic sequences will need to intersect in many elements, so the cyclic order is essential in those constructions. We present such a construction below proving $R(n,m) = \Omega(n^{5/6}m^{1/2})$ for $n^{1/3} < m < n^{2/3}$. This bound is represented in Figure 1 by the dashed line. The area of "uncertainty" is shaded. Even with this construction, the upper and lower bounds for $R(n,m)$ are far apart for $n^{1/3} < m < n$.

Construction. This construction is based on the construction of Gy. Elekes [8] of a set of axis-aligned parabolas and a set of points with a large number of incidences. Observe that axis-aligned parabolas form a collection of pseudo-parabolas: any pair intersects at most twice. For integers $b \geq a \geq 1$ consider the subset $P = \{(i,j) : |i| \leq a, |j| \leq 3a^2b\}$ of the integer grid and consider the collection $\mathcal{C}$ of parabolas (and lines) given by $y = ux^2 + vx + w$ with integers u, v, and w satisfying $|u| \leq b$, $|v| \leq ab$ and $|w| \leq a^2b$. We have $m = |P| = (2a+1)(6a^2b+1) = \Theta(a^3b)$ and $n = |\mathcal{C}| = (2b+1)(2ab+1)(2a^2b+1) = \Theta(a^3b^3)$. Clearly, each curve in $\mathcal{C}$ contains a point in P for each possible x coordinate, a total of $2a+1$ points. For each $p \in P$ we define the linearly ordered list B_p of all the curves in $\mathcal{C}$ passing through p. We order the list B_p according to the slopes of the curves at p (breaking ties arbitrarily). As a result we get m linearly ordered lists of subsets of the set of n symbols. It is easy to verify that these lists are intersection reverse. Their total length is the number of incidences between P and $\mathcal{C}$, which is $\Theta(a^4b^3) = \Theta(n^{5/6}m^{1/2})$.

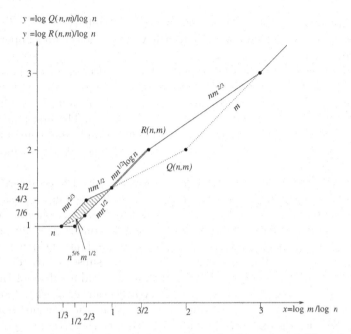

Fig. 1. Bounds and area of uncertainty for $R(n, m)$ and $Q(n, m)$.

Problem 2. Is it possible to find $n^{2/3}$ pairwise intersection reverse cyclic sequences over an alphabet of size n such that their total lengths sum to significantly more than $n^{7/6}$?

Note that for $m = n^{2/3}$ both constructions gives cyclic sequences with total size $\Theta(n^{7/6})$. One of the constructions is based on finite geometry, the other on Euclidean geometry. It seems to be hard to combine these constructions for a better result. The upper bound (provided both by Corollary 7 and the Kővári–T. Sós–Turán Theorem) is $O(n^{4/3})$.

We remark that, although not known if needed in Corollary 7, the $n/\sqrt{m}$ term is meaningful. This is the threshold for a typical pair of symbols appearing together in many cyclic sequences. We need it for our estimate that not many more different than same pairs exist. If a typical pair of symbols appears together in only two cyclic sequences, it is possible that they only contribute different pairs. This happens in the above construction as well; since we construct linearly ordered (rather than cyclic) sequences that are pairwise intersection reverse, no "same pair" ever appears.

One can ask the same extremal question about linearly ordered sequences. Let $Q(n, m)$ stand for the maximum total length of m pairwise intersection reverse sequences over an n element alphabet. In this case two symbols cannot appear together in three sequences. The Kővári–T. Sós–Turán Theorem therefore gives the bounds $Q(n, m) = O(mn^{2/3} + n)$ and $Q(m, n) = O(n\sqrt{m} + m)$. For $m \leq n/\log^2 n$ or $m \geq n^3$ we get the same upper bounds that we did for $R(n, m)$. The upper bound for intermediate values of m is shown by the dotted line in Figure 1. One gets simple constructions of intersection reverse sequences by considering set systems with pairwise intersection limited to singletons. Just as we noted in the case of cyclic sequences, this property ensures that the sequences are pairwise intersection reverse independent of the linear order chosen. The standard construction for such a set system is the set of lines in a finite plane, yielding $Q(n,m) = \Omega(n\sqrt{m})$ for $m \geq n$ and $Q(n,m) = \Omega(m\sqrt{n})$ for $m \leq n$. The bounds $Q(n,m) \geq n$ and $Q(n,m) \geq m$ are trivial. These bounds determine $Q(n,m)$ up to a constant factor for $m \leq n^{1/3}$ and $m \geq n$. Notice that the construction using parabolas in the plane yields pairwise intersection reverse linearly ordered sequences and so we have $Q(n,m) = \Omega(n^{5/6}m^{1/2})$ for $n^{1/3} \leq m \leq n^{2/3}$. Surprisingly, the "area of uncertainty" for $Q(n,m)$ is exactly the same parallelogram as it is for $R(n,m)$. Only when $n < m < n^3$ do the bounds for $Q(n,m)$ and $R(n,m)$ diverge. We do not know if allowing for cyclic sequences can yield longer intersection reverse collections in the $m < n$ case.

Problem 3. Does $R(n,m) = O(Q(n,m))$ hold for $m < n$?

As far as pseudo-circles are concerned, our result is conjectured to be far from optimal. The best known construction is a set of n pseudo-circles that needs $\Omega(n^{4/3})$ cuts before it becomes a collection of pseudo-segments. It is an important open problem to improve either bound on the minimum number of cuts that turn pseudo-circles into pseudo-segments.

Acknowledgments

We would like to thank János Pach for introducing us to this problem and for directing our attention to the many consequences that were shown in Section 3.

References

1. Pinchasi, R., Radoičić, R.: On the number of edges in geometric graphs with no self-intersecting cycle of length 4. In Pach, J., ed.: Towards a Theory of Geometric Graphs. Number 342 in Contemp. Math., Providence, RI, Amer. Math. Soc. (2004)
2. Tamaki, H., Tokuyama, T.: How to cut pseudo-parabolas into pseudosegments. Discrete Comput. Geom. **19** (1998) 265–290
3. Aronov, B., Sharir, M.: Cutting circles into pseudo-segments and improved bounds for incidences. Discrete Comput. Geom. **28** (2002) 475–490
4. Agarwal, P., Nevo, E., Pach, J., Pinchasi, R., Sharir, M., Smorodinsky, S.: Lenses in arrangements of pseudo-circles and their applications. J. ACM **51** (2004) 139–186
5. Chan, T. M.: On levels in arrangements of curves. Discrete and Comput. Geom. **29** (2000) 375–393
6. Chan, T. M.: On levels in arrangements of curves, II: a simple inequality and its consequence. In: Proc. 44th IEEE Symposium on Foundations of Computer Science (FOCS). (2003) 544–550
7. Kővári, T., T. Sós, V., Turán, P.: On a problem of K. Zarankiewicz. Colloq. Math. **3** (1954) 50–57
8. Elekes, Gy.: Sums versus products in number theory, algebra and Erdös geometry. In G. Hálász, et. al., ed.: Paul Erdős and his mathematics II. Based on the conference, Budapest, Hungary, July 4–11, 1999. Volume 11 of Bolyai Soc. Math. Stud., Springer (2002) 241–290

New Exact Results and Bounds
for Bipartite Crossing Numbers of Meshes[*]

Matthew C. Newton[1], Ondrej Sýkora[1], Martin Užovič[2], and Imrich Vrt'o[3]

[1] Department of Computer Science, Loughborough University,
Loughborough, Leicestershire LE11 3TU, UK
{m.c.newton,o.sykora}@lboro.ac.uk
[2] Department of Computer Science, Comenius University,
Mlynská dolina, 842 48 Bratislava, Slovak Republic
[3] Department of Informatics, Institute of Mathematics, Slovak Academy of Sciences,
Dúbravská 9, 841 04 Bratislava, Slovak Republic
vrto@savba.sk

Abstract. The bipartite crossing number of a bipartite graph is the minimum number of crossings of edges when the partitions are placed on two parallel lines and edges are drawn as straight line segments between the lines. We prove exact results, asymtotics and new upper bounds for the bipartite crossing numbers of 2-dimensional mesh graphs. We especially show that $\mathrm{bcr}(P_6 \times P_n) = 35n - 47$, for $n \geq 7$.

1 Introduction

The *planar crossing number* is the minimum number of edge crossings in any drawing of a graph in the plane. This is an important and difficult problem which has been studied in graph theory, as well as in the theory of VLSI [4, 10, 15]. Computing the value of the planar crossing number is NP-hard [6], and exact values are known only for very restricted classes of graphs. In this paper we study a frequent variant of the planar crossing number. Let $G = (V_0, V_1, E)$ be a bipartite graph, where V_0, V_1 is the bipartition of vertices into independent sets. A *bipartite drawing* of G, denoted by $\mathcal{D}(G)$ consists of placing the vertices of V_0 and V_1 into distinct points on two horizontal lines $y = 0$, $y = 1$ in the xy-plane, respectively, and then drawing each edge with one straight line segment which connects the end-vertices. The bipartite crossing number of the drawing, denoted by $\mathrm{bcr}(\mathcal{D}(G))$, is the number of crossing pairs of edges in the drawing. The *bipartite crossing number* of G, denoted by $\mathrm{bcr}(G)$, is the minimum $\mathrm{bcr}(\mathcal{D}(G))$ over all drawings.

A motivation behind studying $\mathrm{bcr}(G)$ comes from the routing of VLSI (see for example [10, 17]). Another motivation appears in the field of graph drawing. It is well known that $\mathrm{bcr}(G)$ is one of the parameters which strongly influences the understanding and aesthetics of drawings of graph-like structures, especially in a hierarchical fashion. For a survey on drawing graphs see [3].

[*] This research was supported by the EPSRC grant GR/R37395/01 and by VEGA grant No. 2/3164/23.

The notion of bcr(G) was first introduced in [7], [8] and [23] where it was stated that bcr$(G) = 0$ iff the graph is caterpillar, and bcr$(C_n) = n/2 - 1$ for an n-vertex cycle (n is even). Some basic observations on bcr(G) were made in [13]. The bipartite crossing number problem is known to be NP-complete [6] but can be solved in polynomial time for bipartite permutation graphs [20], and trees [18]. A great deal of research has been devoted to the design of algorithms and heuristics for solving this problem (see for example [2,5,9,11,12,14,21,22]).

To find exact results of bcr(G) for special families of graphs is of interest from the graph-theoretic point of view, but such results can also be applied in testing heuristics for bipartite drawings. So far the heuristics are mostly compared against each other but not against the optimal drawing, which is typicaly unknown. For this purpose one can use

$$\mathrm{bcr}(S(K_n)) = 4\binom{n}{4} + 2\binom{n}{3},$$

where $S(K_n)$ denotes the complete graph with exactly one new vertex on every edge [8]. They proved a similar formula for the subdivided complete bipartite graph. In [18] we proved for the complete binary tree T_n of depth $n - 1$

$$\mathrm{bcr}(T_n) = \frac{1}{9}((3n - 11)2^n + 2(-1)^n) + 2.$$

For the mesh $P_m \times P_n$, i.e. the graph defined by the Cartesian product of an $m-$vertex path with an $n-$vertex path, where $3 \leq m \leq n$, we found in [19] bcr$(P_3 \times P_n) = 5n - 6$, for $n \geq 3$. To our knowledge these are all known non-trivial exact results for the bipartite crossing number of typical graphs. In [19] we also proved an estimation

$$\frac{3}{4}m(m - 18)n + O(m^3) \leq \mathrm{bcr}(P_m \times P_n) \leq \frac{3}{2}m(m - 1)n.$$

In this paper we prove an asymptotic

$$\lim_{n \to \infty} \frac{\mathrm{bcr}(P_m \times P_n)}{n} = \frac{1}{2}(3m^2 - 7m + 4)$$

and exact results for $m = 4, 5$, and 6. Especially,

$$\mathrm{bcr}(P_6 \times P_n) = \begin{cases} 161, & \text{if } n = 6, \\ 35n - 47, & \text{if } n \geq 7. \end{cases}$$

We conclude the paper with new upper bounds for bipartite crossing numbers of general meshes and some conjectures about exact values.

2   A General Lower Bound

In this section we prove a lower bound for general meshes. Assume that the mesh has m horizontal rows and n vertical columns.

Theorem 1. *For a mesh* $P_m \times P_n, n > 8m$:

$$\mathrm{bcr}(P_m \times P_n) \geq \frac{1}{2}(3m^2 - 7m + 4)n - 4m^3 + \frac{21}{2}m^2 - \frac{7}{2}m.$$

Proof. Let us have a bipartite drawing $\mathcal{D}$ of the mesh. Divide the drawing into three parts: $\mathcal{D}_1, \mathcal{D}_2, \mathcal{D}_3$. Define $\mathcal{D}_1$ as a part of drawing constructed in the following way: starting from the left end of the drawing, take alternate vertices from both partitions until this set of vertices contains vertices from m different columns of the mesh. The convex hull of this set is $\mathcal{D}_1$. Starting from the right end of the drawing we similarly construct the part $\mathcal{D}_3$ until it contains vertices of m different columns (different also from those in the $\mathcal{D}_1$). The rest is $\mathcal{D}_2$. From the Dirichlet principle it follows that the total number of vertices in $\mathcal{D}_1$ and $\mathcal{D}_3$ is at most $2m^2$, and the number of horizontal edges incident to the vertices from $\mathcal{D}_1$ and $\mathcal{D}_3$ is at most $4m^2$. Now, choose m vertices of different columns from $\mathcal{D}_1$ and m vertices of different columns in $\mathcal{D}_3$. Take a bijection between these two groups of vertices and for any pair join its vertices by a path that uses exclusively edges of one row of the mesh and the edges in the columns where the vertices of the pair reside. In Fig. 1 there is such a path, shown by the heavy line. The path can use edges of the mesh as shown in Fig. 2. The m paths are obviously vertex disjoint.

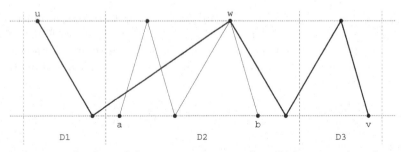

Fig. 1. An example of a path between vertices u and v, which crosses the column $a - b$ in vertex w.

We have m such edge disjoint paths. There are at least $n - 2m$ complete columns in $\mathcal{D}_2$. Now we count the number of crossings of the above m paths with edges of these columns. There are $m - 1$ column edges in one column and at least $m - 2$ paths must cross each of them (two paths may use end-vertices of a column edge). Therefore there are $(n - 2m)(m - 1)(m - 2)$ crossings of this type.

If a row edge lies completely in $\mathcal{D}_2$ it must be crossed by $m - 1$ paths as one path can use the edge. As there are at least $(n - 1)m - 4m^2$ such edges, we have at least $((n - 1)m - 4m^2) \times (m - 1)/2$ crossings of this type (2 in denominator is because a crossing is counted twice), which in total gives the claimed lower bound. $\qquad\square$

In Section 4 we describe an upper bound which coincides with this lower bound up to the second order term for fixed m and $n \gg m$ which implies:

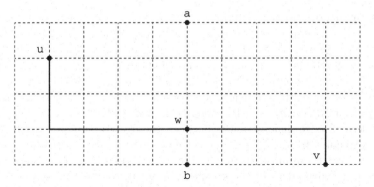

Fig. 2. Example of the path between the vertices u and v in the mesh.

Theorem 2. *For any fixed* $m \geq 3$

$$\lim_{n \to \infty} \frac{\operatorname{bcr}(P_m \times P_n)}{n} = \frac{1}{2}(3m^2 - 7m + 4)$$

3   Three Exact Results

Here we prove exact results on the bipartite crossing numbers for $m = 4, 5, 6$. We will skip the (simpler) proofs for the cases $m = 4, 5$.

Theorem 3. *The bipartite crossing numbers of* $P_4 \times P_n, P_5 \times P_n$ *and* $P_6 \times P_n$ *satisfy*

$$\operatorname{bcr}(P_4 \times P_n) = \begin{cases} 33, & \text{if } n = 4, \\ 12n - 14, & \text{if } n \geq 5. \end{cases}$$

$$\operatorname{bcr}(P_5 \times P_n) = \begin{cases} 81, & \text{if } n = 5, \\ 22n - 28, & \text{if } n \geq 6. \end{cases}$$

$$\operatorname{bcr}(P_6 \times P_n) = \begin{cases} 161, & \text{if } n = 6, \\ 35n - 47, & \text{if } n \geq 7. \end{cases}$$

Proof. The matching upper bounds will be shown in Section 4. It is known that $\operatorname{bcr}(P_6 \times P_6) = 161$ and $\operatorname{bcr}(P_6 \times P_7) = 198$. The values were found by Thomas Odenthal using a branch and bound algorithm [16].

The leftmost 'comb', C, is a sub-graph of the mesh defined as a graph induced by the first two columns without the edges of the second column. Similarly we define the rightmost comb. Let us call the convex hull of the drawing of C the *comb region, R*.

We proceed by induction on n. Let the claim hold for $n - 1 \geq 7$. Consider an optimal drawing $\mathcal{D}(P_6 \times P_n)$. If there are 35 crossings on the leftmost or rightmost comb then by deleting that comb we get a drawing $\mathcal{D}(P_6 \times P_{n-1})$. Hence

$$\operatorname{bcr}(\mathcal{D}(P_6 \times P_n)) \geq 35 + \operatorname{bcr}(\mathcal{D}(P_6 \times P_{n-1})) \geq 35n - 47.$$

However, if both combs contain less than 35 crossings each, our aim is to show that the drawing must be in special forms for which we determine the numbers of crossings directly. Roughly speaking, in this case both combs must be placed in "extreme" positions (left and right) in the drawing. Let Z be the set of vertices which are not in the left comb C. Assume that there are at least two vertices, or one vertex of at least degree 3, of Z to the left of R, and at least two vertices, or one vertex of at least degree 3, of Z to the right of R. Then there exists at least 3 edge disjoint paths between these vertices in the mesh, which are vertex disjoint from C. As C has 11 edges, the 3 paths force 33 crossings on C. Note that there must also be 5 additional crossings on C caused just by the edges of C with edges of the second column. In total we have 38 crossings on C, a contradiction.

WLOG we assume that there is at most one vertex $v \in Z$ of degree two lying to the left of R. Let $S \subseteq Z$ denote the set of vertices lying in the comb region R. We claim that there are at least 17 vertices of Z, which are from the 3rd to the 8th columns, lying to the right of the comb region R. If there are only at most 16 such vertices, then as $n \geq 8$, the graph induced by Z has at least 60 edges and at most 24 of them are in the graph induced by those 16 vertices. Hence the number of edges incident to S is at least 36 and every such edge produces a crossing on C, a contradiction.

Take the set Z' of 11 vertices out of those 17 such that they lie in the 3rd to the 7th columns. One can show that there are 6 internally vertex disjoint paths, forming a set $\mathcal{P}$, between the vertices of the second column and vertices of Z', disjoint from the edges of C and from the rightmost column. It implies that there is no vertex of degree 2 on the paths from $\mathcal{P}$.

If $S = \emptyset$, then we can move the degree two vertex v from the left to the immediate right of R without increasing crossings. We checked by exhaustive search on computer all possible drawings of C (i.e. $(6!)^2$ permutations) and found that if $S = \emptyset$ then the edges of C are crossed at least 32 times. The crossings are caused either by paths from $\mathcal{P}$, or edges of the second column, or by C edges with C edges. Moreover, there is only one drawing of C with precisely 32 crossings, shown in Fig. 3. We call it 'optimal with empty S'.

The notation and type of lines we use for vertices and edges is shown in Fig. 4.

Assume that $S \neq \emptyset$ and that there are less than 35 crossings on the comb, C, i.e. its number of crossings is 33 or 34.

Suppose there is a vertex of degree two in S. As it does not belong to any of the paths of $\mathcal{P}$, it causes at least 2 crossings more on C.

Suppose there is a vertex of degree three (or four) in S. It may belong to one of the paths of $\mathcal{P}$. This implies at least one (or two) more crossings on C. Consequently, if there are two nodes in S and at least one of them is not of degree 3, then there are at least 3 new crossings, i.e. at least 36 crossings on C, a contradiction.

Similarly, we get a contradiction if there is a single degree two vertex of Z to the left of R, and there is a vertex in S.

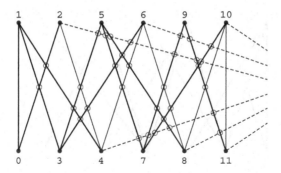

Fig. 3. Optimal drawing of the comb C with empty S.

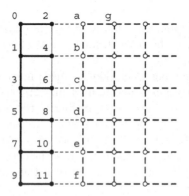

Fig. 4. Notation used for vertices of the comb C.

Let there be exactly one vertex of degree two or four in S. Such a vertex can be moved to the immediate right of the region R without increasing the number of crossings.

Now, assume we have two vertices of degree 3 in S. The two vertices must be either both from the third column or next to each other from the same row (either first or sixth), otherwise we have at least 3 more crossings on C, a contradiction.

By a detailed case analysis one can show that in both situations the vertices can be moved to the immediate right of R without increasing the number of crossings, i.e. all vertices of Z will be to the right of R.

The last case we need to analyse is where there is a single degree 3 vertex in S. We checked by exhaustive search on computer all possible drawings of C (i.e. $7(6!)^2$ situations) and found that if there is one vertex of degree 3 in S then C is crossed at least 33 times, and there is only one such drawing of C with precisely 33 crossings. We call it '*optimal with one vertex in S*' (see Fig. 5).

So we conclude that the drawing of the left comb is either optimal with empty S with 32 crossings or optimal with one vertex in S with 33 crossings. We repeat the above arguments for the right comb.

Let us first have both combs in optimal drawings with empty S. In total they contain 64 crossings. Consider now 6 horizontal paths, forming a set $\mathcal{P}'$,

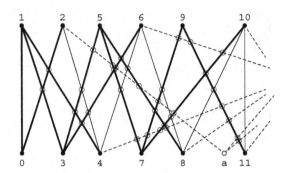

Fig. 5. Optimal drawing of the comb C with one degree 3 vertex in S.

starting in the second column and ending in the $(n-1)$th column. Similarly as in the proof of Theorem 1, the paths from $\mathcal{P}'$ create $20(n-4)$ crossings on the 3rd,...,$(n-2)$nd column.

Now, we count the number of crossings caused mutually on paths from $\mathcal{P}'$. Recall [8] that the bipartite crossing number of a $2p$-vertex cycle is $p-1$. The paths of $\mathcal{P}'$ are divided into 2 groups according to the line on which they start on the left side. Consider any 2 paths from different groups. Join them by 2 new artificial edges on the left and right side, respectively. We get a $(2n-4)$-vertex cycle. As there are no crossings on the new edges the original paths cross at least $n-3$ times and the number of crossings between the paths of different groups is $9(n-3)$.

Consider any 2 paths from the same group. Identify the starting vertices of the paths on the left and right side, respectively. This operation does not increase the crossing number. We get a $(2n-6)$-vertex cycle with $n-4$ crossings. Hence there are $3(n-4)$ crossings between the paths of each group. Altogether the number of mutal crossings on the paths is $15n-51$.

The optimal drawing of the left comb C forces 10 crossings between the edges of the 2nd column and edges joining the 2nd and the 3rd column. A similar argument holds for the right hand side comb. So we have 20 such crossings. In total the number of crossings are

$$\mathrm{bcr}(\mathcal{D}(P_6 \times P_n)) = 64 + 20(n-4) + 15n - 51 + 20 = 35n - 47.$$

If, instead of being *optimal with empty S* a comb is in the form *optimal with one vertex in S*, we have 33 crossings on the comb. We then have one less crossing:

- on edges of the third column with edges between the second and the third columns;
- on the edges between the second and the third columns with themselves; and
- on the edges of the second column with the edges between the second and the third columns.

We have one more crossing on:

– the edges of the second and the third column; and
– on the the edges of the second column with the edges between the third and the fourth columns.

The changes in comparison with 'optimal drawing with empty S' are only local and they give the same number of crossings in total.   □

4   Upper Bounds

First we describe 2 types of bipartite drawings of meshes.

– A mesh $P_m \times P_n$ is drawn in a *diagonal* manner if the vertices are placed in the order shown in Fig. 6.
  We denote it by $\mathcal{D}_d(P_m \times P_n)$.
– We say that the mesh is drawn in a *combined* way if the vertices are placed in the order as shown in Fig. 7.
  The parameter $s < n/2$ denotes the number of left and right columns whose vertices are drawn in diagonal manner. We denote this type of drawing by $\mathcal{D}_c(P_m \times P_n)$.

Note that if $s = 1$, the drawing is just a column by column drawing. One can check that the number of vertices in the above defined types of drawings of meshes is given by the following upper bounds:

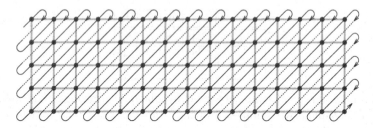

Fig. 6. Vertex order of mesh drawn in diagonal way.

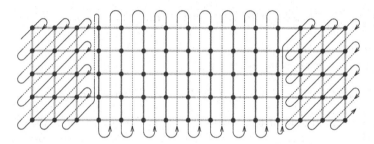

Fig. 7. Vertex order of mesh drawn in combined way.

Theorem 4. *Let $4 \leq m \leq n$. Then*

$$\mathrm{bcr}(\mathcal{D}_d(P_m \times P_n)) = (2m^2 - 6m + 5)n - \frac{1}{3}(2m^3 - 6m^2 + 7m - 3), \quad (1)$$

$$\mathrm{bcr}(\mathcal{D}_c(P_m \times P_n)) = \frac{1}{2}(3m^2 - 7m + 4)n \quad (2)$$

$$- \left\lceil \frac{(4s+3)m^2}{4} \right\rceil + (2s^2 + s + 1)m - \frac{1}{3}s(2s-1)(s+5).$$

□

As a consequence we get matching upper bounds to our lower bounds from Section 3:

$$\mathrm{bcr}(P_4 \times P_n) = \begin{cases} \mathrm{bcr}(\mathcal{D}_d(P_4 \times P_4)) = 33, \\ \mathrm{bcr}(\mathcal{D}_c(P_4 \times P_n)) = 12n - 14, \text{ if } n \geq 5, \text{ with } s = 2. \end{cases}$$

$$\mathrm{bcr}(P_5 \times P_n) = \begin{cases} \mathrm{bcr}(\mathcal{D}_d(P_5 \times P_5)) = 81, \\ \mathrm{bcr}(\mathcal{D}_c(P_5 \times P_n)) = 22n - 28, \text{ if } n \geq 6, \text{ with } s = 2. \end{cases}$$

$$\mathrm{bcr}(P_6 \times P_n) = \begin{cases} \mathrm{bcr}(\mathcal{D}_d(P_6 \times P_6)) = 161, \\ \mathrm{bcr}(\mathcal{D}_c(P_6 \times P_n)) = 35n - 47, \text{ if } n \geq 7, \text{ with } s = 2. \end{cases}$$

At this place we reproduce a table of exact results for small meshes obtained by Thomas Odenthal [16] using a branch and bound method.

m vs. n	4	5	6	7	8
4	33	46	58	70	82
5		81	104	126	148
6			161	198	233

The values for $m = n$ are obtained by diagonal drawings. The values for $n > m$ are obtained by combined drawings. In cases $m = 5, 6, 7$ and $m = 8$ and $n > m$ we use the combined drawing with $s = 2$ and $s = 3$, respectively.

For general square meshes the upper bound (1) gives

$$\mathrm{bcr}(P_m \times P_m) \leq \frac{1}{3}(4m^3 - 12m^2 + 8m + 3),$$

which we believe to be optimal.

For general rectangular meshes with $4 \leq m \leq n$, we have

$$\mathrm{bcr}(P_m \times P_m) \leq \min\{\mathrm{bcr}(\mathcal{D}_d(P_m \times P_n)), \mathrm{bcr}(\mathcal{D}_c(P_m \times P_n))\} \quad (3)$$

The righthand side of the inequality (3) is minimized for a value $s = s_m$, for which as $m \to \infty$, $s_m/m \to 1 - 1/\sqrt{2} = 0.2928932 \cdots$. Comparing the minimum with the value for the diagonal drawing (1) we see that the diagonal drawing is better only in a narrow interval $m \leq n \leq \alpha_m m$, for which if $m \to \infty$, $\alpha_m \to 2(3 - \sqrt{2})/3 = 1.057333 \cdots$.

Acknowledgment

The authors thank Thomas Odenthal who found exact numbers for small size meshes.

References

1. Chung, F.R.K., A conjectured minimum valuation tree, *SIAM Review* **20** (1978), 601–604.
2. Demetrescu, C., Finocchi, I., Removing cycles for minimizing crossings, *ACM Journal of Experimental Algorithms* **6** (2001).
3. Di Battista, J., Eades, P., Tamassia, R., Tollis, I. G., Graph Drawing, Algorithms for the Visualization of Graphs, Prentice Hall, 1999.
4. Erdős, P., Guy, R. P., Crossing number problems, *American Mathematical Monthly* **80** (1973), 52–58.
5. Eades, P., Wormald, N., Edge crossings in drawings of bipartite graphs, *Algorithmica* **11** (1994), 379–403.
6. Garey, M. R., Johnson, D. S., Crossing number is NP-complete, *SIAM Journal on Algebraic and Discrete Methods* **4** (1983), 312–316.
7. Harary, F., Determinants, permanents and bipartite graphs, *Mathematical Magazine* **42** (1969), 146–148.
8. Harary, F., Schwenk, A., A new crossing number for bipartite graphs, *Utilitas Mathematica* **1** (1972), 203–209.
9. Jünger, M., Mutzel, P., 2-layer straight line crossing minimization: performance of exact and heuristic algorithms, *Journal of Graph Algorithms and Applications* **1** (1997), 1–25.
10. Leighton, F. T., Complexity Issues in VLSI, MIT Press, 1983.
11. Martí, R., A tabu search algorithm for the bipartite drawing problem, *European Journal of Operational Research* **106** (1998), 558–569.
12. Matuszewski, C., Schönfeld, R., Molitor, P., Using sifting for *k*-layer crossing minimization, in: *Proc. 7th. Intl. Symposium on Graph Drawing*, Lecture Notes in Computer Science 1731, Springer, Berlin, 2001, 217-224.
13. May, M., Szkatula, K., On the bipartite crossing number, *Control and Cybernetics* **17** (1988), 85–98.
14. Newton, M., Sýkora, O., Vrťo, I., Two new heuristics for two-sided bipartite graph drawings, in: Proc. *10th Intl. Symposium on Graph Drawing*, Lecture Notes in Computer Science 2528, Springer, Berlin, 2002, 312-319.
15. Pach, J., Agarwal, P. K., Combinatorial Geometry, Wiley and Sons, New York, 1995.
16. Odenthal, T., Personal communication, 2002.
17. Sarrafzadeh, M., An Introduction to VLSI Physical Design, McGraw-Hill, New York, 1995.
18. Shahrokhi, F., Sýkora, O., Székely, L. A., Vrťo, I.: On the bipartite drawings and the linear arrangement problem, *5th Intl. Workshop on Algorithms and data Structures*, Lecture Notes in computer Science 1272, Springer, Berlin, 1997, 55-68.
19. Shahrokhi, F., Sýkora, O., Székely, L. A., Vrťo, I., A new lower bound for the bipartite crossing number with applications, *Theoretical Computer Science* **245** (2000), 281–294.

20. Spinrad, J., Brandstädt, A., Stewart, L., Bipartite permutation graphs, *Discrete Applied Mathematics* **19** (1987), 279–292.
21. Sugiyama, K., Tagawa, S., Toda, M., Methods for visual understanding of hierarchical systems structures, *IEEE Transactions on Systems, Man and Cybernetics* **11** (1981), 109–125.
22. Warfield, J., Crossing theory and hierarchy mapping, *IEEE Transactions on Systems, Man and Cybernetics* **7** (1977), 502–523.
23. Watkins, M. E., A special crossing number for bipartite graphs: a research problem, *Annals of New York Academy Sciences* **175** (1970), 405–410.

Drawing Pfaffian Graphs

Serguei Norine

School of Mathematics, Georgia Institute of Technology, Atlanta, Georgia 30332, USA

Abstract. We prove that a graph is Pfaffian if and only if it can be drawn in the plane (possibly with crossings) so that every perfect matching intersects itself an even number of times.

1 Introduction

In this paper we prove a theorem that connects Pfaffian orientations with the parity of the numbers of crossings in planar drawings. The proof is elementary, but it has other consequences and raises interesting questions. Before we can state the theorem we need some definitions.

All graphs considered in this paper are finite and have no loops or multiple edges. For a graph G we denote its edge set by $E(G)$. A *labeled graph* is a graph with vertex-set $\{1, 2, \ldots, n\}$ for some n. If u and v are vertices in a graph G, then uv denotes the edge joining u and v and directed from u to v if G is directed. A *perfect matching* is a set of edges in a graph that covers all vertices exactly once.

Let G be a directed labeled graph and let $M = \{u_1v_1, u_2v_2, \ldots, u_kv_k\}$ be a perfect matching of G. Define the *sign* of M to be the sign of the permutation

$$\begin{pmatrix} 1 & 2 & 3 & 4 \ldots 2k-1 & 2k \\ u_1 & v_1 & u_2 & v_2 \ldots \quad u_k & v_k \end{pmatrix}.$$

Note that the sign of a perfect matching is well-defined as it does not depend on the order in which the edges are written. We say that a labeled graph G is *Pfaffian* if there exists an orientation D of G such that the signs of all perfect matchings in D are positive, in which case we say that D is a *Pfaffian orientation* of G. An unlabeled graph G is *Pfaffian* if it is isomorphic to a labeled Pfaffian graph. It is well-known and also follows from Theorem 1 below that in that case every labeling of G is Pfaffian. The importance of Pfaffian graphs will be discussed in the next section.

By a *drawing* Γ of a graph G we mean an immersion of G in the plane such that edges are represented by homeomorphic images of $[0, 1]$ not containing vertices in their interiors. Edges are permitted to intersect, but there are only finitely many intersections and each intersection is a crossing. For edges e, f of a drawing Γ let $cr(e, f)$ denote the number of times the edges e and f cross. For a perfect matching M let $cr_\Gamma(M)$, or $cr(M)$ if the drawing is understood from context, denote $\sum cr(e, f)$, where the sum is taken over all unordered pairs of distinct edges $e, f \in M$.

J. Pach (Ed.): GD 2004, LNCS 3383, pp. 371–376, 2004.
© Springer-Verlag Berlin Heidelberg 2004

The following theorem is the main result of this paper. The proof will be presented in Section 3.

Theorem 1. *A graph G is Pfaffian if and only if there exists a drawing of G in the plane such that $cr(M)$ is even for every perfect matching M of G.*

The "if" part of this theorem as well as the "if" part of its generalization (Theorem 3) was known to Kasteleyn [4] and was proved by Tesler [14]; however our proof of this part is different. The "only if" part is new.

2 Pfaffian Graphs

Pfaffian orientations have been introduced by Kasteleyn [2–4], who demonstrated that one can enumerate perfect matchings in a Pfaffian graph in polynomial time.

We say that an $n \times n$ matrix $A(D) = (a_{ij})$ is a *skew adjacency matrix* of a directed labeled graph D with n vertices if

$$a_{ij} = \begin{cases} 1 & \text{if } ij \in E(D), \\ -1 & \text{if } ji \in E(D), \\ 0 & \text{otherwise.} \end{cases}$$

Let A be a skew-symmetric $2n \times 2n$ matrix. For each partition

$$P = \{\{i_1, j_1\}, \{i_2, j_2\}, \dots, \{i_n, j_n\}\}$$

of the set $\{1, 2, \dots, 2n\}$ into pairs, define

$$a_P = sgn \begin{pmatrix} 1 & 2 & \dots & 2n-1 & 2n \\ i_1 & j_1 & \dots & i_n & j_n \end{pmatrix} a_{i_1 j_1} \dots a_{i_n j_n}.$$

Note that a_P is well defined as it does not depend on the order of the pairs in the partitions nor on the order in which the pairs are listed. The *Pfaffian* of the matrix A is defined by

$$Pf(A) = \sum_P a_P,$$

where the sum is taken over all partitions P of the set $\{1, 2, \dots, 2n\}$ into pairs. Note that if D is a Pfaffian orientation of a labeled graph G then $Pf(A(D))$ is equal to the number of perfect matchings in G. One can evaluate the Pfaffian efficiently using the following identity from linear algebra: for a skew-symmetric matrix A

$$det(A) = (Pf(A))^2.$$

Thus the number of perfect matchings, and more generally the generating function of perfect matchings of a Pfaffian graph, can be computed in polynomial time.

The problem of recognizing Pfaffian bipartite graphs is equivalent to many problems of interest outside graph theory, eg. the Pólya permanent problem [11], the even circuit problem for directed graphs [15], or the problem of determining which real square matrices are sign non-singular [5], where the latter has applications in economics [13].

The complete bipartite graph $K_{3,3}$ is not Pfaffian. Each edge of $K_{3,3}$ belongs to exactly two perfect matchings and therefore changing an orientation of any edge does not change the parity of the number of perfect matchings with negative sign. One can easily verify that for some (and therefore for every) orientation of $K_{3,3}$ this number is odd.

In fact, Little [6] proved that a bipartite graph is Pfaffian if and only if it does not contain an "even subdivision" H of $K_{3,3}$ such that $G \setminus V(H)$ has a perfect matching.

A structural characterization of Pfaffian bipartite graphs was given by Robertson, Seymour and Thomas [12] and independently by McCuaig [7]. They proved that a bipartite graph is Pfaffian if and only if it can be obtained from planar graphs and one specific non-planar graph (the Heawood graph) by repeated application of certain composition operations. This structural theorem implies a polynomial time algoritheorem for recognition of Pfaffian bipartite graphs.

No satisfactory characterization is known for general Pfaffian graphs. The result of this paper was obtained while attempting to find such a description.

3   Main Theorem

Let Γ be a drawing of a graph G in the plane. We say that $S \subseteq E(G)$ is a *marking* of Γ if $cr(M)$ and $|M \cap S|$ have the same parity for every perfect matching M of G.

Theorem 1 follows from the following more general result.

Theorem 2. *For a graph G the following are equivalent:*
(a) G is Pfaffian;
(b) some drawing of G in the plane has a marking;
(c) every drawing of G in the plane has a marking;
(d) there exists a drawing of G in the plane such that $cr(M)$ is even for every perfect matching M of G.

We say that Γ is *a standard drawing* of a labeled graph G if the vertices of Γ are arranged on a circle in order and every edge of Γ is drawn as a straight line.

The equivalence of conditions (a), (b) and (c) of Theorem 2 immediately follows from the next two lemmas.

Lemma 1. *Let Γ be a standard drawing of a labeled graph G. Then G is Pfaffian if and only if Γ has a marking.*

Proof. Let D be an orientation of G. Let $M = \{u_1v_1, u_2v_2, \ldots, u_kv_k\}$ be a perfect matching of D. The sign of M is the sign of the permutation

$$P = \begin{pmatrix} 1 & 2 & 3 & 4 & \ldots 2k-1 & 2k \\ u_1 & v_1 & u_2 & v_2 & \ldots & u_k & v_k \end{pmatrix}.$$

Let $i(P)$ denote the number of inversions in P, then

$$sgn(M) = sgn(P) = (-1)^{i(P)} = \prod_{1 \le i < j \le 2k} sgn(P(j) - P(i)) =$$

$$= \prod_{1 \le i < j \le k} sgn((u_j - u_i)(v_j - u_i)(u_j - v_i)(v_j - v_i)) \times$$

$$\times \prod_{1 \le i \le k} sgn(v_i - u_i). \qquad (1)$$

In Γ edges $u_i v_i$ and $u_j v_j$ cross if and only if each of the two arcs of the circle containing the vertices of Γ with the ends u_i and v_i contains one of the vertices u_j and v_j, in other words if and only if

$$sgn((u_j - u_i)(v_j - u_i)(u_j - v_i)(v_j - v_i)) = -1.$$

Define $S_D = \{uv \in E(D) | u > v\}$. Note that for every $S \subseteq E(G)$ there exists (unique) orientation D such that $S = S_D$. From (1) we deduce that

$$sgn(M) = (-1)^{cr(M)} \times (-1)^{|M \cap S|}.$$

Therefore M has a positive sign if and only if $cr(M)$ and $|M \cap S|$ have the same parity. It follows that D is a Pfaffian orientation of G if and only if S_D is a marking of the standard drawing of G. $\qquad \square$

Notice that we have in fact shown that there exists a one-to-one correspondence between Pfaffian orientations of a labeled graph and markings of its standard drawing.

Lemma 2. *Let Γ_1 and Γ_2 be two drawings of a labeled graph G in the plane. Then Γ_1 has a marking if and only if Γ_2 has one.*

Proof. For any n and any two sequences $(a_1, a_2, .., a_n)$ and $(b_1, b_2, .., b_n)$ of pairwise distinct points in the plane, there clearly exists a homeomorphic transformation of the plane that takes a_i to b_i for all $1 \le i \le n$. Therefore without loss of generality we assume that the vertices of G are represented by the same points in the plane in both Γ_1 and Γ_2.

It suffices to prove the statement of the lemma for drawings Γ_1 and Γ_2 that differ only in the position of a single edge $e = uv$. Let e_1 and e_2 denote the images of e in Γ_1 and Γ_2 correspondingly. Define $C = e_1 \cup e_2$. The closed curve C separates its complement into two sets P_1 and P_2 with the property that every simple curve with the ends $a \in P_i$ and $b \in P_j$ crosses C even number of times if and only if $i = j$.

Clearly if $e \notin M$ we have

$$cr_{\Gamma_1}(M) = cr_{\Gamma_2}(M). \qquad (2)$$

Let $c = 0$ if both P_1 and P_2 contain an even number of vertices of G and let $c = 1$ otherwise. For two curves C_1 and C_2, let $cr(C_1, C_2)$ denote the total number of times C_1 crosses C_2. For any perfect matching M of G, such that $e \in M$, the following identity holds modulo 2:

$$cr_{\Gamma_1}(M) + cr_{\Gamma_2}(M) = 2 \sum_{\{f,g\} \subseteq M \backslash \{e\}} cr(f,g) + \sum_{f \in M \backslash \{e\}} (cr(f,e_1) + cr(f,e_1))$$

$$= \sum_{f \in M \backslash \{e\}} cr(f,C) = c. (3)$$

Suppose S is a marking of Γ_1. Identities (2) and (3) imply that S is a marking of Γ_2 if $c = 0$, and that $S \triangle \{e\}$ is a marking of Γ_2 if $c = 1$. $\qquad \square$

Since clearly (d) implies (b), to finish the proof of Theorem 2 it remains to show that (b) implies (d). Suppose G satisfies (b) and consider a drawing of G in the plane with a marking S. Suppose there exists $e \in S$. We change the way e is drawn, so that the closed curve C which is composed from the old and the new drawing of e separates one vertex of G from the rest. From the proof of Lemma 2 it follows that $S \setminus \{e\}$ is a marking in the new drawing. By repeating the procedure we produce a drawing of G such that the empty set is a marking, therefore demonstrating that G satisfies condition (d) of Theorem 2.

4   Concluding Remarks

1. The following generalization of Theorem 1 follows from the proof in previous section.

Theorem 3. *Let G be a graph and let $\mathcal{M}$ be the set of all perfect matchings of G. Let $s : \mathcal{M} \to \{-1, 1\}$. Then the following are equivalent:*
(1) there exists an orientation D of G such that for every $M \in \mathcal{M}$ its sign in the corresponding directed graph is equal to $s(M)$;
(2) there exists a drawing of G in the plane such that for every $M \in \mathcal{M}$

$$s(M) = (-1)^{cr(M)}.$$

In [8] I was also able to generalize the methods used in the proof of Theorem 1 to prove a result on the numbers of crossings in "T-joins" in different drawings of a fixed graph.

2. For a labeled graph G, an orientation D of G and a perfect matching M of G, denote the sign of M in the directed graph corresponding to D by $D(M)$. We say that a graph G is k-*Pfaffian* if there exist a labeling of G, orientations $D_1, D_2, \ldots, D_k$ of G and real numbers $\alpha_1, \alpha_2, \ldots, \alpha_k$, such that for every perfect matching M of G

$$\sum_{i=1}^{k} \alpha_i D_i(M) = 1.$$

For surfaces of higher genus the following result was mentioned by Kasteleyn [3] and proved by Galluccio and Loebl [1] and independently by Tesler [14].

Theorem 4. *Every graph that can be embedded on a surface of genus g is 4^g-Pfaffian.*

I was able to prove the following analogue of Theorem 1 for the torus [9].

Theorem 5. *Every 3-Pfaffian graph is Pfaffian. A graph G is 4-Pfaffian if and only if there exists a drawing of G on the torus such that $cr(M)$ is even for every perfect matching M of G.*

Theorems 4 and 5 suggest several questions. For which $k \geq 5$ do there exist graphs that are k-Pfaffian, but not $(k-1)$-Pfaffian? Is it true that a graph G is 4^g-Pfaffian if and only if there exists a drawing of G on a surface of genus g such that $cr(M)$ is even for every perfect matching M of G?

Acknowledgment

I would like to thank my advisor Robin Thomas for his guidance and support in this project. The research was supported in part by NSF under Grant No. DMS-0200595.

References

1. A. Galluccio and M. Loebl, On the theory of Pfaffian orientations. I. Perfect matchings and permanents, Electron. J. combin. 6 (1999), no. 1, Research Paper 6, 18pp. (electronic)
2. P. W. Kasteleyn, The statistics of dimers on a lattice. I. The number of dimer arrangements on a quadratic lattice, Physica 27 (1961), 1209-1225.
3. P. W. Kasteleyn, Dimer statistics and phase transitions, J. Mathematical Phys. 4 (1963), 287-293.
4. P. W. Kasteleyn, Graph Theory and Crystal Physics, Graph Theory and Theoretical Physics, Academic Press, London, 1967, 43-110.
5. V. Klee, R. Lardner and R. Manber, Signsolvability revisited, Linear Algebra Appl. 59 (1984), 131-157.
6. C. H. C. Little, A characterization of convertible (0,1)-matrices, J. Comb. Theory B 18 (1975), 187-208.
7. W. McCuaig, Polya's permanent problem, preprint.
8. S. Norine, Pfaffian graphs, T-joins and crossing numbers, submitted.
9. S. Norine, in preparation.
10. J. Pach and G. Toth, Which crossing number is it anyway?, J. Comb. Theory Ser. B 80 (2000), no. 2, 225–246.
11. G. Pólya, Augfabe 424, Arch. Math. Phys. Ser. 20 (1913), 271.
12. N. Robertson, P. D. Seymour and R. Thomas, Permanents, Pfaffian orientations, and even directed circuits, Ann. of Math. (2) 150(1999), 929-975.
13. P. Samuelson, Foundations of Economic Analysis, Atheneum, New York, 1971.
14. G. Tesler, Matching in graphs on non-orientable surfaces, J. Comb. Theory B 78 (2000), 198-231.
15. V. V. Vazirani and M. Yannakakis, Pfaffian orientations, $0-1$ permanents, and even cycles in directed graphs, Discrete Appl. Math. 25 (1989), 179-190.

3D Visualization of Semantic Metadata Models and Ontologies*

Charalampos Papamanthou[1,2], Ioannis G. Tollis[1,2], and Martin Doerr[2]

[1] Department of Computer Science, University of Crete, Heraklion, Greece
{cpap,tollis}@csd.uoc.gr
[2] Institute of Computer Science, FORTH, Heraklion, Greece
martin@ics.forth.gr

Abstract. We propose an algorithm for the 3D visualization of general ontology models used in many applications, such as semantic web, entity-relationship diagrams and other database models. The visualization places entities in the 3D space. Previous techniques produce drawings that are 2-dimensional, which are often complicated and hard to comprehend. Our technique uses the third dimension almost exclusively for the display of the isa relationships (links) while the property relationships (links) are placed on some layer (plane). Thus the semantic difference between isa links and property links, which should be as vertical or as horizontal as possible respectively, is emphasized. Special reference is made on a certain model, the CIDOC Conceptual Reference Model.

1 Introduction

Semantic graphs (also called models) describe relationships between entities and are very important in many applications [1, 7–9]. They contain a great amount of information and the visualization of such models is very crucial in order to understand the details they incorporate. There have been several attempts to automatically produce drawings of these models. Previous techniques produce drawings that are 2-dimensional, and often complicated and hard to comprehend [5, 6]. We present a technique that produces 3-dimensional drawings that are clear and understandable. In these metadata models there are usually two different types of relationships: the *isa* relationships and the *property* relationships. Our technique uses the third dimension almost exclusively for the display of the isa relationships (links) while the property relationships (links) are placed on some layer (plane). These links have different meaning. In detail, an isa link represents a type of inheritance between the connected entities, whereas the property links correspond to different relationships between the entities. Since it is not always feasible, a few property links may use the third dimension.

Many automated tools for the display of such models have been developed. The tools are naturally categorized into two types: textual and visual [1]. Examples of textual RDFS browsing tools are Protege 2000 [2], Ontoedit [3] and

---

* This work was supported in part by INFOBIOMED code: IST-2002-507585 and the Greek General Secretariat for Research and Technology under Program "ARISTEIA", Code 1308/B1/3.3.1/317/12.04.2002.

Ontomat[1]. However, text-based environments have not proved effective [3, 4]. On the other hand, visualization based tools provide the user with a more general view of the whole model. There are many tools visualizing RDF browsing such as IsaViz [5], FRODO RDFSViz [6], and OntoViz[2].

All these visualization tools produce 2D visualizations of the input model. However, 2D visualizations of these models are complex and confusing, especially when the number of entities of the specific model is large. Furthermore, in two dimensions, it is difficult to understand the meaning of each link, as they are all looked upon under the same perspective. Our approach uses the third dimension for the visualization of the isa links and the input model will be considered as a system of interactions between entities, which push the nodes of the graph towards a specific optimal position. Till now, to the best of our knowledge, no 3D models for the automated visualization of semantic metadata models have been presented.

In this paper, we present an algorithm that produces 3D solutions to the semantic metadata and ontologies visualization problem. The algorithm tries to implement intuitive preference directions (isa links upwards, property links horizontal), that assist the intellectual orientation and understanding of the human spectator. If the number of entities and properties is large, this is virtually impossible to be accomplished in 2D representations. Our algorithm can be applied in order to visualize many ontology models, including the ABC ontology and model described in [7]. We present our results for a conceptual reference model, the CIDOC CRM.

The **CIDOC Conceptual Reference Model (CRM)** provides definitions and a formal structure for describing the implicit and explicit concepts and relationships used in cultural heritage documentation [8, 9]. The CIDOC CRM is a directed graph where nodes represent entities and edges represent property and isa relationships. This graph is isa-disconnected (it contains more than one isa-connected components) and has many interesting characteristics. The complete definition of CIDOC CRM can be found in [9]. An example model of CIDOC CRM is shown in Figure 1.

2   The Problem

Our target is to create a 3D-mapping of a metadata model. The model describing entity relationships is given in various formats, such as XML or RDF.

As the description of the model consists of two kinds of relationships (isa and property links) between entities, we can extract a directed graph $G = (V, E)$ such that V is the set of entities and E is divided into two sets of edges, I, which denotes the set of isa links and P, which denotes the set of property links. Throughout the paper n will denote the number of the original model entities (vertices) whereas m will denote the number of the original model links (edges).

---

[1] http://annotation.semanticweb.org/ontomat
[2] Ontoviz is a visualization plug-in for Protege 2000.

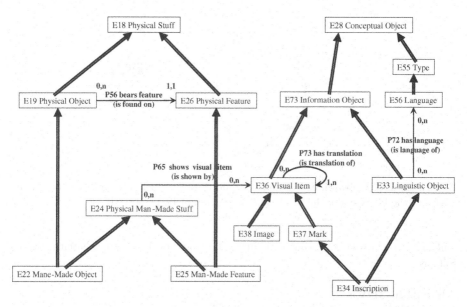

Fig. 1. A CIDOC CRM model (mii) consisting of 15 entities (nodes).

Therefore, we reduce the problem of metadata model 3D visualization to the computation of a 3D embedding of the respective directed graph. Thus, we want to create a 3D visualization of G (i.e., to compute (x, y, z) for all $v \in V$) such that for all edges $(v, u) \in I$ it is $z(v) > z(u)$. The inequality represents a parent-child relationship between two nodes that are connected with isa links. In this way, the importance of isa links is pointed out. Also it is desirable (but clearly not always feasible) to have $z(v) = z(u) + 1$ for each $(v, u) \in I$ (i.e., all the isa-connected nodes lie in adjacent layers). Additionally, we want to minimize the number of edges $(v, u) \in P$ that connect nodes that lie in different layers. This means that our main objective is to place nodes that are connected with property links on the same layer. In this way, an absolute discrimination between the two types of edges is achieved. Finally, minimizing edge crossings per layer is desired. Due to the computation of the (x, y, z) coordinates, crossings between isa links are unlikely to occur, as the probability of independent placed line segments crossing in the 3D space is almost zero. The satisfaction of all three constraints is not always feasible. Thus, in the solution that we will present, some trade-offs have to be made in order to achieve a feasible 3D embedding of the graph. Figure 2 shows an ideal case of a 3D embedding: For all edges $(v, u) \in I$ it is $z(v) = z(u) + 1$ (i.e., all isa links connect nodes lying in adjacent layers). Also, for all edges $(v, u) \in P$ it is $z(v) = z(u)$ (i.e., all property links lie on the same layer). Finally, no crossings between property links per layer and between isa links exist.

The constraints presented are general constraints that have to be taken into consideration in order to produce an aesthetically good and readable drawing. During the description of the algorithm, some more constraints will appear that

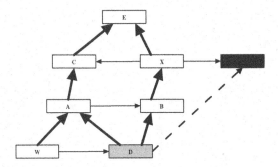

Fig. 2. An ideal embedding. If either a property or an isa link is added between the gray and the black node, the ideal embedding cannot be preserved.

will aim to improve the final drawing. The main body of our algorithm is as follows:

Algorithm 3DVis
Input: A directed graph $G = (V, E = I \cup P)$
Output: A 3D embedding of G

1. **Layer Assignment Stage**
   For each $v \in V$ compute $z(v)$.
2. **Pseudo Nodes and Pseudo Edges Addition Stage**
   Decompose each *isa* link (i, j), such that i, j do not lie in adjacent layers, by introducing *dummy isa* nodes at the layers between i and j.
   Decompose each *property* link (i, j), such that i, j do not lie in the same layer, by introducing one *dummy property* node either at the layer of node i or at the layer of node j.
3. **Coordinate Assignment on Layers Stage**
   Assign (x, y) coordinates to the nodes of the graph (including pseudo nodes).
4. **Return** the computed 3D embedding of $G = (V, I \cup P)$.

In the following sections, we describe the three main stages of the algorithm.

3   Layer Assignment

Layer assignment is equivalent to determining the z-coordinate of each node. The layers are numbered $0, 1, \ldots, L-1$, where L is a positive integer determined by the algorithm. Many layer assignment algorithms have been proposed with the *longest path* layering algorithm being the dominant one [10]. However, the longest path layering, though initially implemented, is not finally used, as it does not take into consideration the property links of each node.

In order to assign layers, both property and isa links should be taken into consideration as we want to place the nodes in a way that minimizes the total number of layers intersected by the edges. Thus the number of pseudo nodes (dummy property and dummy isa nodes) is also minimized. If the graph had only isa links then a technique similar to that proposed in [11] for layered drawing of digraphs could be used (see also [10]).

Let $G = (V, I \cup P)$ be the input graph model to visualize, where V is the set of nodes (entities) of the model, I is the set of isa links and P is the set of the property links. Our first aim is to run an algorithm in order to identify the isa-connected components of the induced undirected graph. This can be easily done, by applying a depth first search that computes the isa connected trees in the depth first search forest.

Suppose now that we have identified the isa DFS trees that correspond to the input model. In the general case, we suppose that we have k such connected subgraphs $G_j = (V_j, E_j)$, $j = 1, \ldots, k$. Let n_j denote the number of nodes of each connected subgraph (i.e., $n_j = |V_j|$). It is clear that $\sum_{j=1}^{k} n_j = n$. We want to compute $z(i)$ for all $i = 1, \ldots, n$. Throughout the paper, we assume that z *increases* by moving downwards. One very important consideration is that the z-coordinates of the nodes must always satisfy the order imposed by the isa hierarchies. The isa hierarchy for all the nodes of the graph is derived from the input model.

Additionally, we would like to impose an upper bound L to the z-values. L could be the number $\max_{i=1,\ldots,k}\{n_i\}$ (i.e., the number of the nodes of the largest DFS tree). Another approach for the definition of L could be the longest path of the directed acyclic graph that is produced by the model if we ignore all the property links. Furthermore, we must define an objective function of each solution. It is desired that a layer assignment is achieved such that the vertical distance between nodes which are connected with property links is minimized. Additionally, we want to ensure that vertical distance between nodes that are connected with isa links is also minimized. Due to the nature of the models we want to visualize, minimizing vertical distance between nodes which are connected with property links is more important than minimizing vertical distance between nodes which are connected with isa links. Thus we minimize the former with priority α and the latter with priority β, where $\alpha > \beta$. Therefore, we define the following integer non-linear optimization problem of the function f:

$$\min f(z) = \alpha \sum_{(i,j) \in P} (z(i) - z(j))^2 + \beta \sum_{(i,j) \in I} (z(i) - z(j))^2$$
$$s.t. \ z(i) > z(j) \ \forall (i,j) \in I; z(i) < L \ \forall i \in V; z(i) \in \mathbb{Z} \ \forall i \in V$$

The first constraint ensures that the isa hierarchy is preserved, the second constraint imposes an upper bound to the values the depths of the nodes can range, while the third one forces an integer solution. Figure 3 shows two legal layer assignments of the same model. The superiority of the second one is clear, since all property links lie on the layers.

The optimization problem, which is at least NP-hard [22], can be solved with integer programming approximation algorithms. However, if we force direction to the property links, it can be solved optimally by using the method proposed in [11]. We have solved the problem using an optimization package, called Lingo (version 8) (see **http://www.lindo.com**).

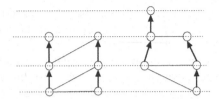

Fig. 3. Two layerings z_1, z_2. The isa hierarchies in both layerings are preserved. Note that $f(z_1) = 2\alpha + 4\beta$, while $f(z_2) = \alpha + 4\beta$. Our layer assignment algorithm will choose the second layout.

4   Pseudo Nodes and Edges Addition

After the layers of the nodes have been computed, we must assign the (x, y) coordinates to each node. Let z_0 be the layout that minimizes f. We have to consider the following two cases:

Case 1 – Property Links

There can exist property links (i, j) such that $z_0(i) \neq z_0(j)$ (i.e., i, j are not placed on the same layer and $f(z_0) > \beta \sum_{(i,j) \in I} (z_0(i) - z_0(j))^2$). Suppose we have p such links. In this case, every one of these p property links will originate from a certain layer and will terminate to another one. Thus we introduce the idea of a *dummy property* node, which will actually appear as a bend in the final visualization. For each pair of nodes i, j that are connected with a property link such that $z_0(i) \neq z_0(j)$, a dummy property node $c(i, j)$ is placed at layer $\min\{z_0(i), z_0(j)\}$. Additionally, it is desired that $c(i, j)$ lies exactly above (in terms of (x, y) coordinates) of the node $v = i$ or $v = j$ such that $z_0(v) = \max\{z_0(i), z_0(j)\}$. A dummy property node $c(i, j)$ is thus defined as a pseudo node of an original node i which is connected to a node j through a property link such that $z_0(i) \neq z_0(j)$. Following this approach, all property links will be drawn as orthogonal lines at the cost of introducing one bend per property link. Additionally, an upward *property tree* is formed (see Figure 4b). In the case that a node j has more than one property destinations which all lie on the same layer, different from $z_0(j)$, then all the created copies are merged into one.

Another approach is to create a *property path* between the respective nodes. In this way, the number of dummy nodes would increase and the semantic difference between the property and isa links would not be so clear. Additionally, this approach is not so useful, as in such metadata models, the number of the property links is much less than the number of isa links.

Case 2 – Isa Links

There can exist isa links (i, j) such that $z_0(i) > z_0(j) + 1$ (i.e., i, j are not placed in adjacent layers and $f(z_0) > \alpha \sum_{(i,j) \in P} (z_0(i) - z_0(j))^2 + \beta|I|)^3$. For each isa link (i, j) such that $z_0(i) - z_0(j) > 1$ we define a set $F^{(i,j)}$ of $z_0(i) - z_0(j) - 1$

---

[3] $|X|$ denotes the cardinality number of set X.

dummy isa nodes that will be placed on layers $z_0(j) + 1, z_0(j) + 2, \ldots, z_0(i) - 1$. Note that $F^{(i,j)} = \emptyset \Leftrightarrow (i,j) \in I \wedge z_0(i) - z_0(j) = 1$.

Another way to describe this is as follows. Suppose we draw a virtual straight line segment between two nodes i and j that are connected with an isa link and lie in different layers. Dummy isa nodes are introduced on each layer pierced by this virtual line segment. These nodes are stored in $F^{(i,j)}$. $F^{(i,j)}$ defines a path of $z_0(i) - z_0(j) - 1$ new isa links $i, i_1, i_2, \ldots, i_r, j$ where $r = z_0(i) - z_0(j) - 1$.

Finally, the total number of added dummy isa nodes is $d = \sum_{(i,j) \in I \wedge F^{(i,j)} \neq \emptyset} |F^{(i,j)}|$. It is clear that $d \leq \sum_{F^{(i,j)} \neq \emptyset} (L - 1)$. In Figure 4, we can see how the addition of the dummy nodes is achieved.

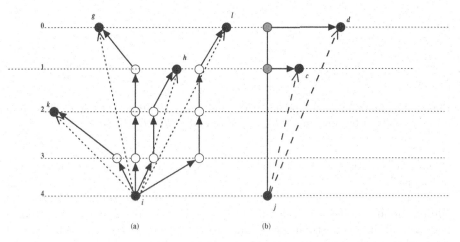

Fig. 4. Dummy nodes and edges addition. In (a), node i has 4 isa links: (i, k), (i, g), (i, h), (i, l). In (b), node j has has 2 property links (j, c) and (j, d).

Therefore, the total number of the inserted dummy nodes is $p + d$. It is easy to prove that no dummy nodes are added if and only if $f(z_0) = \beta |I|$. In this case $p = 0$ and $F^{(i,j)} = \emptyset \; \forall (i,j) \in I$. The existence of dummy nodes is important in the next phase, as they play the role of placeholders and bends in the final drawing.

Finally, one more step has to be performed. As the model behaves as a dynamic system, we must connect each dummy property node $c(i,j)$ of two property connected nodes i, j with a new link between $c(i,j)$ and the original node (i.e., the node copied). This can be done if we introduce a pseudo-isa relationship between each dummy property node and the node from which it has been copied. There will therefore be p pseudo-isa relationships (as many as the dummy property nodes). The final number of isa links will be $|I'| = |I| + p + d$. Thus, after the addition stage is complete, our graph will consist of $|V| + p + d$ nodes, exactly $|P|$ property links and $|I|$ isa links. In the following sections, V, I will denote the *augmented* sets of nodes and isa links respectively. It is important to state that in the final visualization, there would be only $|I'| - p$ isa links, as the pseudo-isa links only contribute to maintain the dynamic aspect of the model.

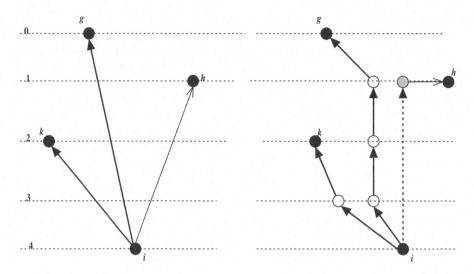

Fig. 5. The function of the pseudo node and edges addition algorithm. The initial model has 3 relationships $ISA(i,k)$, $ISA(i,g)$ and $PROPERTY(i,h)$.

The overall function of the pseudo-node and edges addition algorithm is depicted in Figure 5.

Finally, we could use an alternative approach to this problem. Instead of adding dummy nodes for each isa link (i,j), we can create an *isa tree* from a node i to all its isa destinations, similar to the property tree. The difference of this approach is evident only in the case of a node i having h isa links of type $(i,j_1), (i,j_2), \ldots, (i,j_h)$. This approach reduces the number of the dummy isa nodes. However, the connection between all isa-connected nodes is not very clear. Finally it depends upon the application as to which one of the two alternatives will be chosen. In our case, the first alternative is more satisfying. Figure 6 depicts the dummy isa nodes addition according to the second alternative. Note that the approach of using isa and property trees is trying to mimic the ideal visualization case described in Figure 2.

5   Coordinate Assignment on Layers

In this section, we present the method for the computation of the (x,y) coordinates of each node. Our main objective here is to minimize, with weight α, the 3D distance of all the pairs of nodes that are connected with isa links. In this way, the *isa children* of a node v will be placed symmetrically around v. Furthermore, with lower priority β, we want to minimize the horizontal distances between nodes that are connected with property links. Note, that after the introduction of the dummy property nodes, property links connect nodes lying on the same layers.

This minimization is made under the constraint that nodes do not overlap. Hence, we must ensure that for each layer there is a minimum horizontal distance,

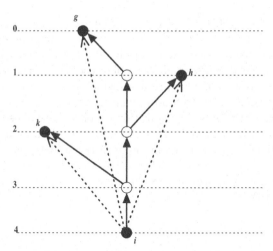

Fig. 6. Forming an ISA tree. Initially, there are 3 isa links: (i,k), (i,g), (i,h).

say 1, between nodes belonging to a certain layer. Additionally, it is desired that for each node i that has a dummy property node j to another layer it is $p(i) = p(j)$ which means i and j have identical x and y coordinates (i.e., it is $x(i) = x(j)$ and $y(i) = y(j)$).

Furthermore, we would like to have the dummy nodes of each isa link (i,j) placed on a straight line. In this way the hierarchy is clearer. For this purpose, we examine two cases: **(a)**, If node i has only **one** isa neighbor j such that $z_0(i) > z_0(j) + 1$, then we want to ensure that

$$p(i) = p(i_1) = p(i_2) = \ldots = p(i_r) \tag{1}$$

where $i_1, i_2, \ldots, i_r$ are the nodes introduced in section 4 and $r = z_0(i) - z_0(j) - 1$. **(b)**, If there are **more** than one isa links originating from i, say h, and we want to ensure that for each isa neighbor j_k, $k = 1, \ldots, h$, the dummy isa nodes that belong in $F^{(i,j_k)}$ are related as follows

$$p(i_1) = p(i_2) = \ldots = p(i_r) \tag{2}$$

where now $r = z_0(i) - z_0(j_k) - 1$, $k = 1, \ldots, h$. Note that only in the first case the origin of the isa link is placed on the same vertical straight line with the dummy isa nodes. This is not feasible in the second case, as we want to separate the multiple isa links originating from i. In both cases, the destination of the isa link is not forced to lie in a specific position. A desired layout for the second case can be seen in Figure 4. This is a technique inspired by polyline drawings [21].

All these observations finally lead to the following non-linear optimization problem:

$$\min g(x,y) = \alpha \sum_{(i,j) \in I} d_{ij} + \beta \sum_{(i,j) \in P} d_{ij}$$

$$s.t. \quad (x(i) - x(j))^2 + (y(i) - y(j))^2 \geq 1 \ \forall (i,j) : z_0(i) = z_0(j) \tag{3}$$

$$p(t) = p(c(i,j)) \ \forall c(i,j), \ t = \arg\max\{z_0(i), z_0(j)\}; C_1; C_2 \tag{4}$$

where $d_{ij} = \sqrt{(x(i) - x(j))^2 + (y(i) - y(j))^2 + (z_0(i) - z_0(j))^2}$ is the 3D Euclidean distance of nodes i, j, C_1, C_2 are the constraints imposed by (1),(2).

Constraint (4) ensures minimum distance per layer, whereas the other constraints ensure that the added nodes are placed on straight lines according to the rules we have presented. This problem, which is also at least NP-hard [22], has also been solved with Lingo 8, in polynomial time on the size of input [12].

Instead of solving the above optimization problem, we can use a force directed algorithm. This can be achieved by using some extra dummy nodes, which will be regarded as fixed. There are many force directed algorithms proposed such as [13–16]. Finally, we can examine if the hierarchical approach [17] or maybe an orthogonal drawing approach [18–20] could be implemented. These are all goals of future work.

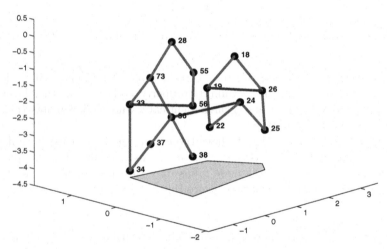

Fig. 7. 3D Representation of the Mark Inscription Information (mii) Model (Figure 1).

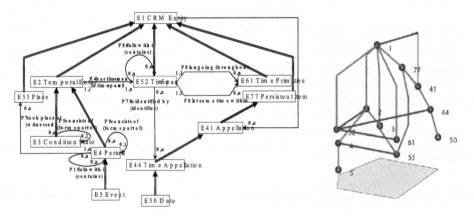

Fig. 8. 3D Representation of the Time Span Information (tsi) Model.

6   Time and Experiments

We have implemented our algorithm and have run it on several examples. The time needed by our algorithm is divided into three main parts: **(1)**, The time needed for minimizing f. Lingo uses efficient approximation algorithms and performs the layer assignment in polynomial time. **(2)**, The time needed for determining dummy nodes. Let L^* be the maximum number of layers needed after the layer assignment stage is complete. This time is obviously $O(mL^*)$, as we need to check all links (isa and property links) in order to decide if the addition of dummy nodes is necessary or not. **(3)**, The time needed for the minimization of g. This part is the most expensive one, as there are many constraints and the size of the problem has increased (due to the introduction of dummy nodes and edges). However, Lingo provides an approximation in reasonable time.

Here, we present some visualization examples produced by our software with reference to two CIDOC CRM models. Isa and property links are depicted with red and blue lines respectively. The time our algorithm needed in order to produce the 3D drawings is 4 seconds in average.

References

1. Alexandru Telea, Flavius Frasincar, Geert-Jan Houben, "Visualizing of RDF(S) based Information" , Proc. IEEE IV '03, IEEE CS Press, 2003
2. N. F. Noy, M. Sintek, S. Decker, M. Crubezy, R. W. Fergerson, and M. A. Musen, "Creating semantic web contents with protege-2000". IEEE Intelligent Systems, 16(2):60–71, 2001
3. Y. Sure, S. Staab, J. Angele. "OntoEdit: Guiding Ontology Development by Methodology and Inferencing" In: Proceedings of the International Conference on Ontologies, Databases and Applications of SEmantics ODBASE 2002
4. S. Card, J. Mackinlay, and B. Shneiderman, "Readings in Information Visualization". M. Kaufmann, 1999
5. E. Pietriga, "Isaviz: a visual environment for browsing and authoring rdf models", The Eleventh International World Wide Web Conference (Developer's day), 2002.
6. Frodo RDFS Visualization Tool, http://www.dfki.uni-kl.de/frodo/RDFSViz/.
7. Lagoze C, Hunter J., "The ABC Ontology and Model" , Proceedings of the International Conference on Dublin Core and Metadata Applications 2001, p. 160–176.
8. Nick Crofts, Martin Doerr, Tony Gill, Stephen Stead, Matthew Stiff (editors), "Definition of the CIDOC Conceptual Reference Model", March 2004 (3.4.10).
9. Doerr M., "The CIDOC CRM – An Ontological Approach to Semantic Interoperability of Metadata", AI Magazine, 4, issue 1 (2003).
10. G. Battista, P. Eades, R. Tamassia, I. Tollis, "Graph Drawing – Algorithms for the Visualization of Graphs" , (1999).
11. E.R. Gausner, E. Koutsofios, S.C. North and K.P. Vo, "A Technique for Drawing Directed Graphs", IEEE Trans. Softw. Eng., 19, 214–230, 1993.
12. Mihir Bellare and Phillip Rogaway, "The complexity of approximating a nonlinear program", Journal of Mathematical Programming B, Vol. 69, No. 3, pp. 429–441
13. P. Eades and X. Lin, "Spring Algorithms and Symmetry", In Proceedings of COCOON 1997, vol 1276 of Lecture Notes in Computer Science, pp. 109–112. Springer.

14. T. Fruchterman and E. Reingold, "Graph Drawing by Force Directed Platcement", Softw.-Pract. Exp., 21, no. 11, 1129–1164, 1991.
15. T. Kamada and S. Kawai, "An Algorithm for Drawing General Undirected Graphs", Inform. Process. Lett., 31, 7–15, 1989.
16. T. Kamada, "Visualizing Abstract Objects and Relations", World Scientific Series in Computer Science, 1989.
17. K. Sugiyama, S. Tagawa and M. Toda, "Methods for visual understanding of hierarchical systems", IEEE Trans. Syst. Man Cybern., SMC-11, no. 2, 109–125, 1981.
18. Achilleas Papakostas, Ioannis G. Tollis, "Algorithms for Incremental Orthogonal Graph Drawing in Three Dimensions", J. Graph Algorithms Appl. 3(4):81–115
19. Achilleas Papakostas, Ioannis G. Tollis, "Algorithms for area-efficient orthogonal drawings", Comput. Geom. 9(1-2):83–110 (1998)
20. Achilleas Papakostas, Ioannis G. Tollis, "Interactive Orthogonal Graph Drawing", IEEE Trans. Computers 47(11):1297–1309 (1998)
21. G. Di Battista, R. Tamassia, and I. G. Tollis, "Area requirement and symmetry display of planar upward drawings", Discrete Comput. Geom., 7:381–401, 1992
22. Michael R. Garey, David S. Johnson, "Computers and Intractability: A Guide to the Theory of NP-Completeness", W H Freeman & Co., 1979

A Note on the Self-similarity
of Some Orthogonal Drawings[*]

Maurizio Patrignani

Università di Roma Tre, Italy
patrigna@dia.uniroma3.it

Abstract. Large graphs are difficult to browse and to visually explore. This note adds up evidence that some graph drawing techniques, which produce readable layouts when applied to medium-size graphs, yield self-similar patterns when launched on huge graphs. To prove this, we consider the problem of assessing the self-similarity of graph drawings, and measure the box-counting dimension of the output of three algorithms, each using a different approach for producing orthogonal grid drawings with a reduced number of bends.

1 Introduction

The picture of Fig. 1 shows a huge planar connected random graph drawn within the orthogonal grid layout standard, where nodes are points with integer coordinates and edges are sequences of horizontal and vertical segments of integer length. The drawing was obtained by using the well-known *topology-shape-metrics* approach [10, 3]. It can be noticed that the drawing can be roughly split into rectangular blocks, some very dense, and some others much more sparse. Also, the same pattern seems to occur at at different zooming levels inside the denser rectangular blocks.

The property of showing the same structure at different scales is referred to as self-similarity and is often associated with the concepts of fractional dimensionality and fractal. These concepts will be more formally defined in Section 2. In the rest of this section we will use the intuitive notion of self-similarity to introduce the main point of this note and the implications of it.

Self-similar structures, characterized by the property of showing similar patterns at all length scales, abound in nature. Lots of examples can be found in disciplines as different as geology (river basins, coastlines, mountains landscapes, etc.), human and animal physiology (blood vessels, nerves, bronchial tubes, etc.), and biology (trail networks for harvesting ants, etc.)

---

[*] Work partially supported by European Commission – Fet Open project COSIN – COevolution and Self-organisation In dynamical Networks – IST-2001-33555, by European Commission – Fet Open project DELIS – Dynamically Evolving Large Scale Information Systems – Contract no 001907, by "Progetto ALINWEB: Algoritmica per Internet e per il Web", MIUR Programmi di Ricerca Scientifica di Rilevante Interesse Nazionale, and by "The Multichannel Adaptive Information Systems (MAIS) Project", MIUR Fondo per gli Investimenti della Ricerca di Base.

J. Pach (Ed.): GD 2004, LNCS 3383, pp. 389–394, 2004.
© Springer-Verlag Berlin Heidelberg 2004

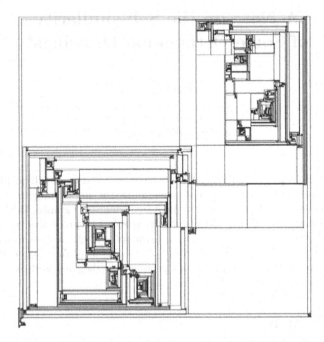

Fig. 1. A drawing with the minimum number of bends of a planar connected graph with 4,700 nodes and 6,155 edges. The number of bends is 2,463.

Although fractals are usually put in relation with chaos, they are sometimes credited with some desirable property caused by their self-similar pattern. However, it is very hard to attribute a fractal structure to a specific optimization problem and such an explicit statement can be searched in vain in the literature. Rigorously defined combinatorial processes, like graph drawing techniques, provide promising examples to study the relationship between fractals and optimization criteria.

This note offers an argument in favor of the hypothesis that self-similarity may be induced by a minimization process. In fact, the line followed by our argument goes the other way around: we start from a well known aesthetic criterion, namely the reduction of the number of bends in orthogonal drawings, and we assess the self-similarity of the produced layouts, a property elusive enough to remain unnoticed for several years.

2   Background

Central to the definition of fractal is that of dimension [4]. The dimension which we generally are familiar with is the so-called topological dimension. The *topological dimension* D_T of a set is always an integer and is zero if the set is totally disconnected, while it is n if the intersection of the set with the boundary of an arbitrarily small neighborhood of one of its points has topological dimension

$n - 1$. Thus, the topological dimension of a collection of points is zero, that of a collection of curves is one, that of a collection of surfaces is two, and so on.

While the topological dimension is based on the concept of intersection, an alternative and more tortuous definition of dimension is based on the concept of cover, and is obtained from the definition of Hausdorff measure. The Hausdorff measure $\mathcal{H}^p(S)$ of a set S is defined parametrically with p, where p is a non negative real. Thus, there are as many Hausdorff measures as many values for p (we will provide a rigorous definition of Hausdorff measure in the next paragraphs.) If there is a value D_H in $[0, \infty]$ such that $\mathcal{H}^p(S) = \infty$ for p in $[0, D_H)$ and $\mathcal{H}^p(S) = 0$ for p in $(D_H, \infty]$, then D_H is the *Hausdorff dimension* of S.

In order to define the Hausdorff measure $\mathcal{H}^p(S)$, used to find the Hausdorff dimension, we need some further definitions. The *diameter* $|U|$ of a set U is the greater distance apart of any pair of its points. A δ-*cover* of a set S is a countable collection of sets $\{U_i\}$ of diameter at most δ whose union contains S. $\mathcal{H}^p_\delta(S)$ is the infimum of the sum of the p-powers of the diameters of a δ-cover of S, i.e., $\mathcal{H}^p_\delta(S) = inf(\sum_{i=0}^\infty |U_i|^p : \{U_i\}$ δ-cover of S). Finally, $\mathcal{H}^p(S) = \lim_{\delta \to 0} \mathcal{H}^p_\delta(S)$.

Fractals are defined by Mandelbrot [6] as geometric sets whose Hausdorff dimension D_H strictly exceeds their topological dimension D_T.

As it could be difficult to directly compute the dimension D_H of a set S, often a variant of the above definition is considered imposing the additional constraint that all the sets in $\{U_i\}$ have equal diameter. With this hypothesis, given an arbitrary value p different from D_H, by studying how $\mathcal{H}^p_\delta(S)$ goes to ∞ (when p is in $[0, D_H)$) or to 0 (when p is in $(D_H, \infty]$) for $\delta \to 0$, we can infer D_H.

In fact, $\mathcal{H}^p_\delta(S) = inf(\sum_{i=0}^\infty |U_i|^p) = inf(\delta^{p-D_H} \sum_{i=0}^\infty |U_i|^{D_H}) = \delta^{p-D_H} \mathcal{H}^{D_H}_\delta$. Assume that the diameter of S is finite and let N be the cardinality of the δ-cover that gives the infimum of $\sum_{i=0}^\infty |U_i|^p$. We have that $\mathcal{H}^p_\delta(S) = N\delta^p = \mathcal{H}^{D_H}_\delta \delta^{p-D_H}$. Thus, for $\delta \to \infty$, $N = C\delta^{-D_H}$, where $C = \mathcal{H}^{D_H}(S)$. It follows that measuring the exponent of the power law that relates the number N of covering sets to their diameter δ gives indirectly the value of the dimension we are searching for. Based on these considerations, a simple method, called *box-counting*, can be devised to compute a dimension D_B, which is usually near to D_H. It consists of partitioning the plane into equal-sized squares of side δ, and of plotting on a log-log scale the number N of squares intersecting S with respect to δ. The slope of this curve gives D_B.

3 Measuring the Self-similarity of Orthogonal Drawings

In order to compute the box-counting dimension of a graph drawing, we use the tool by Leejay Wu and by Christos Faloutsos [11], which in addition to being well known and widely used, is fast and rigorously designed. Since the input of such a tool is a geometric object consisting of a collection of points, we replaced each node with a point and each edge with a dotted line (see Figs 2.a and 2.b).

Fig. 2.c shows the box-counting log-log plot, produced by the package [11]. Between points B and C the -1.64 slope of the curve gives the fractal dimension induced by the self-similarity of the "holes" in the drawing. Note that before point A, as each non empty box covers a single point of the drawing, the dimen-

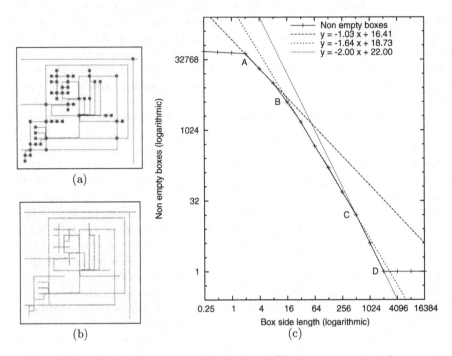

Fig. 2. (a) A portion of the drawing of a graph and (b) the same portion where segments have been replaced with dots (five dots for each grid segment). (c) The box counting log-log plot for a drawing of a graph with 485 vertices.

sion appears to be zero. Between points A and B, since a non empty box rarely intersects more than a single segment, the line with slope -1.03 is measuring the dimension of a collection of segments. Between points C and D, the line with slope -2.00 is measuring the dimension of a continuous surface, since boxes are largest than the largest "holes" of the drawing. Finally, after point D, the whole drawing fits into a single box, and the slope of the line is zero again.

Using Pigale generator [2], we created three test suites of planar connected, biconnected and triconnected graphs ranging from 500 to 3,000 edges, increasing each time by 500 edges, 10 graphs for each type. After the generation multiple edges and self-loops were removed. Fig. 3.a shows the number of edges before and after simplification. For all the three test suites the number of nodes is roughly half of the number of edges before simplification.

Orthogonal layouts were produced by applying three different algorithms corresponding to the three approaches described hereunder.

OFV: The *orthogonal-from-visibility* approach consists of constructing a planar orthogonal grid drawing of a biconnected planar graph starting from a visibility representation of the same graph [3]. Our implementation handles graphs of degree greater than four inserting at most two bends per edge.

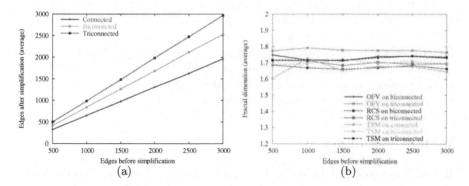

Fig. 3. (a) Average number of edges before and after simplification for the three test suites. (b) The D_B fractal dimensions for orthogonal drawings computed with various approaches.

RCS: The *relative-coordinates scenario* is based on the incremental construction of the drawings. We applied the 'simple algorithm' described in [7] for drawing high degree biconnected graphs. Intersections may be introduced even for planar graphs. Each edge is guaranteed to have a single bend.

TSM: For the *topology-shape-metrics* approach we used the Boyer and Myrvold algorithm [1] to compute a planar embedding. During the subsequent orthogonalization step, the total number of bends was minimized by using the algorithm in [10], modified as described in [5] for handling high degree graphs. Finally, the compaction step was performed with the heuristic based on the rectangularization of the faces [10].

As it can be seen from Fig. 3.b, the D_B fractal dimension computed for the various kinds of graphs is consistently above the D_T of the point sets, which is one. Hence, the drawings are actually fractals, and their fractal dimension is about 1.7 for all types of graphs and for all the three approaches. In all cases the correlation computed by package [11] for the line approximating the curve on the log-log graph is less than -0.999, suggesting a high reliability for the computed D_B values.

4   Conclusions and Open Problems

Although it is well known that random graphs do not exhibit self-similar properties, we wanted to make sure that the input graphs were not self-similar themselves. The self-similarity of graphs and networks has been widely studied. Several real-life networks, for example, are known to show a self-similar node-degree distribution [9, 12]. We plotted the node-degree distribution for the three test-suites and verified on a log-log scale that the curves are not power laws.

Also, we wanted to check whether the planarization step, exploited by both the **OFV** and the **TSM** approaches, folded the faces one into the other, introducing self-similarity. Again, we found that the node-degree distribution of the dual graphs does not follow a power law.

Finally, we tried alternative compaction heuristics for the **TSM** approach, without noticing significant changes of the D_B fractal dimension computed.

Some other issues deserve a deeper investigation: Is there a way to avoid self-similar patterns in orthogonal grid drawings by introducing a few extra bends? Can alternative measures of fractal dimension, like the *correlation dimension* [8], help deepening our understanding of this phenomenon? Do other graph drawing standards also produce self-similar drawings of large graphs?

Acknowledgments

We would like to thank Paolo Branchesi, Giuseppe Di Battista, Maurizio Pizzonia, and Roberto Tamassia for encouragement and interesting discussions.

References

1. J. Boyer and W. Myrvold. Stop minding your P's and Q's: A simplified O(n) planar embedding algorithm. In *10th Annual ACM-SIAM Symposium on Discrete Algorithms*, pages 140–146, 1999.
2. H. de Fraysseix and P. Ossona de Mendez. P.I.G.A.L.E – Public Implementation of a Graph Algorithm Library and Editor. SourceForge project page `http://sourceforge.net/projects/pigale`
3. G. Di Battista, P. Eades, R. Tamassia, and I. G. Tollis. *Graph Drawing*. Prentice Hall, Upper Saddle River, NJ, 1999.
4. K. Falconer. *Fractal Geometry: Mathematical Foundations and Applications*. John Wiley & Sons Ltd, second edition, 2003.
5. U. Fößmeier and M. Kaufmann. Drawing high degree graphs with low bend numbers. In F. J. Brandenburg, editor, *Graph Drawing (Proc. GD '95)*, volume 1027 of *Lecture Notes Comput. Sci.*, pages 254–266. Springer-Verlag, 1996.
6. B. B. Mandelbrot. *The Fractal Geometry of Nature*. W. H. Freeman and Company, New York, 1977.
7. A. Papakostas and I. Tollis. Efficient orthogonal drawings of high degree graphs. *Algorithmica*, 26:100–125, 2000.
8. M. Schroeder. *Fractals, chaos, power laws: minutes from an infinite paradise*. W. H. Freeman and Company, New York, 1991.
9. G. Siganos, M. Faloutsos, P. Faloutsos, and C. Faloutsos. Power laws and the as-level internet topology. In *IEEE/ACM Transactions on Networking (TON)*, volume 11, pages 514–524. ACM Press, 2003.
10. R. Tamassia. On embedding a graph in the grid with the minimum number of bends. *SIAM J. Comput.*, 16(3):421–444, 1987.
11. L. Wu and C. Faloutsos. FracDim, jan 2001. Perl package available at `http://www.andrew.cmu.edu/~lw2j/downloads.html`
12. S.-H. Yook, H. Jeong, and A.-L. Barabasi. Modeling the Internet's large scale topology. In *Proc. PNAS'02*, 2002.

No-Three-in-Line-in-3D[*]

Attila Pór[1] and David R. Wood[1,2]

[1] Department of Applied Mathematics, Charles University, Prague, Czech Republic
por@kam.mff.cuni.cz
[2] School of Computer Science, Carleton University, Ottawa, Canada
davidw@scs.carleton.ca

Abstract. The *no-three-in-line* problem, introduced by Dudeney in 1917, asks for the maximum number of points in the $n \times n$ grid with no three points collinear. In 1951, Erdös proved that the answer is $\Theta(n)$. We consider the analogous three-dimensional problem, and prove that the maximum number of points in the $n \times n \times n$ grid with no three collinear is $\Theta(n^2)$. This result is generalised by the notion of a *3D drawing* of a graph. Here each vertex is represented by a distinct gridpoint in $\mathbb{Z}^3$, such that the line-segment representing each edge does not intersect any vertex, except for its own endpoints. Note that edges may cross. A 3D drawing of a complete graph K_n is nothing more than a set of n gridpoints with no three collinear. A slight generalisation of our first result is that the minimum volume for a 3D drawing of K_n is $\Theta(n^{3/2})$. This compares favourably to $\Theta(n^3)$ when edges are not allowed to cross. Generalising the construction for K_n, we prove that every k-colourable graph on n vertices has a 3D drawing with $\mathcal{O}(n\sqrt{k})$ volume. For the k-partite Turán graph, we prove a lower bound of $\Omega((kn)^{3/4})$.

1 Introduction

In 1917, Dudeney [10] asked what is the maximum number of points in the $n \times n$ grid with no three points collinear? This question, dubbed the *no-three-in-line* problem, has since been widely studied [1, 2, 7, 14, 16–19, 21]. A breakthrough came in 1951, when Erdős [14] proved that for any prime p, the set $\{(x, x^2 \bmod p) : 0 \leq x \leq p-1\}$ contains no three collinear points. If follows that the $n \times n$ grid contains $n/2$ points with no three collinear, and for all $\epsilon > 0$ and $n > n(\epsilon)$, there are $(1 - \epsilon)n$ points with no three collinear. The result has been improved to $(3/2 - \epsilon)n$ by Hall *et al.* [18] using a different construction. These bounds are optimal if we ignore constant factors, since each gridline contains at most two points, and thus the number of points is at most $2n$. Guy and Kelly [17] conjectured that the maximum number of points in the $n \times n$ grid with no three collinear tends to $(2\pi^2/3)^{\frac{1}{3}}n$ as $n \to \infty$.

In this paper we study the *no-three-in-line-in-3D* problem: what is the maximum number of points in the $n \times n \times n$ grid with no three points collinear? The following is our primary result.

---

[*] Research supported by NSERC and COMBSTRU.

J. Pach (Ed.): GD 2004, LNCS 3383, pp. 395–402, 2004.
© Springer-Verlag Berlin Heidelberg 2004

Theorem 1. *The maximum number of points in the $n \times n \times n$ grid with no three collinear is $\Theta(n^2)$.*

Cohen *et al.* [6] generalised the no-three-in-line problem in a similar direction. They proved that for any prime p, the set $\{(x, x^2 \bmod p, x^3 \bmod p) : 0 \leq x \leq p-1\}$ contains no four coplanar points. It follows that the $n \times n \times n$ grid contains at least $n/2$ and $(1 - \epsilon)n$ points with no four coplanar. Each gridplane contains at most three points; thus we have an upper bound of $3n$.

Cohen *et al.* [6] were motivated by three-dimensional graph visualisation. Let G be an (undirected, finite, simple) graph with vertex set $V(G)$ and edge set $E(G)$. A *3D drawing* of G represents each vertex by a distinct point in $\mathbb{Z}^3$ (a *gridpoint*), such that with each edge represented by the line-segment between its endpoints, the only vertices that an edge intersects are its own endpoints. That is, an edge does not 'pass through' a vertex. The *bounding box* of a 3D drawing is the minimum axis-aligned box containing the drawing. If the bounding box has side lengths $X - 1$, $Y - 1$ and $Z - 1$, then we speak of an $X \times Y \times Z$ drawing with *volume* $X \cdot Y \cdot Z$. That is, the volume of a 3D drawing is the number of gridpoints in the bounding box. This definition is formulated so that 2D drawings have positive volume.

Distinct edges in a 3D drawing *cross* if they intersect at a point other than a common endpoint. Based on the observation that the endpoints of a pair of crossing edges are coplanar, Cohen *et al.* [6] proved that the minimum volume for a crossing-free 3D drawing of K_n is $\Theta(n^3)$. The lower bound here is based on the observation that no axis-perpendicular gridplane can contain five vertices, as otherwise there is a planar K_5. Note that it is possible for four vertices to be in a single gridplane, provided that they are not in convex position. Subsequent to the work of Cohen *et al.* [6], crossing-free 3D drawings have been widely studied [4–6, 8, 9, 11, 12, 15, 20, 23]. This paper initiates the study of volume bounds for 3D drawings of graphs, in which crossings are allowed. The following simple observation is immediate.

Observation 1. *A set V of n gridpoints in $\mathbb{Z}^3$ determines a 3D drawing of K_n if and only if no three points in V are collinear.* □

Thus, the following result is a slight strengthening of Theorem 1.

Theorem 2. *The minimum volume for a 3D drawing of K_n is $\Theta(n^{3/2})$.*

A *k-colouring* of a graph G is an assignment of one of k colours to each vertex, so that adjacent vertices receive distinct colours. We say G is *k-colourable*. The *chromatic number* $\chi(G)$ is the minimum k such that G is k-colourable. The Turán graph $T(n, k)$ is the n-vertex complete k-partite graph with $\lceil n/k \rceil$ or $\lfloor n/k \rfloor$ vertices in each colour class. Theorem 2 generalises as follows.

Theorem 3. *Every k-colourable graph on n vertices has a 3D drawing with $\mathcal{O}(n\sqrt{k})$ volume. Moreover, every 3D drawing of the Turán graph $T(n, k)$ has $\Omega((kn)^{3/4})$ volume.*

Note that 2D drawings of k-colourable graphs were studied by Wood [25], who proved an $\mathcal{O}(kn)$ area bound, which is best possible for the Turán graph.

The remainder of this paper is organised as follows. In Section 2 we prove the lower bounds in Theorems 1 and 2, which imply the upper bound in Theorem 1. In Section 3 we prove the upper bounds in Theorems 1 and 2, which imply the lower bound in Theorem 1.

2   Lower Bounds

An axis-parallel line through a gridpoint is called a *gridline*. A gridline that is parallel to the X-axis (respectively, Y-axis and Z-axis) is called an *X-line* (*Y-line* and *Z-line*). An axis-perpendicular plane through a gridpoint is called a *gridplane*.

Lemma 1. *There are at most $2n^2$ points in the $n \times n \times n$ grid with no three collinear.*

Proof. Every X-line contains at most two points, and there are n^2 X-lines.   □

The idea in Lemma 1 can be generalised to give a universal lower bound on the volume of a 3D drawing of a graph.

Lemma 2. *Every 3D drawing of a graph G has at least $\chi(G)^{3/2}/\sqrt{8}$ volume.*

Proof. Say G has an $A \times B \times C$ drawing. The vertices on a single Z-line induce a set of paths, as otherwise an edge passes through a vertex. The set of paths is 2-colourable. Using a distinct pair of colours for each Z-line, we obtain a $2AB$-colouring of G. Thus $\chi(G) \leq 2AB$. Similarly, $\chi(G) \leq 2AC$ and $\chi(G) \leq 2BC$. Thus $8(ABC)^2 \geq \chi(G)^3$, and the volume $ABC \geq \sqrt{\chi(G)^3/8}$.   □

The bound in Lemma 2 is only of interest if $\chi(G) \geq 2n^{2/3}$, since n is a trivial lower bound on the volume of a 3D drawing.

The following lemma proves the lower bound in Theorem 3.

Lemma 3. *For all $n \equiv 0 \pmod{k}$, every 3D drawing of $T(n, k)$ has at least $(kn)^{3/4}/\sqrt{8}$ volume.*

Proof. Consider an $A \times B \times C$ drawing of $T(n, k)$. Let a_i (respectively, b_i and c_i) be the number of X-lines (Y-lines and Z-lines) that contain a vertex in the i-th colour class. Considering the arithmetic and harmonic means of $\{a_i : 1 \leq i \leq k\}$ we have,

$$k^2 \leq \left(\sum_i a_i \right) \left(\sum_i \frac{1}{a_i} \right) .$$

The X- and Y-lines that contain a vertex coloured i intersect in at most $a_i b_i$ gridpoints. There are n/k vertices coloured i. Thus $a_i b_i \geq n/k$, implying $1/a_i \leq kb_i/n$.

Hence,

$$k^2 \leq \left(\sum_i a_i \right) \left(\sum_i \frac{kb_i}{n} \right) .$$

That is,

$$kn \leq \left(\sum_i a_i \right) \left(\sum_i b_i \right) .$$

There are at most two distinct colours represented in each gridline, as otherwise an edge passes through a vertex. There are BC distinct X-lines. Thus $\sum_i a_i \leq 2BC$. Similarly, $\sum_i b_i \leq 2AC$. Thus $kn \leq (2BC)(2AC)$. That is, $ABC^2 \geq kn/4$. By symmetry, $ACB^2 \geq kn/4$ and $BCA^2 \geq kn/4$. Thus $(ABC)^4 \geq (kn/4)^3$, implying that the volume $ABC \geq (kn/4)^{3/4}$.   □

Since $\chi(K_n) = n$ and $K_n = T(n, n)$, Lemmata 2 and 3 both prove the lower bound in Theorem 2.

Corollary 1. *Every 3D drawing of K_n has volume at least $n^{3/2}/\sqrt{8}$.*   □

3   Upper Bounds

The next lemma is the main component in the proof of our upper bounds. For all primes p, define

$$V_p = \left\{ (x, y, (x^2 + y^2) \bmod p) \, : \, 0 \leq x, y \leq p - 1 \right\} .$$

Lemma 4. *For all primes p, the set V_p contains three collinear points if and only if $p \equiv 1 \pmod 4$.*

Proof. The result is trivial for $p = 2$. Now assume that p is odd. Suppose V_p contains three collinear points a, b, and c. Then there exists a vector $\boldsymbol{v} = (v_x, v_y, v_z)$ such that $b = k\boldsymbol{v} + a$ and $c = \ell\boldsymbol{v} + a$, for distinct nonzero integers k and ℓ. (Precisely, $v_x = \gcd(b_x - a_x, c_x - a_x)$, $v_y = \gcd(b_y - a_y, c_y - a_y)$, and $v_z = \gcd(b_z - a_z, c_z - a_z)$.) Since $b \in V_p$,

$$(kv_x + a_x)^2 + (kv_y + a_y)^2 \equiv kv_z + a_z \pmod p .$$

That is,

$$k^2(v_x^2 + v_y^2) + a_x^2 + a_y^2 \equiv kv_z + a_z - 2k(v_x a_x + v_y a_y) \pmod p .$$

Since $a \in V_p$, we have $a_x^2 + a_y^2 \equiv a_z \pmod p$. Since p is a prime and $k \neq 0$,

$$k(v_x^2 + v_y^2) \equiv v_z - 2(v_x a_x + v_y a_y) \pmod p .$$

By the same argument applied to c,

$$\ell(v_x^2 + v_y^2) \equiv v_z - 2(v_x a_x + v_y a_y) \pmod p .$$

Thus,

$$k(v_x^2 + v_y^2) \equiv \ell(v_x^2 + v_y^2) \pmod p .$$

That is,
$$(k - \ell)(v_x^2 + v_y^2) \equiv 0 \pmod{p} \ .$$

Since $k \neq \ell$ and p is a prime,
$$v_x^2 + v_y^2 \equiv 0 \pmod{p} \ .$$

Now v_x and v_y are both not zero, as otherwise a, b and c would be in a single Z-line. Without loss of generality, $v_x \neq 0$. Thus v_x has a multiplicative inverse modulo p, and
$$(v_y v_x^{-1})^2 \equiv -1 \pmod{p} \ .$$

That is, -1 is a quadratic residue. A classical result found in any number theory textbook states that -1 is a quadratic residue modulo an odd prime p if and only if $p \equiv 1 \pmod 4$.

Now we prove the converse. Suppose that $p \equiv 1 \pmod 4$. By the above-mentioned result there is an integer t such that $1 + t^2 \equiv 0 \pmod{p}$. We can assume that $0 \leq t \leq (p-1)/2$ as otherwise $p - t$ would do. Thus $(1, t, 0) \in V_p$ and $(2, 2t, 0) \in V_p$, and the three points $\{(0, 0, 0), (1, t, 0), (2, 2t, 0)\}$ are collinear. $\qquad\square$

To apply Lemma 4 we need primes $p \not\equiv 1 \pmod 4$.

Lemma 5 ([3, 13]).

(a) *For all $t \in \mathbb{N}$, there is a prime $p \not\equiv 1 \pmod 4$ with $t \leq p \leq 2t$.*
(b) *For all $\epsilon > 0$ and $t > t(\epsilon)$, there is a prime $p \equiv 3 \pmod 4$ with $t \leq p \leq (1 + \epsilon)t$.*

Proof. Part (a) is a strengthening of Bertrand's Postulate due to Erdős [13]. Baker *et al.* [3] proved that for all sufficiently large t, the interval $[t, t + t^{0.525}]$ contains a prime. The proof can be modified to give primes $\equiv 3 \pmod 4$ in the same interval [Glyn Harman, personal communication, 2004]. Clearly this implies (b). $\qquad\square$

We can now prove the upper bound in Theorem 2.

Lemma 6. *Every complete graph K_n has a 3D drawing with $(2 + o(1))n^{3/2}$ volume, and for all $\epsilon > 0$ and $n > n(\epsilon)$, K_n has a 3D drawing with $(1 + \epsilon)n^{3/2}$ volume.*

Proof. By Lemma 5 with $t = \lceil \sqrt{n} \rceil$, there is a prime $p \not\equiv 1 \pmod 4$ with $\lceil \sqrt{n} \rceil \leq p \leq 2\lceil \sqrt{n} \rceil$ and $p \leq (1 + \epsilon)\lceil \sqrt{n} \rceil$. By Observation 1 and Lemma 4, the set V_p defines a $p \times p \times p$ drawing of K_{p^2}. By choosing the appropriate vertices, we obtain a $\lceil n/p \rceil \times p \times p$ drawing of K_n. The volume is $(2 + o(1))n^{3/2}$ and $(1 + \epsilon)n^{3/2}$. $\qquad\square$

The same proof gives the lower bound in Theorem 1.

Lemma 7. *There are at least $n^2/4$ points in the $n \times n \times n$ grid with no three collinear. For all $\epsilon > 0$ and $n > n(\epsilon)$, there are at least $(1 - \epsilon)n^2$ points in the $n \times n \times n$ grid with no three collinear.* $\qquad\square$

Lemma 6 generalises to give the following construction of a 3D drawing of $T(n, k)$.

Lemma 8. *Every Turán graph $T(n,k)$ has a 3D drawing with $(2 + o(1))n\sqrt{k}$ volume. For all $\epsilon > 0$ and $k > k(\epsilon)$, $T(n,k)$ has a 3D drawing with $(1 + \epsilon)n\sqrt{k}$ volume.*

Proof. Index the colour classes $\{(x,y) : 0 \leq x,y \leq \lceil\sqrt{k}\rceil - 1\}$. By Lemma 5, there is a prime $p \not\equiv 1 \pmod 4$ with $\lceil\sqrt{k}\rceil \leq p \leq 2\lceil\sqrt{k}\rceil$ and $p \leq (1+\epsilon)\lceil\sqrt{k}\rceil$. For each $1 \leq i \leq \lceil n/k\rceil$, put the i-th vertex in colour class (x,y) at $(x, y, ip + (x^2 + y^2) \bmod p)$. Each colour class occupies its own Z-line. Thus, if an edge passes through a vertex, then three vertices from distinct colour classes are collinear. Observe that for every vertex at (a_x, a_y, a_z), we have $a_x^2 + a_y^2 \equiv a_z \pmod p$. Thus the same argument from Lemma 4 applies here, and no three vertices from distinct colour classes are collinear. Thus no edge passes through a vertex, and we obtain a 3D drawing of $T(n,k)$. The bounding box is $\lceil\sqrt{k}\rceil \times \lceil\sqrt{k}\rceil \times p\lceil n/k\rceil$. The volume is $(1 + o(1))np$, which is $(2 + o(1))n\sqrt{k}$ and $(1 + \epsilon)n\sqrt{k}$.   □

Pach *et al.* [23] proved that every k-colourable graph on n vertices is a subgraph of $T(2n+2k, 2k-1)$. Thus Lemma 8 implies the upper bound in Theorem 3.

Lemma 9. *Every k-colourable graph on n vertices has a 3D drawing with $(4\sqrt{2} + o(1))n\sqrt{k}$ volume. For all $\epsilon > 0$ and $k > k(\epsilon)$, every k-colourable graph on n vertices has a 3D drawing with $(2\sqrt{2} + \epsilon)n\sqrt{k}$ volume.*   □

4   Open Problems

Open Problem 1. Does every k-colourable graph have a crossing-free 3D drawing with $\mathcal{O}(kn^2)$ volume? The best known upper bound is $\mathcal{O}(k^2n^2)$ due to Pach *et al.* [23]. A $\mathcal{O}(kn^2)$ bound would match the $\Theta(n^3)$ bound for the minimum volume of a crossing-free 3D drawing of K_n.

For $1 \leq \ell \leq d - 1$, let $\mathrm{vol}(n, d, \ell)$ be the minimum bounding box volume for n vertices in $\mathbb{Z}^d$, such that no $\ell + 2$ vertices are in any ℓ-dimensional subspace. We have the following lower bound.

Lemma 10. *For $1 \leq \ell \leq d - 1$, $\mathrm{vol}(n, d, \ell) \geq \left(\dfrac{n}{\ell+1}\right)^{d/(d-\ell)}$.*

Proof. Consider n vertices in a d-dimensional box of volume $\mathrm{vol}(n, d, \ell)$, such that no $\ell + 2$ vertices are in any ℓ-dimensional subspace. The box can be partitioned into $\mathrm{vol}(n, d, \ell)^{(d-\ell)/d}$ subspaces of dimension ℓ, each of which have at most $\ell + 1$ vertices by assumption. Thus $n \leq (\ell + 1)\,\mathrm{vol}(n, d, \ell)^{(d-\ell)/d}$, and $\mathrm{vol}(n, d, \ell)$ is as claimed.   □

Open Problem 2. What is $\mathrm{vol}(n, d, \ell)$?

Consider the case of $\mathrm{vol}(n, d, d-1)$. Erdős [14] and Cohen *et al.* [6] proved that $\mathrm{vol}(n, 2, 1) \in \Theta(n^2)$ and $\mathrm{vol}(n, 3, 2) \in \Theta(n^3)$, respectively. Let $V = \{(x, x^2 \bmod p, \ldots, x^d \bmod p) : 0 \leq x \leq n - 1\}$, where p is a prime with $n - 1 \leq p \leq 2n$. The proofs of Erdős [14] and Cohen *et al.* [6] generalise to show that V contains no $d+1$ points in any $(d-1)$-dimensional subspace. Thus $\mathrm{vol}(n, d, d-1) \leq 2^{d-1}n^d$. By Lemma 10, $\mathrm{vol}(n, d, d-1) \in \Theta(n^d)$ for any constant d.

Open Problem 3. What is $vol(n, d, 1)$? Erdős [14] proved that $vol(n, 2, 1) \in \Theta(n^2)$. Theorem 2 proves that $vol(n, 3, 1) \in \Theta(n^{3/2})$. This problem is unsolved for all constant $d \geq 4$. Note that for $d \geq \log_2 n$ the problem becomes trivial. Just place the vertices at $\{(x_1, \ldots, x_d) : x_i \in \{0, 1\}\}$, and $vol(n, d, 1) \in \Theta(n)$.

Open Problem 4. What is $vol(n, d, 2)$? This case is interesting as it relates to crossing-free drawings. Cohen *et al.* [6] proved $vol(n, 3, 2) \in \Theta(n^3)$. Wood [24] proved that for $d = 2 \log n + \mathcal{O}(1)$, we have $vol(n, d, 2) \in \mathcal{O}(n^2)$. In particular, K_n has a $2 \times 2 \times \cdots \times 2$ crossing-free d-dimensional drawing with $\mathcal{O}(n^2)$ volume. What is the minimum volume for a crossing-free drawing of K_n, irrespective of dimension, is of some interest.

References

1. MICHAEL A. ADENA, DEREK A. HOLTON, AND PATRICK A. KELLY. Some thoughts on the no-three-in-line problem. In *Proc. 2nd Australian Conf. on Combinatorial Mathematics*, vol. 403 of *Lecture Notes in Math.*, pp. 6–17. Springer, 1974.
2. DAVID BRENT ANDERSON. Update on the no-three-in-line problem. *J. Combin. Theory Ser. A*, 27(3):365–366, 1979.
3. ROGER C. BAKER, GLYN HARMAN, AND JÁNOS PINTZ. The difference between consecutive primes. II. *Proc. London Math. Soc.*, 83(3):532–562, 2001.
4. PROSENJIT BOSE, JUREK CZYZOWICZ, PAT MORIN, AND DAVID R. WOOD. The maximum number of edges in a three-dimensional grid-drawing. *J. Graph Algorithms Appl.*, 8(1):21–26, 2004.
5. TIZIANA CALAMONERI AND ANDREA STERBINI. 3D straight-line grid drawing of 4-colorable graphs. *Inform. Process. Lett.*, 63(2):97–102, 1997.
6. ROBERT F. COHEN, PETER EADES, TAO LIN, AND FRANK RUSKEY. Three-dimensional graph drawing. *Algorithmica*, 17(2):199–208, 1996.
7. D. CRAGGS AND R. HUGHES-JONES. On the no-three-in-line problem. *J. Combinatorial Theory Ser. A*, 20(3):363–364, 1976.
8. EMILIO DI GIACOMO. Drawing series-parallel graphs on restricted integer 3D grids. In LIOTTA [22], pp. 238–246.
9. EMILIO DI GIACOMO AND HENK MEIJER. Track drawings of graphs with constant queue number. In LIOTTA [22], pp. 214–225.
10. HENRY ERNEST DUDENEY. *Amusements in Mathematics*. Nelson, Edinburgh, 1917.
11. VIDA DUJMOVIĆ, PAT MORIN, AND DAVID R. WOOD. Layout of graphs with bounded tree-width. *SIAM J. Comput.*, to appear.
12. VIDA DUJMOVIĆ AND DAVID R. WOOD. Three-dimensional grid drawings with subquadratic volume. In JÁNOS PACH, ed., *Towards a Theory of Geometric Graphs*, vol. 342 of *Contemporary Mathematics*, pp. 55–66. Amer. Math. Soc., 2004.
13. PAUL ERDŐS. A theorem of Sylvester and Schur. *J. London Math. Soc.*, 9:282–288, 1934.
14. PAUL ERDŐS. Appendix, in KLAUS F. ROTH, On a problem of Heilbronn. *J. London Math. Soc.*, 26:198–204, 1951.
15. STEFAN FELSNER, GIUSSEPE LIOTTA, AND STEPHEN WISMATH. Straight-line drawings on restricted integer grids in two and three dimensions. *J. Graph Algorithms Appl.*, 7(4):363–398, 2003.
16. ACHIM FLAMMENKAMP. Progress in the no-three-in-line problem. II. *J. Combin. Theory Ser. A*, 81(1):108–113, 1998.

17. RICHARD K. GUY AND PATRICK A. KELLY. The no-three-in-line problem. *Canad. Math. Bull.*, 11:527–531, 1968.

18. RICHARD R. HALL, TERENCE H. JACKSON, ANTHONY SUDBERY, AND K. WILD. Some advances in the no-three-in-line problem. *J. Combinatorial Theory Ser. A*, 18:336–341, 1975.

19. HEIKO HARBORTH, PHILIPP OERTEL, AND THOMAS PRELLBERG. No-three-in-line for seventeen and nineteen. *Discrete Math.*, 73(1-2):89–90, 1989.

20. TORU HASUNUMA. Laying out iterated line digraphs using queues. In LIOTTA [22], pp. 202–213.

21. TORLEIV KLØVE. On the no-three-in-line problem. III. *J. Combin. Theory Ser. A*, 26(1):82–83, 1979.

22. GUISEPPE LIOTTA, ed., *Proc. 11th International Symp. on Graph Drawing* (GD '03), vol. 2912 of *Lecture Notes in Comput. Sci.* Springer, 2004.

23. JÁNOS PACH, TORSTEN THIELE, AND GÉZA TÓTH. Three-dimensional grid drawings of graphs. In BERNARD CHAZELLE, JACOB E. GOODMAN, AND RICHARD POLLACK, eds., *Advances in discrete and computational geometry*, vol. 223 of *Contemporary Mathematics*, pp. 251–255. Amer. Math. Soc., 1999.

24. DAVID R. WOOD. Drawing a graph in a hypercube. Manuscript, 2004.

25. DAVID R. WOOD. Grid drawings of k-colourable graphs. *Comput. Geom.*, to appear.

Visual Navigation of Compound Graphs

Marcus Raitner

University of Passau, D-94032 Passau, Germany
`Marcus.Raitner@Uni-Passau.De`

Abstract. This paper describes a local update scheme for the algorithm of Sugiyama and Misue (IEEE Trans. on Systems, Man, and Cybernetics **21** (1991) 876–892) for drawing views of compound graphs. A view is an abstract representation of a compound graph; it is generated by contracting subgraphs into meta nodes. Starting with an initial view, the underlying compound graph is explored by repeatedly expanding or contracting meta nodes. The novelty is a totally local update scheme of the algorithm of Sugiyama and Misue. It is more efficient than redrawing the graph entirely, because the expensive steps of the algorithm, e. g., level assignment or crossing minimization, are restricted to the modified part of the compound graph. Also, the locality of the updates preserves the user's mental map: nodes not affected by the expand or contract operation keep their levels and their relative order; expanded edges take the same course as the corresponding contracted edge.

1 Introduction

A well-established technique to deal with huge graphs is to partition them recursively into a hierarchy of subgraphs; this leads to *compound graphs* [1] or *clustered graphs* [2]. Within both models, abstract representations of the graph, so-called *views*, can be defined [3, 4]. Intuitively, a view is generated by contracting subgraphs not needed in detail into *meta nodes*. Edges from within the contracted subgraph to nodes outside become edges from the meta node to the outside node. Views can be used for interactively exploring a large graph: one can choose which subgraphs to *contract* into meta nodes and which to *expand*. To this end, it is important that the drawing of the current view can be adjusted efficiently after these operations such that the user's mental map [5] is preserved. In this paper, an update scheme for the compound graph drawing algorithm of Sugiyama and Misue [1] is presented. Particular emphasis is laid on the locality of the updates: every expand and contract has only a local effect. The drawing of the new graph is computed only on the manipulated subgraph. This is a significant improvement for the time consuming phases of the algorithm, such as level assignment or crossing minimization. The user's mental map of the old view is preserved by keeping all uninvolved nodes on their levels and in the same relative order. Furthermore, expanded edges take the same course as the corresponding contracted edge.

J. Pach (Ed.): GD 2004, LNCS 3383, pp. 403–413, 2004.

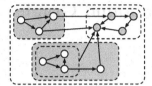
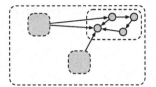

Fig. 1. A compound digraph; the hierarchy tree is depicted by the inclusion of the dashed rectangles.

Fig. 2. The view having as leaves the darker shaded nodes of the compound digraph in Fig. 1.

1.1   Problem Description

A *compound digraph* $D = (V, E, F)$ consists of nodes V, *inclusion edges* E, and *adjacency edges* F. It is required that the *inclusion digraph* $T = (V, E)$ is a rooted tree, and no adjacency edge connects a node to one of its descendants or ancestors; see Fig. 1. For convenience, we adopt the terminology of [1]: for a node $v \in V$, let $\mathrm{Ch}(v)$ denote the set of all *children* of v and $\mathrm{Pa}(v)$ the *parent* of v in T. The *descendants* of v, $\mathrm{De}(v)$, are all nodes in the subtree rooted at v (including v). The *depth* of v, $\mathrm{depth}(v)$, is the number of nodes on the path from the root of T to v. A *view* of D is a compound digraph $D[U] = (U, E[U], F[U])$ given by the nodes $U \subseteq V$, such that $T[U] = (U, E[U])$ with $E[U] = \{(u, v) \in E \mid u, v \in U\}$ is connected and contains the root of T; additionally, the leaves of $T[U]$ must *cover* the leaves of T, i.e., for each leaf u of T, (exactly) one of its ancestors is a leaf in $T[U]$; thus, $T[U]$ is a subtree of T from the root. The adjacency edges $F[U]$ comprise all edges $(u, v) \in F$ with $u, v \in U$ as well as *induced edges*: two *leaves* $u, v \in U$ are connected by an induced edge if and only if there are nodes $u' \in \mathrm{De}(u)$ and $v' \in \mathrm{De}(v)$ such that u' and v' are connected by an adjacency edge $(u', v') \in F$. Intuitively, given the designated set of leaves of $T[U]$, a view is constructed by shrinking each such leaf and all its descendants in T into a single meta node; see Fig. 2.

For the visual navigation of the underlying compound digraph D, the following operations shall be performed on a view $D[U]$:

– expand(v), where v is a leaf in $T[U]$; refines the view at v, i.e., the result is the view $D[U']$ with $U' = U \cup \mathrm{Ch}(v)$,
– contract(v), where $\mathrm{Ch}(v)$ are leaves in $T[U]$; coarsens the view at v, i.e., the result is the view $D[U']$ with $U' = U \setminus \mathrm{Ch}(v)$.

This paper concentrates on *visualizing* the above operations. We start with an initial layout of some view $D[U]$, and the user iteratively applies expand or contract operations; after each operation the previous layout has to be adjusted. Clearly, the obvious solution is redrawing the complete graph; unfortunately, it is neither efficient nor will it preserve the users mental map [5], which can be seen by comparing Figs. 3 and 5. In this context, our notion of preserving the mental map is that all old nodes U stay on their levels with their relative order unchanged; furthermore, expanded edges should take the same course as the corresponding contracted edge.

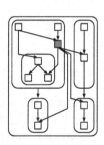 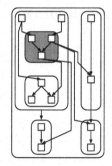 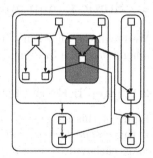

Fig. 3. Before expanding the shaded node.   **Fig. 4.** Result of our update scheme.   **Fig. 5.** Result of redrawing.

Our solution is an update scheme for the algorithm of Sugiyama and Misue [1] for drawing compound graphs. This algorithm, briefly recalled in Sect. 2, consists of four steps, which produce intermediate results. The initial view is laid out with the original algorithm; after each **expand** or **contract** operation, the intermediate results of the previous run are adjusted with our update scheme, as described in Sect. 3.

1.2   Related Work

There are several algorithms for drawing compound graphs, most notably the one of Sander [6], which differs from the algorithm of Sugiyama and Misue [1], among other things, by its *global layering*. While producing more compact – and supposedly more pleasant layouts –, this global layering is more difficult to update than the *local layering* of Sugiyama and Misue.

Dynamic or online graph drawing concentrates on updating layouts of ordinary graphs subject to insertions and deletions of nodes and edges [7]. The client-server model for online hierarchical graph drawing of North and Woodhull [8] allows insertion and removal of subgraphs, but only for ordinary DAGs and not for compound graphs. By using a *clan-based hierarchical decomposition*, the incremental drawing approach for DAGs of Shieh and McCreary [9] restricts the adjustments of the layout to the modified part. Visual navigation of compound (or clustered) graphs by expanding and contracting nodes has been introduced by Sugiyama and Misue [1], but they seem to implement them through reapplying their algorithm. Huang and Eades [10] briefly describe a system for handling huge clustered graphs visually, but their layout method is force-directed.

The visualization system of Abello and Korn [11] supports the interactive exploration of very large clustered graphs. They also use views as an abstraction of the underlying graph and provide methods for expanding *edges*. This expansion, however, restricts the view to the expanded edge, whereas our method preserves the relations of the expanded part to the remainder.

An efficient data structure supporting **expand** and **contract**, also known as *graph view maintenance problem* [3, 4, 12], is indispensable for an efficient implementation of the update method proposed in this paper. To this end, the software architecture of [13] should be considered as well.

2   Static Compound Graph Drawing

We recall the algorithm of Sugiyama and Misue [1] briefly, because our update scheme uses the intermediate results of these steps.

2.1   Step I: Hierarchization

Input of this step is the original compound graph $D = (V, E, F)$; the result is the *assigned compound graph* $D_A = (V, E, F_A, \text{clev})$, where $\text{clev} : V \to \mathbb{N}^+$ maps each node to its *compound level* and F_A are the adjacency edges F oriented from lower to higher level. This step internally uses the *derived compound graph* $D_D = (V, E, F_D, \text{type})$; the edges F_D with their types $\text{type} : F_D \to \{<, \leq\}$ are derived from F by replacing every adjacency edge $(u, v) \in F$ with edges between those ancestors of u and v having equal depth. The deepest such edge is of type $<$; all others of type $\leq$; see Fig. 6.

After resolving cycles in D_D, compound levels are assigned to the resulting cycle-free graph $D_F = (V, E, F_F, \text{type})$. The root is placed on level (1); then children of already placed nodes are treated recursively. Note that the children of all nodes on the same level are always evaluated in a common recursive call. The local level of the children is determined by a standard level assignment algorithm that takes into account the two types of edges: $\text{type}(u, v) = <$ enforces that the level of u is *less than* the level of v; with type $\leq$ the levels may also be *equal*. The compound level of a child v, $\text{clev}(v)$, is built by appending its local level to its parent's compound level, $\text{clev}(\text{Pa}(v))$. Finally, the adjacency edges F are oriented from lower to higher level. Let the complexity of this step be $\mathcal{O}(\mathcal{S}_I(n))$, where n is the size of the input D.

2.2   Step II: Normalization

In this step all adjacency edges of the assigned compound digraph D_A are made *proper*: an edge (u, v) is proper if $\text{clev}(\text{Pa}(u)) = \text{clev}(\text{Pa}(v))$ and $\text{tail}(\text{clev}(u)) = \text{tail}(\text{clev}(v)) + 1$, i.e., the parents lie on the same level and the children's levels differ by one. This is achieved by replacing each improper edge (u, v) with a linear compound graph as in Fig. 7. The result of this step is the *assigned proper compound graph* $D_P = (V_P, E_P, F_P, \text{clev})$. Let the complexity of this step be $\mathcal{O}(\mathcal{S}_{II}(n))$, where n is the size of the input D_A.

2.3   Step III: Vertex Ordering

Given the assigned proper compound graph D_P, this step calculates the relative order of the nodes on each level, so as to minimize edge crossings. To this end, the local order of all inner nodes' children on their levels is determined. The result of this step is the *ordered compound graph* $D_O = (V_P, E_P, F_P, \text{clev}, \sigma)$, where for each inner node $u \in V_P$, σ describes the local order of u's children.

The vertex ordering algorithm works depth-first: at an inner node u the compound graph induced by $\text{De}(u)$ is reduced to an ordinary DAG, the *local*

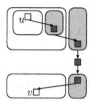

Fig. 6. The two dashed edges are derived from the solid one.

Fig. 7. Dummy nodes replace the improper edge (u, v).

hierarchy, by shrinking each child of u into a single node. This leads to two types of edges in the local hierarchy: edges between adjacent levels and those connecting nodes on the same level. A child u' of u or a descendant of u' may be adjacent to a node $v \notin \mathrm{De}(u)$. By the definition of a proper edge, it follows that u and $\mathrm{Pa}(v)$ lie on the same level. Since the algorithm has already ordered the children of all ancestors of u, it is known whether v lies to the left or to the right of u. Therefore, each child u' is annotated with two values $\lambda(u')$ and $\rho(u')$ counting the edges going to the left and to the right, respectively.

The crossing minimization for the local hierarchy starts with a preprocessing step – the so-called *splitting-method* – pinning the children u' with $\lambda(u') - \rho(u') > 0$ to the left end and those with $\lambda(u') - \rho(u') < 0$ to the right end of their level; the larger $|\lambda(u') - \rho(u')|$ is, the nearer to the end the node is placed. For the remaining nodes the crossings are minimized with a modified bary center heuristic that takes into account the horizontal edges. Let the complexity of this step be $\mathcal{O}(\mathcal{S}_{\mathrm{III}}(n))$, where n is the size of the input D_P.

2.4 Step IV: Metric Layout

This step assigns coordinates and dimensions to the nodes of the ordered compound graph D_O. A recursive algorithm assigns local coordinates to the children relative to their parents position; for an inner node u it is applied to all children of u first, thus determining their width and height. The local coordinates are optimized with the so-called *priority method* on the *metrical local hierarchy*, which is basically the local hierarchy from the previous step without the horizontal edges. Similar to the bary center heuristic for minimizing crossings, the priority method improves the nodes positions by moving them – as far as possible without changing the order on the level – to their respective (metrical) bary centers. A final depth-first traversal calculates the absolute coordinates. The first phase is done with a recursive algorithm. Let the complexity of this step be $\mathcal{O}(\mathcal{S}_{\mathrm{IV}}(n))$, where n is the size of the input D_O.

3 Update Scheme

Let $D[U] = (U, E[U], F[U])$ be a view of a compound graph $D = (V, E, F)$, where $D[U]$ already has been drawn with the standard algorithm; node $v \in U$ shall be expanded, resulting in a new view $D[U'] = (U', E[U'], F[U'])$, with

$U' = U \cup \mathrm{Ch}(v)$. It is assumed that, given $D[U]$, the *structure* of $D[U']$ can be determined efficiently, which is where the data structures for maintaining hierarchical graph views [3, 4, 12] come into play.

3.1   Step I: Hierarchization

In this step the assigned compound digraph has to be updated. In order to preserve the mental map, all old nodes $u \in U$ stay on their levels; only for the children of v appropriate levels are determined. In other words, the level assignment function clev $: U \to \mathbb{N}^+$ has to be extended to the set U'. Expanding v in the cycle-free graph $D_F[U]$ inherits the direction of edges incident to v to all corresponding expanded edges, i. e., if an edge was reversed during the cycle removal of the previous run, all corresponding expanded edges are reversed as well. In this setting, it is obvious that new cycles entirely consist of newly added edges, and the cycle removal can be restricted to the children of v.

From the definition of the derived graph, it follows that all edges adjacent to a leaf are of type $<$, and in the level assignment algorithm an edge of type $<$ causes the target node to be placed on a higher level than the source node. Therefore, v is not connected to any node on its level; neither is any child of v in the updated cycle-free graph $D_F[U']$. Consequently, the level assignment does not need to take into account children of nodes on the same level as v, but can be restricted to the subgraph of $D_F[U']$ induced by the children of v. After the level assignment clev has been extended to U', updating the assigned compound digraph $D_A[U]$ to $D_A[U']$ is just a matter of adding $\mathrm{Ch}(v)$ (and the corresponding inclusion edges) and inserting the new adjacency edges directed from lower to higher levels; induced adjacency edges incident with v are removed.

How much does the updated assigned compound digraph differ from the one the hierarchization algorithm of Sect. 2.1 applied to $D[U']$ would have produced? Since all old nodes stay on their level, it is not possible to place ancestors of v's neighbors on the same level as v. Compare, for instance, Fig. 9, which shows the level assignment produced by our update scheme, and Fig. 10, which would be the result of the hierarchization algorithm applied anew. The adjacency edge (u, v) in Fig. 8 leads to a type $<$ derived edge $(\mathrm{Pa}(u), v)$, which results in $\mathrm{Pa}(u)$ being placed on a level above v, where it is bound to stay during our update. If the derived graph would be built anew, this edge were of type $\leq$, and $\mathrm{Pa}(u)$ and v were placed on the same level, as shown in Fig. 10.

Property 1. *Let k denote the number of elements added to $D[U]$ by expanding v. The complexity of updating the assigned compound digraph $D_A[U]$ is $\mathcal{O}(\mathcal{S}_I(k))$, compared to $\mathcal{O}(\mathcal{S}_I(n + k))$ for reapplying step I to $D[U']$, where n is the size of $D[U]$. The user's mental map is supported by keeping all old nodes U on their level.*

3.2   Step II: Normalization

The assigned proper compound graph $D_P[U]$ is updated in two steps: first the node v is expanded and then the new improper adjacency edges are made proper.

Fig. 8. Before expanding v.   **Fig. 9.** Update of the de-   **Fig. 10.** Building the derived graph.   rived graph anew.

Expanding v means to add $\mathrm{Ch}(v)$ with appropriate inclusion edges, (induced) adjacency edges between two children of v, and those between a child of v and some other node $u \notin \mathrm{Ch}(v)$. For the latter the old induced edge connecting v and u has to be removed; if this edge has been made proper all the associated dummy nodes and edges are removed as well. Since the levels of the old nodes U are unchanged, the only improper edges are adjacent to at least one child of v. They are made proper exactly as described in Sect. 2.2. In the worst case this has to be done for every *new* adjacency edge, whereas for the other adjacency edges the construction from the previous proper compound digraph $D_P[U]$ is reused. Clearly, the result is exactly the same as if the normalization had been applied to the updated assigned compound digraph $D_A[U']$ as a whole.

Property 2. *Let k denote the number of elements added to $D_A[U]$ by expanding v. The complexity of updating the proper, assigned compound digraph $D_P[U]$ is $\mathcal{O}(\mathcal{S}_{\mathrm{II}}(k))$, compared to $\mathcal{O}(\mathcal{S}_{\mathrm{II}}(n+k))$ for reapplying step II to $D_A[U']$, where n is the size of $D_A[U]$.*

3.3   Step III: Vertex Ordering

In this step preserving the mental map means to keep the order of nodes that are common in the old and the new graph: only for the nodes that have been added during the update from $D_P[U]$ to $D_P[U']$ a position in the relative order on the respective levels has to be determined. These nodes are either children of v – including dummy nodes for edges between two children of v – or dummy nodes that belong to an edge between a child of v and a node $u \notin \mathrm{Ch}(v)$. As described in Sect. 2.3, the vertex ordering algorithm recursively calculates local orders for the children of an inner node on their levels; hence, determining the order of the children of v is just a matter of applying the algorithm to the subtree rooted at v. A precondition, however, is that the children of all ancestors of v already have been ordered. Therefore, the positions of dummy nodes that are not children of v have to be fixed prior to ordering the children of v.

Consider an induced edge $(v, u) \in F[U]$ (and symmetrically for an edge (u, v)); after expanding v, children $v_1, \ldots, v_k$ inherit this edge. Clearly, the subgraph inserted to make any of the (v_i, u) proper is – except for an extra dummy node complex on the level of v – identical to the one for edge (v, u). The idea is to reuse the positions of dummy nodes of the edge (v, u) for the dummy nodes of the edges $(v_1, u), \ldots, (v_k, u)$. The dummy nodes of these edges are treated as one block; the position of the block is the position of the respective dummy node

 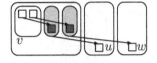 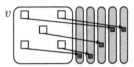

Fig. 11. Two proper non-local edges (v, u) and (v, w) before expanding v.

Fig. 12. After expanding v, the two new dummy node complexes inherit the order of u and w.

Fig. 13. The order of the dummy nodes is determined by the order of v's children.

of the edge (v, u). This reusing has the effect that the expanded edges take the *same course* as the old edge.

Since the positions of the old dummy nodes are reused, it can be assumed without loss of generality that the edge (v, u) is proper; if it is not, it suffices to expand the edge from v to the first dummy node of (v, u), which is proper. Expanding the proper edge (v, u) results in improper edges $(v_1, u), \ldots, (v_k, u)$; see Figs. 11 and 12. Each edge (v_i, u) is made proper with a dummy node complex consisting of nodes p_i and c_i with $\mathrm{Pa}(c_i) = p_i$ and edges (v_i, c_i) and (p_i, u); the nodes p_i are siblings of v and lie on the same level as v. We distinguish two types of proper edges (v, u): a *local* edge has $\mathrm{Pa}(v) = \mathrm{Pa}(u)$ and a *non-local* edge $\mathrm{Pa}(v) \neq \mathrm{Pa}(u)$. The reason is that if $\mathrm{Pa}(v) \neq \mathrm{Pa}(u)$, the definition of proper demands that $\mathrm{Pa}(v)$ and $\mathrm{Pa}(u)$ lie on the same level. Since the relative position of old nodes must be preserved, it is known whether $\mathrm{Pa}(u)$ is to the left or to the right of $\mathrm{Pa}(v)$. Clearly, this determines whether the dummy nodes p_i are to the left or to the right of v. For a local edge, the position of the new dummy node can be *anywhere* on the level of v.

All dummy nodes p_i belonging to the same expanded edge are treated as one block; therefore, one representant p is sufficient. In the local hierarchy induced by $\mathrm{Pa}(v)$'s children, a representant p for expanded edges belonging to a non-local edge (v, u) has $\lambda(p) - \rho(p) = \pm 1$, depending on whether $\mathrm{Pa}(u)$ lies to the left $(+1)$ or to the right (-1) of $\mathrm{Pa}(v)$. The splitting method (cf. Sect. 2.3) puts p to the left or right end of the level, with the exact position determined by the $\lambda(p) - \rho(p)$ value. Let q be the representant for the expanded edges of another non-local edge (v, w) such that $\lambda(p) - \rho(p) = \lambda(q) - \rho(q)$. Then p and q are indistinguishable in the splitting method; they are pinned to one end in *arbitrary* relative order. This order, however, should be the same as for the nodes u and w, which have both the same level $\mathrm{clev}(v) + 1$. This problem, incidentally, is immanent to the original algorithm of Sugiyama and Misue [1]. It can be solved by taking the relative order of the end nodes of the expanded edges as secondary sorting criterion in the splitting method. For our update scheme this means that a representant p is placed into the old order σ according to $\lambda(p) - \rho(p)$, and if there are more nodes with the same value, the order of the respective end nodes determines their order. Consider, for instance, the two dummy nodes on v's level in Fig. 12; having value -1 they are all placed to the right end. The order derived from the order of u and w shown in Fig. 12 clearly is the best choice.

After the splitting method, all representants for expanded edges of non-local edges are fixed; it remains to do the same for local edges. If (v, u) and (v, w) are two proper local edges, then u and w lie on the same level and thus determine the relative order of the representants p and q of the expanded edges for (v, u) and (v, w) respectively. Essentially, the only degree of freedom is whether to place p or q right or left of v. It makes no sense to have some non-dummy node x between p and v: otherwise the edges that p represents would cross x. In the local hierarchy induced by Pa(v)'s children, dummy nodes like p and q have only one outgoing edge; hence, their bary center is identical to the position of u and w. The bary centers are used to decide whether representants are placed left or right of v.

It remains to determine the relative order of the expanded edges *within* their respective block. Since this order depends on the positions of v's children, the crossing reduction algorithm of Sect. 2.3 is applied to the local hierarchy of v first. Note that this is possible without knowing the exact order of the expanded edges: the representant already determines on which side they leave v; this information is sufficient for the λ and ρ values of v's children. Consider the edge (v, u) with its expanded edges $(v_1, u), \ldots, (v_k, u)$. The order of the dummy nodes $p_1, \ldots, p_k$ within the block represented by p is determined as follows: if p lies to the right of v, and if $v_{\sigma(1)}, \ldots, v_{\sigma(k)}$ is the order of v's children from bottom to top and within the same level from left to right, then $p_{\sigma(1)}, \ldots, p_{\sigma(k)}$ is the order of the dummy nodes from left to right; see Fig. 13. The case that p is right to v as well as the two cases for an incoming edge (u, v) are symmetric.

Property 3. *Let k denote the number of elements added to $D_P[U]$ by expanding v. Then the complexity of updating the local order σ is $\mathcal{O}(\mathcal{S}_{\mathrm{III}}(k))$, compared to $\mathcal{O}(\mathcal{S}_{\mathrm{III}}(n + k))$ for reapplying step III to $D_P[U']$, where n is the size of $D_P[U]$. The relative order of all old nodes U is preserved and expanded edges take the same course as the corresponding contracted edge.*

3.4   Step IV: Metric Layout

Expanding v changes the width and height of v, which leads to adjustments of the local coordinates for v's siblings; this, in turn, changes the width and height of Pa(v), and so on up to the root. On the other hand, the local coordinates of children of a node that is no ancestor of v are not affected. As described in Sect. 2.4, the metric layout consists of two steps: computing local coordinates followed by a traversal of the hierarchy to sum them up to absolute coordinates. Hence, we adjust the local coordinates at v and all its ancestors and then use the second step unalteredly. For the updates of the local coordinates basically the same recursive procedure as in Sect. 2.4 is used; the only difference is that recursive calls are made only for ancestors of v. The local coordinates in the subtrees rooted at nodes that are no ancestors are reused from the previous layout.

Property 4. *Let n denote the size of $D_O[U']$. In the worst case, the complexity of updating the coordinates is $\mathcal{S}_{\mathrm{IV}}(n)$. The final depth-first traversal to sum up the absolute coordinates is completely applied in any case; the local coordinates are adjusted only for ancestors of the expanded node.*

4   Contraction

Contracting a node v that has been expanded with the above update scheme is straightforward: in the contracted view $D[U']$ all nodes are old, i.e., $U' \subseteq U$; hence, the level assignment clev and the vertex order σ just need to be restricted to U'. The position of the dummy nodes for a new induced edge incident to v is given by the position of the blocks of the corresponding expanded edges. Since the width and height of v has changed, the metric layout has to be updated as described in Sect. 3.4. This has the side-effect that expanding and contracting are also *visually inverse*, i.e., the drawing after expanding and contracting a node v is the same as before expanding.

Why is contraction more complicated for nodes v that have not been expanded with our update scheme? Consider a child v' of v with an edge (u, v') such that v and $Pa(u)$ lie on the same level, e.g., as in Fig. 10. As pointed out in Sect. 3.1, this cannot happen if v has been expanded before, yet it is possible in the layout of the initial view. Note that because of the deepest derived edge being of type $<$, for each edge (x, y) the compound levels clev(x) and clev(y) differ – after a common start sequence – by at least one position (cf. Sect. 2.1). The induced edge (u, v), representing (u, v') after contracting v, would violate this invariant, because clev(v) would be a subsequence of clev(u). This problem manifests itself in the derived graph: before contracting v the deepest edge, the one of type $<$, was adjacent to v' and is removed; therefore, the type of the derived edge between v and the ancestor of u at the same depth as v would have to be adjusted from $\leq$ to $<$, which could lead to substantial changes in the level assignment and thus to the user's mental map; see Fig. 9.

The easiest way to deal with this problem is to allow contraction only for nodes that have been expanded before, i.e., no node of the initial view can be contracted. Another way is to modify the algorithm of Sugiyama and Misue [1] used for the initial view such that all edges in the derived graph are of type $<$. The consequence is that the initial layout is less compact, because nodes with descendants that are connected never lie on the same level.

Property 5. *Let k denote the number of elements removed from $D_P[U]$; then the elements removed from the other (intermediate) compounds graphs are at most k. Updating the drawing after contracting a node v that has been expanded with our update scheme takes $\mathcal{O}(k)$ for steps I to III. Step IV is the same as after expanding; see Property 4.*

5   Summary

The proposed update scheme for the algorithm of Sugiyama and Misue [1] supports efficient visual navigation of compound graphs through expand and contract operations. The locality of our update scheme makes it much more efficient than redrawing the entire new view. For expanding a node the complexity of updating the drawing essentially is determined by applying each of the steps I to III to the *modified part* of the compound graph, followed by step IV adjusting the coordinates. The user's mental map of the old view is preserved well: old nodes stay on their levels in the same relative order and expanded edges take the same course as the corresponding contracted edge.

References

1. Sugiyama, K., Misue, K.: Visualization of structural information: Automatic drawing of compound digraphs. IEEE Trans. on Systems, Man, and Cybernetics **21** (1991) 876–892
2. Eades, P., Feng, Q.W.: Multilevel visualization of clustered graphs. In: Proc. 4th GD. Vol. 1190 of LNCS. (1996) 101–112
3. Buchsbaum, A.L., Westbrook, J.R.: Maintaining hierarchical graph views. In: Proc. 11th SODA. (2000) 566–575
4. Raitner, M.: Dynamic tree cross products. In: Proc. 15th ISAAC. LNCS. (2004)
5. Misue, K., Eades, P., Lai, W., Sugiyama, K.: Layout adjustment and the mental map. Journal of Visual Languages and Computing **6** (1995) 183–210
6. Sander, G.: Graph layout for applications in compiler construction. TCS **217** (1999) 175–214
7. Branke, J.: Dynamic graph drawing. In Kaufmann, M., Wagner, D., eds.: Drawing Graphs – Methods and Models. Vol. 2025 of LNCS. Springer (2001) 228–246
8. North, S.C., Woodhull, G.: Online hierarchical graph drawing. In: Proc. 9th GD. Vol. 2265 of LNCS. (2001) 232–246
9. Shieh, F.S., McCreary, C.L.: Clan-based incremental drawing. In: Proc. 8th GD. Vol. 1984 of LNCS. (2000) 384–395
10. Huang, M.L., Eades, P.: A fully animated interactive system for clustering and navigating huge graphs. In: Proc. 6th GD. Vol. 1547 of LNCS. (1998) 374–383
11. Abello, J., Korn, J.: MGV: A system for visualizing massive multigraphs. IEEE Trans. on Visualization and Computer Graphics **8** (2002) 21–38
12. Buchsbaum, A.L., Goodrich, M.T., Westbrook, J.R.: Range searching over tree cross products. In: Proc. 8th ESA. Vol. 1879 of LNCS. (2000) 120–131
13. Raitner, M.: HGV: A library for hierarchies, graphs, and views. In: Proc. 10th GD. Vol. 1528 of LNCS. (2002) 236–243

Layout Volumes of the Hypercube*

Lubomir Torok[1] and Imrich Vrt'o[2]

[1] Institute of Mathematics and Computer Science,
Severná 5, 974 01 Banská Bystrica, Slovak Republic
[2] Institute of Mathematics, Slovak Academy of Sciences,
Dúbravská 9, 841 04 Bratislava, Slovak Republic

Abstract. We study 3-dimensional layouts of the hypercube in a 1-active layer and a general model. The problem can be understood as a graph drawing problem in $3D$ space and was addressed at Graph Drawing 2003 [5]. For both models we prove general lower bounds which relate volumes of layouts to a graph parameter called cutwidth. Then we propose tight bounds on volumes of layouts of N-vertex hypercubes. Especially, we have $\mathrm{VOL}_{1-AL}(Q_{\log N}) = \frac{2}{3}N^{\frac{3}{2}}\log N + O(N^{\frac{3}{2}})$, for even $\log N$ and $\mathrm{VOL}(Q_{\log N}) = \frac{2\sqrt{6}}{9}N^{\frac{3}{2}} + O(N^{4/3}\log N)$, for $\log N$ divisible by 3. The 1-active layer layout can be easily extended to a 2-active layer (bottom and top) layout which improves a result from [5].

1 Introduction

The research on three-dimensional circuit layouts started in seminal works [15, 17] as a response to advances in VLSI technology. Their model of a 3-dimensional circuit was a natural generalization of the 2-dimensional model [18]. Several basic results have been proved since then which show that the 3-dimensional layout may essentially reduce material, measured as volume [6, 12]. The problem may be also understood as a special 3-dimensional orthogonal drawing of graphs, see e.g., [9]. In all these models the degrees of vertices of a graph, which represents the circuit, are at most 6. There exist only a few papers investigating layouts of graphs of arbitrary degrees, e.g., [4, 5]. There are two basic variants of the layout model. In a *1-active layer model*, a vertex of degree d is represented by a square of side d and is placed in the bottom layer of the 3-dimensional grid. In the *general model*, a vertex of degree d is represented by a cube of side d and can lie anywhere in the grid. In both models edges are routed as edge disjoint paths in the underlying grid. The variants are straightforward extensions of linear layouts [13] and 2-dimensional layouts [18, 19], respectively, and generalizations of 3-dimensional layouts for bounded degree graphs mentioned above.

In this paper we solve an open problem posed by Calamoneri and Massini [5] concerning the layout volume of the hypercube. We solve the problem for both models. First we prove general lower bounds on volumes of any graph G of cutwidth $\mathrm{cw}(G)$:

* This research was partially supported by the VEGA grant No. 2/3164/23.

J. Pach (Ed.): GD 2004, LNCS 3383, pp. 414–424, 2004.

$$\mathrm{VOL}_{1-AL}(G) \geq \mathrm{cw}(G)\sqrt{\sum_{v \in V} \deg^2(v)},$$

$$\mathrm{VOL}(G) \geq (\mathrm{cw}(G) - \sqrt{2\mathrm{cw}(G)})^{\frac{3}{2}}.$$

The lower bounds represent 3-dimensional analogues of similar estimations of the area of linear or 2-dimensional layouts [13, 16] and are of special interest. Then we propose optimal layouts in both models for the $\log N$-dimensional hypercube. Especially, for even $\log N$ we have

$$\mathrm{VOL}_{1-AL}(Q_{\log N}) = \frac{2}{3}N^{\frac{3}{2}}\log N + O(N^{\frac{3}{2}}),$$

$$\mathrm{VOL}(Q_{\log N}) = \frac{2\sqrt{6}}{9}N^{\frac{3}{2}} + O(N^{\frac{4}{3}}\log N),$$

for $\log N$ divisible by 3. Recall that 2-dimensional layouts of hypercubes were studied in several papers [3, 8, 11, 14], but only recently an asymptotically exact result has appeared [10], for even $\log N$.

Our layout can be easily extended to the 2-active layer model, where vertices represented as rectangles are placed on the bottom and top layer of a grid. For such a model, Calamoneri and Massini [5] designed nearly optimal layout for the hypercube. We prove an optimal layout for the hypercube in this model.

2   Preliminaries

The 3-dimensional grid consists of all points (a, b, c) of integer coordinates in the standard xyz-coordinate system. Two points are joined by a straightline segment if, and only if, their Euclidean distance is 1. The lines parallel to axes and going through the points are called tracks. At the same time we consider the 3-dimensional grid as an infinite grid graph whose vertices corresponds to points and edges are the segments of lengths 1. The 3-dimensional 1-active layer layout of a graph G is a mapping of G into the grid such the following conditions are satisfied:

- A vertex of degree d is represented by a square of integer coordinates of side d lying in the *basic plane* given by $z = 0$. The sides of the square are parallel to the x and y axes.
- Two vertices (squares) do not touch.
- Edges are represented as edge disjoint paths in the grid graph in the halfspace above the basic plane. A path touches only two squares which represent the endvertices of the corresponding edge.

The 3-dimensional general layout of a graph G is a mapping of G into the grid such the following conditions are satisfied:

- A vertex of degree d is represented by a cube of integer coordinates of side edge length d. The edges of the cube are parallel to the axes.
- Two vertices (cubes) do not touch.

– Edges are represented as edge disjoint paths in the grid graph. A path touches only two cubes which represent the endvertices of the corresponding graph edge.

The volume of the 3-dimensional 1-active layer (general) layout of G, denoted by $\text{VOL}_{1-AL}(G)$ (denoted by $\text{VOL}(G)$) is defined as the volume of the smallest box containing the layout. From technical reasons we assume that the coordinates of the box vertices are of the form $(a + 1/2, b + 1/2, c + 1/2)$, for some integers a, b, c.

Remark. The models we described, differs from the so called multilayer model [7], in which vertices are represented as boxes lying on the bottom layer and edges are routed in layers parallel to the basic plane.

Let $\phi : V \to 1, 2, 3, ..., |V|$ be a 1-1 labelling of vertices of a graph $G = (V, E)$. Define

$$cw(G, \phi) = \max_i \{|\{uv \in E : \phi(u) \leq i < \phi(v)\}|\}.$$

The cutwidth of the graph G is defined as

$$cw(G) = \min_\phi \{cw(G, \phi)\}.$$

The cutwidth is strongly related to linear layouts and, roughly saying, represents the largest edge cut in a graph which is embedded in a line.

Let $\deg(v)$ denote the degree of a vertex v. If N is a power of two, the $\log N$-dimensional hypercube graph has a vertex set consisting of all binary strings of length $\log N$. Two vertices are joined by an edge if the corresponding strings differ in precisely one position.

3   Lower Bounds

In this section we prove a lower bound which is of special interest for its general use and easy applicability.

Theorem 1. *The volume of the 3-dimensional 1-active layer layout of any graph G of cutwidth $cw(G)$ satisfies*

$$\text{VOL}_{1-AL}(G) \geq cw(G)\sqrt{\sum_{v \in V} \deg^2(v)}.$$

Proof. Assume we have an optimal 1-active layer layout of G of the volume $\text{VOL}_{1-AL}(G)$. The layout is put into a box of the width W, length L and height H, where W, L, H are measured along the x, y, z axes, respectively. Thus

$$\text{VOL}_{1-AL}(G) = WLH. \tag{1}$$

As two vertices do not touch we have

$$WL \geq \sum_{v \in V} (\deg(v) + 1)^2. \tag{2}$$

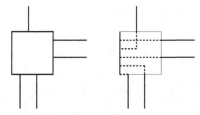

Fig. 1. Replacing a 5×5 square vertex by a straight-line segment.

Consider now the first layer. For every vertex (square) remove its sides except for the left one. It is easy to see that the edges which ended on those 3 sides can be prolonged and connected to the left side using the original tracks and such that the edges are edge disjoint. See Fig. 1

We obtain a new layout in which the square vertices are replaced by straight-line segments. Now we use a similar idea as in earlier paper [16] (Lemma 2.1), for estimating the area of 2-dimensional layouts. For the sake of completeness we repeat the argument. Any such segment is identified by its coordinates given by the position of the lower end of the segment. Sort the segments according to their coordinates lexicographically. Label the segments by $1, 2, ..., |V|$, according to the lexicographic order. Let ϕ denote this labelling. Take the first i segments in this labelling. We can find a surface, normal to xy-plane, which separates precisely i segments from the rest, as in Fig. 2.

The area of the surface is $(L+1)H$. Clearly the number of edges between the two parts of the layout is given by

$$\{|\{uv \in E : \phi(u) \leq i < \phi(v)\}|\} \leq (L+1)H$$

Maximizing the left hand side over all i's and then minimizing it over all ϕ's we get the cutwidth of G on the left side. By rearranging the terms we have

$$(L+1)H \geq \mathrm{cw}(G). \tag{3}$$

We may repeat the above argument by changing the role of L and W. We get

$$(W+1)H \geq \mathrm{cw}(G). \tag{4}$$

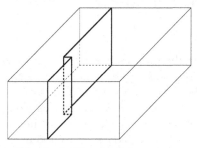

Fig. 2. A 3D cut of the volume.

Combining relations (1),(2),(4) and (3) we have

$$\mathrm{VOL}_{1-AL}(G) \geq (\mathrm{cw}(G) - H)\sqrt{\sum_{v \in G}(\deg(v) + 1)^2}$$

$$= \left(\mathrm{cw}(G) - \frac{\mathrm{VOL}_{1-AL}(G)}{WL}\right)\sqrt{\sum_{v \in G}(\deg(v) + 1)^2}$$

$$\geq \left(\mathrm{cw}(G) - \frac{\mathrm{VOL}_{1-AL}(G)}{\sum_{v \in G}(\deg(v) + 1)^2}\right)\sqrt{\sum_{v \in G}(\deg(v) + 1)^2}$$

$$\geq \mathrm{cw}(G)\sqrt{\sum_{v \in G}(\deg(v) + 1)^2} - \frac{\mathrm{VOL}_{1-AL}(G)}{\sqrt{\sum_{v \in G}(\deg(v) + 1)^2}}.$$

By eliminating $\mathrm{VOL}_{1-AL}(G)$ and some algebraic manipulations we get

$$\mathrm{VOL}_{1-AL}(G) \geq \mathrm{cw}(G)(\sqrt{\sum_{v \in V}(\deg(v) + 1)^2} - 1) \geq \mathrm{cw}(G)\sqrt{\sum_{v \in V}\deg^2(v)}.$$

$\square$

Remark. The idea of cutting of the 3-dimensional layout into special parts was used in [6] in a model for bounded degree graphs, where they considered the so called *special bisection width* of a graph which is however in general a smaller quantity than the cutwidth.

Corollary 1. *The optimal volume of the 3-dimensional 1-active layer layout of the N-vertex hypercube $Q_{\log N}$ satisfies*

$$\mathrm{VOL}_{1-AL}(Q_{\log N}) \geq \frac{2}{3}N^{\frac{3}{2}}\log N + O(\sqrt{N}\log N).$$

Proof. Several papers proved that $\mathrm{cw}(Q_{\log N}) = \lfloor 2N/3 \rfloor$, e.g., [1,2]. $\square$

Theorem 2. *The optimal volume of the 3-dimensional layout of any graph $G = (V, E)$ satisfies*
$$\mathrm{VOL}(G) \geq (\mathrm{cw}(G) - \sqrt{2\mathrm{cw}(G)})^{\frac{3}{2}}.$$

Proof. Consider an optimal 3-dimensional layout of G. Assume the layout is put into a box of sizes W, L, H such that

$$W \geq L \geq H. \tag{5}$$

Take any vertex (a cube of sizes $n \times n \times n$) and choose one of its vertical "edges" (a straightline segment). Delete the cube except for that segment. Observe that the graph edges originally attached to the cube can be prolonged and connected to the segment using edge disjoint paths along the tracks. Repeating this operation for all vertices we get a new layout in which the vertices are replaced by segments of length n and edges are routed in the edge disjoint manner. Similarly as in the proof of Theorem 1, we can find a surface, see Fig. 3, which

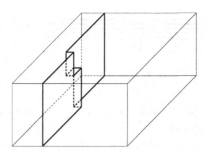

Fig. 3. A 3D cut of the volume.

separates the segments into two parts such that there are at least $\mathrm{cw}(G)$ edges between the segments.

The edges must cross the surface. The area of the surface is $HL + H + 1$. It follows

$$HL + H + 1 \geq \mathrm{cw}(G)$$

Then

$$HL \geq \mathrm{cw}(G) - 1 - H. \tag{6}$$

The relation (5) implies

$$LW \geq \mathrm{cw}(G) - 1 - H. \tag{7}$$

$$WH \geq \mathrm{cw}(G) - 1 - H. \tag{8}$$

Multiplying (6)(7) and (8) and taking the cube root we have

$$\mathrm{VOL}^{\frac{2}{3}}(G) \geq \mathrm{cw}(G) - 1 - H \geq \mathrm{cw}(G) - 1 - \mathrm{VOL}^{\frac{1}{3}}(G).$$

Solving this quadratic inequality for $\mathrm{VOL}^{\frac{1}{3}}(G)$ we get the lower bound.   □

Corollary 2. *The optimal volume of the 3-dimensional layout of the N-vertex hypercube $Q_{\log N}$ satisfies*

$$\mathrm{VOL}(Q_{\log N}) \geq \frac{2\sqrt{6}}{9} N^{\frac{3}{2}} + O(N).$$

4   One-Active Layer Layout of Hypercubes

Theorem 3. *The optimal volume of the 1-active layer layout of the $\log N$-dimensional hypercube satisfies*

$$\mathrm{VOL}_{1-AL}(Q_{\log N}) \leq \frac{2^{\frac{i}{2}+1}}{3} N^{\frac{3}{2}} \log N + O(N^{\frac{3}{2}}),$$

where $i = \log(N \bmod 2)$.

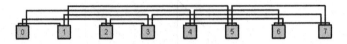

Fig. 4. Linear layout of Q_3.

Proof. Assume $\log N$ is divisible by 2. The second case is similar. Our basic building block is a linear layout of the m-dimensional hypercube Q_m using $cw(Q_m) = \lfloor 2^{m+1}/3 \rfloor$ horizontal tracks, while every vertex is represented by a square of size m. See Figure 4 for the case $m = 3$. Such a layout can be obtained by placing the vertices on the line in the natural order and assigning the edges to tracks properly. This was observed in several papers, e.g., [8, 10, 11]. We utilize a fact that $Q_{\log N}$ can be represented as a cartesian product $Q_{\frac{\log N}{2}} \times Q_{\frac{\log N}{2}}$. Using the linear layout minimizing the cutwidth for $Q_{\frac{\log N}{2}}$, one can easily design a 2-dimensional layout of $Q_{\log N}$ as shown in Figure 5, for N=64.

Now rotating every edge in the angle $\pi/2$ around the line defined by its endpoints and compacting the vertices in the natural way we get a 1-active layer layout of the volume

$$\left\lfloor \frac{2}{3} N^{\frac{1}{2}} \right\rfloor \times N(\log N + 1)^2.$$

The first term stands for the height of the layout and the second one for the occupied area on the 1st layer. We decrease the height of the layout by a factor of $\log N$ in the following way. Consider the layout of any $Q_{\frac{\log N}{2}}$ - the sublayout of the global layout of $Q_{\log N}$. Assume the vertices of $Q_{\frac{\log N}{2}}$ are aligned along

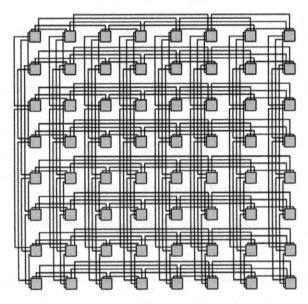

Fig. 5. 2-dimensional layout of $Q_6 = Q_3 \times Q_3$.

the x-axis. The edges occupy $\lfloor 2N^{1/2}/3 \rfloor$ tracks parallel to x. Divide the edges evenly into $\log N$ groups according to the distance of the corresponding tracks from the basic plane. I.e., the first group occupy the first $t = \lceil \lfloor 2N^{1/2}/3 \rfloor / \log N \rceil$ tracks when counting from the basic plane, the second group occupy the second t tracks, and so on. Reattach the edges to their corresponding vertices such that:

  i) The endpoints of edges of the same group has the same y coordinate.
 ii) The endpoints of edges of different groups have different y coordinate.
iii) The endpoints of edges lies in the first or the third "quadrant" of the square vertex. This can be viewed as a "shifting" of edges in y direction on a proper place on its square endvertices. Finally, decrease the height h of every track to $(h \bmod t) + 1$ and reroute correspondigly the edges. See Figure 6.

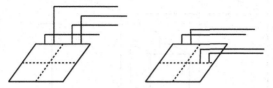

Fig. 6. Decreasing the height of the layout.

We repeat this operation for all $Q_{\frac{\log N}{2}}$'s in x and y directions. For the y direction, the endpoits of edges lie in the second and fourth quadrants of square vertices. This avoids overlapping of vertical parts of two edges starting in the same square vertex. The new height of the layout is t, which implies the claimed volume.   □

5   Layouts of Hypercubes in the General Model

Theorem 4. *The optimal volume of the general 3-dimensional layout of the* $\log N$-*dimensional hypercube satisfies*

$$\mathrm{VOL}(Q_{\log N}) \leq \frac{2^{\frac{i}{2}+1}\sqrt{6}}{9} N^{\frac{3}{2}} + O(N^{\frac{4}{3}} \log N),$$

where $i = \log(N \bmod 3)$.

Proof. Let $\log N$ be divisible by 3. The other 2 cases are similar. Consider a cube of edge length $n = \log N$ in the $3D$ grid. Consider another cube of edge length $s + n$ such that the first cube is positioned "centrally" in the second one and

$$s = \left\lceil \sqrt{\left\lceil \frac{2N^{\frac{1}{3}}}{3} \right\rceil} \right\rceil.$$

Place $N^{1/3}$ copies of the second cube (with the first one inside) along the x axis with unit spacing such that they form a box C of size $(s+n+1)N^{1/3} \times (s+n) \times (s+n)$. Note that in the box C, there are at least

$$(s+n)^2 - n^2 \geq \left\lceil \frac{2N^{\frac{1}{3}}}{3} \right\rceil$$

tracks, parallel to the x-axis, which do not cross the cubes of sides n. Now let the $N^{1/3}$ small cubes be vertices of a $Q_{\frac{\log N}{3}}$ placed along the x-axis in the natural order. Thus having cutwidth $\lfloor 2N^{1/3}/3 \rfloor$, if we assume that the edges are drawn as in the linear layout in one plane. It is easy to redraw the edges of $Q_{\frac{\log N}{3}}$ such that they lie in the box C and if two edges shared the same track in the linear layout they will share the same track in C. Moreover the edges are attached to the opposite sides of a vertex only. See Figure 7.

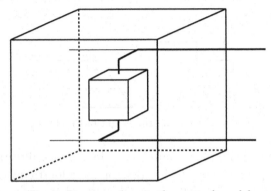

Fig. 7. Routing edges in the general model.

Finally, we use again the fact that $Q_{\log N} = Q_{\frac{\log N}{3}} \times Q_{\frac{\log N}{3}} \times Q_{\frac{\log N}{3}}$. Repeating the above construction for all $Q_{\frac{\log N}{3}}$'s in all 3 dimensions we get a layout for $Q_{\log N}$. One can check that the layout satisfies the assumptions of the model. The total volume is

$$(s+n+1)^3 N = \frac{2\sqrt{6}}{9} N^{\frac{3}{2}} + O(N^{\frac{4}{3}} \log N). \qquad \square$$

6   A 2-Active Layers Layout for Hypercubes

Calamoneri and Massini [5] proposed a 2-active layer layout model for bipartite graphs. In this model, all vertices are represented as rectangles and lie on two opposite sides of the bounding box of the layout volume. The vertices are distributed evenly between the two layers. The other properties of the model are as in the 1-active layer layout model. Particularly, they studied layouts of $\log N$-dimensional hypercubes, assuming that each vertex is represented as a $1 \times \lceil \frac{\log N}{2} - 1 \rceil$ rectangle, and proved

$$\text{VOL}_{2-AL}(Q_{\log N}) = \Omega(N^{\frac{3}{2}} \log^{\frac{1}{2}} N),$$
$$\text{VOL}_{2-AL}(Q_{\log N}) = O(N^{\frac{3}{2}} \log N).$$

We follow the same model but we represent every vertex of the hypercube as a square of side $\log N$. We have

Theorem 5. *The optimal volume of the 3-dimensional 2-active layer layout of the N-vertex hypercube $Q_{\log N}$ satisfies*

$$\mathrm{VOL}_{2-AL}(Q_{\log N}) = \Theta(N^{\frac{3}{2}} \log N).$$

Proof. The lower bound is proved in a similar way as in Theorem 1. The matching upper bound is obtained by placing two $(\log N - 1)$-dimensional hypercubes on the two opposite layers using the 1-active layout from Theorem 3 and adding the edges of the $\log N$-th dimension as straight-line segments between the layers.
□

References

1. Bel Hala, A., Congestion optimale du plongement de l'hypercube $H(n)$ dans la chaine $P(2n)$, *ITA* **27** (1993), 465-481.
2. Bezrukov, S., Chavez, J.D., Harper, L.H., Röttger, M., Schroeder, U.-P., The congestion of n-cube layout on a rectangular grid, *Discrete Mathematics*, **213**, (2000), 13-19.
3. Brebner, G., Relating routing graphs and two-dimensional grids, in: Proc. *VLSI: Algorithms and Architectures*, North Holland, 1985.
4. Biedl, T., Thiele, T., Wood, D.R., Three-dimensional orthogonal graph drawing with optimal volume, in: Proc. *11th Intl. Symposium on Graph Drawing*, Lecture Notes in Computer Science 1984, Springer, Berlin, 2000, 284-295.
5. Calamoneri, T., Massini, A., Nearly optimal three-dimensional layout of hypercube networks, in: Proc. *11th Intl. Symposium on Graph Drawing*, Lecture Notes in Computer Science 2912, Springer, 2003, 247-258.
6. Calamoneri, T., Massini, A., Optimal three-dimensional layout of interconnection networks, *Theoretical Computer Science* **255** (2001), 263-279.
7. Chi-Hsiang Yeh, Varvarigos, E.A., Parhami, B., Multilayer VLSI layout for interconnection networks, in: Proc. *Intl. Conf. on Parallel Processing*, 2000, 21-24.
8. Chi-Hsiang Yeh, Varvarigos, E.A., Parhami, B., Efficient VLSI Layouts of hypercubic networks, in: Proc. *Frontiers of Massivelly Parallel Computation*, 1999, 98-105.
9. Eades, P., Symvonis, A., Whitesides, S., Two algorithms for three-dimensional graph drawing, in: Proc. *4th Intl. Symposium on Graph Drawing*, Lecture Notes in Computer Science 1190, Springer Berlin 1996, 139-154.
10. Even, S., Kupershtok, R., Layout area of the hypercube, *Journal of Interconnection Networks* **4** (2003), 395-417.
11. Greenberg, R.I., Lee Guan, On the area of hypercube networks, *Information Processing Letters* **41** (2002), 41-46.
12. Leighton, F.T., Rosenberg, A.L., Three-dimensional circuit layouts, *SIAM Journal on Computing* **15** (1986), 793-813.
13. Leiserson, C.E., Area-efficient graph layouts (for VLSI), in: Proc. *21st Annual IEEE Symposium on Foundation of Computer Science*, IEEE Computer Science Press, 1980, 270-281
14. Patel, A., Kusalik, A, McCroskey, C., Area-efficient layouts for binary hypercubes, *IEEE Transactions on Computers* **49** (2000), 160-169.

15. Preparata, F.P., Optimal three-dimensional VLSI layouts, *Mathematical Systems Theory* **16** (1983), 1-8.
16. Raspaud, A., Sýkora, O., Vrt'o, I., Cutwidth of the de Bruijn graph, *RAIRO* **26** (1995), 509-514.
17. Rosenberg, A.L., Three-dimensional VLSI: A case study, *Journal of the ACM* **30** (1983), 397-416.
18. Thompson, C.D., Area-time complexity for VLSI, in: Proc. *11th Annual ACM Symposium on Theory of Computing*, 1979, 81-88.
19. Ullman, J.D., Computational Aspects of VLSI, Comp. Sci. Press, Rockville, 1983.

New Theoretical Bounds of Visibility Representation of Plane Graphs[*]

Huaming Zhang and Xin He

Department of Computer Science and Engineering,
SUNY at Buffalo, Buffalo, NY, 14260, USA

Abstract. In a *visibility representation* (VR for short) of a plane graph G, each vertex of G is represented by a horizontal line segment such that the line segments representing any two adjacent vertices of G are joined by a vertical line segment. Rosenstiehl and Tarjan [6], Tamassia and Tollis [7] independently gave linear time VR algorithms for 2-connected plane graph. Afterwards, one of the main concerns for VR is the size of VR. In this paper, we prove that any plane graph G has a VR with height bounded by $\lfloor \frac{5n}{6} \rfloor$. This improves the previously known bound $\lceil \frac{15n}{16} \rceil$. We also construct a plane graph G with n vertices where any VR of G require a size of $(\lfloor \frac{2n}{3} \rfloor) \times (\lfloor \frac{4n}{3} \rfloor - 3)$. Our result provides an answer to Kant's open question about whether there exists a plane graph G such that all of its VR require width greater that cn, where $c > 1$.

1 Introduction

The first simple linear time VR algorithm was given in [6, 7] for a 2-connected plane graph G. One of the main concerns afterwards for VR is the size of the VR, i.e., the height and width of VR. Some work has been done to reduce the size of VR. We summarize related previous results in the following table:

References	Plane graph G	4-Connected plane graph G
[6, 7]	Width of VR $\leq (2n - 5)$	Height of VR $\leq (n - 1)$
[2]	Width of VR $\leq \lfloor \frac{3n-6}{2} \rfloor$	
[5]	Width of VR $\leq \lfloor \frac{22n-42}{15} \rfloor$	
[3]		Width of VR $\leq (n - 1)$
[8]	Height of VR $\leq \lceil \frac{15n}{16} \rceil$	
[9]	Width of VR $\leq \lfloor \frac{13n-24}{9} \rfloor$	Height of VR $\leq \lceil \frac{3n}{4} \rceil$

In this paper, we obtain the following main results:

(1) We prove that every plane graph G has a VR with height bounded by $\lfloor \frac{5n}{6} \rfloor$, which can be obtained in linear time.

---

[*] Research supported in part by NSF Grant CCR-0309953.

J. Pach (Ed.): GD 2004, LNCS 3383, pp. 425–430, 2004.

(2) We give a plane graph G with n vertices such that any VR of G require a size of $(\lfloor \frac{2n}{3} \rfloor) \times (\lfloor \frac{4n}{3} \rfloor - 3)$. This answers Kant's open question about whether there exists a plane graph G such that all of its VR require width greater that cn, where $c > 1$ [2].

2 Preliminaries

In this section, we give definitions and preliminary results. We abbreviate the words "counterclockwise" and "clockwise" as ccw and cw respectively.

An *orientation* of a graph G is a digraph obtained from G by assigning a direction to each edge of G. We will use G to denote both the resulting digraph and the underlying undirected graph unless otherwise specified. For a 2-connected plane graph G and an exterior edge (s, t), an orientation of G is called an *st-orientation* if the resulting digraph is acyclic with s as the only source and t as the only sink. Note that, for every face f of G, its boundary cycle consists of two directed paths. The path on its left (right, resp.) side is called the *left (right, resp.) path* of f. There is exactly one source (sink, resp.) vertex on the boundary of f, it is called the source (sink, resp.) of f.

Let G be a 2-connected plane graph and (s, t) an exterior edge. An *st-numbering* of G is a one-to-one mapping $\xi : V \rightarrow \{1, 2, \cdots, n\}$, such that $\xi(s) = 1$, $\xi(t) = n$, and each vertex $v \neq s, t$ has two neighbors u, w with $\xi(u) < \xi(v) < \xi(w)$, where u (w, resp.) is called a *smaller neighbor (bigger neighbor, resp.)* of v. Lempel et. al. [4] showed that for every 2-connected plane graph G and an exterior edge (s, t), there exists an st-numbering. Given an st-numbering ξ of G, we can orient G by directing each edge in E from its lower numbered end vertex to its higher numbered end vertex. The resulting orientation is called the *orientation derived from* ξ which is an st-orientation of G. On the other hand, if $G = (V, E)$ has an st-orientation $\mathcal{O}$, we can define an 1-1 mapping $\xi : V \rightarrow \{1, \cdots, n\}$ by topological sort. Thus, we can interchangeably use the term an st-numbering of G and the term an st-orientation of G, where each edge of G is directed accordingly. The following lemma was given in [6, 7]:

Lemma 1. *Let G be a 2-connected plane graph. Let $\mathcal{O}$ be an st-orientation of G. A VR of G can be obtained from $\mathcal{O}$ in linear time. The height of the VR is the length of the longest directed path in $\mathcal{O}$.*

Let G be a plane triangulation, $v_1, v_2, \cdots, v_n$ an ordering of the vertices of G where v_1, v_2, v_n are the three exterior vertices of G in ccw order. Let G_k be the subgraph of G induced by $v_1, v_2, \cdots, v_k$ and H_k the exterior face of G_k. Let $G - G_k$ be the subgraph of G obtained by removing $v_1, v_2, \cdots, v_k$.

Definition 1. [1] *An ordering $v_1, \cdots, v_n$ of a plane triangulation G is canonical if the following hold for every $k = 3, \cdots, n$:*

1. *G_k is 2-connected, and its exterior face H_k is a cycle containing the edge (v_1, v_2).*

2. *The vertex v_k is on the exterior face of G_k, and its neighbors in G_{k-1} form a subinterval of the path $H_{k-1}-(v_1, v_2)$ with at least two vertices. Furthermore, if $k < n$, v_k has at least one neighbor in $G - G_k$. (Note that the case $k = 3$ is degenerated, and $H_2 - (v_1, v_2)$ is regarded as the edge (v_1, v_2) itself.)*

A canonical ordering of G can be viewed as an order in which G is reconstructed from a single edge (v_1, v_2) step by step. At step k, when v_k is added to construct G_k, let $c_l, , \cdots, c_r$ be the lower ordered neighbors of v_k from left to right on the exterior face of G_{k-1}. We call (v_k, c_l) the *left edge* of v_k, (v_k, c_r) the *right edge* of v_k, and the edges (c_p, v_k) with $l < p < r$ the *internal edges* of v_k. The collection T of the left edges of the vertices v_j for $3 \le j \le n$ plus the edge (v_1, v_2) is a spanning tree of G and is called a *canonical ordering tree* of G [1].

3   Compact Visibility Representation of Plane Graphs

Let G be a plane triangulation with n vertices, v_1, v_2, v_n be its exterior vertices in ccw order. Let T be a canonical ordering tree of G rooted at v_n with at least $\lceil \frac{n+1}{2} \rceil$ leaves [8]. We will construct the st-numbering ξ_T. (see Figure 1 for an illustration. Only part of the tree is drawn. Tree edges are drawn in solid lines, non-tree edges are drawn in dotted lines, dashed lines represent a path in the tree.)

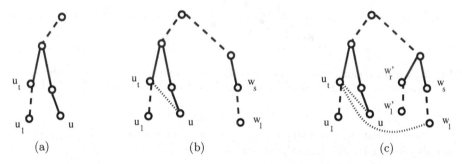

Fig. 1. (a) No edge between u_t and u. (b) An edge between u_t and u, no edge between u_t and w_1. (c) Edges between u_t and u, u_t and w_1. No edge between u_1 and w_r'.

Traveling from the leftmost unassigned leaf of T by **ccw postordering** with respect to T. (The first visited vertex is v_2). Assume we begin from the leaf u_1, continue to $u_1, u_2, \cdots, u_t$, then we reach the next leaf u. There are three cases to be considered for each step:

Case 1: There is no edge between u_t and u. We then assign the remaining numbers from $1, 2, \cdots, n$ to $u_1, u_2, \cdots, u_t$ by **ccw postordering** with respect to T, continue to leaf u, continue on, stop before a new leaf encountered. Continue our assignment if there are leaves left unassigned. See Figure 1 (a).

Case 2: There is an edge between u_t and u, but no edge between u_t and w_1, where w_1 is the rightmost unassigned leaf of T. We then assign the remaining numbers from $1, 2, \cdots, n$ to $u_1, u_2, \cdots, u_t$ by **ccw postordering**. Then jump to leaf w_1, continue on by **cw postordering**, stop before a new leaf encountered. Continue our assignment if there are leaves left unassigned. See Figure 1 (b).

Case 3: There is an edge between u_t and u, and an edge between u_t and w_1. Then Starting from the leaf w_1, assign the remaining numbers to T by **cw postordering** with respect to T. Assume that vertices assigned are $w_1, w_2, \cdots, w_s$, then to next leaf, denote it by w'_1, keep assign numbers to $w'_1, w'_2, \cdots, w'_r$ until a new leaf encountered. Jump back to u_1, keep assign numbers to $u_1, u_2, \cdots, u_t$ by **ccw postordering** with respect to T. Stop before a new leaf encountered. Continue our assignment if there are leaves left unassigned. See Figure 1 (c).

There are at most one or two leaves left at last, then assign remaining numbers to them by **ccw postordering**. (or **cw postordering**). Figure 2 shows such a construction.

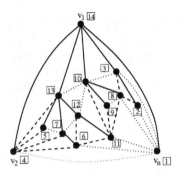

Fig. 2. A plane triangulation G, a canonical ordering tree T of G (drawn in solid lines). The st-numbering ξ_T of G constructed from T as described above.

Observe that, for each step above, at least one vertex has to be bypassed by any directed path. We have the following:

Theorem 1. *Let G be a plane triangulation with n vertices, T a canonical ordering of G with at least $\lceil \frac{n+1}{2} \rceil$ leaves. Let ξ_T be constructed as above. Then:*

1. *ξ_T is an st-numbering of G. The longest directed path of G according to ξ_T is bounded by $\lfloor \frac{5n}{6} \rfloor$.*
2. *G has a VR **with the same exterior face** with height bounded by $\lfloor \frac{5n}{6} \rfloor$, which can be obtained in linear time.*

Next, we have the following theorem regarding the lower bound of VR:

Theorem 2. *There is an n-vertex plane triangulation G such that any VR of G **with the same exterior face** requires a size of $(\lfloor \frac{2n}{3} \rfloor) \times (\lfloor \frac{4n}{3} \rfloor - 3)$.*

Proof. Suppose that G_k is the graph of k nested triangles with $n = 3k$ vertices as shown in Figure 3 (a). We want to show that any VR of G_k requires a grid size of $(\lfloor \frac{2n}{3} \rfloor) \times (\lfloor \frac{4n}{3} \rfloor - 3) = 2k \times (4k - 3)$.

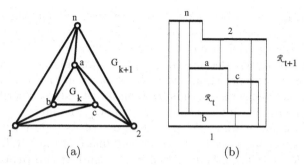

Fig. 3. (a) Nested triangles, (b) VR of nested triangles.

First, we want to use induction on k to prove the height bound. When $k = 1$, it is true. Suppose that it is true for $k = t$. Now, consider $k = t + 1$:

Given any VR $\mathcal{R}_{t+1}$ of G_{t+1}, by removing the horizontal lines segments representing the exterior vertices of G_{t+1} and the vertical line segments representing the edges adjacent to the exterior vertices of G_{t+1}, the resulting representation is a VR of G_t, denoted by $\mathcal{R}_t$. Applying induction hypothesis, the height of $\mathcal{R}_t$ is at least $2t$.

Observe the topmost and the lowest horizontal line segments in $\mathcal{R}_{t+1}$, they have to represent the exterior vertices of G_{t+1}. Thus, the height of $\mathcal{R}_{t+1}$ is at least 2 plus the height of $\mathcal{R}_t$, which is at least $2(t + 1)$. (See Figure 3 (b)) This finishes the induction for the height bound.

Next, we want to prove the width bound. Given any VR $\mathcal{R}_k$ of G_k. It is easy to see that we can obtain an st-numbering ξ of G_k from $\mathcal{R}_k$. We assign numbers from $1, 2, \cdots, n$ to the vertices of G_k such that the lower its corresponding horizontal line segment in $\mathcal{R}_k$ is, the smaller its assigned number is. (If two or more vertices have the same level of horizontal line segments, then arbitrarily assign consecutive numbers to them.) Denote the vertices by $v_1, v_2, \cdots, v_{n=3k}$, where $\xi(v_i) = i$. Without loss of generality, we may assume that the vertical line segment representing the edge $(v_1, v_{n=3k})$ is the leftmost vertical line in $\mathcal{R}_k$. G_k can be directed according to ξ. And ξ induces its dual st-orientation ξ^* of R_k^*.

Claim: The width of the VR $\mathcal{R}_k$ is greater or equal to the length of the longest directed paths in ξ^* of G_k^* .

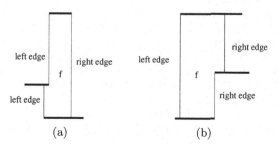

Fig. 4. (a) Two left edges and one right edge, (b) one left edge and two right edges.

Proof of Claim: Let P be a longest directed path of G_k^* in the st-orientation ξ^*. Obviously, it starts from its source s^*, i.e. an interior face of G_k. And it ends at its sink t^*, the exterior face of G_k. We try to trace P in $\mathcal{R}_k$. It starts from the leftmost interior face. Because G_k is a plane triangulation, each face f of G_k only has two possible representations in $\mathcal{R}_k$ as shown in Figure 4. Its right edges in $\mathcal{R}_k$ always have bigger x-coordinates than its left edges. Therefore, when P passes through one face of G_k (i.e. one vertex in G_k^*), it enters a face f from one of its left edges, and it walks out of f from one of its right edges. Thus, each edge on P has to add at least 1 to the x-coordinate of the VR $\mathcal{R}_k$. Therefore, the width of $\mathcal{R}_k$ is at least the length of P.   **End of the proof of Claim.**

Now, we only need to show that for any st-numbering of G_k, the length of its longest directed paths in its dual st-orientation is no short than $\lfloor \frac{4n}{3} \rfloor - 3 = 4k - 3$. We want to show this by induction: When $k = 1$, this is trivially true. Assume that it is true when $k = t$. Consider the case of $k = t+1$: Given any st-numbering ξ_{t+1} of G_{t+1}, it induces an st-numbering ξ_t of G_t. According to the induction hypothesis, a longest directed path P_t in its dual orientation ξ_t^* is no shorter than $4t - 3$. Consider P_t in ξ_{t+1}^*, it can be extended to a directed path P_{t+1} in ξ_{t+1}^* from its source to its sink. Because of the way G_{t+1} is nested, P_{t+1} has to pass through at least 2 vertices (2 faces of G_{t+1}) before it can share edges with the path P_t. (see Figure 3) After it passes through P_k, it also has to pass through at least 2 vertices (2 faces of G_{t+1}) before it can reach its sink. Therefore, the length of P_{t+1} has to be at least 4 plus the length of P_t. Therefore, the length of P_{t+1} has to be at least $4(t + 1) - 3$. This finishes our proof.

References

1. H. de Fraysseix, J. Pach and R. Pollack, How to draw a planar graph on a grid. *Combinatorica* 10 (1990), 41-51.
2. G. Kant, A more compact visibility representation. *International Journal of Computational Geometry and Applications* 7 (1997), 197-210.
3. G. Kant and X. He, Regular edge labeling of 4-connected plane graphs and its applications in graph drawing problems. *Theoretical Computer Science* 172 (1997), 175-193.
4. A. Lempel, S. Even and I. Cederbaum, An algorithm for planarity testing of graphs, in *Theory of Graphs* , pp. 215-232, Rome, 1967.
5. C.-C. Lin, H.-I. Lu and I-F. Sun, Improved compact visibility representation of planar graph via Schnyder's realizer, in: Proc. STACS'03, LNCS, Vol. 2607, (Springer-Verlag, Berlin, 2003) 14-25.
6. P. Rosenstiehl and R. E. Tarjan, Rectilinear planar layouts and bipolar orientations of planar graphs. *Discrete Comput. Geom.* 1 (1986), 343-353.
7. R. Tamassia and I.G.Tollis, An unified approach to visibility representations of planar graphs. *Discrete Comput. Geom.* 1 (1986), 321-341.
8. H. Zhang and X. He, Compact Visibility Representation and Straight-Line Grid Embedding of Plane Graphs, in: Proc. WADS'03, LNCS, Vol. 2748, (Springer-Verlag Heidelberg, 2003) 493-504.
9. H. Zhang and X. He, On Visibility Representation of Plane Graphs, in: Proc. STACS'04, LNCS, Vol. 2996, (Springer-Verlag Heidelberg, 2004) 477-488.

Visualizing Large Graphs
with Compound-Fisheye Views and Treemaps[*]

James Abello[1], Stephen G. Kobourov[2], and Roman Yusufov[2]

[1] DIMACS Center,
Rutgers University
abello@dimacs.rutgers.edu
[2] Department of Computer Science,
University of Arizona
{kobourov,ryusufov}@cs.arizona.edu

Abstract. Compound-fisheye views are introduced as a method for the display and interaction with large graphs. The method relies on a hierarchical clustering of the graph, and a generalization of the traditional fisheye view, together with a treemap representation of the cluster tree.

1 Introduction

Many of the challenges in visualization today arise from the volume of data. As the volume of data grows, so too does our desire to visualize the data. Often the data contain relationships between objects, and can be represented as a graph. A great deal of research and investment has gone into developing better display systems, high-resolution screens, and visualization walls. However, no matter how good our display systems get and how many pixels per square inch can be obtained, there will always be graphs that are too large to be fully displayed and too complex to comprehend as a whole.

Graphs with hundreds of thousands of nodes and millions of edges are commonplace in many of today's applications, such as telecommunications, software engineering, and databases. Recent graph drawing algorithms allow us to compute layouts for large graphs in reasonable times. However, exploring and interacting with such graphs in their entirety is likely to be ineffective.

A visualization technique that relies on *fisheye views, clustering,* and *treemaps* is introduced in order to provide a way to explore and interact with large graphs. In this context, clustering implies any hierarchical decomposition of the graph. The cluster computation in turn yields level-views of the input graph at different levels of detail. Navigation from one level of the hierarchy to the next is provided by partial refinement and/or coarsening of different parts of the view. These operations correspond to zooming in and zooming out. *Compound-fisheye views* are introduced as a technique that provides high level of detail in the focus area while also providing a global view of the graph, through distortion of the view in

---

[*] This work is partially supported by the NSF under grant ACR-0222920 and by ITCDI under grant 003297.

J. Pach (Ed.): GD 2004, LNCS 3383, pp. 431–441, 2004.

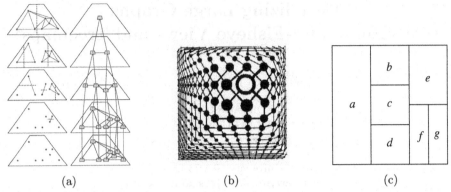

(a)   (b)   (c)

Fig. 1. Different graph views: (a) Multi-level view of a geometrically clustered graph; (b) Fisheye view of a grid-like graph; (c) Treemap for a small tree.

areas away from the focus. The compound-fisheye view combines clusters from different levels of the cluster-hierarchy by showing high detail clusters in the area of interest and progressively lower detail clusters away from the focus. A treemap view of the cluster tree is also used to provide a global view of the original graph.

1.1 Related Work

Multi-level display algorithms are described in the context of visualization for clustered graphs in [9]. Compound and clustered graphs are studied in [10, 18]. Multi-level views [8, 9] show large graphs at multiple abstraction levels. A natural realization of such multiple level representations is a 3D drawing with each level drawn on a plane at a different z-coordinate, and with the clustering structure drawn as a tree in 3D; see Fig. 1(a). The related concept of a graph sketch is introduced in [1] and is used in the MGV system [2].

The above algorithms assume that the clustering of the graph is given. In the case where the input graph has no clustering information, hierarchical clustering algorithms based on the structural properties of the graph can be used [4, 15]. Alternatively, geometric graph clustering based on binary space partitions can also be used to display large graphs, as described in [8]. The quality of the resulting multi-level drawings depends on the initial embedding of the graph in the plane. Hence, a good initial embedding of a large graph is a prerequisite for this method. Recently, a number of efficient algorithms for layout of large graphs have been developed, based on multi-scale, high-dimensional and spectral methods [12–14, 20]. Data structures supporting cluster-graph operations (such as cluster-expand and cluster-collapse) have been studied in [6, 7].

Fisheye views are introduced in the context of viewing and filtering computer programs [11]. Fisheye views show one area of interest quite large and in detail and show other areas successively smaller and in less detail by using a distortion function; see Fig. 1(b). 2-D fisheye view graph drawings with position and size distortions are studied in [16]. Finally, treemaps [17] have been studied for over a decade as an efficient space-filling layout of tree-like structures; see Fig. 1(c). The nodes of the tree are displayed as nested rectangles in the treemap. The children of a node are within the rectangle of the parent. Squarified treemaps ensure good aspect ratio for the rectangles [5] and ordered treemaps keep related items spatially close to each other in the map [3].

1.2   Our Contribution

A visualization technique called a *compound-fisheye view* is described. It makes possible to extend the effective use of the traditional fisheye view to larger graphs. The technique relies on creating a clustered graph from the original graph, via a hierarchical clustering algorithm. The resulting cluster tree is then shown as a treemap and is also used to navigate the compound-fisheye view.

Compound-fisheye views allow the exploration of an area of interest in detail, by providing an interactive view of the graph, while still capturing the global context. When a focus node is selected from the current view, the corresponding subgraph at the next level is depicted and the view is updated. Similarly, less details can be requested about a particular node in the current view, which results in the replacement of the node (and its siblings in the cluster tree) with its parent, and the subsequent update of the view. These operations provide the ability to zoom in and out with respect to the current view.

Moreover, the compound-fisheye view has clusters from different levels in the cluster tree, depending on how close they are to the area of interest. In a way similar to traditional fisheye views, when more detail is requested in a particular area, the areas farthest away from the focus are automatically reduced in detail. Unlike traditional fisheye views, however, the reduction in the detail is achieved by replacing parts of the graph far away from the focus with coarser representations from the cluster tree (rather than just shrinking the area allocated to these parts, via distortion).

To aid the comprehension of the overall structure, a treemap view of the cluster tree provides global context. A prototype of the visualization system has been implemented and tested with graphs of varying sizes, up to 10,000 nodes. The screen-shots in Fig 7 show the system in action.

2   Hierarchical Clustering

A graph clustering algorithm is *geometric* if nodes are clustered according to their spacial locality, given an initial embedding of the entire graph. Similarly, a graph clustering algorithm is *structural* if nodes are clustered based on structural features of the original graph (such as connectivity and density). Any clustering algorithm can be used for the purpose of compound-fisheye view navigation, provided that the clustering is hierarchical. One structural clustering algorithm and one geometric clustering algorithm have been implemented as a part of the prototype.

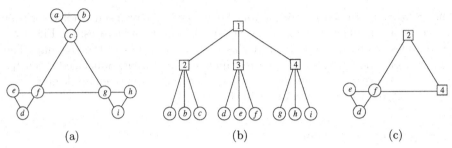

Fig. 2. (a) Input graph G; (b) Cluster-tree T: lettered nodes represent input graph nodes and numbered nodes represent clusters; (c) Compound-fisheye view of $C = (G, T)$.

The structural clustering algorithm implemented in the system is a Markov clustering algorithm [19]. Markov clustering, also known as MCL, uses a random walk in the graph to identify densely connected components. MCL is also general enough to allow weighted graphs (including negative weights) that can be directed or undirected.

The geometric clustering algorithm implemented in the system is a Binary Space Partition (BSP) algorithm, similar to that in [8]. Starting with a 2D layout of the entire graph, a k-d tree recursive partition is used to obtain the clustering. The initial embedding is obtained using a high-dimensional multi-level method, similar to that in [12].

2.1   Clustered Graphs

Whether structural or geometric, the clustering algorithm produces a cluster tree. A leaf node in this tree represents a node from the original graph. An internal node represents a cluster of nodes, which consists of all the nodes in its subtree. A cluster may contain leaf nodes and/or other clusters. The cluster tree data structure is the interface between the clustering algorithm, the compound-fisheye view, and the treemap.

Fig. 2(a-b) show an example of an input graph G and the recursive clustering defined by a tree T. Together the input graph and the cluster tree make up the clustered-graph $C = (G, T)$. All the nodes of T at a given depth i represent the clusters of that level. A *view at level i*, $G_i = (V_i, E_i)$, consists of the nodes of depth i in T and a set of representative edges. The edge $(u, v) \in E_i$ if there is an edge between a and b in G, where a is in the subtree of u and b is in the subtree of v. Fig. 2(c) shows a compound-fisheye view of the clustered graph. The compound-fisheye view is initialized with the root of the tree and interaction is accomplished by means of the two clustered-graph operations: `cluster-expand` and `cluster-collapse`. The compound-fisheye view may contain a combination of nodes from different levels in T.

Consider the `cluster-expand` operation. This operation takes a node in the compound-fisheye view, called a *cluster-node*, replaces it with its children in the cluster tree, and performs the necessary updates to the edges in the graph.

Cluster-nodes do not exist in the original graph, but are created as part of a clustering tree. In Fig. 2(c), all numbered nodes are cluster-nodes and all the nodes from the input graph are leaves in the cluster tree. When expanding a cluster, determining which nodes need to be added to the graph is straightforward, since they will always be the children in the clustering tree of the cluster-node being expanded. The challenge arises when trying to determine what new edges should be added to the current view. An edge exists between two cluster-nodes only if some member from one cluster is adjacent to a member of the other cluster in the original graph. In Fig. 2 an edge exists between cluster nodes 2 and 4 because nodes c and g are adjacent in the original unclustered graph.

The `cluster-collapse` operation is simpler. When a cluster is to be collapsed, one of its children in the clustering tree must be selected, since the cluster-node representing the cluster will not be in the graph. In Fig. 2(b), the cluster-node 3 can only be collapsed by selecting one of its children, d, e, or f from the compound-fisheye view in Fig. 2(c). Thus, in order to collapse a cluster, the children of the cluster-node (siblings of the selected node in the cluster tree) must be identified. Next, the set of nodes that are adjacent to any of those children is identified. Finally, an edge is added between the collapsed cluster-node and each adjacent node.

3   Compound-Fisheye Views

Fisheye views of graph drawings allow a user to understand the structure of a graph near a specific set of nodes (local detail), and at the same time they display the graph's overall structure (global context). Such views achieve smooth integration of both local detail and global context by repositioning and resizing nodes and edges in the graph. However, even for graphs with a few hundred nodes, the benefits of this approach are lost as the areas away from the focus become too congested to comprehend.

The fisheye view idea is applied on a hierarchically clustered graph to obtain a *compound-fisheye view*; see Fig. 3 (a-b). The compound-fisheye view makes it possible to extend the effective use of the traditional fisheye view to larger graphs. Conceptually, the nodes in the compound-fisheye view of a clustered graph can be obtained by taking the intersection of an inverted cone with the level-views of the clustered graph; see Fig. 3(c). The cone is centered at the area of interest in the original graph (at the deepest level in the multi-level view). The farther away from the point of the cone, the coarser are the views that it intersects. Thus, parts of the graph that are far away from the focus are represented using clusters at higher levels in the cluster tree.

The compound-fisheye view allows the user to navigate it, modify it, and interact with it. The underlying representation provides an adaptive view with the look and feel of a normal graph. Interaction with the compound-fisheye view is accomplished through the `cluster-expand` and `cluster-collapse` operations. These operations correspond to zooming in and out with respect to the current view. When a focus area is selected, the corresponding subgraph at the next (higher or lower) level is depicted and the current view is updated.

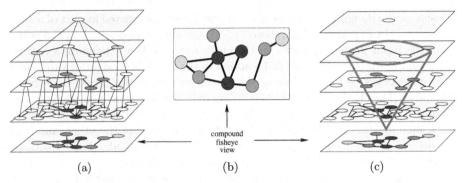

Fig. 3. Compound-fisheye view of a clustered graph: (a) A multi-level view of a clustered graph with highlighted nodes that make up the compound-fisheye view at the bottom; (b) The compound-fisheye view is made of clusters from three different levels of the hierarchy; (c) Conceptual view: the intersection of the multi-level view with an inverted cone.

Once the nodes in the current compound-fisheye view have been identified, the edges connecting them can be determined. Similar to the edges in the level-views, the edges in a compound-fisheye view are easily defined: the edge (u, v) is in the compound-fisheye view if there is an edge between a and b in G, where a is in the subtree of u and b is in the subtree of v. Unlike in level-views, the nodes in the compound-fisheye view are made of clusters from different levels in the cluster tree. This makes the edge computations more challenging, especially for large graphs. While efficient data structures have been designed for this problem [6, 7], a simple node and edge hashing scheme are employed in the implementation of the prototype system.

To ensure that the compound-fisheye view does not become too cluttered, the parts of the view away from the focus are automatically collapsed. With standard fisheye views, it is not difficult to support more than one focus. This idea generalizes to compound-fisheye views as well, although this functionality has not yet been implemented.

4   Treemaps

Treemaps are a space-filling graph visualization technique first introduced in [17]. An important feature of treemaps is that they make very efficient use of display space. Thus it is possible to display large trees with many hierarchical levels in a minimal amount of space. Fig. 4(a) shows a sample tree structure and Fig. 4(b) shows the corresponding treemap. The algorithm used to partition the display space is known as the "slice-and-dice algorithm" and functions like a k-d tree space partition. The positioning of tree nodes in a treemap is a recursive process. First, the children of the root are placed across the display area horizontally, where each node's area is directly proportional to its weight. Then, for each node n already displayed, each of n's children is placed across vertically within n's display area. This process is repeated, alternating between horizontal and vertical placement until all nodes have been displayed.

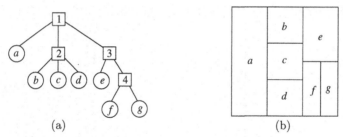

(a)   (b)

Fig. 4. A cluster tree (a) and its treemap representation (b).

Treemaps can be especially helpful when dealing with large clustered graphs. While the compound-fisheye view combines detailed local information and a global context, treemaps lend themselves naturally to showing the information encapsulated in the clustering tree. When viewing a graph at some level of abstraction, the viewer is really looking at nodes belonging to some level in the cluster tree. A treemap can display the whole structure of a cluster tree, thus allowing the user to place the current view in context.

In the standard treemap of Shneiderman [17] the nodes are represented as rectangles of various shapes. This makes a visual comparison of their importance (as determined by area) difficult, especially as the rectangles vary in orientation as well. In squarified treemaps [5] the *aspect ratio* (the ratio between the width and height of a rectangle) is taken into account when placing nodes in the treemap. The resulting treemaps contains squarish elements, making it easier to visually compare their areas.

Squarified treemaps with a modified visual appearance are implemented in the prototype of the compound-fisheye view visualization system. In traditional treemaps, only the leaf nodes of a tree are displayed as rectangular areas. It is often difficult to determine the nesting depth of the treemap structure as can be seen in Fig. 5(a). For the purpose of navigating the compound-fisheye view it is important to show depth information about the clustering tree, so the first step is to display the nesting information as shown in Fig. 5(b). In order to make the nested representation easier to view, progressively thinner borders for deeply nested elements are used. The thickness of the border around an element is inversely proportional to its level in the original tree. Thus, the root node has the thickest border and the leaf nodes have the thinnest borders as shown in Fig. 5(c).

5   Visualization Technique

Our visualization technique provides two views: one of the compound-fisheye view and another of the treemap defined on the cluster tree. On their own, neither of these approaches is powerful enough to represent and navigate a large graph. The treemap algorithm applies only to trees, and while it could be applied to a spanning tree of the graph, it does not show connectivity well. The compound-

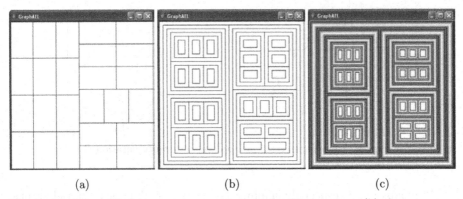

(a)   (b)   (c)

Fig. 5. (a) A squarified treemap representation of a 25-node tree; (b) the same tree with nested rectangles; the same tree with frames.

fisheye view abstracts a great deal of the graph information, which can only be recovered by recursive expansion of clusters. While showing connectivity and local details well, it only shows an abstraction of the overall structure, making navigational decisions difficult.

Together, the compound-fisheye view of the graph and the treemap of the cluster tree offer a better approach to showing both local details and global context. The combined view is shown in Fig. 6. One of the main shortcomings of a compound-fisheye view is that nodes that are clustered become invisible in the display and the viewer cannot deduce information about the structure underneath the cluster. By using a combined view displaying both the compound-fisheye view and the treemap, information about the subtree rooted at the cluster-node can be better conveyed. In Fig. 6, the red node (rightmost node) has been selected by the user and its corresponding rectangle in the treemap is highlighted in blue (top left). It is easy to see that the selected cluster-node is

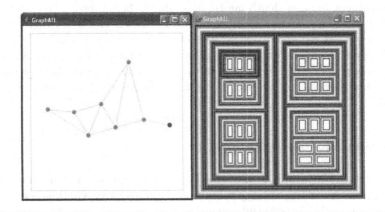

Fig. 6. Compound-fisheye view and treemap: the red cluster-node (rightmost) has been selected and its corresponding representation (top left) is highlighted in the treemap.

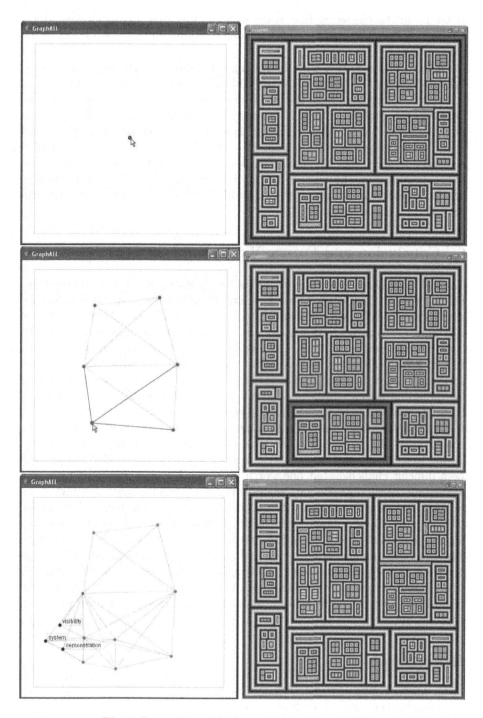

Fig. 7. Interaction with the system: steps 1, 2, and 3.

at level 4 in the clustering tree, contains 3 children that are also leaves, and if the user chooses to expand the cluster-node it will be replaced by 3 nodes in the current view.

When a user selects a node by placing the mouse pointer over it, it is highlighted along with all of its cluster tree siblings. This reveals the branching factor and permits the user to see the set of nodes that will be collapsed, should he choose to collapse the cluster (via a right mouse-click). During a collapse operation, the nodes being collapsed are highlighted in red, and whenever an expand operation is performed (via a left mouse-click), the cluster-node being expanded is marked in green. Once a cluster has been fully expanded, the resulting nodes are the nodes from the original graph and are colored in black. If all clusters are expanded, then the resulting graph is identical to the original graph and will contain only black nodes.

6   Conclusion and Future Work

The technique described in this paper has been implemented in a Java prototype. Fig. 7 shows snapshots of the exploration of the GD literature topic graph for the years 1994-2000, with 332 nodes and 1,338 edges. In step 1 the cluster tree root and its treemap are shown. In step 2 the root is expanded and one of the cluster-nodes has been selected (highlighting its corresponding treemap region). In step 3, the selected node has been expanded and labeled leaves of the tree appear in the view. For this example, the combined computation time needed for the initial layout, clustering, treemap, and rendering take under 5 seconds. Interaction with the compound-fisheye view is in real-time.

While the system can deal with larger graphs (with up to 10,000 nodes) the computation times are not nearly as good. Incorporating this prototype into a fully functional graph visualization system that can handle large graphs will be a difficult challenge, but one worth pursuing. Efficient data structures and algorithms, to support expand/collapse operations will become important if one requires real-time interaction with graphs with hundreds of thousands of nodes.

The current system uses cluster-node positions that are set by the clustering algorithm and are never modified. A natural alternative is to apply a layout algorithm to the compound-fisheye view and reapply it after an expand/collapse operation is performed. Since the view changes as nodes are being added/removed through expand and collapse operations, the layout algorithm must preserve the mental map between consecutive layouts and make smooth transitions between such layouts.

References

1. J. Abello, I. Finocchi, and J. Korn. Graph sketches. In *IEEE Proc. Information Visualization*, pages 67–71, 2001.
2. J. Abello and J. Korn. Mgv: A system to visualize massive multi-digraphs. *IEEE Transactions on Computer Graphics and Visualization*, pages 1–16, 2002.

3. B. Bederson, B. Shneiderman, and M. Wattenberg. Ordered and quantum treemaps: Making effective use of 2D space to display hierarchies. *ACM Transactions on Graphics*, 21(4):833–854, 2002.
4. U. Brandes, M. Gaertler, and D. Wagner. Experiments on graph clustering algorithms. In *11th Euro. Symp. on Algorithms (ESA)*, pages 568–579, 2003.
5. M. Bruls, K. Huizing, and J. J. van Wijk. Squarified treemaps. In *Proc. Joint Eurographics/IEEE TVCG Symp. Visualization, VisSym*, pages 33–42, 2000.
6. A. L. Buchsbaum, M. T. Goodrich, and J. R. Westbrook. Range searching over tree cross products. In *8th Euro. Symp. on Algorithms (ESA)*, pages 120–131, 2000.
7. A. L. Buchsbaum and J. R. Westbrook. Maintaining hierarchical graph views. In *Proceedings of the 12th ACM-SIAM Symposium on Discrete Algorithms (SODA)*, pages 566–575, 2000.
8. C. A. Duncan, M. T. Goodrich, and S. G. Kobourov. Balanced aspect ratio trees and their use for drawing large graphs. *Journal of Graph Algorithms and Applications*, 4:19–46, 2000.
9. P. Eades and Q. Feng. Multilevel visualization of clustered graphs. In *Proceedings of the 4th Symposium on Graph Drawing (GD)*, pages 101–112, 1996.
10. P. Eades, Q. Feng, and X. Lin. Straight-line drawing algorithms for hierarchical graphs and clustered graphs. In *Proceedings of the 4th Symposium on Graph Drawing (GD)*, pages 113–128, 1996.
11. G. W. Furnas. Generalized fisheye views. In *Proceedings of ACM Conference on Human Factors in Computing Systems (CHI '86)*, pages 16–23, 1986.
12. P. Gajer and S. G. Kobourov. GRIP: Graph dRawing with Intelligent Placement. *Journal of Graph Algorithms and Applications*, 6(3):203–224, 2002.
13. R. Hadany and D. Harel. A multi-scale algorithm for drawing graphs nicely. *Discrete Applied Mathematics*, 113(1):3–21, 2001.
14. D. Harel and Y. Koren. A fast multi-scale method for drawing large graphs. *Journal of graph algorithms and applications*, 6:179–202, 2002.
15. R. Sablowski and A. Frick. Automatic graph clustering. In *Proceedings of the 4th Symposium on Graph Drawing (GD)*, pages 395–400, 1996.
16. M. Sarkar and M. H. Brown. Graphical fisheye views. *Communications of the ACM*, 37(12):73–84, 1994.
17. B. Shneiderman. Tree visualization with treemaps: a 2-d space-filling approach. Technical report, HCI Lab, University of Maryland, Mar. 1991.
18. K. Sugiyama and K. Misue. Visualization of structural information: Automatic drawing of compound digraphs. *IEEE Transactions on Systems, Man, and Cybernetics*, 21(4):876–892, 1991.
19. S. van Dongen. *Graph Clustering by Flow Simulation*. PhD thesis, University of Utrecht, 2000.
20. C. Walshaw. A multilevel algorithm for force-directed graph drawing. In *Proceedings of the 8th Symposium on Graph Drawing (GD)*, pages 171–182, 2000.

A Compound Graph Layout Algorithm
for Biological Pathways

Uğur Dogrusoz[1,2,*], Erhan Giral[1], Ahmet Cetintas[1],
Ali Civril[1], and Emek Demir[1]

[1] Center for Bioinformatics, Bilkent University, Ankara 06800, Turkey
ugur@cs.bilkent.edu.tr
[2] Tom Sawyer Software, Oakland, CA 94612, USA

Abstract. We present a new compound graph layout algorithm based
on traditional force-directed layout scheme with extensions for nesting
and other application-specific constraints. The algorithm has been successfully implemented within PATIKA, a pathway analysis tool for drawing complicated biological pathways with compartmental constraints and
arbitrary nesting relations to represent molecular complexes and pathway abstractions. Experimental results show that execution times and
quality of the produced drawings with respect to commonly accepted
layout criteria and pathway drawing conventions are quite satisfactory.

1   Introduction

The notion of compound graphs has been used in the past to represent more complex types of relations or varying levels of abstractions in data [10, 8]. One such
application is in bioinformatics; PATIKA (Pathway Analysis Tool for Integration
and Knowledge Acquisition) is a software tool providing an integrated, multi-
user environment for visualizing and manipulating network of cellular events [4].

There has been a great deal of work done on general graph layout [5] but
considerably less on layout of compound graphs [14, 12, 2, 6], which has mostly
focused on layout of hierarchical graphs.

There have been a few studies done specifically for layout of biological pathways as well, focusing on metabolic pathways [11, 1, 13]. Certain tools such as
PATIKA enforce a more restricted ontology to represent signaling pathways whose
underlying graph structure can be arbitrarily more complicated and irregular.

A layout algorithm for signaling pathways was proposed and implemented
within PATIKA earlier [9]. However neither this algorithm nor any of the previously proposed ones address advanced pathway representations including nested
drawings, intergraph relations, and application-specific constraints such as compartmental constraints at the same time. In this paper we describe a new algorithm for layout of compound pathway graphs.

2   Pathway Model

The structure of pathway graphs highly depend on the type of pathways (e.g.,
metabolic or signaling) and the model or ontology used to represent the biological

---

* To whom correspondence should be addressed.

J. Pach (Ed.): GD 2004, LNCS 3383, pp. 442–447, 2004.

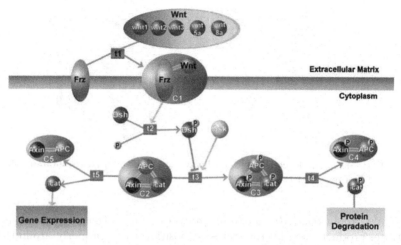

Fig. 1. Canonical wnt pathway represented by PATIKA ontology, including molecular complexes (e.g., C2) and various abstractions (e.g., Wnt and Protein Degradation).

phenomenon. We assume the ontology described in [3], which represents a cellular process in the form of a directed compound pathway graph (Figure 1).

3   Layout Algorithm

A force-directed layout algorithm with constraints to satisfy general drawing conventions in compound pathway graphs has been chosen. Basically, it is a virtual dynamic system in which nodes are assumed to be physical objects with a certain "electrical charge", connected via "springs" of a pre-specified desired length. Objects pull or repel each other depending on current lengths of any connected springs. In addition, relatively minor repulsion forces act on any pair of objects that are too close to each other to avoid node-to-node overlaps. Furthermore, each nested graph including the root of the nesting hierarchy is assumed to have a dynamic (with respect to the graph bounds) center of gravity. Thus the optimal layout is regarded as the state of this system in which total energy is minimal [7]. The following additions are made to this basic model (Figure 2):

- An expanded node and its associated nested graph are represented as a single entity, similar to a "cart" which can move freely in every direction. Multiple levels of nesting is modeled with smaller carts on top of larger ones.
- The nodes and edges of a nested graph are set in motion on this cart, confined within its bounds. Each cart is assumed to be surrounded by a material, elastic enough to adapt to the current bounds of the associated nested graph. Thus, as nodes of the nested graph are pushed out, expanding the nested graph, the expanded node adjusts its bounds accordingly.
- To avoid overlaps of variably sized nodes, desired edge lengths are calculated using parts of edges in between borders of end-nodes.
- Intergraph edges are treated specially; their desired lengths are set to be proportional to the nesting depth of the graphs their end nodes belong to.

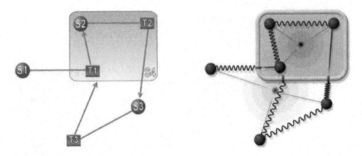

Fig. 2. Part of a sample compound pathway (left) and the corresponding physical model used by our algorithm (right).

We also apply *relativity constraint forces* or simply *relativity forces* on each substrate, product and effector states to position them properly around associated transition(s). The convention is to align substrates and products of a transition on opposite sides of the transition to form a certain flow direction. When calculating relativity forces, we first determine a flow, called *orientation*, for each transition by simply looking at current, relative positions of their associated substrates and products. Then each associated state of the transition is applied a relativity force to respect this orientation (Figure 3).

Another important constraint is due to cellular locations of biological nodes (compartments) represented by rectangular regions. The layout algorithm must keep each biological node within the bounds of the associated compartment and must enlarge or shrink it as required by the geometry of the pathway.

The algorithm is composed of three phases preceded by initialization:

Initialization: Initial node and compartment sizes, and threshold values for convergence are calculated as well as initial random positions of nodes. In addition, for efficiency and quality reasons parts of the given pathway that are

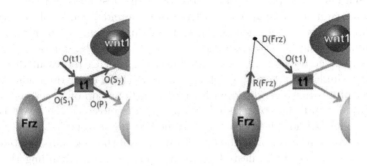

Fig. 3. An example of how the orientation of a transition is determined shown on transition t1 of Figure 1 (left) and used to calculate the relativity force on one of its substrates, Frz (right). O(t1), R(Frz), and D(Frz) respectively denote orientation of t1, relativity force on Frz due to t1, and desired location of Frz to obey this force, where magnitude of R(Frz) is equal to that of O(t1), and the distance of D(Frz) from t1 is equal to the desired edge length.

trees are temporarily removed. The remaining part of the pathway forms the "skeleton" of the pathway graph.

Phase 1: In this phase the skeleton graph is laid out using the spring embedder model described earlier but relativity and gravitational forces are disabled.

Phase 2: Trees reduced earlier on are introduced back level by level in this phase, also taking relativity and gravitational forces into account.

Phase 3: This phase is the stabilization phase where we "polish" the layout.

algorithm COMPOUNDLAYOUT()
(1) **call** INITIALIZATION()
(2) set **phase** to 1
(3) **if** layout type is incremental **then**
(4)   increment **phase** to 3
(5) **while phase** ≤ 3 **do**
(6)   set **step** to 1, **error** to 0
(7)   **while** (**step** < maxIterCount(**phase**) **and**
                 error > errorThreshold(**phase**)) **or** !allTreesGrown **do**
(8)   **call** APPLYSPRINGFORCES()
(9)   **call** APPLYREPULSIONFORCES()
(10)   **if phase** ≠ 1 **then**
(11)   **call** APPLYGRAVITATIONFORCES()
(12)   **call** APPLYRELATIVITYFORCES()
(13)   **call** CALCNODEPOSITIONSANDSIZES()
(14)   **call** UPDATECOMPARTMENTBOUNDS()
(15)   **if phase** = 2 **and** !allTreesGrown **and step** % growStep = 0 **then**
(16)   **call** GROWTREESONELEVEL()
(17)   increment **step** by 1
(18)   increment **phase** by 1

A quick analysis reveals that the running time of layout is $O(k \cdot n^2)$ where n is the total number of nodes in the compound pathway, and k is the number of iterations required to reach an energy minimal state.

4   Implementation

The algorithm has been implemented within the PATIKA editor built on top of Tom Sawyer Software's GET for Java ver. 5.5. The results have been found satisfactory as far as the general graph drawing criteria such as number of crossings and total area are concerned. In addition, application-specific constraints such as compartmental constraints and relative positioning constraints seem to be highly satisfied. Figures 4 and 5 show sample pathway drawings produced.

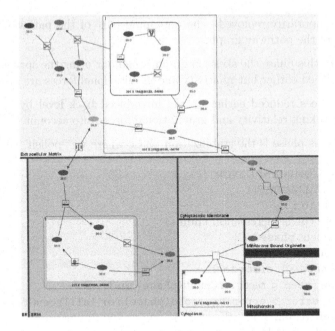

Fig. 4. Sample pathway from the PATIKA editor laid out by our algorithm.

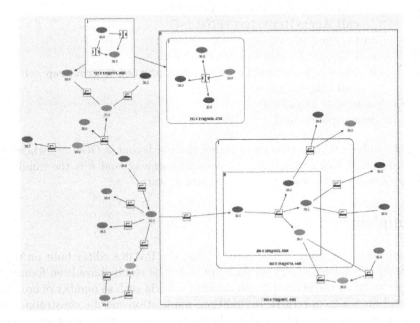

Fig. 5. Sample pathway from the PATIKA editor laid out by our algorithm.

From the theoretical analysis given earlier, a quadratic behavior of execution time versus number of nodes is expected, assuming k does not grow in the order of the graph size. The experiments validate this argument.

References

1. M. Y. Becker and I. Rojas. A graph layout algorithm for drawing metabolic pathways. *Bioinformatics*, 17:461–467, 2001.
2. F. Bertault. Force-directed algorithm that preserves edge crossing properties. In J. Kratochvil, editor, *GD '99*, volume 1731 of *Lecture Notes in Computer Science*, pages 351–358. Springer-Verlag, 1999.
3. E. Demir, O. Babur, U. Dogrusoz, A. Gursoy, A. Ayaz, G. Gulesir, G. Nisanci, and R. Cetin-Atalay. An ontology for collaborative construction and analysis of cellular pathways. *Bioinformatics*, 20(3):349–356, 2002.
4. E. Demir, O. Babur, U. Dogrusoz, A. Gursoy, G. Nisanci, R. Cetin-Atalay, and M. Ozturk. PATIKA: An integrated visual environment for collaborative construction and analysis of cellular pathways. *Bioinformatics*, 18(7):996–1003, 2002.
5. G. Di Battista, P. Eades, R. Tamassia, and I. G. Tollis. *Graph Drawing, Algorithms for the Visualization of Graphs.* Prentice-Hall, 1999.
6. P. Eades, Q. Feng, and X. Lin. Straight-line drawing algorithms for hierarchical graphs and clustered graphs. In S. North, editor, *GD '96*, volume 1190 of *Lecture Notes in Computer Science*, pages 113–128. Springer-Verlag, 1997.
7. T. M. J. Fruchterman and E. M. Reingold. Graph drawing by force-directed placement. *Software Practice and Experience*, 21(11):1129–1164, 1991.
8. K. Fukuda and T. Takagi. Knowledge representation of signal transduction pathways. *Bioinformatics*, 17(9):829–837, 2001.
9. B. Genc and U. Dogrusoz. A constrained, force-directed layout algorithm for biological pathways. In G. Liotta, editor, *GD '03*, volume 2912 of *LNCS*, pages 349–356, 2003.
10. K. Sugiyama and K. Misue. A Generic Compound Graph Visualizer/Manipulator: D-ABDUCTOR. In F. J. Brandenburg, editor, *GD '95*, volume 1027 of *Lecture Notes in Computer Science*, pages 500–503. Springer-Verlag, 1995.
11. P. D. Karp and S. Paley. Automated drawing of metabolic pathways. In *Third International Conference on Bioinformatics and Genome Research*, pages 225–238, Tallahassee, Florida, June 1994.
12. G. Sander. Layout of compound directed graphs. Technical Report A/03/96, University of Saarlandes, CS Dept., Saarbrücken, Germany, 1996.
13. F. Schreiber. High quality visualization of biochemical pathways in BioPath. *In Silico Biology*, 2(2):59–73, 2002.
14. K. Sugiyama and K. Misue. Visualization of structural information: Automatic drawing of compound digraphs. *IEEE Transactions on Systems, Man and Cybernetics*, 21(4):876–892, 1991.

Curvilinear Graph Drawing
Using the Force-Directed Method

Benjamin Finkel[1] and Roberto Tamassia[2]

[1] MIT Lincoln Laboratory
finkel@ll.mit.edu
[2] Brown University
rt@cs.brown.edu

Abstract. We present a method for modifying a force-directed graph drawing algorithm into an algorithm for drawing graphs with curved lines. Our method is based on embedding control points as dummy vertices so that edges can be drawn as splines. Our experiments show that our method yields aesthetically pleasing curvilinear drawing with improved angular resolution. Applying our method to the GEM algorithm on the test suite of the "Rome Graphs" resulted in an average improvement of 46% in angular resolution and of almost 6% in edge crossings.

1 Introduction

Curvilinear drawings of graphs give a significant amount of flexibility to a layout algorithm, creating the potential for improved aesthetics. Such drawings are ideally suited for several applications (e.g., flight maps). However, the literature on curvilinear drawing algorithms is not as extensive as that on straight-line and orthogonal drawings. In this paper, we present a methodology for computing curvilinear drawings using force-directed methods and we report on the result of experiments showing that our technique yields aesthetically pleasing curvilinear drawings with improved angular resolution and number of crossings.

The idea of the force-directed method is to use physical simulations to lay out a graph. Forces are calculated, applied to vertices, and recalculated over many iterations. In the pioneering "spring-embedder" algorithm [10], the edges are modeled as stretchable springs of different length, which oscillate until the system reaches equilibrium. Effective force-directed techniques include subatomic forces [12] and simulated annealing [8]. An experimental comparison of force-directed methods is presented in [3]. Recent work includes [13, 19].

Previous work on curvilinear drawings of graphs has focused on planar drawings and on edge-routing methods. Early systems for layered drawings of digraphs use heuristics that transform a drawing with polygonal chains into a curvilinear drawing by replacing polygonal chains with splines [14–16]. The routing of a curvilinear edge that is being added to an existing drawing is investigated in [9]. In [17, 18], algorithms for drawing planar graphs using cubic Bézier curves for the edges are given. In [7], an algorithm is presented for drawing planar graphs

J. Pach (Ed.): GD 2004, LNCS 3383, pp. 448–453, 2004.

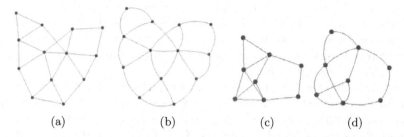

Fig. 1. Layout improvements obtained by our method: (a) straight-line drawing produced by the KK algorithm; (b) associated curvilinear drawing; (c) straight-line drawing produced by the GEM algorithm; (d) associated curvilinear drawing.

with asymptotically optimal area and angular resolution such that the edges are sequences of two circular arcs. Brandes and Wagner show how to use Bézier curves to draw graphs with vertices at fixed locations [6], with applications to the display of geographic networks. Related work on 3D curvilinear drawings of geographic networks is presented in [4, 5, 21].

Imagine an architect who uses graph drawing software to design a sculpture garden. In the graph, vertices represent sculptures and edges represent paths. Using a standard force-directed algorithm, the architect would probably get a drawing with a good spread of sculptures, but a mediocre layout of paths. The architect might want the paths themselves to be more separated, and to arrive at the sculptures at more distinct angles. These properties correspond to the aesthetic criteria of edge separation and angular resolution. It seems plausible that by introducing curved paths the architect could improve the aesthetic layout with respect to these two qualities. The architect might also prefer smooth, curved lines to the rigid lines connecting the sculptures in the layout generated by the standard force-directed algorithm. We show in Fig. 1 how drawings produced by the KK [20] and GEM [11] algorithms can be smoothed and aesthetically improved using the method set forth in this paper.

The fundamental function of force-directed methods is to find a layout for the vertices, which are the only objects subject to forces. Edges merely influence these forces. In many circumstances, this is acceptable. However, one could imagine a situation in which the layout of the edges is also important and should be considered by the algorithm. Thus, we would like to give edges a "mass-like" quality so they can also be pushed and pulled and acted upon by forces.

2   Force-Directed Curvilinear Drawings

For the curvilinear drawings studied in this paper, we consider three relevant aesthetic criteria: angular resolution, edge separation, and number of crossings. The *angular resolution* refers to the angles formed by pairs of edges incident on a vertex. In general, small angles are not desirable. We measure the angular resolution as the difference between the smallest angle and the optimal angle at a vertex v ($360/\deg(v)$), averaged over all vertices. Thus, lower values indicate better layouts. The *edge separation* refers to the distance between an edge and another non-incident, non-intersecting edge. In this paper, we measure the edge separation as the average distance between pairs of non-incident, non-intersecting edges. Hence, larger values are desirable. Finally, edge crossings are undesirable and should be minimized.

Our algorithm for curvilinear graph drawing using the force-directed method is based on the following simple idea (see Fig. 2a–b). Given a graph G, we insert a new fictitious vertex C along each edge (A, B), thus replacing (A, B) with a path of two edges (A, C) and (C, B). Let G' be the resulting graph. We compute a straight-line drawing of G' using a force-directed algorithm. Finally, we transform the drawing of G' into a drawing of G by replacing each polygonal chain (A, C, B), where C is a fictitious vertex, with a curve that joins A and B and uses C as a control point. More generally, we can embed several fictitious vertices (control points) on each edge to magnify the curving effect. In practice, one or two vertices will produce good results. Our approach generalizes the one used by Brandes and Wagner [6] for the layout of geographic networks, where the vertex locations are fixed and forces operate only in local neighborhoods.

Adding vertices and edges to a graph may increase the number of crossings. In theory, one could prevent the addition of new crossings by imposing boundaries on the control vertices. However, most force-directed algorithms, while evaluating a modified graph, will sometimes generate new crossings (see Fig. 2c). In practice, many of these crossings occur among edges incident on the same vertex and can be easily detected.

Any number of algorithmic heuristics could be used to unwind these crossings. We use a binding technique, inspired from [6], which was designed to be simple and algorithm independent: its only mechanism is adding hidden edges. To bind a vertex, connect the closest control vertices around that vertex in the order of the "real" (non-control) endpoints. The points are sorted based on polar coordinates, similar to a Graham Scan. The vertices are connected with edges that will not be drawn, but will pull the control points into place. Convex angles are also excluded, so as to preserve good angular resolution.

Binding a vertex will generally fix these added local intersections. It should be noted, however, that binding is specific to each embedding of a graph. Thus, it can be done only after the algorithm has run and then it requires the algorithm to be repeated from the current position with the new hidden edges. In this way, the bindings can be done in several passes, or all at once, yielding different results.

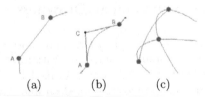

(a)   (b)   (c)

Fig. 2. (a–b) Inserting a fictitious vertex C, serving as a control point, on edge (A, B). (c) Crossing between edges incident on the same vertex.

3   Implementation and Experimental Results

We have implemented our force-directed curvilinear graph drawing method in Java using the JDSL library [1] and the VGJ tool [2]. Experiments were performed on 1.5 GHz AMD Athlon workstations with 512MB RAM running Linux. The test suite was made from the "Rome Graphs" obtained from graphdrawing.org. The force-directed algorithms chosen were GEM [11] and KK [20] because of their effectiveness, speed and simplicity. Indeed, GEM and KK were shown to be all-around performers in [3]. The implementations of GEM and KK were adapted from the JDSL library and VGJ tool, respectively.

We use a Bézier curve because of its attractive shape and geometric properties: it is contained inside the convex hull of its endpoints and control points, and its slope starts directly towards a control point. Edges with single and double control points use quadratic and cubic Bézier curves, respectively. For binding, there are two successive passes. We also test the effect of binding all vertices.

A step-by-step example of the execution of the modified GEM method using quadratic Bézier curves and the "bind all" heuristic is shown in Fig. 3.

Each one of the Rome Graphs was drawn twice, with and without the curvilinear method, starting from the same initial random vertex placement. Fictitious control vertices were evenly spaced on the edges. The above process was repeated with three different binding heuristic options. First with no binding, simply executing the algorithm on the modified graph and measuring the aesthetics. Second, with the binding heuristic on only vertices with local intersections. In this case we ran the algorithm on the graph as before, then bound vertices

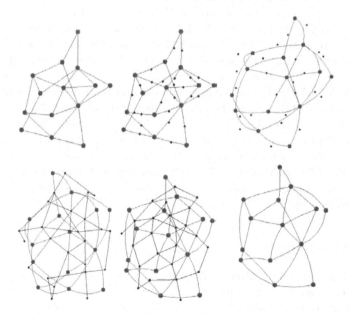

Fig. 3. Step-by-step example of execution of the modified GEM method using quadratic Bézier curves and the "bind all" heuristic.

Table 1. Averages of the aesthetic properties using our curvilinear method with the GEM force-directed layout algorithm on the Rome Graphs.

edge type, type of binding	angular resolution	edge separation	# crossings
straight edges (without method)	46.175257	0.043419	30.718649
quadratic Bézier, none	34.447798	0.044492	29.652424
quadratic Bézier, crossed	32.612817	0.044642	30.316422
quadratic Bézier, all	31.748013	0.045090	30.393090
cubic Bézier, none	31.233901	0.044849	29.209033
cubic Bézier, crossed	28.199770	0.045294	29.216517
cubic Bézier, all	24.814479	0.045931	28.986946

with local intersections, executed, bound again, and executed a final time before making the measurements. Finally, a test was done where the algorithm was executed on the graph, then all vertices were bound, and then the algorithm was run again.

Our experimental results, summarized in Table 1, show that the curvilinear drawing method significantly improves the angular resolution. Binding all vertices, using cubic Bézier curves led to an overall average angular resolution 46% closer to optimal and average decrease of more than one crossing per graph.

It should be noted that binding in general was not very effective when using quadratic Bézier edges. This is because two vertices share each control point; binding from one vertex can adversely affect the other. For cubic Bézier curves, the results show that binding is an extremely effective heuristic; it is best when used on all vertices, not just the ones with local intersections.

Unlike the GEM algorithm, the KK algorithm did not perform well with our curvilinear method. Since the method increases the number of vertices and edges, it affects the running time of the drawing algorithm. Because of its $O(n^3)$ running time per iteration, compared to GEM's $O(n^2)$, KK did not scale well and was prohibitively slow on large graphs. In addition, the KK algorithm uses graph-theoretic distances, and thus the presence of the fictitious vertices hampers the effectiveness of the algorithm itself, creating far too many new crossings to justify any improvement in angular resolution or edge separation.

Future research should investigate the effect of this process on other force-directed methods, such as [8, 13, 19]. It would also be interesting to investigate the use of interpolating splines (e.g., Catmull-Rom), which go through their control points, instead of Bézier curves.

Acknowledgements

We would like to thank Mike Goodrich and Franco Preparata for useful discussions. Our research was supported in part by National Science Foundation grants CCR–0098068 and DUE–0231202. This work originates from Benjamin Finkel's Undergraduate Honors Thesis and was performed while he was at Brown University.

References

1. JDSL: the data structures library in Java. http://jdsl.org.
2. VGJ: Visualizing graphs with Java. http://www.eng.auburn.edu/department/cse/research/graph_drawing/graph_drawing.html.
3. F. J. Brandenburg, M. Himsolt, and C. Rohrer. An experimental comparison of force-directed and randomized graph drawing algorithms. In *Graph Drawing (Proc. GD 1995)*, *LNCS* 1027, pp. 76–87. 1996.
4. U. Brandes, G. Shubina, and R. Tamassia. Improving angular resolution in visualizations of geographic networks. In *Proc. Joint Eurographics — IEEE TCVG Symposium on Visualization (VisSym '00)*, pp. 23–32, 2000.
5. U. Brandes, G. Shubina, R. Tamassia, and D. Wagner. Fast layout methods for timetable graphs. In *Graph Drawing (Proc. GD 2000)*, *LNCS* 1984, pp. 127-138, 2001.
6. U. Brandes and D. Wagner. Using graph layout to visualize train interconnection data. In *Graph Drawing (Proc. GD 1998)*, *LNCS* 1547, pp. 44–56, 1998.
7. C. C. Cheng, C. A. Duncan, M. T. Goodrich, and S. G. Kobourov. Drawing planar graphs with circular arcs. In *Graph Drawing (Proc. GD 1999)*, *LNCS* 1731, pp. 117–126, 1999.
8. R. Davidson and D. Harel. Drawing graphics nicely using simulated annealing. *ACM Trans. Graph.*, 15(4):301–331, 1996.
9. D. P. Dobkin, E. R. Gansner, E. Koutsofios, and S. C. North. Implementing a general-purpose edge router. In *Graph Drawing (Proc. GD 1997)*, *LNCS* 1353, pp. 262–271, 1997.
10. P. Eades. A heuristic for graph drawing. *Congr. Numer.*, 42:149–160, 1984.
11. A. Frick, A. Ludwig, and H. Mehldau. A fast adaptive layout algorithm for undirected graphs. In *Graph Drawing (Proc. GD 1994)*, *LNCS* 894, pp. 388–403, 1995.
12. T. Fruchterman and E. Reingold. Graph drawing by force-directed placement. *Softw. – Pract. Exp.*, 21(11):1129–1164, 1991.
13. P. Gajer, M. T. Goodrich, and S. G. Kobourov. A fast multi-dimensional algorithm for drawing large graphs. *Graph Drawing (Proc. GD 2000)*, *LNCS* 1984, pp. 211–221, 2000.
14. E. R. Gansner, E. Koutsofios, S. C. North, and K. P. Vo. A technique for drawing directed graphs. *IEEE Trans. Softw. Eng.*, 19:214–230, 1993.
15. E. R. Gansner, S. C. North, and K. P. Vo. DAG – A program that draws directed graphs. *Softw. – Pract. Exp.*, 18(11):1047–1062, 1988.
16. A. Garg and R. Tamassia. GIOTTO3D: A system for visualizing hierarchical structures in 3D. In *Graph Drawing (Proc. GD 1996)*, *LNCS* 1190, pp. 193–200, 1997.
17. M. T. Goodrich and C. G. Wagner. A framework for drawing planar graphs with curves and polylines. In *Graph Drawing (Proc. GD 1998)*, *LNCS* 1547, pp. 153–166, 1998.
18. C. Gutwenger and P. Mutzel. Planar polyline drawings with good angular resolution. In *Graph Drawing (Proc. GD 1998)*, *LNCS* 1547, pp. 167–182, 1998.
19. D. Harel and Y. Koren. Graph drawing by high-dimensional embedding. *Graph Drawing (Proc. GD 2002)*, *LNCS* 2528, pp. 207–219, 2002.
20. T. Kamada and S. Kawai. An algorithm for drawing general undirected graphs. *Inform. Process. Lett.*, 31:7–15, 1989.
21. T. Munzner, E. Hoffman, K. Claffy, and B. Fenner. Visualizing the global topology of the MBone. In *Proc. IEEE Symp. on Information Visualization*, pp. 85–92, 1996.

Graphael: A System for Generalized Force-Directed Layouts*

David Forrester, Stephen G. Kobourov, Armand Navabi,
Kevin Wampler, and Gary V. Yee

Department of Computer Science,
University of Arizona
{forrestd,kobourov,navabia,wamplerk,gyee}@cs.arizona.edu

Abstract. The `graphael` system implements several traditional force-directed layout methods, as well as several novel layout methods for non-Euclidean geometries, including hyperbolic and spherical. The system can handle large graphs, using multi-scale variations of the force-directed methods. Moreover, `graphael` can layout and visualize graphs that evolve though time, using static views, animation, and morphing. The implementation includes a powerful interface that allows the user to put together existing algorithms and visualization techniques, and to easily add new ones. The system is written in Java and is available as a downloadable program or as an applet at `http://graphael.cs.arizona.edu`.

1 Introduction

As researchers in the graph drawing community develop new algorithms and visualization techniques it is natural for the creation of new graph drawing tools to follow. It is often the case, however, that the implementation of an algorithm is accompanied by time consuming tasks that have little to do with the algorithm itself. Researchers who would like to test a new layout algorithm should only have to concern themselves with the details of the algorithm itself rather than with graphics packages, file parsers, or user interface design.

In this paper we present `graphael`: yet another graph drawing system designed to provide the necessary structure and flexibility for force-directed graph drawing research. Our system is built with the following design considerations: (1) *Plug-and-Play:* it should be easy to integrate new algorithms and visualization methods; (2) *User Friendliness:* the user interface should be easy to use, but also powerful and versatile; (3) *Portability:* the system should run on any computational platform.

The `graphael` system attempts to meet these goals by providing a set of core algorithms and visualization routines, as well as an interface that allows the user to combine different algorithms and visualization methods, and to easily add new ones. In addition, the system contains several novel algorithms and visualization techniques, such as force-directed methods in non-Euclidean geometries, and techniques for dealing with graphs that evolve through time.

---

* This work is supported in part by the NSF under grant ACR-0222920 and ITCDI under grant 003297.

1.1   Related Work

A number of automated graph drawing systems have been developed over the last few years; see [9] for a survey. The GraphServer [3] is an online service that allows users to draw graphs and translate graph descriptions between multiple formats. Tulip is a framework built to facilitate large graph drawing research [1]. WilmaScope [4] is a Java application designed specifically for 3D visualization. yFiles [14] is a commercial library of Java classes developed to provide building blocks for graph drawing applications. Pajek [2] is a Windows program designed to handle large graphs for social networks analysis. TGRIP [6] is an extension on the GRIP system [8] and efficiently draws large temporal graphs using intelligent placement. The GraphAEL [5] system extracts three types of evolving graphs from a custom-built graph drawing literature database and creates 2D and 3D animations of the evolutions.

1.2   Our Contributions

In addition to sharing all the letters with GraphAEL [5], the system described in this paper was inspired by it. We wanted to provide a graph visualization framework that can easily be coupled with the bibliographic database to provide visualizations of the co-citation, collaboration, and topic graphs, produced from the database. This led to the development of the current system, which is equipped with a core package of force-directed algorithms and visualization tools. In addition to putting together well-known algorithms and visualization methods, graphael contains several novel features. Among these features are support for temporal graphs, interactive graph visualization, multi-scale layout algorithms for large graphs, and embedding graphs in non-Euclidean spaces, such as hyperbolic space and spherical space. The current system also includes an interactive Control Flow (CF) Graph, used to put together different combinations of layout algorithms, projections and visualizations, while offering a visual representation of the process.

2   System Overview

2.1   Force-Directed Layouts

Force-directed layout algorithms are a powerful and practical graph drawing heuristic. They rely on an objective function that maps a particular graph layout to an energy value. Typically such algorithms start with a random drawing of the graph and utilize standard optimization methods to minimize the energy function. The algorithms define functions in which low energies are associated with layouts where adjacent vertices are near some preferred distance from each other, and non-adjacent vertices are well-spaced. The main difference between force-directed algorithms is the choice of energy function and the methods for its minimization.

We have implemented two traditional force-directed algorithms in graphael. The first one is the Fruchterman-Reingold [7] algorithm. It defines an attractive force function for adjacent vertices and a repulsive force function for non-adjacent vertices. For a vertex v, $F_{FR}(v) = F_{a,FR} + F_{r,FR}$, where the attractive force is defined as $F_{a,FR} = \sum_{u \in Adj(v)} \frac{\text{dist}_{R^n}(u,v)^2}{\text{edgeLength}^2}(\text{pos}[u] - \text{pos}[v])$ and the repulsive force is defined as $F_{r,FR} = \sum_{u \in Adj(v)} s \cdot \frac{\text{edgeLength}^2}{\text{dist}_{R^n}(u,v)^2} \cdot (\text{pos}[u] - \text{pos}[v])$.

The second force-directed method is the Kamada-Kawai [10] layout algorithm. In this method each pair of vertices connected by a path has forces proportional to the length of the path. The displacement of a vertex v of G is calculated by: $F_{KK}(v) = \sum_{u \in N_i(v)} \left(\frac{\text{dist}_{R^n}(u,v)^2}{\text{dist}_G(u,v) \cdot \text{edgeLength}^2} - 1 \right)(\text{pos}[u] - \text{pos}[v])$.

In the above equations, $\text{dist}_{R^n}(u,v)$ is the Euclidean distance between $\text{pos}[u]$ and $\text{pos}[v]$, $\text{dist}_G(u,v)$ is the graph distance between u and v along a shortest path, edgeLength is the unit edge length, $\text{Adj}(v)$ is the set of vertices adjacent to v, and s is a small scaling factor.

2.2 Multi-scale Graph Drawing

The effectiveness of force-directed methods rapidly decreases as the input graphs get larger. This is mainly due to the increased difficulty of getting out of local minima and to the runtime complexity, typically quadratic, or cubic in the size of the graph. Multi-scale graph drawing methods address both of these problems by filtering the graph into different levels, called filtration levels, each containing a subset of the initial graph. The levels are laid out from least to most complex. The multi-scale methods rely on good filtrations, good initial placement of the vertices, and on local refinement on each level.

Filtrations: The effectiveness of the multi-scale method depends on each successive filtration level containing a constant fraction of the vertices from the previous level. Thus, good filtrations have $\Theta(\lg n)$ depth and can be quickly computed. In `graphael` we currently provide three filtration methods: Maximal Independent Set Filtration, Random Graph Filtration, and Cores Filtration:

1. *Maximal Independent Set Filtration:* A filtration $V = V_0 \supset V_1 \supset \ldots \supset V_k \supset \emptyset$ of the vertex set V of G is called a maximal independent set filtration if V_1 is a maximal independent set of G, and each V_i is a maximal subset of V_{i-1} so that the graph distance between any pair of its elements is at least $2^{i-1} + 1$. Maximal Independent Set filtrations have depth $O(\lg n)$ and can be computed in near-linear time [8].
2. *Random Graph Filtration:* Random filtrations are created by repeatedly removing half of the vertices, chosen at random, starting with the original vertex set V of G. The depth of this filtration is also $O(\lg n)$ and the computation time required is linear in the size of V. Although simple, this method produces reasonable layouts for large graphs.
3. *Cores Graph Filtration:* Graph cores are described in [13]. Given a graph $G = (V, E)$, a subgraph $H_k = (W; E|W)$ induced by the set W is a k-core, or a core of order k if $\forall v \in W : deg_H(v) \geq k$, and H_k is the maximum

subgraph with this property. The core of maximum order is also called the main core. Graph cores can be computed in linear time [2]. If the number of cores is a small constant compared to the size of the graph, we augment the filtration induced by the cores to depth $O(\lg n)$ using the peeling process inherent in the core computation.

Initial Placement and Refinement: The main idea of good initial placement is to add vertices to the current drawing one at a time at a carefully computed position, rather than a random one [8]. For simplicity we describe the process in 2D, but in practice this is done in arbitrary Euclidean, and even some non-Euclidean, spaces. Assume that the highest filtration level has exactly 3 vertices. These vertices are placed at the endpoints of a triangle with sides proportional to the graph distances between the points in the original graph. Vertices in subsequent filtration levels are placed based on their graph distances from already placed vertices from previous filtration levels. The intuition is that if we place the vertices close to their optimal positions initially, the refinement phase will only need a few iterations of a local force-directed calculations to reach a minimal energy state. In **graphael**'s implementation, we use the "3-closest-vertices" strategy. Using this method we place the vertex t at the barycenter of u, v, and w, the three vertices closest to t from the previous filtration level. Once all the vertices at the current filtration level have been placed, we apply a local force-directed refinement. The refinement stage is local as for a given vertex v in the current filtration, only a small neighborhood of vertices $N_i(v)$ is considered in the force computation.

2.3   Graphs That Evolve Through Time

We have also implemented algorithms for visualization of graphs that evolve through time based on techniques described in [5, 6]. The algorithms are modifications of the standard force-directed algorithms that allow us to deal with vertex-weighted and edge-weighted graphs. Graphs that evolve through time are converted to vertex-weighted and edge-weighted graphs, by treating each instance of the graphs as a timeslice, and connecting neighboring timeslices. The edges connecting different timeslices are called inter-timeslice edges. By changing the weights of these edges, we are able to balance the individual graph readability with the overall mental map preservation between consecutive graphs. Making the inter-timeslice edges heavy, results in fixing the vertex positions in each graph instance. Alternatively, making the inter-timeslice edges light, results in nearly independent layouts of each graph instance.

Weighted Graphs: We modify the force-directed equations for calculating the force vectors to include edge weights and vertex weights so as to place heavy vertices well away from each other and to place vertices connected by heavy edges closer to each other. The unit edge length is modified to $\sqrt{w_u \cdot w_v}/w_e$ for an edge of weight w_e, connecting vertices u, v of weight w_u, w_v, respectively.

The Kamada-Kawai method relies on the notion of graph distance between pairs of vertices. It is easy to generalize this notion to weighted graphs, but because of the computational and space overhead associated with calculating the shortest path between all pairs of vertices in the graph, we use an approximation. Let $p_1, p_2, \ldots, p_n$ be the sequence of vertices in the shortest unweighted path in G connecting two vertices, u and v. The modified Kamada-Kawai force vector is given by $F_{KK}(v) =$

$$\sum_{u \in N_i(v)} \left(\frac{2 \cdot \text{dist}_{R^n}(u,v)^2}{\text{optDist}_G(u,v)^2 \cdot \text{edgeLength}^2 + \text{dist}_{R^n}(u,v)^2} - 1 \right) (\text{pos}[u] - \text{pos}[v]),$$

where $\text{optDist}_G(u,v) = \sum_{i=2}^{n} \frac{\sqrt{w_{p_i} \cdot w_{p_{i-1}}}}{w_{e_{p_i p_{i-1}}}}$. Similarly, we modify the Fruchterman-

Reingold forces as follows: $F_{a,FR} = \sum_{u \in Adj(v)} \frac{w_e \cdot \text{dist}_{R^n}(u,v)^2}{\text{edgeLength}^2} (\text{pos}[u] - \text{pos}[v])$

and $F_{r,FR} = \sum_{u \in Adj(v)} s \cdot \left(\frac{\text{edgeLength}^2 \cdot \sqrt{w_u \cdot w_v}}{\text{dist}_{R^n}(u,v)^2} \right) (\text{pos}[u] - \text{pos}[v])$.

Timeslice Attribute: To visualize a series of graphs embodying the evolution of a set of relationships over time, we associate a timeslice attribute with each vertex. The timeslice of a vertex is just a label identifying which graph instance the vertex belongs to. We use the timeslice attribute to partition the vertices of a graph into groups by time. Additional modifications to the force-directed algorithms are needed to accommodate timeslice information.

For the Kamada-Kawai layout method, the function $\text{optDist}_G(u,v)$ is modified so that for two vertices u, v with timeslice indices of t_u and t_v, respectively, is given by: $\text{optDist}_G(u,v) = \sum_{i=2}^{n} \delta_{t_u t_v} \cdot \frac{\sqrt{w_{p_i} \cdot w_{p_{i-1}}}}{w_e}$, where $p_1, p_2, \ldots, p_n$ is the shortest unweighted path in G connecting two vertices, u and v and $\delta_{t_u t_v}$ is 1 if $t_u = t_v$ and 0 otherwise.

The modifications needed for the Fruchterman-Reingold calculations are similar. Repulsive forces are simply eliminated between vertices in different timeslices, $F_{r,w,t,FR} = \delta \cdot F_{r,w,FR}$ while the attractive forces remain unchanged, $F_{a,w,t,FR} = F_{a,w,FR}$.

2.4   Visualizing Evolving Graphs

The timeslice information alone is not enough to nicely layout evolving graphs; we must also arrange edges between timeslices so that the layouts can be used for animation. The most straightforward method to animate is simply to use a series of "snapshots" of a graph taken at some interval over a period of time. When visualizing an evolving graph, we would ideally like the graphs of each timeslice to have high readability (i.e. have a pleasing layout) and for consecutive timeslices to be similar, that is, the mental map should be preserved. To meet these constraints the timeslices are combined into a single graph by connecting vertices with the same labels from adjacent timeslices.

Because of the modified optimal distance function, corresponding vertices in different timeslices have no repulsive force on each other, but they still have

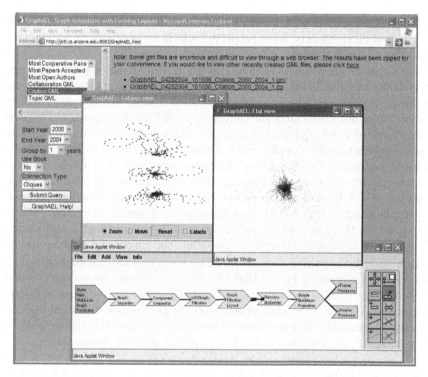

Fig. 1. graphael in cooperation with GraphAEL. The graphs shown are the column view and flat 2D view of a citation graph from 2000 to 2003 by 1 year increments.

attractive forces due to the inter-timeslice edges. In graphael the balance between readability and mental map preservation can be controlled by changing the weights of the inter-timeslice edges.

Once the layout of the evolving graph has been computed, graphael offers different methods for visualizing the graphs. Each timeslice can be drawn in a restricted 2D view, or the graphs can be drawn in 3D with individual graphs arranged on top of each other (column-view). The column view lays out each timeslice on a separate plane, allowing the user to view the changes in the graph over time; see Fig. 1. The inter-timeslice edges can be hidden or displayed. Smoothly stepping through the evolving graphs using linear interpolation of vertex positions, colors, and weights yields visually pleasing animations.

2.5   Graph Drawing in Non-Euclidean Spaces

A novel feature in graphael is the option to layout graphs in non-Euclidean spaces, in particular, in hyperbolic space and in spherical space [11]. Existing force-directed algorithms are restricted to calculating a graph layout in Euclidean geometry. Euclidean space has a very convenient structure for force-directed methods. It is easy to define distances and angles, and the relationship between the vector representing the net force on an object and the appropriate motion of that object are quite straightforward. Certain non-Euclidean geometries, specifically hyperbolic geometry, have properties which are particularly well suited to the layout and visualization of large classes of graphs [12].

With this in mind we have implemented a generalization of force-directed methods to non-Euclidean geometries that relies on mappings between non-Euclidean geometries and corresponding tangent spaces. While a non-Euclidean geometry does not afford all of the conveniences of Euclidean geometry, there is a straightforward way to define distances and angles, provided we restrict ourselves to geometries which are smooth. Such geometries are known as Riemannian geometries, and while they have less convenient structure than Euclidean geometry, they retain many of the characteristics which are useful for force-directed graph layouts. A Riemannian manifold M has the property that for every point $x \in M$, the tangent space $T_x M$ is an inner product space. This means that for every point on the manifold, we can define local notions of length and angle.

Using a local notion of length we can define the length of a continuous curve $\gamma : [a, b] \to M$ by $\text{length}(\gamma) = \int_a^b \|\gamma'\| dt$. This leads to a natural generalization of the concept of a straight line to that of a *geodesic*, where the geodesic between two points, $u, v \in M$ is defined as a continuously differentiable curve of minimal length between them. In Euclidean geometry the geodesics are straight lines, and in spherical geometry they are arcs of great circles. Hence, the distance between two points, $d(x, y)$ is defined as the length of the geodesic between them.

Hyperbolic Geometry: Hyperbolic geometry is particularly well suited to graph layout because it has "more space" than Euclidean geometry – in the same sense that spherical geometry has "less space". Unlike in Euclidean geometry, where the relationship between the radius and circumference of a circle in two-dimensional geometry is linear with a factor of 2π, and constant in a spherical geometry, in hyperbolic geometry the circumference of a circle increases exponentially with its radius. The applicability of this geometric property to graph layout is well-illustrated with the example of a tree. In hyperbolic space, it is possible to layout a tree structure with a uniform distribution of the vertices and with uniform edge lengths despite the fact that the number of vertices at a certain depth in the tree increases exponentially with the depth.

In order to visualize a layout in hyperbolic geometry it is necessary to map the layout into the (2D) Euclidean geometry of a computer monitor. The method used in `graphael` is the Poincare projection, that maps hyperbolic space onto the open unit disk. The projection compresses the space near the boundary of the unit disk, giving the impression of a fish-eye view. This naturally provides a useful focus+context technique for visualizing the layouts of graphs. This model preserves angles, but distorts lines. A line in hyperbolic space is mapped to a circular arc which intersects the unit circle at right angles. The Poincare disk progressively distorts the graph view as we move away from the center of projection. In Fig. 2(a) we show a drawing of a graph obtained in hyperbolic geometry and displayed in 2D Euclidean space.

Spherical Geometry: Using the same ideas, we can generalize force-directed methods to spherical space. Spherical geometry, like hyperbolic geometry, has

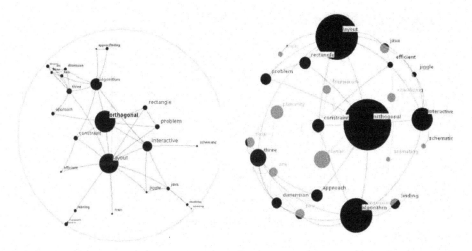

Fig. 2. Layouts of a title-word graph, obtained in Hyperbolic space and in Spherical space. The graph has 27 vertices and 50 edges and the vertices correspond to title-words from papers in the 1999 Graph Drawing conference. The size of a vertex is determined by its frequency and edges are placed between two vertices if they co-occur in at least one paper.

a constant curvature and the equations for mapping to and from the tangent space can be calculated analytically. Each point on a sphere is given a longitude and latitude. The sphere can then be embedded in 3D Euclidean space by a simple parameterization. In Fig. 2(b) we show a drawing of a graph obtained in spherical geometry and displayed in three-dimensional Euclidean space.

Multi-scale Graph Drawing in Non-Euclidean Space: Since we are able to utilize tangent space mapping to use existing force-directed methods for graph layouts, we can also generalize the multi-scale method for drawing large graphs to non-Euclidean spaces. Of the tree stages in the multi-scale method (filtration, initial placement, and refinement) the only stage that requires further consideration is the initial placement stage.

In the initial placement stage we place each vertex one at a time in the barycenter of its neighbors. In Euclidean space we simply take the average of each dimension and place the vertex at that point. For non-Euclidean points we use the mapping to and from a tangent space. Specifically, we map the non-Euclidean points that correspond to the location of the neighbors to a tangent space. From there we are able to calculate the barycenter. We map that back into the non-Euclidean manifold and place the vertex at that location.

2.6   Graph Editor

There are several ways to experiment with the `graphael` system. Loading one of the sample graph files, loading a new file, or creating a new graph. When a user wants to create a graph manually, they have the option of using a basic

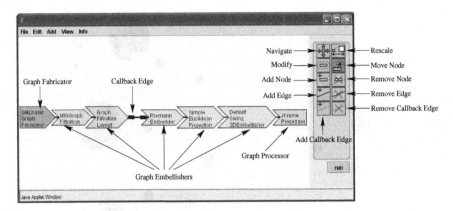

Fig. 3. A screenshot of the `graphael` CF-graph.

graph editor that can be accessed from within **graphael**. The editor is simple, but useful for users that do not have data to generate a graph from. The graph editor is especially helpful in the testing of new components, since simple cases can be modeled easily in the editor. With a point/click/drag user interface, vertices and edges can be added, deleted, or moved.

3   Control Flow Graph

Here we describe another novel features of **graphael**, the Control Flow (CF) graph[1]. The CF-graph allows the user to put together different combinations of layout algorithms, projections and visualizations, while offering a visual representation of the derived graph's production process; see Fig. 3. CF-graphs contain CF-nodes (such as layout algorithms) that act to generate and refine a derived graph, and CF-edges that represent the channels of input and output, internally passed between the CF-nodes.

CF-graphs are created by adding new CF-nodes and connecting them with edges. Once a complete chain of appropriate CF-nodes has been completed, the "run" button activates all the graph fabricators (described below). Modifying how an existing CF-graph produces derived graphs can be done by manipulating the composition of the current CF-graph or by changing the internal properties of the production units of CF-nodes.

3.1   CF-Nodes

There are three different types of production units: fabricators, embellishers, and processors. Each of them is briefly described below.

1. *Graph Fabricators* are graph production units that take no input from other CF-graph entities. They act as the starting point for the generation of derived

---

[1] Note: For clarity we shall call a Control Flow graph a CF-graph and a graph produced by a CF-graph a "derived" or "production" graph.

graphs since are not dependent on a derived graph as input. Many of the current graph fabricators available in **graphael** create the most raw form of a derived graph (i.e., vertex and edge declarations) by reading input files.

2. *Graph Embellishers* are methods that require a single, derived graph element as input from within the CF-graph and output a newly augmented, derived graph. In many cases, graph embellishers are used to add or modify properties of the derived graph they receive as input. For example, an embellisher could take a weighted graph as input and produce a weighted graph in which the color property of the heaviest vertices makes them stand out.

3. *Graph Processors* are methods in the final stage in any CF-graph, since they do not pass derived graphs to other production units. These methods typically output the final derived graph in the form of a picture, or a file.

3.2 Callback Edges

In a CF-graph with normal edges, once a production unit is finished with its input, the graph is passed to the next unit until it reaches the end of the CF-graph. However, there are cases when this is not desirable. If we wish, for example, to show a graph layout in a series of iterations (as opposed to just the final product), we would require the use of **graphael**'s callback edges. These edges allow the source CF-node to suspend its execution and pass the graph to the remainder of the CF-graph, starting at the target of the callback edge. Once this finishes, execution resumes where the source CF-node left off. Using callbacks, we implemented features such as animation. Specifically, a layout that needs to iterate over the graph multiple times can suspend itself to let the resulting graph from each iteration reach the processor and be displayed on the screen. After the processor finishes, the layout runs the next iteration. Callback edges can be identified as thick, dotted lines.

3.3 CF-Node Property Management

Whereas the panel in Fig. 3 offers different ways to manipulate the CF-graph, the individual CF-nodes can be manipulated as well. Recall that one of the two ways to modify how a derived graph is produced is to change CF-node properties. While we do allow for customized property managers, we have implemented an automatic GUI generator to minimize the amount of work required to make additions to the **graphael** library. The GUI generator is implemented using Java's reflection capabilities. This allows **graphael** to dynamically examine methods and data members of Java classes that have been added to its library. The system detects which properties can be modified by looking for a pair of *getter* and *setter* methods that meet certain conditions.

4 Conclusion

The **graphael** system has been implemented in Java, and is can be used to draw static graphs and evolving graphs online or offline. It can also be used as visualization platform for tools that generate graphs as output. For example, a database that produces graphs, such as the one described in [5] can be coupled with `graphael` to provide visual interaction with the graphs. Our system currently supports the (graph markup language) file format. Simple modifications to the standard format accommodate vertex-weights and edge-weights, as well as timeslice information.

References

1. D. Auber. Tulip - a huge graph visualization framework. In M. Jünger and P. Mutzel, editors, *Graph Drawing Software*, pages 105–126. Springer-Verlag, 2003.
2. V. Batagelj and A. Mrvar. Pajek - analysis and visualization of large networks. In M. Jünger and P. Mutzel, editors, *Graph Drawing Software*, pages 77–103. Springer-Verlag, 2003.
3. S. S. Bridgeman, A. Garg, and R. Tamassia. A graph drawing and translation service on the www. *International Journal on Computational Geometry and Application*, 9(4–5):419–446, 1999.
4. T. Dwyer and P. Eckersley. Wilmascope - a 3d graph visualization system. In M. Jünger and P. Mutzel, editors, *Graph Drawing Software*, pages 55–75. Springer-Verlag, 2003.
5. C. Erten, P. J. Harding, S. G. Kobourov, K. Wampler, and G. Yee. GraphAEL: Graph animations with evolving layouts. In *11th Symposium on Graph Drawing*, pages 98–110, 2003.
6. C. Erten, P. J. Harding, S. G. Kobourov, K. Wampler, and G. Yee. Exploring the computing literature using temporal graph visualization. In *Visualization and Data Analysis*, pages 45–56, 2004.
7. T. M. J. Fruchterman and E. M. Reingold. Graph drawing by force-directed placement. *Softw. Pract. Exper.*, 21(11):1129–1164, 1991.
8. P. Gajer and S. G. Kobourov. GRIP: Graph drawing with intelligent placement. *Journal of Graph Algorithms and Applications*, 6(3):203–224, 2002.
9. M. Jünger and P. Mutzel, editors. *Graph Drawing Software*. Springer-Verlag, 2003.
10. T. Kamada and S. Kawai. An algorithm for drawing general undirected graphs. *Inf. Process. Lett.*, 31(1):7–15, 1989.
11. S. G. Kobourov and K. Wampler. Non-Euclidean spring embedders. In *10th Annual IEEE Symposium on Information Visualization (InfoVis)*. To appear in 2004.
12. J. Lamping, R. Rao, and P. Pirolli. A focus+context technique based on hyperbolic geometry for visualizing large hierarchies. In *Proceedings of the SIGCHI conference on Human factors in computing systems*, pages 401–408, 1995.
13. S. B. Seidman. Network structure and minimum degree. *Social Networks*, 5:269–287, 1983.
14. R. Wiese, M. Eiglsperger, and M. Kauffmann. yfiles - visualization and automatic layout of graphs. In M. Jünger and P. Mutzel, editors, *Graph Drawing Software*, pages 173–192. Springer-Verlag, 2003.

QUOGGLES: Query On Graphs –
A Graphical Largely Extensible System

Paul Holleis[1] and Franz J. Brandenburg[2]

[1] University of Munich, 80333 Munich, Germany
[2] University of Passau, 94030 Passau, Germany

Abstract. We describe the query and data processing language *QUOGGLES* which is particularly designed for the application on graphs. It uses a pipeline-like technique known from command line processing, and composes its queries as directed acyclic graphs. The main focus is on the extensibility and the ease of use. The language permits queries that select a distinguished subgraph, e.g., the set of all green nodes with degree at least d or the set of edges whose endnodes have a neighbor which has exactly one neighbor. It is `SQL` complete, however, it cannot describe paths of arbitrary length; otherwise NP-hard problems like Hamilton path could directly be expressed. *QUOGGLES* also enables the user to concatenate queries with algorithms, e.g. with graph drawing algorithms, which are then applied to the selected subgraph.

1 Introduction

Graphs are frequently used to represent discrete data with objects as nodes and (binary) relations represented as edges. A relational database can be seen as a graph with n-ary relations, which can be modelled by hyperedges or by bipartite graphs. Often, a user has a special view on the data. In terms of graphs this means a distinguished subgraph, which is described by a collection of nodes, edges and attributes. This is particularly true for huge graphs such as the WWW or communication networks, from which the user selects a particular section.

This scenario coincides with the theme of Category C of the 10th Graph Drawing Contest 2002 [1]. The initiator Joe Marks had posted example graphs and wanted on-line answers on questions like "what is the largest wheel of green and blue nodes". At GD 2002, nobody could answer this question on-line. This was the starting point for quoggles ("QUeries On Graphs: A Graphical Largely Extensible System"), a plug-in for the graph visualization toolkit Gravisto developed at the University of Passau [2]. It is fully described in [3]. Gravisto associates graph elements with a hierarchy of attributes. These are addressed by queries and used for further computations like comparisons and sorting. The query language is capable of simulating relational algebra and `SQL`. However, it cannot express transitive closures and the existence of paths of arbitrary length. *QUOGGLES* itself is fully graphical and composes its queries in terms of directed acyclic graphs, which are automatically drawn by a simple algorithm.

Here we give a short description of the language and illustrate its use by some examples. For details we refer to [3] and [2].

J. Pach (Ed.): GD 2004, LNCS 3383, pp. 465–470, 2004.

2   Description of the Language

The query language of *QUOGGLES* consists of a set of fundamental operations which can be combined to form more complex operations. Every operation has i inputs, p parameters and o outputs. For maximal generality these numbers can depend on the values of the parameters. The resulting language has the full power of relational algebra and SQL (the proof can be found in [3]); however it cannot express paths of arbitrary length.

QUOGGLES is fully graphical. Every operation has a **box** as graphical representation as shown in Fig. 1. The box includes the name of the operation, values for its parameters and is numbered consecutively in the order of the creation within the query. It has i incoming lines on the left hand side for the input and o outgoing lines on the right for the output. These act as connection points to other boxes.

In the graphical representation, queries are composed from operations as directed acyclic graphs, combining inputs and outputs of boxes of operations in an appropriate way, possibly observing intermediate results using Output boxes. See, e.g., Fig. 2 for an example which is explained later in more detail.

For the evaluation of a query, the pipeline idea is used, which is well known from Unix command line and batch file processing. *QUOGGLES* applies and extends this general approach in information processing to graphs.

The set of graph elements from a graph acts as a source for each query. Data is processed as it flows through the pipelined operations. Since the notion of one single linear pipeline is quite restrictive, a directed acyclic graph can be built instead. It is constructed from operations with any finite number of inputs and outputs. Data flows along the edges of such a query graph. A query can then be evaluated using a topological ordering. A graphical user interface helps to create, change, execute and debug query graphs.

2.1   Basic Operations

In this section we describe a set of basic operations necessary to generate a sensible range of queries. This includes input operations, filters, general purpose and

Fig. 1. Graphical representations of a general and a sample operation.

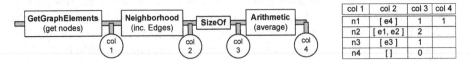

col 1	col 2	col 3	col 4
n1	[e4]	1	1
n2	[e1, e2]	2	
n3	[e3]	1	
n4	[]	0	

Fig. 2. This query computes the average degree of the nodes of a graph.

graph specific operations. Every operation receives one element or a collection of elements as input, checks and transforms the element(s) and outputs its result. Such elements can be nodes and edges, numbers and strings, e.g., nodes and edge labels, or tuples of such elements. In the implementation a collection is a list of Java objects. The following categories of operations are available:

Input Operations. Input boxes have no inputs and are used to create specific constants, such as text labels or numerical values or access external information like saved graphs.

Graph Specific Operations. For the navigation through graphs it suffices to provide an operation that accesses the neighborhood of a graph element and one that returns associated attribute(s):

The Neighborhood operation accesses elements in the graph theoretical neighborhood from input graph elements. Possible parameter values include *neighbors*, *incoming* or *outgoing edges* and *source* or *target nodes*.

The GetAttributeValue operation is used to retrieve attributed information from graph elements, such as node and edge identifiers or their labels. Graphical attributes like shape and size can also be queried.

General Query Operations. Most of the data processed by *QUOGGLES* is present as lists of elements. Hence, operations on collections are common. These include *flatten*, which converts a nested to a flat collection, *reverse*, which reverses lists and *make distinct*, which removes duplicates and *sort* for sorting a list using a string representation of the objects. The *union* and *intersection* boxes take two input lists and compute the set union and intersection, respectively.

Further general operations use the textual representation of elements in the input collection, count elements, compute the average, do arithmetic, comparisons and boolean operations or check the type of an input. Figure 3 shows the CompareTwoValues operation that compares its two inputs according to some specified relation.

There is a special TwoSplitConnector operation that duplicates the input and thus enables producing queries beyond that of simple linear pipelining.

Since it is not always clear that the sinks of the query graph are the (only) places that should contribute to the query result, Output boxes specify which part of the data present somewhere in the query pipeline should contribute to its result. They can also be used to check intermediate results.

Figure 2 shows an example query. The table on the right shows its output if the small graph displayed in Fig. 4 is used as input to the query. The data in the first column is retrieved from the first output box ('col 1'). It shows a list of

Fig. 3. Comparing two inputs using the $\leq$ relation and two different orders.

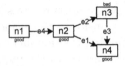

Fig. 4. A small graph used as sample input to queries.

all nodes of the graph. The `Neighborhood` operation produces a list that holds
(lists of) all incident edges of the corresponding nodes. The third columns shows
the sizes of these edge lists in the second column, i.e., the number of incident
edges of each node. The `Arithmetic` box then computes the average degree of all
nodes of the input graph ('col 4'). The selection of elements according to some
predicate is the most frequently used operation in database systems. Here, the
`Filter` operation retrieves those objects from an input collection of arbitrary
objects that meet the condition specified by a predicate subquery. The predicate
can be an arbitrary query. Its output will be interpreted as a boolean value.
Empty collections or the value zero will for example be converted to *false*.

Figure 5 shows a query that filters all graph elements that have an attribute
called *value* with value equal to *good*. The table on the right shows the result for
the example graph displayed in Fig. 4. Since edges do not have such an attribute,
column one displays a "-" for them. Nodes *n1, n2, n4* match the criterium (which
can be verified by examining their attribute value in column three).

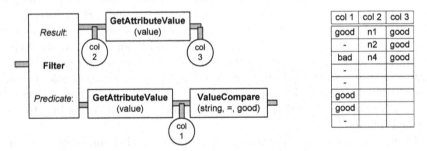

Fig. 5. Find all nodes from the small graph on the left that have outgoing edges.

To ensure reusability of queries and a reasonable size of queries, it is important that any query or part thereof can be saved as a subquery for later use. This
is shown in Fig. 6 where the calculation of the average node degree of a graph is
saved. This subquery can then be used to get all nodes that have a degree larger
than average. The query and the result ('col 3') is shown in Fig. 7.

3   Application in Graph Drawing

Good layouts of graphs can often not be achieved by generic graph drawing
algorithms alone. A certain degree of interactivity can be necessary or at least
helpful. However, tweaking algorithms for those special uses can be time con-

Fig. 6. The query on the left is saved and can later be used in a Subquery box.

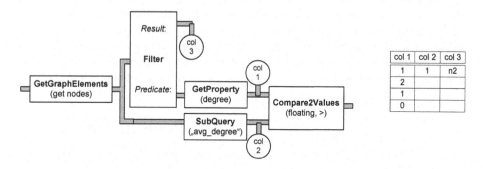

Fig. 7. Using a Subquery box, nodes with high degree ($n2$) can easily be found.

suming or even impossible if third-party programs are used. *QUOGGLES* enables the user to layout different parts of the graph differently.

Subgraphs that should be drawn in some special way can be retrieved by using queries. Then these sets can be further processed by applying layout algorithms on them. This renders it extremely easy to, e.g., quickly test which type of centrality best serves to find a good drawing. Nodes with a high centrality value can be drawn, e.g., more central than others.

Figure 8 shows a generic example query that finds a certain set of 'important' nodes. This might be done, e.g., by using the query shown in Fig. 7 as a subquery. This set is drawn using a spring embedder algorithm ('spring') with a small value as parameter ('10') indicating that nodes will be close together. All other nodes (retrieved using the *set minus* operation) are drawn using an algorithm that places nodes on a circle with a rather large radius ('75'). The algorithms directly work on the data structure. The result of the query (as specified by the circular box titled 'col 1') is the set of special nodes. This helps to manually adjust the relative placement of the two layouted subgraphs. This query has been applied to a random placement of nodes of a small graph producing the layout shown in Fig. 8.

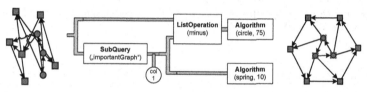

Fig. 8. A query used to apply two layout algorithms to different parts of the graph.

4   Conclusion

The *QUOGGLES* system is an implementation of a query language specifically designed to retrieve information from graphs. It combines general and graph specific operations using an extended pipeline principle. It can been shown that the system is relational complete and even provides similar functionality as SQL 92. An intuitive user interface is provided. Its extensibility renders it especially useful for semi-automatic graph processing. As an example, we showed how to apply its feature to include algorithms in query processing to address layouting and graph drawing problems. Fig. 9 shows a screen dump of the system.

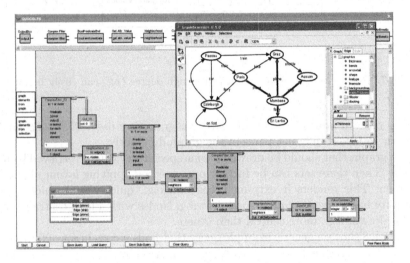

Fig. 9. A screen dump of Gravisto with an example graph and the *QUOGGLES* system after executing the introductory query: "Get the set of edges whose endnodes have a neighbor which has exactly one neighbor".

References

1. Brandenburg, F.J.: Graph-drawing contest report. Proceedings Graph Drawing 2002, LNCS 2528 (2002) 376–379
2. University of Passau: Gravisto. http://www.gravisto.org/ (2002)
3. Holleis, P.: Design and implementation of an extensible query language on graphs. Diploma thesis (2004)

Visualisation of Large and Complex Networks Using `PolyPlane`*

Seok-Hee Hong[1] and Tom Murtagh[2]

[1] National ICT Australia; School of Information Technologies,
University of Sydney, Australia
`shhong@it.usyd.edu.au`
[2] School of Information Technologies,
University of Sydney, Australia
`tfm@it.usyd.edu.au`

Abstract. This paper discusses a new method for visualisation of large and complex networks in three dimensions. In particular, we focus on visualising the *core tree structure* of the large and complex network. The algorithm uses the concept of *subplanes*, where a set of subtrees is laid out. The subplanes are defined using *regular polytopes* for easy navigation. The algorithm can be implemented to run in linear time. We implemented the algorithm and the experimental results show that it produces nice layouts of large trees with up to ten thousand nodes. We further discuss how to extend this method for more general case.

1 Introduction

Recent technological advances produce a lot of data, and have led to many large and complex network models in many domains; examples include social networks, biological networks and webgraphs.

Visualization can be an effective tool for analysis of such networks. Good visualisation reveals the hidden structure of the networks and amplifies human understanding, thus leading to new insights, new findings and possible prediction.

However, recent advances in technology have made available data on networks with millions of nodes; visualization of such large and complex networks is very challenging. Current methods exhibit at least one the following problems: poor scalability, lack of good navigation methods, poor integration with analysis methods, and lack of good 3D visualisation.

In this paper, we present a new method for visualisation of large and complex networks in three dimensions. In particular, we focus on visualising the *core tree*

---

* This research has been supported by a SESQUI grant from the University of Sydney, a research grant from the School of Information Technologies, Special Study Leave Program of the University of Sydney, and NICTA Summer Vacation Scholarship. Animated drawings are available from http://www.cs.usyd.edu.au/~shhong/3dtreedraw.htm. National ICT Australia is funded by the Australian Government's Backing Australia's Ability initiative, in part through the Australian Research Council.

structure of the large and complex network to reduce both cognitive overload and visual complexity.

The tree is one of the most common relational structure. Many applications can be modeled as trees. Examples include family trees, hierarchical information, BFS tree of WWW graphs and phylogenetic trees. There are many layout methods for trees in two and three dimensions. However, most existing methods focus on two dimensions. Layout methods for trees in three dimensions are not well investigated.

We present a new drawing algorithm for trees in three dimensions. The algorithm uses the concept of the *subplanes*, where a set of subtrees are laid out. The main reason to use subplanes is to reduce visual complexity and for easy navigation. Note that 3D visualisation may suffer from occlusion and navigation problem. However, using subplanes defined by *regular polytopes*, the drawing is easy to navigate. Further, the algorithm can be implemented to run in linear time.

We implemented the algorithm and the experimental results show that it produces nice layouts of large trees with up to ten thousands nodes. Figure 1(a) shows a drawing of a tree with 6929 nodes using our algorithm. Here, we use the Icosahedron polytope to define 30 subplanes.

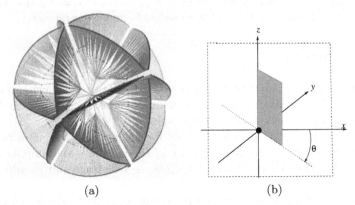

(a)   (b)

Fig. 1. (a) Example output of the algorithm (b) example of a subplane.

This paper is organized as follows. In Section 2, we review previous drawing algorithms for trees. The main results of the paper is in Section 3: here we present a new drawing algorithm for drawing trees in three dimensions. An implementation and experimental results are described in Section 4. In Section 5, we discuss how to extend this method for more general case.

2   Related Work

There are many tree drawing algorithms and systems are available [2, 3, 7–9, 11–13]. For a survey, see [5]. These can be classified as 2D visualisation [2, 3,

7, 8, 11, 12] or 3D visualisation [9, 13], rooted tree [7, 12, 13] or free (unrooted) tree [3, 8, 11], binary tree [12] or general tree [3, 7, 8, 11, 13], and their aesthetics or optimization goal, such as efficient use of spaces [7, 11].

First we consider 2D tree drawing algorithms. The *radial drawing* algorithm is suitable for drawing free trees in two dimensions [3]. It uses concentric circles and then recursively draw each subtree in a wedge of the circle. However, there is some unused space in order to guarantee no edge crossings. For rooted binary trees, one can use *Tidier Drawing* algorithm [12].

Treemaps use a space filling technique for the visualisation of the tree in two dimensions [7]. It uses all available space, but it may be difficult to understand the relationship between the nodes[11]. The *hyperbolic tree browser* uses hyperbolic geometry for layout in two dimensions and classical *focus+context* techniques [8]. It produces nice layouts and can be used to visualise large hierarchies, up to a thousand or so nodes.

Recently, the *space optimization tree* was presented for viewing very large hierarchies in two dimensions [11]. The method uses the space in an optimized way and can display trees with up to 55000 nodes.

In three dimensions, *cone trees* are the best known [13]. They allow *focus+context* view and provide rotation operations. However, it uses only the surface of the cone and there is some unused space in 3D. They are able to display trees with thousands of nodes. The *H3* method uses hyperbolic geometry in three dimensions [9]. It produces a three dimensional layout for the spanning trees of large directed graphs, and also provides *focus+context* view.

However, in general, 3D drawing algorithms for trees are not well investigated compared to the methods in two dimensions. In this paper, we present a new drawing algorithm for trees in three dimensions.

3   The PolyPlane Algorithm

The algorithm uses the concept of *subplanes*, which are defined by regular polytopes. Roughly speaking, we choose the root of the tree and then partition the subtrees of the root. Then we assign each set of subtrees to a subplane and we draw each set on the subplane. An example of a subplane is shown in Figure 1(b).

The algorithm described in this paper treats the input tree as a rooted tree. The deletion of the root results in *subtrees* $T_1, T_2, \ldots, T_m$. The drawing algorithm draws the root of T at the origin o, and distributes the subtrees $T_1, T_2, \ldots, T_m$ onto disjoint subplanes P_j, which are equally spaced around the z axis.

The algorithm uses a two dimensional drawing algorithm, Draw2D, as a subroutine. This algorithm draws a rooted tree in a subplane. For the purposes of this paper there are no specific requirements for Draw2D; there are many linear time algorithms available (see [2]). Thus, the main algorithm can be described as follows.

Algorithm PolyPlane
1. Compute the core tree structure T of a graph G.
2. Choose the root r of the tree T.

3. Choose the regular polytope which defines the subplanes. Let j be the number of subplanes defined from the regular polytope.
4. Choose a partitioning $\mathcal{S} = S_1 \cup S_2 \cup \ldots \cup S_j$ of the set $\{T_i : 1 \leq i \leq m\}$ of subtrees rooted at r.
5. For all i, $1 \leq i \leq j$, consider the subtrees in S_i to be a single tree T_i' with a common root r. Use Draw2D to draw T_i' in the subplane P_i.

It is clear that the drawings in the subplanes have no edge crossings. Further, as long as we use a linear time algorithm Draw2D, and compute the partitioning at step 3 in linear time, the whole algorithm takes linear time.

Note that the algorithm is very *flexible*, as there are many steps at which an arbitrary choice can be made at each step. We now explain each step in details.

3.1   Computation of the Core Tree Structure

We first discuss how to compute the core tree structure of the large and complex networks. One can use a spanning tree, Steiner tree, BFS (Breath-First-Search) tree, or DFS (Depth-First-Search) tree based on the application domain.

For example, for weighted graphs, one may use the maximum weight spanning tree or minimum weight spanning trees. These can be computed in polynomial time. For the simplest, one can use a BFS or DFS Tree which can be computed in linear time.

3.2   Choice of the Root Node

The choice of the root may depend on the application domain. For example, the input tree may already have a designated root from hierarchy.

For free (unrooted) trees, one may choose the *center* of the tree as root. Every tree T has a center c, that is, a vertex such that the maximum distance from c to the leaves of the tree is minimized. Further it can be found in linear time.

In practice, one may choose *the (domain dependent) most important vertex* as the root. For example, highest degree vertex based on degree centrality, or betweenness centrality in social network analysis.

3.3   Choice of the Regular Polytopes

We use regular polytopes to define subplanes. The main reason to use regular polytopes is to provide an easy navigation method. There are many regular polytopes. However, all are variations on just three types: pyramids, prisms, and the Platonic solids. We now explain each polytope in detail.

Regular Pyramid A *regular pyramid* is a pyramid with a regular g-gon as its base. There is only one *g-fold rotation axis*, called the *vertical rotation axis*, passing through the apex and the center of its base. The axis defines the g-fold rotational symmetries of the regular pyramid polytope. An example is shown in Figure 2(a). Also, there are g reflection planes, called *vertical reflection* planes, each containing the principal axis.

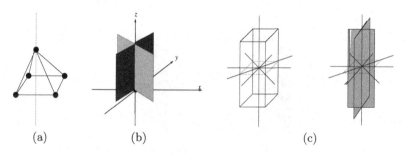

Fig. 2. (a) pyramid polytope (b) subplanes of pyramid polytope (c) prism polytope and subplanes.

We can define the g reflection planes, each of which is a rotation of $2\pi i/g$, $i = 0, 1, \ldots, g - 1$. The basic idea is to construct a regular pyramid drawing of a tree by placing the center of the tree at the apex of the pyramid, and the g partitioned subtrees on the subplanes that contain the side edges of the pyramid. An example is illustrated in Figure 2(b).

Regular Prism A *regular prism* has a regular g-gon as its top and bottom face. There are $g + 1$ rotation axes and they can be divided into two classes. The first one, called the *principal* axis or *vertical rotation* axis, is a g-fold rotation axis which passes through the centers of the two g-gon faces. The second class, called *secondary* axes or *horizontal rotation* axes, consists of g 2-fold rotation axes which lie in a plane perpendicular to the principal axis. Also, there are g reflection planes, called *vertical reflection* planes, each containing the principal axis, and another reflection plane perpendicular to the principal axis, called *horizontal reflection*. Figure 2(c) shows the rotation axes in the prism polytope.

We can define the $2g$ reflection planes, each of which is a rotation of $2\pi i/g$, $i = 0, 1, \ldots, g - 1$ in both directions of z coordinates. The basic idea is to construct a regular prism drawing of a tree by placing the center of the tree at the centroid of a prism, and $2g$ partitioned subtrees on the subplanes that contain the side edges of the prism.

An example is illustrated in Figure 3. In fact, there are two variations. One can define $2g$ subplanes as illustrated in Figure 3(a), or $3g$ subplanes as in Figure 3(b), which include g subplanes on the xy plane.

The Platonic Solids Basically, we use the rotation axes of the regular polytopes to define the subplanes. Using the regular g-gon pyramid polytope, we can define g subplanes, and using the regular g-gon prism polytope, we can define either $2g$ or $3g$ subplanes. For the Platonic solids, we can define more subplanes, as there are more rotation axes.

The tetrahedron has four 3-fold rotation axes and three 2-fold rotation axes. It has 12 rotational symmetries and 24 symmetries in total. The cube has three 4-fold rotation axes, four 3-fold axes, and six 2-fold rotation axes. It has 24 rotational symmetries and a full symmetry group of size 48. The icosahedron has six 5-fold rotation axes, ten 3-fold rotation axes, and fifteen 2-fold rotation

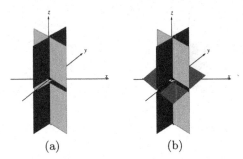

Fig. 3. Example of the subplanes of the regular prism polytope.

axes. It has 60 rotational symmetries and a full symmetry group of size 120. Note that the cube and the octahedron are dual solids, and the dodecahedron and the icosahedron are dual.

As an example, we consider the case of the cube. We can define 24 subplanes using the cube. The cube has three 4-fold axes, and these can define six half-axes. We can use each of the half-axis, to define 4-subplanes. This means that we can define 24 subplanes in total.

Similarly, we can define 12 subplanes using the tetrahedron, 3 subplanes each around four 3-fold axes. We can also define 60 subplanes using the icosahedron, as we can define 5 subplanes each around six 5-fold axes and these six 5-fold axes define 12 half-axes.

3.4   The Partitioning Algorithm

Once we have chosen the regular polytope and fixed the number g of the subplanes, we then need to divide the subtrees into g subsets.

For this step, we need to find a balanced partitioning of the subtrees. This problem can be formulated as a traditional bin-packing problem. Note that the bin-packing problem is NP-hard [4]. However, many heuristics and approximation algorithms are available [1,4]. For our implementation, we use first-fit and best-fit [4]. One main advantage for using these heuristics is that they run in linear time. One may use other well-known approximation algorithms for more sophisticated balancing. For details, see [1].

Note that in many applications, the partitioning may be given based on clustering or analysis of the large and complex networks. For example, in social network analysis, clustering can be defined using centrality or status measure. In biological networks, clustering can be defined by users or functionality.

3.5   The 2D Drawing Algorithm

Once we have chosen the regular polytope and computed a partitioning of the subtrees into g subsets, we then choose a 2D drawing algorithm for trees to draw each subset in a *subplane*. Formally, a subplane can be defined as a maximal simply connected open subset of a reflection plane that does not intersect any other reflection plane.

Many linear time algorithms are available to implement `Draw2D`(see [2]). For our implementation, we choose the radial drawing algorithm [3] to create the drawings in the subplanes as wedges. To guarantee no edge crossings, we allow a small space between each pair of subplanes.

4   Implementation and Experimental Results

We implemented the new layout algorithm as a part of the system *3DTree-Draw* [10]. In fact, the system 3DTreeDraw implements two 3D tree drawing algorithms.

The first one is a symmetric drawing algorithm [6], which finds the maximum number of symmetries in a tree and then constructs a maximally symmetric drawing of trees in three dimensions. The second algorithm is the algorithm `PolyPlane`, which is presented in this paper.

The system also provides simple zoom in and zoom out functions, as well as rotation of the 3D drawing. This rotation function is sufficient for navigation, as the subplanes were defined using regular polytopes which make the drawing easy to navigate. It also provides a function that you can save the result as a `bmp` file.

We use two different data sets, regular data sets and real world data sets from software engineering, webgraph and social network domains. We now present experimental results in details.

Firstly, we use randomly generated regular data sets, from a few hundred up to a hundred thousands nodes. The experimental results show that it produces aesthetically pleasing drawings of trees with up to ten thousands nodes. For the regular data sets, the drawings produce balanced appearance. Figure 4(a) shows a tree with 8613 nodes, using a regular 3-gon prism polytope with 6 subplanes.

One can define more subplanes based on variations of the Platonic solids. For example, in Figure 1, we can define three planes in each of the triangular shape space recursively. This improves resolution. However, we observed that too many planes make navigation a bit difficult. See Figure 4 (b) for an example. The tree has 483 nodes and it was drawn using the icosahedron polytopes.

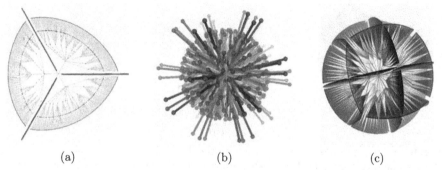

(a)	(b)	(c)

Fig. 4. (a) tree with 8613 nodes drawn with prism polytope (6 subplanes) (b) tree with 483 nodes drawn with the icosahedron polytope (c) tree with 6929 nodes drawn with the cube and the octahedron polytopes (36 subplanes).

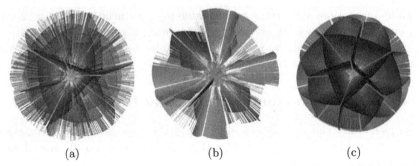

Fig. 5. Trees drawn the dodecahedron and the icosahedron polytopes (90 subplanes) (a) tree with 22001 nodes (b) tree with 59732 nodes (c) tree with 139681 nodes.

Figure 4(c) shows a tree with 6929 nodes, using a variation of the cube and the octahedron polytopes with 36 subplanes. Figure 5(a) shows a tree with 22001 nodes, using a variation of the dodecahedron and the icosahedron polytopes with 90 subplanes. Figure 5(b) shows a tree with 59732 nodes, using a variation of the dodecahedron and the icosahedron polytopes with 90 subplanes. Figure 5(c) shows a tree with 139681 nodes, using a variation of the dodecahedron and the icosahedron polytopes with 90 subplanes. Figure 6(a) shows a tree with 6929 nodes, using the dodecahedron polytope with 30 subplanes.

Finally, we apply our algorithm to visualise large and complex network from real world data. Figure 6(b) shows a home directory with 1385 nodes, using the icosahedron polytope with 30 subplanes.

Figure 6(c) shows a BFS tree of the School of IT, University of Sydney webgraph, with 4485 nodes, using the cube polytope with 12 subplanes. This is rooted at the main home page, and the core tree structure is a BFS tree. Figure 7(a) shows a tree of depth-limited search of the it.usyd.edu.au website with 146716 nodes, using a variation of the dodecahedron polytope with 30 subplanes.

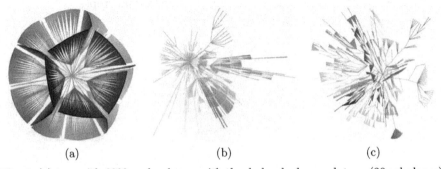

Fig. 6. (a) tree with 6929 nodes drawn with the dodecahedron polytope (30 subplanes) (b) a home directory with 1385 nodes drawn with the icosahedron polytope (30 subplanes) (c) BFS tree of School of IT website with 4485 nodes drawn with the cube polytope (12 subplanes).

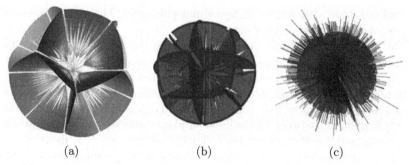

(a)   (b)   (c)

Fig. 7. (a) tree of depth-limited search of website with 146716 nodes drawn with the dodecahedron polytope (30 subplanes) (b) Erdos Number visualisation using the icosahedron polytope (30 subplanes) (c) Kevin Bacon number visualisation using the prism.

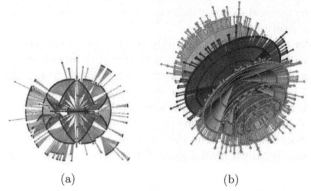

(a)   (b)

Fig. 8. Kevin Bacon number visualisations using (a) the icosahedron polytope with 30 subplanes (b) multiple planes defined by concentric cones.

Figure 7(b) shows Erdos number visualisation of mathematician collaboration network. We root at Erdos and compute a BFS tree. We then use the icosahedron polytope with 30 subplanes. Note that the current database keeps a record up to Erdos number 2, hence it displays a balanced appearance.

Figure 7(c), 8(a), and 8(b) show "Kevin Bacon number visualisations" of Hollywood movie actor collaboration network. We root at Kevin Bacon and then compute a BFS tree to visualise Kevin Bacon Number. We use three different polytopes: Figure 7(c) uses the prism polytope, Figure 8(a) uses the icosahedron polytope with 30 subplanes, and Figure 8(b) uses a variation of the `PolyPlane` algorithm with multiple planes defined by concentric cones.

In summary `PolyPlane` produces nice layouts. In particular, the effective use of subplanes reduces both visual and cognitive complexity and provides easy navigation.

5   Conclusion and Current Work

In this paper, we present a simple method to visualise the core tree structure of a large and complex network in three dimensions. The algorithm uses the concept of subplanes which are defined using regular polytopes. The algorithm is easy to implement and runs in linear time.

The algorithm is flexible, as one can choose the regular polytope and the 2D drawing algorithm for their own purpose. For example, for rooted trees, the pyramid polytope is more suitable. For dense trees with small diameter, the prism polytope or one of the Platonic solids is preferred. To improve resolution, one can define more subplanes using the method described in Section 4. However, there is a trade off between the number of planes and the navigation problem.

In summary, `PolyPlane` has the following advantages. It is flexible, easy to implement and can run in linear time. It can scale very well and is suitable for visualising a tree with high degree nodes, short diameter, or short average path length.

Our current work is to implement good navigation methods for `PolyPlane` and extend the method to cover more general cases. That is, to draw the whole network using multiple planes.

We consider many different variations as extensions. One is to use two parallel planes, to draw an *important subgraph* of the network on the top plane and draw the remaining subgraph of the network in the bottom plane. The important subgraph can be a set of vertices, a set of edges, or a small subgraph of the network. Note that this method can be generalised using up to k planes and a method to draw hierarchical graphs in three dimensions.

Another extension is to extend this multi plane idea to draw clustered graphs in three dimensions. Both involve some fundamental problems that need to be solved.

References

1. E. G. Coffman, M. R. Garey and D. S. Johnson, Approximation Algorithms for Bin Packing: A Survey, Approximation Algorithms for NP-Hard Problems, D. Hochbaum (editor), PWS Publishing, Boston, pp. 46-93. 1997.
2. G. Di Battista, P. Eades, R. Tamassia and I. G. Tollis, *Graph Drawing: Algorithms for the Visualization of Graphs*, Prentice-Hall, 1998.
3. P. Eades, Drawing Free Trees, Bulleting of the Institute of Combinatorics and its Applications, pp. 10-36, 1992.
4. M. R. Garey and D. S. Johnson, Computers and Intractability: A Guide to the Theory of NP Completeness, *Freeman*, 1979.
5. I. Herman, G. Melançon G, M. Marshall, Graph Visualization in Information Visualization: a Survey, IEEE Transactions on Visualization and Computer Graphics, 6, pp. 24-44, 2000.
6. S. Hong and P. Eades, Drawing Trees Symmetrically in Three Dimensions, *Algorithmica*, vol. 36, no. 2, 2003.
7. B. Johnson and B. Shneiderman, Tree-maps: A Space-Filling Approach to the Visualization of Hierarchical Information Structures, Proc. of IEEE Visualization'91, IEEE, Piscataway, NJ, pp. 284-291, 1991.
8. J. Lamping, R. Rao and P. Piroli, A Focus + Context Technique Based on Hyperbolic Geometry for Visualizing Large Hierarchies, Proc. of ACM CHI'95 Conference: Human Factors in Computing Systems, ACM, New York, NY, pp. 401-408, 1995.

9. T. Munzner, H3: Laying Out Large Directed Graphs in 3D Hyperbolic Space, Proc. of the 1997 IEEE Symposium on Information Visualization, pp. 2-10, 1997.
10. T. Murtagh and S. Hong, 3DTreeDraw: A Three Dimensional Tree Drawing System, Proc. of SoCG, pp. 380-382, 2003.
11. Q. V. Nguyen and M. Huang, A Space-Optimized Tree Visualization, Proc. of IEEE Symposium on Information Visualization (InfoVis2002), pp. 85-92, 2002.
12. E. Reingold and J. Tilford, Tidier Drawing of Trees, IEEE Transactions on Software Engineering, 7, pp 223-228, 1981.
13. G. Roberston, J. Mackinlay and S. Card, Cone Trees: Animated 3D Visualizations of Hierarchical Information, Proc. of SIGCHI'91, pp. 189-194, 1991.

The Metro Map Layout Problem*

Seok-Hee Hong[1], Damian Merrick[1], and Hugo A.D. do Nascimento[2]

[1] National ICT Australia**; School of Information Technologies,
University of Sydney, Australia
{shhong,dmerrick}@it.usyd.edu.au
[2] Instituto de Informatica-UFG, Brazil
hadn@inf.ufg.br

Abstract. We initiate a new problem of *automatic* metro map layout. In general, a metro map consists of a set of lines which have intersections or overlaps. We define a set of aesthetic criteria for good metro map layouts and present a method to produce such layouts automatically. Our method uses a variation of the spring algorithm with a suitable preprocessing step. The experimental results with real world data sets show that our method produces good metro map layouts quickly.

1 Introduction

A metro map is a simple example of a geometric network that appears in our daily life. An example of such, the Sydney Cityrail NSW train network, is shown in Figure 1 (a) [13]. Furthermore, the *metro map metaphor* has been used successfully for visualising *abstract* information, such as the "train of thought" network in Figure 1 (b) [8], website networks [9], and networks of related books in Figure 5 (a) [14].

In general, a metro map can be modeled as a graph, and automatic visualisation of graphs has received a great deal of interest from visualisation researchers over the past 10 years. However, automatic visualisation of metro maps is a very challenging problem, as already observed by Beck [4] and Tufte [12]. Note that existing metro maps are produced manually. Hence, it would be interesting to know how far automatic visualisation methods can go towards achieving the quality of the hand drawn pictures.

In this paper, we address this new problem of metro map layout. We define a set of aesthetic criteria for good metro map layouts and present a method to produce such layouts automatically. Our method uses a variation of the spring algorithm with a suitable preprocessing step. The experimental results with real world data sets show that our method produces good metro map layouts quickly.

In the next section, we define the problem. We present our metro map layout methods in Section 3 and discuss metro map labeling methods in Section 4. In Section 5, we present the experimental results and Section 6 concludes.

---

* A preliminary version of this paper was published in [5]. For a version of this paper with full-size colour images, see [6].

** National ICT Australia is funded by the Australian Government's Backing Australia's Ability initiative, in part through the Australian Research Council.

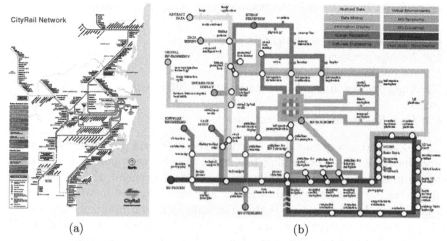

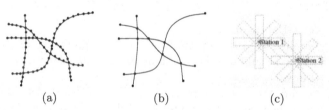

Fig. 1. (a) Sydney Cityrail NSW network (b) "Train of thought" network.

2   The Metro Map Layout Problem

A *metro map graph* consists of a graph G and a set of paths that cover all the vertices and edges of G. Some vertices and edges may appear in more than one path, but each occurs in at least one path. See Figure 2 (a) for an example.

Fig. 2. (a) Example of a metro map graph (b) Simplified metro map graph (c) Eight label positions for each vertex.

A *layout* of a metro map consists of a drawing of the graph. Thus, the main problem of this paper can be formally defined as follows.

The Metro Map Layout Problem
Input: a metro map graph G with a set of lines, each being a sequence of stations.
Output: a good layout L of G.

We now need a definition of a *good* layout of a metro map graph. For this purpose, we have studied existing *hand-drawn* metro maps from all over the world. For example, a detailed study of the London metro map by Beck can be found in [4]. From these manually produced layouts, we derive the following criteria for a good metro map layout.

C1: Each line drawn as straight as possible.
C2: No edge crossings.
C3: No overlapping of labels.
C4: Lines mostly drawn horizontally or vertically, with some at 45 degrees.
C5: Each line drawn with unique color.

We have designed layout methods based on these criteria. It should be noted that producing layouts conforming to exact geometry or topology is not the primary aim of our project. This is partly because, in general, the metro map metaphor can be used for visualisation of abstract information which has no fixed geometry. Another reason is that the most common usage of the metro map is for *navigation*, that is, to find out how to get to a specific destination. For example, consider the situation where a visitor to London (who does not know the exact geometry of London) uses the metro map for navigation.

3   The Layout Methods

We have tried five different layout methods using various combinations of spring algorithms. The tools that we use are GEM [2], a *modified version* of PrEd [1] and a magnetic spring algorithm [11]. In summary, each method can be briefly described as follows.

1. `Method 1`: The GEM algorithm.
2. `Method 2`: Simplify the metro map graph using a preprocessing step described in Section 3.1 and use the GEM algorithm *with* edge weight (details are explained later).
3. `Method 3`: Simplify the metro map graph and use the GEM algorithm *without* edge weight. Then we use the modified PrEd algorithm *with* edge weight.
4. `Method 4`: Simplify the metro map graph and use the GEM algorithm without edge weight. Then we use the modified PrEd algorithm with edge weight, plus *orthogonal* magnetic spring algorithm.
5. `Method 5`: Simplify the metro map graph and use the GEM algorithm without edge weight. Then we use the modified PrEd algorithm with edge weight, orthogonal magnetic spring algorithm, plus *45 degree* magnetic field forces.

We now describe each method in detail. *Method 1* simply uses GEM, a generic spring embedder. We use *Method 1* mainly as a baseline for comparison.

3.1   Method 2

First, we explain the *preprocessing step*. Note that there are many vertices of degree two in the metro map graph G. However, they do not contribute to the *embedding*, that is the overall topology, of the graph. This motivates us to remove these vertices and define a simplified graph G'. The resulting graph only contains intersection vertices and vertices with degree one. For example, the metro map graph in Figure 2 (a) can be simplified as in Figure 2 (b). Note that special care is needed to handle self loop and multiple edge cases.

After drawing the simplified graph G', we need to reinsert those removed vertices to get a layout of G. This requires space; hence we assign edge weights, according to the number of removed vertices, to edges of G' and produce a layout of G' which reflects those edge weights. Thus, Method 2 can be described as follows.

Method 2
1. Compute a simplified metro map graph G' by removing degree two vertices from the metro map graph G.
2. Produce a layout L' of G' using the GEM algorithm with edges weighted according to the number of vertices removed at Step 1.
3. Produce a layout L of G by reinserting the removed vertices, spaced evenly along the edges in the layout L'.

3.2   Method 3

Method 3 uses the preprocessing step, the GEM algorithm and the PrEd algorithm [1]. The PrEd algorithm is a special force directed method that preserves the topology of its input layout. For our purpose, we modified the PrEd algorithm to take into account edge weights.

In our modification, the x-components of the attraction force $F^a(u,v)$ and the repulsion force $F^r(u,v)$ between two vertices u and v are defined as follows:

$$F_x^a(u,v) = \frac{d(u,v)}{\delta(u,v)}(x(v) - x(u)), \quad F_x^r(u,v) = \frac{-\delta(u,v)^2}{d(u,v)^2}(x(v) - x(u)) \quad (1)$$

where $x(u)$ and $x(v)$ are the x coordinates of vertices u and v respectively, $d(u,v)$ is the distance between vertices u and v, and $\delta(u,v)$ is the ideal distance between the two vertices defined as $\delta(u,v) = L \times min(W, weight(u,v))^2$, where L, W are positive constants and $weight(u,v)$ is the weight of the edge between u and v. In the case of multiple edges between u and v, the maximum weight of that set of edges is used. The y-components $F^a y(u,v)$ and $F_y^r(u,v)$ of the force vectors are computed similarly.

Node-edge repulsion forces are computed in an identical manner to the original PrEd algorithm, that is:

$$F_x^e(v,(a,b)) = -\frac{(\gamma - d(v,i_v))^2}{d(v,i_v)}(x(i_v) - x(v)) \quad (2)$$

if $i_v \in (a,b)$, $d(v,i_v) < \gamma$, otherwise $F_x^e(v,(a,b)) = 0$.

As in the PrEd algorithm, the total force acting on a vertex v is calculated by summing all attraction and repulsion forces on that vertex, that is:

$$F_x(v) = \sum_{(u,v)\in E} F_x^a(u,v) + \sum_{u\in V} F_x^r(u,v)$$

$$+ \sum_{(a,b)\in E} F_x^e(v,(a,b)) - \sum_{u,w\in V,(v,w)\in E} F_x^e(u,(v,w)) \quad (3)$$

We now describe Method 3.

Method 3
1. Compute a simplified metro map graph G' by removing degree two vertices from the metro map graph G.
2. Produce an initial layout L' of G' using the GEM algorithm (with no edge weights).
3. Produce a better layout L'' of G' using the PrEd algorithm, modified to include edge weights.
4. Produce a layout L of G by reinserting the removed vertices, spaced evenly along the edges in the layout L''.

3.3 Method 4

Method 4 uses the preprocessing step, GEM and PrEd with orthogonal magnetic springs, that is, with horizontal and vertical aligning forces.

Forces are calculated as for Method 3, but with the addition of magnetic field forces acting on each edge. Equal and opposite forces are applied to each vertex of an edge to attempt to align that edge with a horizontally or vertically directed vector [11].

The magnitude of a force from an individual force field vector on the edge connecting vertices u and v is determined by a similar calculation to that for a magnetic spring:

$$F^m(u, v) = c_m bd(u, v)^\alpha \theta^\beta \qquad (4)$$

where b represents the strength of the magnetic field, θ is the angle between the edge (u, v) and the magnetic force vector, and $c_m, \alpha, \beta > 0$ are model-tuning constants.

Four force field vectors are used - left, right, up and down directed vectors. At any instant only a single magnetic force is applied to a given edge. The magnitude of the force applied is calculated according to the above equation for the force field vector to which the edge has the lowest angle θ. The direction of the force applied is perpendicular to the direction of the edge. To effect the desired rotational force on the edge, a force of magnitude $F^m(u, v)$ is applied to one vertex of the edge and a force of magnitude $-F^m(u, v)$ is applied to the other.

With the addition of the magnetic field, the total force applied to a vertex v becomes:

$$F_x(v) = \sum_{(u,v)\in E} F_x^a(u, v) + \sum_{u \in V} F_x^r(u, v) + \sum_{(a,b)\in E} F_x^e(v, (a, b)) \qquad (5)$$

$$- \sum_{u,w \in V, (v,w)\in E} F_x^e(u, (v, w)) + \sum_{(u,v)\in E} F_x^m(u, v)$$

Method 4 can be described similarly to Method 3, except at Step 3: Produce a better layout L'' of G' using the PrEd algorithm, modified to include edge weights and orthogonal magnetic field forces.

3.4   Method 5

Method 5 uses the preprocessing step, the GEM algorithm and the PrEd algorithm with orthogonal magnetic springs and 45 degree magnetic forces. Forces are calculated as for Method 4, but with the addition of four diagonal magnetic force field vectors.

Vectors running bottom-left to top-right, bottom-right to top-left, top-right to bottom-left and top-left to bottom-right are added to the set of magnetic field forces used. Magnetic field forces are calculated as described for Method 4, and the equation for the total force acting on a vertex v does not change.

Method 5 can be described similarly to Method 3, except at Step 3: Produce a better layout L'' of G' using the PrEd algorithm, modified to include edge weights and orthogonal magnetic field forces and 45 degree magnetic forces.

Note that the production of a metro map layout which preserves geographical constraints can be achieved with a small modification. Instead of using GEM at Step 2, we can assign real geographical coordinates to vertices according to the real world latitude and longitude of their associated train stations. This ensures that the geographical embedding of the graph remains unchanged throughout the layout process. As a result, the relative ordering of edges and their crossings are preserved.

4   Metro Map Labeling

The second part of this project is to produce a good labeling for metro map layout. We use a well known combinatorial approach for labeling map features. In this approach, a predefined set of label positions is assigned to every feature and a subset of these positions is chosen for producing an overlap-free label placement.

The first step of the approach is to specify the predefined label positions. We define an *eight position model*, orthogonal and diagonal, for metro map labels as illustrated in Figure 2 (c). Note that the labeling of other types of maps typically uses only a four position model.

The next step is to construct a *conflict graph* that describes all overlaps between label positions. The conflict graph has a vertex for every label position, and an edge linking every pair of label positions that overlap in the map. Moreover, the set of label positions assigned to each feature forms a clique in the graph. Each vertex can also have a cost value indicating the preference of using its associated label position on the map.

A labeling solution is then generated by computing the *maximum independent set with minimum cost* of the conflict graph. If a vertex is included in the independent set then its associated label position is used for label placement. Features which have no label positions appearing in the set are left unlabeled. Note that the maximum independent set represents label positions that do no overlap.

We have labeled the metro maps using the LabelHints system with eight position model. LabelHints implements the automatic approach described above, and offers several other tools for interactively exploring map labeling solutions [7]. It uses a simulated annealing algorithm and a greedy heuristic for producing an initial labeling. Whenever the users are not satisfied with the computer-generated result, they can improve the solution by directly changing the conflict graph, re-executing the algorithms, and/or modifying the pre-computed independent set. Such interactions allow the users to include important domain knowledge that was not considered in the automatic process.

Through experiments done with the system and in studies of other metro maps we observed that labels with diagonal orientation (45 degrees) are visually more pleasing. We have, therefore, decided to use mostly this orientation in our layouts.

5   Implementation and Experimental Results

The layout algorithms were implemented as a plugin to jjGraph [3]. The tests were executed on a single processor 3.0GHz Pentium 4 machine with 1GB of RAM, and the code was run under the Sun Microsystems Java(TM) 2 Runtime Environment, Standard Edition.

Metro map data is stored in a custom text file format describing the sequence of stations along each line in the network. These files are read by the metro map plugin, which then lays out the network and displays the resulting graph layout in jjGraph. jjGraph allows the user to navigate and modify this graph layout, as well as providing save and image export functionality. The metro map plugin was later made to export a complete layout to another format to be loaded into LabelHints [7].

We used real world data sets with several hundred vertices. Let $G = (V, E)$ be the original metro map graph and $G' = (V', E')$ be the reduced metro map graph. Details of the data sets are as follows. *Sydney*: $|V| = 319, |E| = 897, |V'| = 41, |E'| = 178$, *Barcelona*: $|V| = 101, |E| = 111, |V'| = 22, |E'| = 32$, *Tokyo*: $|V| = 224, |E| = 292, |V'| = 62, |E'| = 122$, *London*: $|V| = 271, |E| = 745, |V'| = 92, |E'| = 317$, *Train of thought network*: $|V| = 76, |E| = 120, |V'| = 36, |E'| = 67$, *O'Reilly book network*: $|V| = 116, |E| = 137, |V'| = 44, |E'| = 65$.

In summary, the results are comparable to hand drawn metro maps. Our method produces a good metro map layout very quickly. First, we present results of the Sydney Cityrail network. The methods gradually improve both in terms of the running time and the quality of the layout. Details of the running time of each method are as follows. *Method 1*: 12, *Method 2*: 0.6, *Method 3*: 1.9, *Method 4*: 2.1, and *Method 5*: 2.3 seconds.

Note that Method 2 significantly reduces the running time over Method 1. Methods 3, 4 and 5 take slightly longer than Method 2, due to use of two spring algorithms; however, they are still significantly faster than Method 1.

Each method's results for Cityrail are illustrated in Figures 3 and 4. Note that each successive Figure improves over the previous one; and Method 5 is clearly the best. It satisfies most of the criteria that we wanted to achieve.

Since Method 5 produces the best result, we chose this layout for labeling. The results for the Barcelona, Sydney Cityrail NSW, Tokyo and London metro

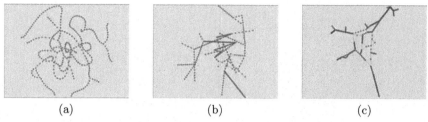

(a)   (b)   (c)

Fig. 3. Sydney Cityrail NSW network produced by (a) method 1 (b) method 2 (c) method 3.

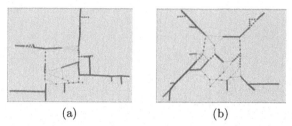

(a)   (b)

Fig. 4. Sydney Cityrail NSW network produced by (b) method 4 (c) method 5.

map layouts are shown in Figures 5 (b), 7 (a), 8 (a) and 8 (b) respectively. The running times are: *Barcelona*: 0.2, *Sydney Cityrail NSW*: 2.3, *Tokyo*: 9.2, and *London*: 22 seconds. Note that London is the most complex of these networks.

Figure 6 shows two examples of metro map metaphor visualisation; the train of thought network and the book network. Figure 7 (b) shows an example of the Sydney Cityrail NSW network with *fixed embedding*. The running times are: *Sydney*: 7.6, *Train of thought*: 3.3, and *O'Reilly book network*: 3.8 seconds.

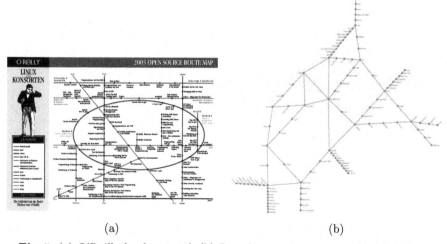

(a)   (b)

Fig. 5. (a) O'Reilly book network (b) Barcelona city metro map with labeling.

An Interactive Multi-user System
for Simultaneous Graph Drawing*

Stephen G. Kobourov[1] and Chandan Pitta[2]

[1] Department of Computer Science,
University of Arizona
kobourov@cs.arizona.edu
[2] Department of Electrical and Computer Engineering,
University of Arizona
chandanp@ece.arizona.edu

Abstract. In this paper we consider the problem of simultaneous drawing of two graphs. The goal is to produce aesthetically pleasing drawings for the two graphs by means of a heuristic algorithm and with human assistance. Our implementation uses the DiamondTouch table, a multi-user, touch-sensitive input device, to take advantage of direct physical interaction of several users working collaboratively. The system can be downloaded at http://dt.cs.arizona.edu where it is also available as an applet.

1   Introduction

Simultaneous drawings of multiple graphs are a useful visualization technique when different relationships are defined on the same set of objects, or when a relationship evolves through time. The objects are represented by graph nodes and the relationships are represented by graph edges. In simultaneous drawings, the placement of the graph nodes is the same in all the drawings, in order to preserve the viewer's mental map. Thus, it is more difficult to obtain good node placement for simultaneous drawings of two or more graphs, compared to the case when only one graph is to be displayed.

Even in the case when only two graphs are given, and individually they are planar, it is not always possible to find consistent node positions that realize plane drawings for each graph. It is not known whether pairs of graphs from a large number of classes allow simultaneous, straight-line, crossing-free embeddings. To aid in the design of algorithms for simultaneous plane drawings for certain classes of graphs and also to help in finding counter-examples (pairs of graphs that cannot be realized) we designed an interactive, multi-user system for manipulating simultaneous drawings of pairs of graphs.

Motivation for this problem comes from applications where it is often necessary to visually compare two relationships. Evolutionary trees on the same set of

* This work is partially supported by the NSF under grant ACR-0222920 and by ITCDI under grant 003297.

J. Pach (Ed.): GD 2004, LNCS 3383, pp. 492–501, 2004.

species are often constructed in computational biology. Biologists spend count-less hours pouring over tree drawings to determine the most likely evolutionary branches. The problem is particularly difficult when the drawings of different trees are laid out independent of each other.

1.1 Related Work

The problem of simultaneous embedding of planar graphs was introduced in [2], where it is shown that pairs of paths, cycles, and caterpillars can always be realized, while for general planar graphs and even outerplanar graphs this is not always possible. Modified force-directed methods are used to visualize general graphs simultaneously such that the mental map is preserved in [6]. Conceptually, the problem of simultaneously embedding graphs is the reverse of the geometric thickness problem [5].

The TreeJuxtaposer is a system designed to support the comparison task for large trees [13]. A tool for visualizing large numbers of evolutionary trees on the same set of species is presented in [10].

Traditional informal definitions of aesthetically pleasing graph drawings in-clude features such as straight-line segments for edges, few if any crossings, and display of symmetries. In crossing minimization, the problem is to find a draw-ing with the minimum number of crossings. The problem is NP-Complete [7] but there has been a great deal of research on heuristic algorithms [8]. Graph planarization [12] is often used together with careful reinsertion of edges.

The Human Guided Search (HuGS) framework described in [1,9] is an inter-active, or human-in-the-loop, optimization system. It leverages people's abilities in areas in which they outperform computers, such as visual and strategic think-ing. Users can steer interactive optimization systems towards solutions which satisfy real-world constraints. HuGS has been applied to graph drawing prob-lems in [11]. The DiamondTouch table is introduced in [4] and it has been used for an interactive, multi-player game [3] and for gestural interaction [14].

1.2 Our Contributions

We present an interactive multi-user system for simultaneous graph drawing. The system uses the DiamondTouch table, and allows for collaborative work of up to four users. We also provide a heuristic algorithm that attempts to minimize the number of crossings. The algorithms can be used on the entire graphs or on subsets of nodes. The users can stop the algorithm, move nodes around and restart it with the updated positions. Thus, the users can help the algorithm move out of a local minimum, or guide the algorithm towards a more aesthetically appealing solution. Alternatively, if the users get stuck in a local minimum, the algorithm can be started from a random position that may lead to a better solution. Finally, our system works not only with the DiamondTouch table, but also as a Java desktop application, or as a Java applet. The system is operational at http://dt.cs.arizona.edu.

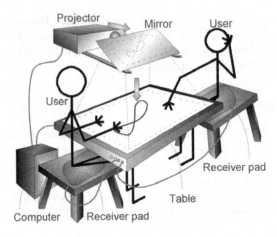

Fig. 1. Conceptual DiamondTouch table setup (from [3]).

2   The DiamondTouch Table

The DiamondTouch table [4] from Mitsubishi Electric Research Laboratories (MERL) is a desktop device that allows up to four users to simultaneously manipulate virtual objects. Users can move objects around on the table by touching and dragging them with their fingers. The purpose of the table is to allow several people to interact with a program at the same time and to do so using their hands rather than more common input devices such as mice. The conceptual setup is shown in Fig. 1.

The DiamondTouch table not only detects multiple users, but also identifies which user is touching where on the table. The table is physically large at 32" x 24" and allows several users to work together comfortably; see Fig. 2. Under the surface of the table, there is a grid of antennae. Each antenna transmits a signal to the computer that corresponds to the strength of the capacitance between the user and table. The capacitance is greatest when the user is in direct contact with a particular antenna: a circuit is completed from the antenna, through the user's body, through the receiver pad on which the user is sitting or standing, and back into the table.

The table is designed to be used with an ordinary desktop PC or laptop. It sends the data from the antennae to the DiamondTouch SDK drivers through the USB port, allowing the software to examine the data and to determine where on the table the user's fingers are located. The table is not a touch-screen: it has no ability to display output. Instead all images which would normally appear on the display monitor are routed to a video projector which projects them onto the surface of the table with the aid of a mirror and some painstaking calibration.

3   Our System

The input to the system consists of two graphs $G_1 = (V_1, E_1)$ and $G_2 = (V_2, E_2)$ defined on the same set of nodes, $V_1 = V_2$, or a subset of a larger common set,

Fig. 2. Physical setup of the DiamondTouch table setup with two users untangling graphs.

$V_1 \subseteq V$ and $V_2 \subseteq V$. The goal is to obtain aesthetically pleasing simultaneous layouts for both graphs.

In the case where G_1 and G_2 are planar, the goal is to obtain a node configuration that realizes plane drawings for each graph. That is, we are looking for a point set P and bijective function $m : V \rightarrow P$ that maps the set of nodes to points in $\mathcal{R}^2$ such that: (1) in a straight-line drawing of G_1 on P, using the mapping m, there are no crossings; and (2) in a straight-line drawing of G_2 on P, using the mapping m, there are no crossings; see Fig. 4.

In the case when the two graphs cannot be realized simultaneously as plane drawings, the goal is to obtain symmetric straight-line drawings with as few edge crossings as possible. Note that edge crossings are acceptable if in each pairwise edge crossing one of the edges is from E_1 and the other from E_2.

The system overview is shown in Fig. 3. The system requires an input file, which contains node and edge information about the two graphs. The graphs are then displayed on the table and users can interact with the system in various ways. Some of the interactions possible are:

- loading and storing graphs via input/output files;
- selecting single-view or split-view;
- selecting drawings to show in the view (G_1, G_2, or both);

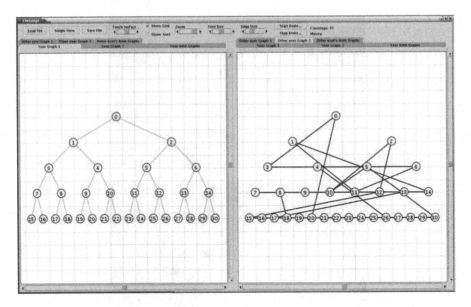

Fig. 3. System interface with split view.

- calling a heuristic crossing-minimization algorithm on both graphs;
- calling the same algorithm on selected parts of the graphs;
- interrupting the algorithm and manually repositioning nodes;
- zooming in and out, or scrolling across larger areas;
- changing colors and sizes of nodes and edges.

3.1 Examples

If the union graph, $G = (V_1 \cup V_2, E_1 \cup E_2)$, of the input pair of graphs, $G_1 = (V_1, E_1)$ and $G_2 = (V_2, E_2)$ is a planar graph, then our problem has a trivial solution. Since G is planar, there exists a plane drawing of it, and hence for each of G_1 and G_2 independently. However, if the union graph G is not a planar graph, a solution may or may not exist.

Consider the pair of graphs in Fig. 4(a-b). Both G_1 and G_2 are simple cycles on 5 nodes. Their union is K_5 as seen in Fig. 4(c). However, it is easy to find node locations that realize each of the two graphs with straight-lines and no-crossings; see Fig. 4(d-e). The only crossing in Fig. 4(f) is between edges of different graphs.

While pairs of paths, cycles, and caterpillars are easy to simultaneously draw without crossings and using straight-line edges, this is not the case for all pairs of planar graphs. In fact, it is not known whether two trees can be simultaneously drawn without crossings and using straight-line edges. With this in mind, we experimented with different classes of trees. It is not difficult to construct a pair of trees, such that their union contains a subdivision of K_n for any n. For the cases when $n \leq 4$ it is fairly straight-forward to obtain by hand straight-line, crossing-free simultaneous drawings. The pen-and-paper solution is difficult to find for K_5 and K_6. For these two cases our system helped us greatly; see Fig. 5-6.

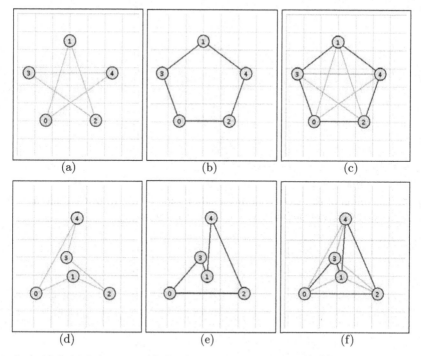

Fig. 4. (a-b) Initial drawings of G_1 G_2 with crossings in G_1; (c) Combined view of both drawings; (e-f) Crossing-free drawings of G_1 and G_2; (f) Combined view of both drawings.

It is also known that there exist pairs of outerplanar graphs that cannot be realized simultaneously [2]; Thus, while it is not possible to design an algorithm for simultaneously realizing pairs of general planar graphs, in many cases solutions do exist. Moreover, no polynomial time algorithms are known for determining whether two planar graphs have a simultaneous embedding or not. Our system can be helpful in gaining insight into the problem and in bridging the gap between the classes of graphs for which algorithms for simultaneous embeddings exist and those for which such embeddings are not possible.

3.2 Different Graph Views

Our system offers several different ways to view the input graphs. The main choice in selecting a view is whether it will be a single-view or a split-view. Regardless of the choice, the views can show graph G_1, or graph G_2, or both graphs at the same time. The split-view with G_1 in one and G_2 in the other seems the most useful for the purpose of untangling graphs. This view is useful when two groups of people simultaneously work on untangling the two graphs. When a node (and its adjacent edges) is moved in one of the drawings, it also moves in the other drawing. Showing both drawings at the same time allows a user to see the impact of the move in both drawings. To aid the user in identifying the two graphs, the edges of G_1 are colored red and the edges of G_2 are colored blue.

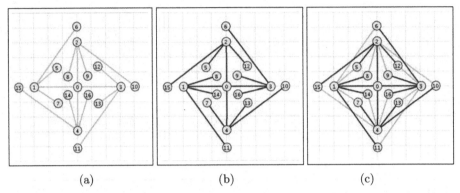

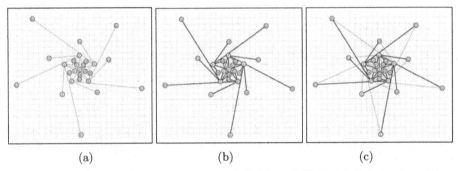

Fig. 5. The union of two trees contains a subdivision of K_5. (a-b) Crossing-free drawings of G_1 and G_2; (c) Combined view of both drawings.

Fig. 6. The union of two trees contains a subdivision of K_6. (a-b) Crossing-free drawings of G_1 and G_2; (c) Combined view of both drawings.

3.3   Heuristic Crossing Removal

Given a crossing pair of edges from the same graph, $e_1 = (p, q)$ and $e_2 = (r, s)$ we employ a crossing removal strategy consisting of three node-manipulating operations: `flip`, `shrink`, and `rotate` (FSR strategy). We briefly describe the three operations in the FSR strategy below.

The `flip` operation consists of flipping the positions of two nodes that are not endpoints of the same edge. This implies that given crossing pair of edges $e_1 = (p, q)$ and $e_2 = (r, s)$, there are 4 possible flips. Without loss of generality, consider the case where p and r are flipped; see Fig. 7. It is easy to see that flipping the position of two nodes that are not endpoints of the same edge removes the crossing.

The `shrink` operation is performed on edges. It is attempted for each endpoint of each of the edges in the crossing edge pair $e_1 = (p, q)$ and $e_2 = (r, s)$. Without loss of generality, consider the case where the operation is performed on node p; see Fig. 8. Let d_1 (d_2) be the distance from p (q) to the intersection point of e_1 and e_2. The shrink operation for e_1 at node p results in moving p along the edge e_1 in the direction of q for a of distance $d_1 + k * d_2$ to its new position p', where k is a parameter in the range $0 < k < 1$.

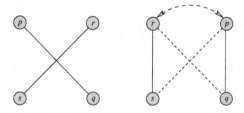

Fig. 7. The `flip(e_1, e_2)` operation.

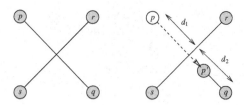

Fig. 8. The `shrink(e_1, e_2)` operation.

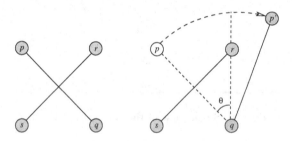

Fig. 9. The `rotate(e_1, e_2)` operation.

The **rotate** operation is attempted for each node in the crossing edge pair $e_1 = (p, q)$ and $e_2 = (r, s)$. Again, consider the case when the operation is performed on node p. Let θ be the angle determined by the intersection of the lines passing thorough the points (p, q) and (p, r); see Fig. 9. We rotate p around q at an angle $\theta + \epsilon$ to its new position p', where $\theta \leq \epsilon \leq 2\pi$.

Each of the operations in the FSR strategy can be executed a number of times on a particular crossing. Some of them are also parametrized by k and ϵ for **shrink** and **rotate**, respectively. The three operations are attempted on all of the undesirable crossings until either they are all removed or we have reached a local minimum.

4   Conclusion and Future Work

We have presented an interactive multi-user system for drawing graphs simultaneously. While the system is designed for the DiamondTouch table, it is also available as a Java application, and as a Java applet at `http://dt.cs.arizona.`

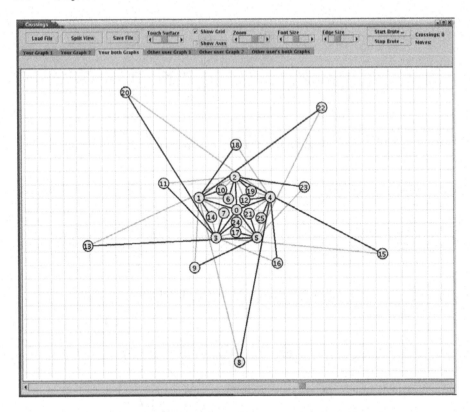

Fig. 10. Single view of the two trees in Fig. 6.

edu. With the aid of our system we were able to untangle many pairs of graphs that had stumped us in the past. We also used the system successfully, to come up with counter-examples for the cases where simultaneous embedding is not possible. However, there are many other examples that we have neither realized simultaneously, nor proved that they cannot be simultaneously realized. Some of these example graphs are available at the URL above and we hope that with the help the PC-version of our system their status will be determined.

Currently the heuristic algorithm for minimizing the crossings in the simultaneous drawings of the two graphs relies on simple heuristics. We would like to explore better heuristic algorithms, or leverage algorithms and heuristics from traditional crossing-minimization. Finally, we would like to design a brute-force algorithm which can be used to implement a fully functioning HuGS system.

Acknowledgments

We would like to thank Joe Marks and Kathy Ryall of MERL for supplying us with the DiamondTouch table and for their great help with all the information we needed about it. We would also like to thank Christian Duncan, for coming up with a challenging class of trees (in Fig. 5 and Fig. 6) for our system, and Yuhong Liu, Quanfu Fan, and Joe Schlecht for countless hours of untangling graphs.

References

1. D. Anderson, E. Anderson, N. Lesh, J. Marks, K. Perlin, D. Ratajczak, and K. Ryall. Human-guided simple search: Combining information visualization and heuristic search. In *Proceedings of the Workshop on New Paradigms in Information Visualization and Manipulation*, pages 21–25, 1999.
2. P. Brass, E. Cenek, C. A. Duncan, A. Efrat, C. Erten, D. Ismailescu, S. G. Kobourov, A. Lubiw, and J. S. B. Mitchell. On simultaneous graph embedding. In *8th Workshop on Algorithms and Data Structures*, pages 243–255, 2003.
3. C. Collberg, S. G. Kobourov, S. Kobes, B. Smith, S. Trush, and G. Yee. Tetratetris: An application of multi-user touch-based human-computer interaction. In *9th International Conference on Human-Computer Interaction (INTERACT)*, pages 81–88, 2003.
4. P. Dietz and D. Leigh. Diamondtouch: A multi-user touch technology. In *14th ACM Symposium on User Interface Software and Technology*, pages 219–226, 2001.
5. M. B. Dillencourt, D. Eppstein, and D. S. Hirschberg. Geometric thickness of complete graphs. *Journal of Graph Algorithms and Applications*, 4(3):5–17, 2000.
6. C. Erten, S. G. Kobourov, A. Navabia, and V. Le. Simultaneous graph drawing: Layout algorithms and visualization schemes. In *11th Symposium on Graph Drawing (GD)*, pages 437–449, 2003.
7. M. R. Garey and D. S. Johnson. Crossing number is NP-complete. *SIAM J. Algebraic Discrete Methods*, 4(3):312–316, 1983.
8. C. Gutwenger and P. Mutzel. An experimental study of crossing minimization heuristics. In *Proceedings of the 11th Symposium on Graph Drawing (GD)*, pages 13–24, 2003.
9. G. W. Klau, N. B. Lesh, J. W. Marks, M. Mitzenmacher, and G. T. Schafer. The hugs platform: A toolkit for interactive optimization. In *Advanced Visual Interfaces (AVI)*, 2002.
10. J. Klingner and N. Amenta. Case study: Visualization of evolutionary trees. In *IEEE Symposium on Information Visualization (INFOVIS)*, pages 71–74, 2002.
11. N. Lesh, J. Marks, and M. Patrignani. Interactive partitioning. In *Proceedings of the Symposium on Graph Drawing (GD)*, pages 31–36, 2000.
12. A. Liebers. Planarizing graphs - a survey and annotated bibliography. *Journal of Graph Algorithms and Applications*, 5(1):1–74, 2001.
13. T. Munzner, F. Guimbretiere, S. Tasiran, L. Zhang, and Y. Zhou. Treejuxtaposer: scalable tree comparison using focus+context with guaranteed visibility. *ACM Transactions on Graphics*, 22(3):453–462, 2003.
14. M. Wu and R. Balakrishnan. Multi-finger and whole hand gestural interaction techniques for multi-user tabletop displays. In *ACM UIST Symposium on User Interface Software and Technology*, pages 192–202, 2003.

Gravisto: Graph Visualization Toolkit

Christian Bachmaier[1], Franz J. Brandenburg[1],
Michael Forster[1], Paul Holleis[2], and Marcus Raitner[1]

[1] University of Passau, 94030 Passau, Germany
{bachmaier,brandenb,forster,raitner}@fmi.uni-passau.de
[2] Ludwig Maximilian University Munich, 80333 Munich, Germany
paul.holleis@ifi.lmu.de

Abstract. *Gravisto*, the Graph Visualization Toolkit, is more than a (Java-based) editor for graphs. It includes data structures, graph algorithms, several layout algorithms, and a graph viewer component. As a general toolkit for the visualization and automatic layout of graphs it is extensible with plug-ins and is suited for the integration in other Java-based applications.

Overview: *Gravisto* is a new approach towards an extensible graph visualization toolkit. Entirely written in Java, *Gravisto* runs on all Java2 1.4 platforms, including Linux, Solaris, MacOS X, and Microsoft Windows. *Gravisto* can be obtained under the terms of the GNU General Public License (GPL) from [2].

Architecture: *Gravisto* consists of three layers; see Fig.1. The basic layer contains the graph data structures. The editor layer uses the basic data structures and provides managers for easy extension and customization. The top layer comprises all plug-ins, either delivered with *Gravisto* or from third-parties.

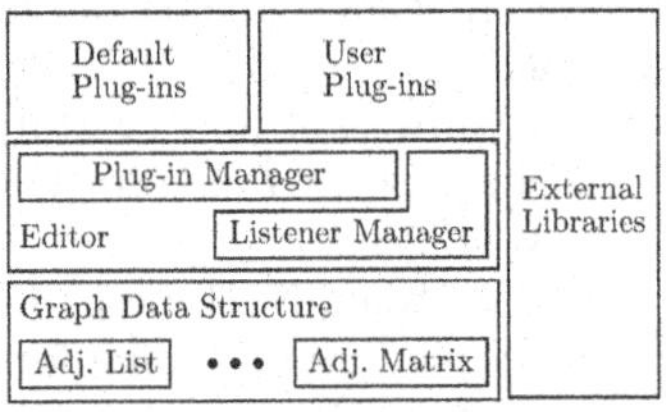

As several components must be notified about changes of the data structure, e. g., a view component in a plug-in, *Gravisto* employs the Observer Design Pattern: the *Event Manager* allows a component to register as a special type of event handler, depending on the events it likes to receive.

Fig. 1. System architecture.

Interfaces: *Gravisto* provides a powerful plug-in mechanism with a comfortable plug-in manager. Most non-core functionality is realized as plug-in, including algorithms, node and edge attributes, graphical user interface components, input and output serializers, attribute inspectors, node and edge shapes, tools, or entire views. The idea behind this paradigm is to facilitate extensions of *Gravisto*; thus encouraging people to contribute to the project. Furthermore, this concept allows easy customization of the editor for different application scenarios, e. g., a biologist, working with biochemical pathways, does not need most functionality of the standard graph editor, but, for instance, the nodes and edges must be linked to additional information in a data base; thus all unnecessary plug-ins

J. Pach (Ed.): GD 2004, LNCS 3383, pp. 502–503, 2004.
© Springer-Verlag Berlin Heidelberg 2004

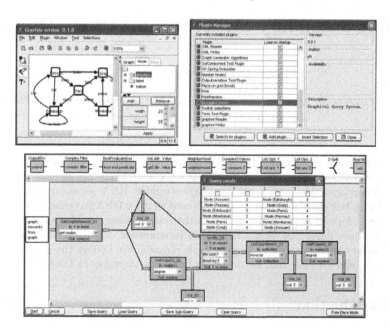

Fig. 2. *Gravisto*'s editing view, its plug-in manager, and the *Quoggles* interface.

can be excluded and a custom plug-in for the data base connection may be added. Of course, the toolkit includes all plug-ins for the basic functionality and many more. Plug-ins for the widely used GML and GraphML file formats allow data exchange with non-Java applications and other graph drawing tools.

Queries on Graphs: An innovative feature is *Quoggles*, [3], a plug-in implementing an extensible, graphical query system for graph properties. The idea for *Quoggles* arose from the 10th Graph Drawing Contest, 2002, Category C, [1]. The query itself is composed into a graph. The input to the query are the set *GE* of all nodes and edges of the queried graph. For example, the query "*GE* → *GetGraphElements(nodes)* → *GetProperty(degree)* → *Arithmetic(avg)*" calculates the average degree of all nodes in the graph. The query shown in Fig. 2 sorts the nodes according to their degree.

References

[1] F. J. Brandenburg. Graph-drawing contest report. In M. T. Goodrich and S. G. Kobourov, editors, *Proc. Graph Drawing, GD 2002*, volume 2528 of *LNCS*, pages 376–379. Springer, 2002.
[2] *Gravisto.* http://www.gravisto.org/. University of Passau.
[3] P. Holleis. Design and implementation of an extensible query system for graphs. diploma thesis, University of Passau, 2004.

DNA Secondary Structures for Probe Design*

Yanga Byun and Kyungsook Han

School of Computer Science and Engineering, Inha University, Inchon 402-751, Korea
quaah@hanmail.net, khan@inha.ac.kr

Abstract. Visualizing DNA secondary structures is essential to fast
and efficient design of probes for DNA chips. There are several programs available for visualizing single-stranded RNA secondary structures, but these programs cannot be used to draw DNA secondary structures formed by several hundred to thousand primers and target genes.
We have developed an algorithm and program for visualizing DNA secondary structures formed by multiple strands. We believe the program
will be a valuable tool for designing primers and probes in DNA chips.

1 Introduction

The high-throughput analysis of genes using DNA chips has a great impact
on modern biological research. Several thousand different primers are required
for a DNA chip [1]. DNA primers are DNA sequence fragments consisting of 4
types of nucleotides: adenine (A), guanine (G), cytosine (C), and tymine (T).
Target genes and primers form secondary structures by hydrogen bonds between
complementary base pairs. The secondary structure is an important criterion for
the selection of the primer since interaction between primers should be avoided
to conserve the maximum sensitivity of the primer and the spot on a DNA chip.

Several programs are available for drawing RNA secondary structures [2–4],
but none of these can be used to draw DNA secondary structures because the
programs are intended for drawing single-stranded RNA. DNA itself is a double-
stranded molecule and primer design should be able to consider secondary structures formed by multiple primers and target genes. We have developed a program
called DNAdraw for fast and accurate selection of primers to be used in DNA
chips. The input for the program is the DNA or cDNA sequence(s) of the target
gene, candidate primer sequences, and their secondary structures. Experimental
results demonstrate that DNAdraw is capable of automatically producing a clear
and aesthetically appealing drawing of DNA secondary structures. This paper
describes an algorithm and its implementation.

2 Algorithm

In the structure data, a pair of parentheses represents a base pair. The parenthe-
sis pairs used in DNAdraw are '()', '[]', and '{}'. In visualizing DNA structure, we

---

* This work was supported by the Korea Science and Engineering Foundation
(KOSEF) under grant R01-2003-000-10461-0.

J. Pach (Ed.): GD 2004, LNCS 3383, pp. 504–507, 2004.

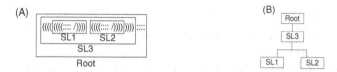

Fig. 1. (A) Example of a structure with 2 simple stem-loops (SL1 and SL2) and a composite stem-loops (SL3). (B) Tree structure of Fig. 1A.

call a structure element enclosed by matching parentheses a *stem-loop* (Fig. 1A). A *simple stem-loop* corresponds to a single hairpin loop-stem, and a *composite stem-loop* contains one or more other stem-loops. From the standpoint of graph theory, a drawing of DNA secondary structures can be considered as a tree with simple stem-loops as leaf nodes of the tree (Fig. 1B). Computation starts with a leaf node.

The algorithm of DNAdraw is outlined as follows: (1) stem-loops are identified from the input structure data; (2) the position and shape of a simple stem-loop are computed; and (3) the position and shape of a composite stem-loop are computed.

Base pairs of a stem in a simple stem-loop are stacked on the y-axis. In Fig. 2A, n represents the number of bases in the loop region plus 2 (for the base pair at the end of a stem). If the loop region contains a terminal base either at $5'$ or $3'$ end of a strand, 10 is added to n to make space between the base and other parts. L represents the distance between adjacent bases of a loop. L is also the distance between a pair of bases of a stem. Then, the angle a is π/n and the radius R of the loop is $L/2sin(a)$.

To determine the loop center, we first compute the midpoint P_m of points P_1 and P_2. If we use N to represent the unit vector directed toward the loop center C from a point P_m, vector N can be obtained by rotating the vector $P_2 - P_1$ clockwise with respect to P_m and then by normalizing the vector. The distance d between C and P_m is determined by equation (1). From the distance d, vector N, and the position vector P_m, we can compute the position vector C representing the loop center.

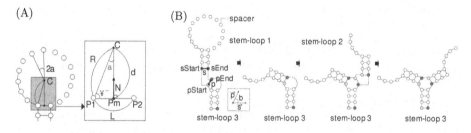

Fig. 2. (A) The radius, angle, and center of a loop in a simple stem-loop. (B) Example of inserting two simple stem-loops into a composite stem-loop by rotating and translation the simple stem-loops.

$$\theta = \pi/2 - a, \quad d = \|\overrightarrow{C} - \overrightarrow{P_m}\| = R \cdot \sin(\theta), \quad \overrightarrow{C} = d \cdot \overrightarrow{N} + \overrightarrow{P_m} \qquad (1)$$

Consider a composite stem-loop pSL containing a simple stem-loop sSL. In Fig. 2B we use $sStart$ and $sEnd$ to represent the position of the first and the last base of sSL before being enclosed in pSL; $pStart$ and $pEnd$ to represent the position at which the first and the last base of sSL to be located in pSL. Let s be the unit vector in the direction of $sEnd - sStart$ and p the unit vector in the direction of $pEnd - pStart$. The simple stem-loop sSL can be inserted into the composite stem-loop pSL by rotating sSL by the angle between s and p with respect to $sStart$ and then translating it by the vector of $pEnd - sStart$. Fig. 2B shows an example of enclosing two simple stem-loops in a composite stem-loop.

3   Results and Discussion

DNAdraw is written in Microsoft Visual C#, and is executable on any Windows system (see Figs. 3 and 4). DNAdraw takes as input the DNA sequence with its structure data in bracket view. In the input data below, 5 and 3 denote the start and termination of the DNA sequence, respectively.

Input Format 1: Both ends of the DNA sequence are denoted either by 5 or 3.

```
# primer1  // optional sequence name
3-ATGCCGTAGGTA-5
5-TAGGTGAGCCAT-3
3-CTCAGCATTGCA-5
3-(((((((((:::::-5
5-))))((((:::::-3
3-)))))))))::::-5
```

Input Format 2: The sequence in each line is ended by a slash ("/") character, with the sequence direction from the 5′ end to 3′ end.

```
# primer2  // optional sequence name
ATGCCGTAGGTA/
TAGGTGAGCCAT/
CTCAGCATTGCA
(((((((((:::/
))))((((:::/
)))))))))::::
```

In summary, we have developed a new algorithm for visualizing DNA secondary structures with multiple DNA strands and have implemented the algorithm in a web-based program called DNAdraw. For given secondary structures, DNAdraw identifies all simple stem-loops and composite stem-loops enclosing other stem-loops. Stem-loops are inserted into their enclosing composite stem-loops by rotation, and/or translation operations. The DNAdraw algorithm is the first capable of automatically drawing DNA structures with multiple strands. DNAdraw will be a valuable tool for fast and accurate design of primers and probes in DNA chips.

(A) 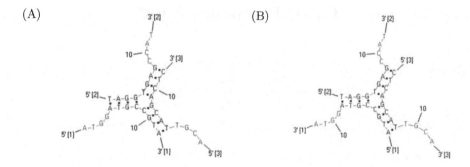 (B)

Fig. 3. (A) Structure drawing for input data format 1. (B) Structure drawing for input data format 2.

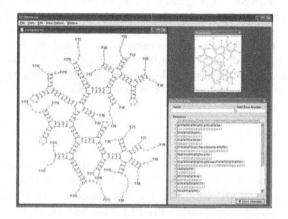

Fig. 4. Hypothetical DNA secondary structure with 15 DNA strands.

References

[1] Lander, E. S.: Array of hope, Nature Genetics **21** (1999) 3-4
[2] De Rijk, P., Wuyts J., De Wachter, R.: RnaViz 2: an improved representation of RNA secondary structure. Bioinformatics **19** (2003) 299-300
[3] Han, K., Kim, D., Kim, H.-J.: A vector-based method for drawing RNA secondary structure. Bioinformatics **15** (1999) 286-197
[4] Matzura, O. and Wennborg, A.: RNAdraw: an integrated program for RNA secondary structure calculation and analysis under 32-bit Microsoft Windows. Computer Applications in the Biosciences **12** (1996) 247-249

Open Problems Wiki*

Marcus Raitner

University of Passau, D-94032 Passau, Germany
Marcus.Raitner@Uni-Passau.De

Abstract. This project was inspired by the last year's paper on *Selected Open Problems in Graph Drawing* by Brandenburg et al. (Proc. 11th GD. Vol. 2919 of LNCS. (2003) 515–539). While being a very good start, a paper is inherently static and will become out-dated. For dynamic content, what open problems (hopefully) are, a web-site is more appropriate. Keeping such a site up-to-date, however, is time consuming and requires good knowledge of recent work. In projects like the free encyclopedia *Wikipedia* these obstacles are overcome with a collaborative approach: everyone is allowed, and even requested, to contribute his knowledge to the site. The Open Problems Wiki makes use of this paradigm to provide a forum for collecting open problems in graph drawing.

Introduction

Recently, collaboratively edited projects on the WWW were impressively successful; by the time of writing, the free encyclopedia *Wikipedia*, for instance, has more than 340 000 articles in the English version (with translations in over 80 languages in steady progress), edited by over 100 000 users. The Wikipedia project is a so-called *Wiki*, a simple form of content management system where *everyone* not only is allowed to but also is requested to create new pages or edit existing pages. Meaning "super fast", the Hawaiian word "wiki wiki" was used for this kind of software, because the pages are written in a very simple, yet powerful, markup language; the wiki software then stores the source code in a database, from where it is read and transformed to standard HTML.

It seems that a Wiki is an appropriate framework for collecting and discussing open problems in graph drawing. The task of editing is balanced among the whole community; therefore, it tends to be more comprehensive and more up-to-date than other solutions. This wiki shall become the primary site for the open problems in graph drawing, a site everyone knows and contributes to.

Features

The Open Problems Wiki uses the *MediaWiki* software, which was developed for the Wikipedia project and is by far the most advanced wiki software available. It already has proved its qualities; it is well documented and supported, and

---

* http://problems.graphdrawing.org

J. Pach (Ed.): GD 2004, LNCS 3383, pp. 508–509, 2004.

freely available under the terms of the GNU General Public License (GPL). The following is only a small fraction of the features; a more detailed description can be found in the "Help" section of the wiki.

Users. Although it is not required to register or log in – not even for editing pages – there are many reasons to do so. A registered user can pick a username; all edits, made while being logged in, will be assigned to that name, giving the user full credit for each contribution in the page history (when not logged in, the edits are just assigned to the respective IP address). When logged in, all own contributions are accessible via the "My contributions" link. Many features that are only available to registered users: for example, registered users can mark edits as *minor*. Minor edits can be filtered from the list of "Recent changes". One very important feature, which active contributors will likely use a lot, are *watchlists*. When logged in a new link "Watch this page" is shown on every page. That link adds a page to the user's watchlist, which thus becomes basically a filtered view of the "Recent changes" page, showing changes recently made to items in the watchlist. Only registered users are allowed to rename pages, a feature that is very important to maintain structure and consistency. Also, images can be uploaded only by registered users.

Basic Markup. As the pages are stored in a database rather than as files on the server, links between the pages are established via the names of the pages. For instance, `[[Graph Drawing]]` in the source code results in a link to a page with title "Graph Drawing"; if this page does not yet exist, the link leads to a new page with a text field for editing. Most pages are written in a simple markup language, which is sufficient for standard editing tasks like headings (`== heading ==`), highlighting text (`''italic''` and `'''bold'''`), or even simple tables. Also, standard HTML can be used for more advanced tasks.

Formulas. The main reason for choosing the same software as in the Wikipedia project was its integrated support for TEX-formulas. TEX-source within special tags (`<math></math>`) is rendered on the server and included as Portable Network Graphics (PNG) picture.

History. Since everyone is allowed to edit a page, it is very important to have a sophisticated version control system. The software used for the Open Problems Wiki stores the complete editing history of every page and allows to rollback a page to any previous state.

Acknowledgments

I would like to thank Ulrik Brandes for supporting the project with the problems subdomain of www.graphdrawing.org.

References

1. Brandenburg, F.J., Eppstein, D., Goodrich, M.T., Kobourov, S., Liotta, G., Mützel, P.: Selected open problems in graph drawing. In Liotta, G., ed.: Proc. 11th Intl. Symposium on Graph Drawing (GD). Volume 2919 of LNCS. (2003) 515–539

Visualization and ILOG CPLEX

Georg Sander and Adrian Vasiliu

ILOG SA, 9 rue de Verdun – BP 85, 94253 Gentilly Cedex, France
{sander,vasiliu}@ilog.fr
http://www.ilog.com

1 Introduction

Graph layout methodologies often solve difficult subproblems in order to satisfy the aesthetic constraints: finding the maximal planar subgraph, minimizing crossings, minimizing area, maximizing symmetries, etc. One standard approach is to translate the subproblem into a linear optimization problem (LOP) and to use a standard mathematical tool to find the solution. The standard solving tool for such LOPs is ILOG CPLEX.

ILOG CPLEX delivers high-performance, robust, flexible optimizers for solving linear, mixed-integer and quadratic programming problems (including mixed integer quadratic constrained problems). It is a component that includes C, C++ and Java API, and it is integrated via the Concert Technology into the ILOG Optimization Suite.

With ILOG JViews, we also offer a visualization suite that includes sophisticated graph layout algorithms. Our recent investigations focused on cross-product development and the question, how ILOG CPLEX can be used to help graph layout and how ILOG JViews can be used to help the LOP solving.

2 How ILOG CPLEX Helps Graph Layout

ILOG CPLEX can be used to solve various subproblems in graph layout. The following is a collection of algorithms found in the literature that make extensive use of ILOG CPLEX:

- Link Crossing – Jünger and Mutzel (GD'95, LNCS 1027, p. 337ff) compare a CPLEX implementation of the LOP for the 2-layer straightline crossing minimization problem with various other heuristics.
- Labeling Problem – Binucci e.a. (GD'02, LNCS 2528, p. 66ff) show an LOP algorithm to compute optimal label positions in an orthogonal drawing.
- Detecting Symmetries – Buchheim and Jünger (GD'01, LNCS 2265, p. 178ff) use ABACUS and CPLEX to detect automorphisms and symmetries in arbitrary graphs.
- Graph Layering Problem – Healy and Nikolov (GD'01, LNCS 2265, p. 16ff) partition a DAG into layers as needed for the hierarchical layout algorithm. Again, CPLEX was used as reference implementation.

J. Pach (Ed.): GD 2004, LNCS 3383, pp. 510–511, 2004.

- Orthogonal Layout Compaction – Klau and Mutzel (IPCO'99, LNCS 1610, p. 304ff) compress an orthogonal layout so that the the edges length are minimized by applying a branch and cut approach that can be implemented with CPLEX.
- and many more.

3   How ILOG JViews Graph Layout Helps LOP Solving

ILOG CPLEX uses nonvisual algorithms on a mathematical model of the problem to be solved. Visualization is used to help CPLEX users to model their problems, and to detect or debug the internal behavior of the CPLEX routines.

- ILOG OPL Studio – This is the modeling tool for the ILOG Optimization Suite. It has a graphical GUI to visualize scheduling problems and their solutions.
- Branch and Bound Tree – Mixed integer problem are often solved by branch and bound. The branching structure is essentially a binary tree. Graph layout technology helps to visualize this tree. The RINS algorithm (Relaxation Induces Neighborhood Search) is a local search for feasible solutions of mixed integer problems and can be visualized this way.
- Precedence graph of linear equations – Many subroutines of CPLEX solve linear equation systems with sparse matrix. A precedence graph helps to analyze in which order the variables must be calculated. A reduction mechanism of the precedence graph leaded to a major performance boost of CPLEX. The precedence graph can be visualized.
- Parallel CPLEX – CPLEX can be executed on multiple processor machines. The debug trace of parallel CPLEX can be visualized, which is essentially a graph.

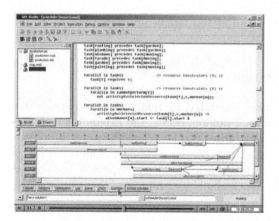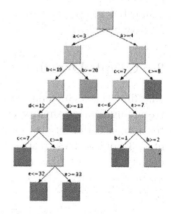

Fig. 1. Left: ILOG OPL Studio, Right: Branch and Bound Tree

Graph-Drawing Contest Report

Franz J. Brandenburg[1], Christian A. Duncan[2],
Emden R. Gansner[3], and Stephen G. Kobourov[4]

[1] Universität Passau 94030 Passau, Germany
brandenb@informatik.uni-passau.de
[2] University of Miami, Coral Gables FL 33124, USA
duncan@cs.miami.edu
[3] AT&T Research, FLORHAM PARK, NJ 07932, USA
erg@research.att.com
[4] University of Arizona, Tucson, AZ 85721, USA
kobourov@cs.arizona.edu

Abstract. This report describes the Eleventh Annual Graph Drawing
Contest, held in conjunction with the 2004 Graph Drawing Symposium in
New York, USA. The purpose of the contest is to monitor and challenge
the current state of the graph-drawing technology.

1 Introduction

This year's graph drawing contest had two distinct tracks: the graph drawing
challenge and the free-style contest. The graph drawing challenge took place
during the conference. The challenge was straight-line crossing minimization of
10 graphs with 20-100 vertices. The contestants were given one hour and were free
to use custom designed software or a provided program for manual graph editing,
GraphMan. The free-style submission offered the opportunity for participants to
present their best graph visualizations. Ten teams of one to three participants
submitted graphs to the challenge, and eleven teams submitted entries to the
free-style contest.

2 Graph Drawing Challenge

The first three challenge graphs were generated by taking randomly-generated
trees, augmenting them with additional random edges, and expanding each vertex into a clique of size 3-6. The next three challenge graphs were the unions
of 2, 3, and 4 trees, respectively. The next two were random graphs with edge
density 10% and 20%, respectively. The ninth challenge graph was the contest
graph from 1998, category C. The tenth challenge graph was the contest graph
from 1999, category B.

The team's submissions to the challenge were measured objectively by assigning scores between 0 and 10 to each participating team as follows: each of
the 10 submitted graphs G_i was assigned a score: $s_i = \frac{min_i}{cur_i}$, where min_i was the
minimal number of crossings found for the i-th graph and cur_i was the number

J. Pach (Ed.): GD 2004, LNCS 3383, pp. 512–516, 2004.

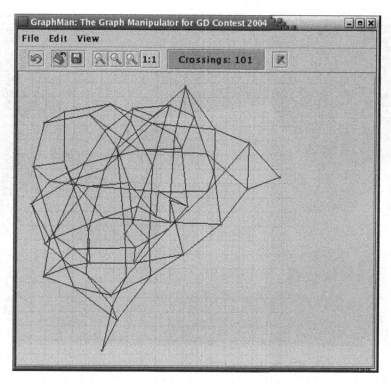

Fig. 1. One of the drawings submitted by the winning team (the 10th challenge graph).

of crossings in G_i. The individual scores for the 10 graphs were added to obtain the total score for each team. The contest committee awarded one first prize and two honorable mentions.

The first place winner was the team of Andrei Grecu and Gunnar Klau; see Fig. 1. They used a custom-built tool called Grapla. Their tool is based on an evolutionary algorithm with four mutation operators, no recombination, and modified tournament selection. The mutation operators take a single node and move it to a new location on a fixed grid. Depending on the specific operator, the new location is either chosen randomly, with a probability according to the distance from the original position, randomly within a local window around the old position, or within a window whose size exponentially decreases with time. The population is realized by a multilevel hierarchy of different solutions whose constant size is maintained by a selection operator that moves solutions between different hierarchy levels or removes them. The tool gives visual feedback of the quality of the population and allows for interactively manipulating the operators as well as their parameter settings.

Two honorable mentions were given to Chandan Pitta and the team of Josh Cooper, Rob Ellis, and Reid Andersen. Both used a combination approach of automated and manual crossing minimization.

Fig. 2. Spectral Dance (original in color).

3   Free-Style Contest

Eleven submissions for the free-style category were received. The submissions consisted of individual drawings, SVG files, VRML files, papers, and movies. The contest committee awarded one first prize and three honorable mentions.

The winning submission was "Spectral Dance" by Ulrik Brandes, Daniel Fleischer, and Thomas Puppe; see Fig. 2. Each frame of this one-minute animation is a three-dimensional spectral layout of a square grid graph. Continuously changing weights on the edges yield continuous changes in the layout, so that no interpolation between layouts was necessary. The animation was used as background in the trailer for "Language of Networks", a conference during the 2004 Ars Electronica Festival in Linz, Austria. This trailer was projected onto the facade of the "Museum of the Future".

An honorable mention was awarded to Kim Hansen and Stephen Wismath for their submission "Arrangement for an Upright Bass"; see Fig. 3. The submission was an animation showing a representation of an arrangement graph in three dimensions, induced by seven planes. Arrangement graphs in 2D have been widely studied, but 3D arrangement graphs have been largely ignored as they are difficult to visualize. Viewing the structure of such graphs in 3D is best accomplished by means of animation. "Arrangement for an Upright Bass" explores the problem of arranging planes to create graphs in a modern computing context.

Fig. 3. Arrangement for an Upright Bass (original in color).

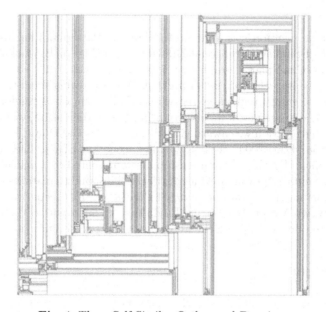

Fig. 4. Three Self-Similar Orthogonal Drawings.

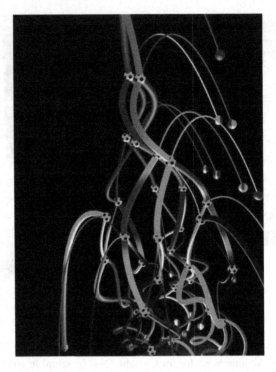

Fig. 5. Euro 2004 Virtual Reality Scene (original in color).

An honorable mention was awarded to Maurizio Patrigniani for his submission "Three Self-Similar Orthogonal Drawings"; see Fig. 4. By looking at the drawing of a maximal planar graph with 5,000 nodes and 14,994 edges, its self-similarity may be appreciated. Small portions of the drawing seem to show the same patterns of empty regions as the whole. This intuition is reinforced by a more rigorous measurement of its box-counting fractal dimension, which is 1.70.

An honorable mention was awarded to Ulrik Brandes and Daniel Fleischer for their submission "Euro 2004", a virtual reality scene; see Fig. 5. The matches of the 2004 European Football Championship are represented by a walkable three-dimensional layout of a graph, in which vertices correspond to matches and edges represent teams moving from one match to the next. The scene was prepared for a CAVE (Cave Automated Virtual Environment) and open to visitors of the "Museum of the Future" during the above mentioned conference.

Acknowledgments

We wish to thank Chandan Pitta for the first version of the GraphMan program, the sponsors of the symposium, and all the contestants.

Fast Algorithms for Hard Graph Problems: Bidimensionality, Minors, and Local Treewidth

Erik D. Demaine and Mohammad Taghi Hajiaghayi

MIT Computer Science and Artificial Intelligence Laboratory,
32 Vassar Street, Cambridge, MA 02139, USA,
{edemaine,hajiagha}@mit.edu

Abstract. This paper surveys the theory of bidimensional graph problems. We summarize the known combinatorial and algorithmic results of this theory, the foundational Graph Minor results on which this theory is based, and the remaining open problems.

1   Introduction

The newly developing theory of bidimensional graph problems, developed in a series of papers [DHT, DHN+04, DFHT, DH04a, DFHT04b, DH04b, DFHT04a, DHT04, DH05b, DH05a], provides general techniques for designing efficient fixed-parameter algorithms and approximation algorithms for NP-hard graph problems in broad classes of graphs. This theory applies to graph problems that are *bidimensional* in the sense that (1) the solution value for the $k \times k$ grid graph (and similar graphs) grows with k, typically as $\Omega(k^2)$, and (2) the solution value goes down when contracting edges and optionally when deleting edges. Examples of such problems include feedback vertex set, vertex cover, minimum maximal matching, face cover, a series of vertex-removal parameters, dominating set, edge dominating set, R-dominating set, connected dominating set, connected edge dominating set, connected R-dominating set, and unweighted TSP tour (a walk in the graph visiting all vertices).

Bidimensional problems have many structural properties; for example, any graph in an appropriate minor-closed class has treewidth bounded above in terms of the problem's solution value, typically by the square root of that value. These properties lead to efficient – often subexponential – fixed-parameter algorithms, as well as polynomial-time approximation schemes, for many minor-closed graph classes. One type of minor-closed graph class of particular relevance has *bounded local treewidth*, in the sense that the treewidth of a graph is bounded above in terms of the diameter; indeed, such a bound is always at most linear.

The bidimensionality theory unifies and improves several previous results. The theory is based on algorithmic and combinatorial extensions to parts of the Robertson-Seymour Graph Minor Theory, in particular initiating a parallel theory of graph contractions. The foundation of this work is the topological theory of drawings of graphs on surfaces.

J. Pach (Ed.): GD 2004, LNCS 3383, pp. 517–533, 2004.

This survey is organized as follows. Section 2 defines the various graph classes of increasing generality to which bidimensionality theory applies. Section 3 describes several structural properties of graphs in these classes, in particular from Graph Minor Theory, that form the basis of bidimensionality. Section 4 defines bidimensional parameters and problems and gives some examples. Section 5 describes one of the main structural properties of bidimensionality, namely, that the treewidth is bounded in terms of the parameter value. Sections 6–10 describe several consequences of bidimensionality theory: separator theorems, bounds on local treewidth, fixed-parameter algorithms, and polynomial-time approximation schemes. Section 11 discusses the main remaining open problems in this area.

2   Graph Classes

In this section, we introduce several families of graphs, each playing an important role in both the Graph Minor Theory and the bidimensionality theory. Refer to Figure 1. All of these graph classes are generalizations of planar graphs, which are well-studied in algorithmic graph theory. Unlike planar graphs and map graphs, every other class of graphs we consider can include any particular graph G; of course, this inclusion requires a bound or excluded minor large enough depending on G. This property distinguishes this line of research from other work considering exclusion of particular minors, e.g., $K_{3,3}$, K_5, or K_6.

2.1   Definitions of Graph Classes

The first three classes of graphs relate to embeddings on surfaces. A graph is *planar* if it can be drawn in the plane (or the sphere) without crossings. A graph has *genus* at most g if it can be drawn in an orientable surface of genus g without

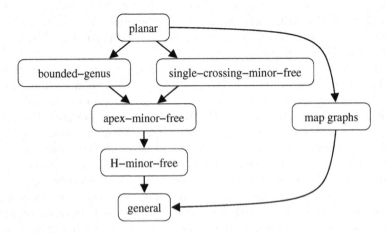

Fig. 1. Interesting classes of graphs. Arrows point from more specific classes to more inclusive classes.

crossings[1]. A class of graphs has *bounded genus* if every graph in the class has genus at most g for a fixed g.

Given an embedded planar graph and a two-coloring of its faces as either *nations* or *lakes*, the associated *map graph* has a vertex for each nation and an edge between two vertices corresponding to nations (faces) that share a vertex. The dual graph is defined similarly, but with adjacency requiring a shared edge instead of just a shared vertex. Map graphs were introduced by Chen, Grigni, and Papadimitriou [CGP02] as a generalization of planar graphs that can have arbitrarily large cliques. Thorup [Tho98] gave a polynomial-time algorithm for constructing the underlying embedded planar graph and face two-coloring for a given map graph, or determining that the given graph is not a map graph.

We view the class of map graphs as a special case of taking fixed powers of a family of graphs. The *kth power* G^k of a graph G is the graph on the same vertex set $V(G)$ with edges connecting two vertices in G^k precisely if the distance between these vertices in G is at most k. For a bipartite graph G with bipartition $V(G) = U \cup W$, the *half-square* $G^2[U]$ is the graph on one side U of the partition, with two vertices adjacent in $G^2[U]$ precisely if the distance between these vertices in G is 2. A graph is a map graph if and only if it is the half-square of some planar bipartite graph [CGP02]. In fact, this translation between map graphs and half-squares is constructive and takes polynomial time.

The next three classes of graphs relate to excluding minors. Given an edge $e = \{v, w\}$ in a graph G, the *contraction* of e in G is the result of identifying vertices v and w in G and removing all loops and duplicate edges. A graph H obtained by a sequence of such edge contractions starting from G is said to be a *contraction* of G. A graph H is a *minor* of G if H is a subgraph of some contraction of G. A graph class $\mathcal{C}$ is *minor-closed* if any minor of any graph in $\mathcal{C}$ is also a member of $\mathcal{C}$. A minor-closed graph class $\mathcal{C}$ is *H-minor-free* if $H \notin \mathcal{C}$. More generally, we use the term "H-minor-free" to refer to any minor-closed graph class that excludes some fixed graph H.

A *single-crossing graph* is a minor of a graph that can be drawn in the plane with at most one pair of edges crossing. Note that a single-crossing graph may not itself be drawable with at most one crossing pair of edges; see [DHN+04]. Such graphs were first defined by Robertson and Seymour [RS93]. A minor-closed graph class is *single-crossing-minor-free* if it excludes a fixed single-crossing graph.

An *apex graph* is a graph in which the removal of some vertex leaves a planar graph. A graph class is *apex-minor-free* if it excludes some fixed apex graph. Such graph classes were first considered by Eppstein [Epp95, Epp00], in connection to the notion of bounded local treewidth as described in Section 7.

The next section describes strong structural properties of the last three classes of graphs (minor-excluding classes) in terms of the first two classes of graphs (embeddable on surfaces) and other ingredients.

---

[1] This definition also includes graphs that can be drawn in non-orientable surfaces of low genus, because if a graph has non-orientable genus g, then it has orientable genus at most $2g$.

3   Structural Properties

Graphs from single-crossing-minor-free and H-minor-free graph classes have powerful structural properties from the Graph Minor Theory. First we need to define treewidth, pathwidth, and clique sums.

3.1   Background

The notion of treewidth was introduced by Robertson and Seymour [RS86a]. To define this notion, first we consider a representation of a graph as a tree, called a tree decomposition. Precisely, a *tree decomposition* of a graph $G = (V, E)$ is a pair (T, χ) in which $T = (I, F)$ is a tree and $\chi = \{\chi_i \mid i \in I\}$ is a family of subsets of $V(G)$ such that

1. $\bigcup_{i \in I} \chi_i = V$;
2. for each edge $e = \{u, v\} \in E$, there exists an $i \in I$ such that both u and v belong to χ_i; and
3. for all $v \in V$, the set of nodes $\{i \in I \mid v \in \chi_i\}$ forms a connected subtree of T.

To distinguish between vertices of the original graph G and vertices of T in the tree decomposition, we call vertices of T *nodes* and their corresponding χ_i's *bags*. The *width* of the tree decomposition is the maximum size of a bag in χ minus 1. The *treewidth* of a graph G, denoted $\mathbf{tw}(G)$, is the minimum width over all possible tree decompositions of G. A tree decomposition is called a *path decomposition* if $T = (I, F)$ is a path. The *pathwidth* of a graph G, denoted $\mathbf{pw}(G)$, is the minimum width over all possible path decompositions of G.

The notion of clique sums goes back to characterizations of $K_{3,3}$-minor-free and K_5-minor-free graphs by Wagner [Wag37] and serves as an important tool in the Graph Minor Theory. Suppose G_1 and G_2 are graphs with disjoint vertex sets and let $k \geq 0$ be an integer. For $i = 1, 2$, let $W_i \subseteq V(G_i)$ form a clique of size k and let G'_i be obtained from G_i by deleting some (possibly no) edges from the induced subgraph $G_i[W_i]$ with both endpoints in W_i. Consider a bijection $h : W_1 \to W_2$. We define a *k-sum* G of G_1 and G_2, denoted by $G = G_1 \oplus_k G_2$ or simply by $G = G_1 \oplus G_2$, to be the graph obtained from the union of G'_1 and G'_2 by identifying w with $h(w)$ for all $w \in W_1$. The images of the vertices of W_1 and W_2 in $G_1 \oplus_k G_2$ form the *join set*. Note that each vertex v of G has a corresponding vertex in G_1 or G_2 or both. It is also worth mentioning that $\oplus$ is not a well-defined operator: it can have a set of possible results.

3.2   Structure of Single-Crossing-Minor-Free Graphs

The structure of single-crossing-minor-free graphs can be described as follows:

Theorem 1 ([RS93]). *For any fixed single-crossing graph H, every H-minor-free graph can be obtained by a sequence of k-sums, $0 \leq k \leq 3$, of planar graphs and graphs of bounded treewidth, where the bound on treewidth depends on H.*

This theorem generalizes characterizations of $K_{3,3}$-minor-free and K_5-minor-free graphs [Wag37]. A graph is $K_{3,3}$-minor-free if and only if it can be obtained by k-sums, $0 \leq k \leq 2$, of planar graphs and K_5. A graph is K_5-minor-free if and only if it can be obtained by k-sums, $0 \leq k \leq 3$, of planar graphs and V_8 (the length-8 cycle C_8 together with eight edges joining diametrically opposite vertices).

This structural property of single-crossing-minor-free graphs has since been strengthened to ensure that the summands are minors of the original graph and to provide algorithms for finding the decomposition:

Theorem 2 ([DHN+04]). *For any fixed single-crossing graph H, there is an $O(n^4)$-time algorithm to compute, given an H-minor-free graph G, a decomposition of G as a sequence of k-sums, $0 \leq k \leq 3$, of planar graphs and graphs of bounded treewidth (where the bound on treewidth depends on H), each of which is a minor of G.*

3.3   Structure of H-Minor-Free Graphs

The structure of H-minor-free graphs is described by a deep theorem of Robertson and Seymour [RS03]. Intuitively, their theorem says that, for every graph H, every H-minor-free graph can be expressed as a "tree structure" of pieces, where each piece is a graph that can be drawn in a surface in which H cannot be drawn, except for a bounded number of "apex" vertices and a bounded number of "local areas of non-planarity" called *vortices*. Here the bounds depend only on H.

Roughly speaking, we say that a graph G is *h-almost embeddable* in a surface S if there exists a set X of size at most h of vertices, called *apex vertices* or *apices*, such that $G - X$ can be obtained from a graph G_0 embedded in S by attaching at most h graphs of pathwidth at most h to G_0 within h faces in an orderly way. More precisely, a graph G is *h-almost embeddable* in S if there exists a vertex set X of size at most h (the *apices*) such that $G - X$ can be written as $G_0 \cup G_1 \cup \cdots \cup G_h$, where

1. G_0 has an embedding in S;
2. the graphs G_i, called *vortices*, are pairwise disjoint;
3. there are faces $F_1, \ldots, F_h$ of G_0 in S, and there are pairwise disjoint disks $D_1, \ldots, D_h$ in S, such that for $i = 1, \ldots, h$, $D_i \subset F_i$ and $U_i := V(G_0) \cap V(G_i) = V(G_0) \cap D_i$; and
4. the graph G_i has a path decomposition $(\mathcal{B}_u)_{u \in U_i}$ of width less than h, such that $u \in \mathcal{B}_u$ for all $u \in U_i$. The sets $\mathcal{B}_u$ are ordered by the ordering of their indices u as points along the boundary cycle of face F_i in G_0.

An h-almost embeddable graph is *apex-free* if the set X of apices is empty.

Now, the deep result of Robertson and Seymour is as follows:

Theorem 3 ([RS03]). *For every graph H, there exists an integer $h \geq 0$ depending only on $|V(H)|$ such that every H-minor-free graph can be obtained by at most h-sums of graphs that are h-almost-embeddable in some surfaces in which H cannot be embedded.*

In particular, if H is fixed, any surface in which H cannot be embedded has bounded genus. Thus, the summands in the theorem are h-almost-embeddable in bounded-genus surfaces.

3.4   Structure of Apex-Minor-Free Graphs

Apex-minor-free graph classes are an important subfamily of H-minor-free graph classes. The general structural theorem for H-minor-free graphs applies in this context as well. However, reductions developed in [DH04b] suggest that the decomposition can be restricted to a particular form in the apex-minor-free case:

Conjecture 1 ([DH04b]). For every graph H, there is an integer $h \geq 0$ depending only on $|V(H)|$ such that every H-minor-free graph can be obtained by at most h-sums of graphs that are h-almost-embeddable in some surfaces in which H cannot be embedded and whose apices are connected via edges only to vertices within vortices.

3.5   Grid Minors

The $r \times r$ grid is the canonical planar graph of treewidth $\Theta(r)$. In particular, an important result of Robertson, Seymour, and Thomas [RST94] is that every planar graph of treewidth w has an $\Omega(w) \times \Omega(w)$ grid graph as a minor. Thus every planar graph of large treewidth has a grid minor certifying that its treewidth is almost as large (up to constant factors). Recently, this result has been generalized to any H-minor-free graph class:

Theorem 4 ([DH05b]). *For any fixed graph H, every H-minor-free graph of treewidth w has an $\Omega(w) \times \Omega(w)$ grid as a minor.*

Thus the $r \times r$ grid is the canonical H-minor-free graph of treewidth $\Theta(r)$ for any fixed graph H. This result is also best possible up to constant factors. Section 11 discusses the remaining issue of bounding the constant factor and its dependence on H.

A similar but weaker bound plays an important role in the Graph Minor Theory [RS86b]: for any fixed graph H and integer $r > 0$, there is an integer $w > 0$ such that every H-minor-free graph with treewidth at least w has an $r \times r$ grid graph as a minor. This result has been re-proved by Robertson, Seymour, and Thomas [RST94], Reed [Ree97], and Diestel, Jensen, Gorbunov, and Thomassen [DJGT99]. Among these proofs, the best known bound on w in terms of r is that every H-minor-free graph of treewidth larger than $20^{5|V(H)|^3 r}$ has an $r \times r$ grid as a minor [RST94]. Theorem 4 therefore offers an exponential (and best possible) improvement over previous results.

Theorem 4 cannot be generalized to arbitrary graphs: Robertson, Seymour, and Thomas [RST94] proved that some graphs have treewidth $\Omega(r^2 \lg r)$ but have grid minors only of size $O(r) \times O(r)$. The best known relation for general graphs is that having treewidth more than 20^{2r^5} implies the existence of an $r \times r$ grid minor [RST94]. The best possible bound is believed to be closer to $\Theta(r^2 \lg r)$ than $2^{\Theta(r^5)}$, perhaps even equal to $\Theta(r^2 \lg r)$ [RST94].

4   Bidimensional Parameters/Problems

Bidimensionality has been introduced and developed in a series of papers [DHT, DHN$^+$04, DFHT, DH04a, DFHT04b, DH04b, DFHT04a, DHT04, DH05b, DH05a]. Although implicitly hinted at in [DHT, DHN$^+$04, DFHT, DH04a], the first use of the term "bidimensional" was in [DFHT04b].

First we define "parameters" as an alternative view on optimization problems. A *parameter* P is a function mapping graphs to nonnegative integers. The *decision problem associated with* P asks, for a given graph G and nonnegative integer k, whether $P(G) \leq k$. Many optimization problems can be phrased as such a decision problem about a graph parameter P.

Now we can define bidimensionality. A parameter is $g(r)$-*bidimensional* (or just *bidimensional*) if it is at least $g(r)$ in an $r \times r$ "grid-like graph" and if the parameter does not increase when taking either minors ($g(r)$-*minor-bidimensional*) or contractions ($g(r)$-*contraction-bidimensional*). The exact definition of "grid-like graph" depends on the class of graphs allowed and whether we are considering minor- or contraction-bidimensionality. For minor-bidimensionality and for any H-minor-free graph class, the notion of a "grid-like graph" is defined to be the $r \times r$ *grid*, i.e., the planar graph with r^2 vertices arranged on a square grid and with edges connecting horizontally and vertically adjacent vertices. For contraction-bidimensionality, the notion of a "grid-like graph" is as follows:

1. For planar graphs and single-crossing-minor-free graphs, a "grid-like graph" is an $r \times r$ grid partially triangulated by additional edges that preserve planarity.
2. For bounded-genus graphs, a "grid-like graph" is such a partially triangulated $r \times r$ grid with up to genus(G) additional edges ("handles").
3. For apex-minor-free graphs, a "grid-like graph" is an $r \times r$ grid augmented with additional edges such that each vertex is incident to $O(1)$ edges to nonboundary vertices of the grid. (Here $O(1)$ depends on the excluded apex graph.)

Contraction-bidimensionality is so far undefined for H-minor-free graphs (or general graphs)[2].

Examples of bidimensional parameters include the number of vertices, the diameter, and the size of various structures such as feedback vertex set, vertex cover, minimum maximal matching, face cover, a series of vertex-removal parameters, dominating set, edge dominating set, R-dominating set, connected dominating set, connected edge dominating set, connected R-dominating set, and unweighted TSP tour (a walk in the graph visiting all vertices). (See [DFHT04b, DFHT04a] for arguments of either contraction- or minor-bidimensionality for the above parameters.) We also say that the corresponding optimization problems based on these parameters, e.g., finding the minimum-size dominating

---

[2] For the parameters to which we have applied bidimensionality, contraction-bidimensionality does not seem to extend beyond apex-minor-free graphs, but perhaps a suitably extended definition could be found in the context of different applications or a "theory of graph contractions".

set, are *bidimensional*. With the exception of diameter, all of these bidimensional problems are $\Theta(r^2)$-bidimensional, which is the most common case (and in some papers used as the definition of bidimensionality). Diameter is the main exception, being only $\Theta(r)$-contraction-bidimensional for planar graphs, single-crossing-minor-free graphs, and bounded-genus graphs, and only $\Theta(\lg r)$-contraction-bidimensional for apex-minor-free graphs.

5   Parameter-Treewidth Bounds

The genesis of bidimensionality was in fact the notion of a parameter-treewidth bound. A *parameter-treewidth bound* is an upper bound $f(k)$ on the treewidth of a graph with parameter value k. In many cases, $f(k)$ can even be shown to be sublinear in k, often $O(\sqrt{k})$. Parameter-treewidth bounds have been established for many parameters and graph classes; see, e.g., [ABF+02, KP02, FT03, AFN04, CKL01, KLL02, GKL01, DFHT, DHN+04, DHT, DFHT04a, DH04b, DFHT04b]. Essentially all of these bounds can be obtained from the theory of bidimensional parameters. Thus bidimensionality is the most powerful method so far for establishing parameter-treewidth bounds, encompassing all such previous results for H-minor-free graphs.

   The central result in bidimensionality that generalizes these bounds is that every bidimensional parameter has a parameter-treewidth bound, in its corresponding family of graphs as defined in Section 4. More precisely, we have the following result:

Theorem 5 ([DH05b, DFHT04a]). *If the parameter P is $g(r)$-bidimensional, then for every graph G in the family associated with the parameter P, $\mathbf{tw}(G) = O(g^{-1}(P(G)))$. In particular, if $g(r) = \Theta(r^2)$, then the bound becomes $\mathbf{tw}(G) = O(\sqrt{P(G)})$.*

   This theorem is based on the grid-minor bound from Theorem 4 and the proof of a weaker parameter-treewidth bound, $\mathbf{tw}(G) = (g^{-1}(P(G)))^{O(g^{-1}(P(G)))}$, established in [DFHT04a]. The stronger bound of $\mathbf{tw}(G) = O(g^{-1}(P(G)))$ was obtained first for planar graphs [DFHT], then single-crossing-minor-free graphs [DHT, DHN+04], then bounded-genus graphs [DFHT04b, DHT04], and finally apex-minor-free graphs for contraction-bidimensional parameters and H-minor-free graphs for minor-bidimensional parameters [DH05b] (Theorem 5 above).

   We can extend the definition of $g(r)$-minor-bidimensionality to general graphs by again defining a "grid-like graph" to be the $r \times r$ grid. Still we can obtain a parameter-treewidth bound [RST94, DH04c], but the bound is weaker: $\mathbf{tw}(G) = 2^{O(g^{-1}(k))^5}$.

6   Separator Theorems

If we apply the parameter-treewidth bound of Theorem 5 to the parameter of the number of vertices in the graph, which is minor-bidimensional with $g(r) = r^2$, then we immediately obtain the following (known) bound on the treewidth of an H-minor-free graph:

Theorem 6 ([AST90, Proposition 4.5; Gro03, Corollary 24; DH05b]).
For any fixed graph H, every H-minor-free graph G has treewidth $O(\sqrt{|V(G)|})$.

A consequence of this result is that every vertex-weighted H-minor-free graph G has a vertex separator of size $O(\sqrt{|V(G)|})$ whose removal splits the graph into two parts each with weight at most $2/3$ of the original weight [AST90, Theorem 1.2]. This generalization of the classic planar separator theorem has many algorithmic applications; see e.g. [AST90, AFN03]. Also, this result shows that the structural properties of H-minor-free graphs given by Theorem 3 are powerful enough to conclude that these graphs have small separators, which we expect from such a strong theorem.

Section 11 discusses the issue of how tight a lead constant can be obtained in such a result.

7 Local Treewidth

Eppstein [Epp00] introduced the *diameter-treewidth property* for a class of graphs, which requires that the treewidth of a graph in the class is upper bounded by a function of its diameter. He proved that a minor-closed graph family has the diameter-treewidth property precisely if the graph family excludes some apex graph. In particular, he proved that any graph in such a family with diameter D has treewidth at most $2^{2^{O(D)}}$. (A simpler proof of this result was obtained in [DH04a].)

If we apply the parameter-treewidth bound of Theorem 5 to the diameter parameter, which is contraction-bidimensional with $g(r) = \Theta(\lg r)$ [DH04a], then we immediately obtain the following stronger diameter-treewidth bound for apex-minor-free graphs:

Theorem 7 ([DH05b]). *For any fixed apex graph H, every H-minor-free graph of diameter D has treewidth $2^{O(D)}$.*

This theorem is not the best possible. In some sense it is necessarily limited because it still does not exploit the full structure of H-minor-free graphs from Theorem 3. The difficulty is that, in a grid-like graph, the $O(1)$ edges from a vertex to nonboundary vertices can accumulate to make the diameter small. However, it is possible to show that, effectively, not too many vertices can have such edges. This fact comes from the property that there are a bounded number of apices in the clique-sum decomposition of Theorem 3, and in an apex-minor-free graph, each apex cannot have more than a bounded number of edges to "distant" vertices. Based on this fact, a complicated proof establishes the following even stronger diameter-treewidth bound in apex-minor-free graphs:

Theorem 8 ([DH04b]). *For any fixed apex graph H, every H-minor-free graph of diameter D has treewidth $O(D)$.*

This diameter-treewidth bound is the best possible up to constant factors. Thus this theorem establishes that, in minor-closed graph families, having any diameter-treewidth bound is equivalent to having a linear diameter-treewidth bound. As mentioned before, no minor-closed graph families beyond apex-minor-free graphs can have any diameter-treewidth bound. Theorem 8 is therefore the ultimate characterization of diameter-treewidth bounds in minor-closed graph families (up to constant factors).

The proof of Theorem 8 is the basis for Conjecture 1. In fact, Theorem 8 would not be hard to prove assuming Conjecture 1.

The diameter-treewidth property has been used extensively in a slightly modified form called the *bounded-local-treewidth property*, which requires that the treewidth of any connected subgraph of a graph in the class is upper bounded by a function of its diameter. For minor-closed graph families, which is the focus of most work in this context, these properties are identical. Graphs of bounded local treewidth have many similar properties to both planar graphs and graphs of bounded treewidth, two classes of graphs on which many problems are substantially easier. In particular, Baker's approach for polynomial-time approximation schemes (PTASs) on planar graphs [Bak94] applies to this setting. As a result, PTASs are known for hereditary maximization problems such as maximum independent set, maximum triangle matching, maximum H-matching, and maximum tile salvage; for minimization problems such as minimum vertex cover, minimum dominating set, minimum edge-dominating set; and for subgraph isomorphism for a fixed pattern [DHN+04, Epp00, HN02]. Graphs of bounded local treewidth also admit several efficient fixed-parameter algorithms. In particular, Frick and Grohe [FG01] give a general framework for deciding any property expressible in first-order logic in graphs of bounded local treewidth. Theorem 8 substantially improves the running time of these algorithms, in particular improving the running time of the PTASs from $2^{2^{2^{O(1/\varepsilon)}}} n^{O(1)}$ to $2^{O(1/\varepsilon)} n^{O(1)}$, where n is the number of vertices in the graph.

8   Subexponential Fixed-Parameter Algorithms

A *fixed-parameter algorithm* is an algorithm for computing a parameter $P(G)$ of a graph G whose running time is $h(P(G)) n^{O(1)}$ for some function h. The exponent $O(1)$ must be independent of G; thus the exponentiality of the algorithm is bounded by the parameter $P(G)$, and the dependence on n is only polynomial. A typical function h for many fixed-parameter algorithms is $h(k) = 2^{O(k)}$. In the last five years, several researchers have obtained *exponential speedups* in fixed-parameter algorithms in the sense that the h function reduces exponentially, e.g., to $2^{O(\sqrt{k})}$. For example, the first fixed-parameter algorithm for finding a dominating set of size k in planar graphs [AFF+01] has running time $O(8^k n)$; subsequently, a sequence of subexponential algorithms and improvements have been obtained, starting with running time $O(4^{6\sqrt{34k}} n)$ [ABF+02], then $O(2^{27\sqrt{k}} n)$ [KP02], and finally $O(2^{15.13\sqrt{k}} k + n^3 + k^4)$ [FT03]. Other subexponential algorithms for other domination and covering problems on planar graphs have also been obtained [ABF+02, AFN04, CKL01, KLL02, GKL01].

All subexponential fixed-parameter algorithms developed so far are based on showing a sublinear parameter-treewidth bound and then using an algorithm whose running time is singly exponential in treewidth and polynomial in problem size. As mentioned in Section 5, essentially all sublinear treewidth-parameter bounds proved so far can be obtained through bidimensionality. Theorem 5 and the techniques of [DFHT04a] yield the following general result for designing subexponential fixed-parameter algorithms:

Theorem 9 ([DH05b, DFHT04a]). *Consider a $g(r)$-bidimensional parameter P that can be computed on a graph G in $h(w)\,n^{O(1)}$ time given a tree decomposition of G of width at most w. Then there is an algorithm computing P on any graph G in P's corresponding graph class, with running time $[h(O(g^{-1}(k))) + 2^{O(g^{-1}(k))}]\,n^{O(1)}$. In particular, if $g(r) = \Theta(r^2)$ and $h(w) = 2^{o(w^2)}$, then this running time is subexponential in k.*

In particular, this result gives subexponential fixed-parameter algorithms for many bidimensional parameters, including feedback vertex set, vertex cover, minimum maximal matching, a series of vertex-removal parameters, dominating set, edge dominating set, R-dominating set, clique-transversal set, connected dominating set, connected edge dominating set, connected R-dominating set, and unweighted TSP tour.

For minor-bidimensional parameters, these algorithms apply to all H-minor-free graphs. The next section describes to what extent these algorithms can be extended to general graphs.

For contraction-bidimensional parameters, these algorithms apply to apex-minor-free graphs. On the other hand, subexponential fixed-parameter algorithms can be obtained for dominating set, which is contraction-bidimensional, on H-minor-free graphs [DFHT04b], map graphs [DFHT], and fixed powers of planar graphs (or even fixed powers of H-minor-free graphs) [DFHT, DFHT04b]. These algorithms are necessarily more complicated than those produced from Theorem 9, because apex-minor-free graphs are precisely the minor-closed graph classes for which domatinating set has a parameter-treewidth bound [DFHT04a]. An intriguing open question is whether these techniques can be extended to other contraction-bidimensional problems than dominating set, for fixed powers of H-minor-free graphs and/or other classes of graphs.

9   Fixed-Parameter Algorithms for General Graphs

As mentioned in Section 5, minor-bidimensionality can be defined for general graphs as well. In this section we show how the bidimensionality theory in this case leads to a general class of fixed-parameter algorithms.

A major result from the Graph Minor Theory (in particular [RS95, RS]) is that every minor-closed graph property is characterized by a finite set of forbidden minors. More precisely, for any property P on graphs such that a graph having property P implies that all its minors have property P, there is a finite set $\{H_1, H_2, \ldots, H_h\}$ of graphs such that a graph G has property P if and only if G does not have H_i as a minor for all $i = 1, 2, \ldots, h$. The algorithmic consequence of this result is that there exists an $O(n^3)$-time algorithm to decide any fixed minor-closed graph property, by finitely many calls to an $O(n^3)$-time minor test [RS95]. This consequence has been used to show the existence of polynomial-time algorithms for several graph problems, some of which were not previously known to be decidable [FL88].

However, all of these algorithmic results (except the minor test) are non-constructive: we are guaranteed that efficient algorithms exist, but are not told what they are. The difficulty is that we know that a finite set of forbidden minors exists, but lack "a means of identifying the elements of the set, the cardinality of the set, or even the order of the largest graph in the set" [FL88]. Indeed, there is a mathematical sense in which any proof of the finite-forbidden-minors theorem must be nonconstructive [FRS87].

We can apply these graph-minor results to prove the existence of algorithms to compute parameters, provided the parameters never increase when taking a minor. For any fixed parameter and any fixed $k \geq 0$, there is an $O(n^3)$-time algorithm that decides whether a graph has parameter value $\leq k$. Unfortunately, the existence of these algorithms does not necessarily imply the existence of a single fixed-parameter algorithm that works for all $k \geq 0$, because the algorithms for individual k (in particular the set of forbidden minors) might be uncomputable. We do not even know an upper bound on the running time of these algorithms as a function of n and k, because we do not know the dependence of the size of the forbidden minors on k.

In [DH04c], fixed-parameter algorithms are constructed for nearly all parameters that never increase when taking a minor, with explicit time bounds in terms of n and k. Essentially, by assuming a few very common properties of the parameter, we obtain the generalized form of minor-bidimensionality.

Theorem 10 ([DH04c]). *Consider a parameter P that is positive on some $g \times g$ grid, never increases when taking minors, is at least the sum over the connected components of a disconnected graph, and can be computed in $h(w) n^{O(1)}$ time given a width-w tree decomposition of the graph. Then there is an algorithm that decides whether P is at most k on a graph with n vertices in $\left[2^{2^{O(g\sqrt{k})^5}} + h(2^{O(g\sqrt{k})^5}) \right] n^{O(1)}$ time.*

As mentioned in [DH04c], a conjecture of Robertson, Seymour, and Thomas [RST94] would improve the running time to $h(O(k \lg k)) n^{O(1)}$, which is $2^{O(k \lg k)} n^{O(1)}$ for the typical case of $h(w) = 2^{O(w)}$. This conjectured time bound almost matches the fastest known fixed-parameter algorithms for several parameters, e.g., feedback vertex set, vertex cover, and a general family of vertex-removal problems [FL88].

10   Polynomial-Time Approximation Schemes

Recently, the bidimensionality theory has been extended to obtain polynomial-time approximation schemes (PTASs) for essentially all bidimensional parameters, including those mentioned above [DH05a]. These PTASs are based on techniques that generalize and in some sense unify the two main previous approaches for designing PTASs in planar graphs, namely, the Lipton-Tarjan separator approach [LT80] and the Baker layerwise decomposition approach [Bak94]. The PTASs apply to H-minor-free graphs for minor-bidimensional parameters and to apex-minor-free graphs for contraction-bidimensional parameters. To achieve this level of generality, [DH05a] uses the sublinear parameter-treewidth bound of Theorem 5 as well as a recent $O(1)$-approximation algorithm for treewidth in H-minor-free graphs [FHL04].

Before we can state the general theorem for constructing PTASs, we need to define a few straightforward required conditions, which are commonly satisfied by most bidimensional problems. The theorem considers families of problems in which we are given a graph and our goal is to find a minimum-size set of vertices and/or edges satisfying a certain property. Such a problem naturally defines a parameter and therefore the notion of bidimensionality. A minor-bidimensional problem has the *separation property* if it satisfies the following three conditions:

1. If a graph G has k connected components $G_1, G_2, \ldots, G_k$, then an optimal solution for G is the union of optimal solutions for each connected component G_i.
2. There is a polynomial-time algorithm that, given any graph G, given any vertex cut C whose removal disconnects G into connected components $G_1, G_2, \ldots, G_k$, and given an optimal solution S_i to each connected component G_i of $G - C$, computes a solution S for G such that the number of vertices and/or edges in S within the induced subgraph $G[C \cup \bigcup_{i \in I} V(G_i)]$ consisting of C and some connected components of $G - C$ is $\sum_{i \in I} |S_i| \pm O(|C|)$ for any $I \subseteq \{1, 2, \ldots, k\}$. In particular, the total cost of S is at most $\mathrm{OPT}(G - C) + O(|C|)$.
3. Given any graph G, given any vertex cut C, and given an optimal solution OPT to G, for any union G' of some subset of connected components of $G - C$, $|\mathrm{OPT} \cap G'| = |\mathrm{OPT}(G')| \pm O(|C|)$.

For contraction-bidimensional problems, the exact requirements on the problem are slightly different but similarly straightforward. The main distinction is that the connected components are always considered together with the cut C. As a result, the merging algorithm in Condition 2 must take as input a solution to a generalized form of the problem that does not count the cost of including all vertices and edges from the cut C. We omit the exact definition of the separation property in this case in the interest of space.

Theorem 11 ([DH05a]). *Consider a bidimensional problem satisfying the separation property. Suppose that the problem can be solved on a graph G with n vertices in $f(n, \mathbf{tw}(G))$ time. Suppose also that the problem can be approximated within a factor of α in $g(n)$ time. For contraction-bidimensional problems, suppose further that both of these algorithms also apply to the generalized form of the problem. Then there is a $(1+\varepsilon)$-approximation algorithm whose running time is $O(nf(n, O(\alpha^2/\varepsilon)) + n^3 g(n))$ for the corresponding graph class of the bidimensional problem.*

This result shows a strong connection between subexponential fixed-parameter tractability and approximation algorithms for combinatorial optimization problems on H-minor-free graphs. In particular, this result yields a PTAS for the following minor-bidimensional problems in H-minor-free graphs: feedback vertex set, face cover (defined just for planar graphs), vertex cover, minimum maximal matching, and a series of vertex-removal problems. Furthermore, the result yields a PTAS for the following contraction-bidimensional problems in apex-minor-free graphs: dominating set, edge dominating set, R-dominating set, connected dominating set, connected edge dominating set, connected R-dominating set, and clique-transversal set.

11   Open Problems

Several combinatorial and algorithmic open problems remain in the theory of bidimensionality and related concepts.

One interesting direction is to generalize bidimensionality to handle general graphs, not just H-minor-free graph classes. As mentioned in Section 5, the natural generalization of minor-bidimensionality still yields a parameter-treewidth bound, but it is very large. This direction essentially asks for the size of the largest grid minor guaranteed to exist in any graph of treewidth w. Robertson, Seymour, and Thomas [RST94] proved that every graph of treewidth larger than 20^{2r^5} has an $r \times r$ grid as a minor, but that some graphs of treewidth $\Omega(r^2 \lg r)$ have no grid larger than $O(r) \times O(r)$, conjecturing that the right requirement on treewidth for an $r \times r$ grid is closer to the $\Theta(r^2 \lg r)$ lower bound. If this conjecture is correct, we would obtain nearly as good parameter-treewidth bounds for minor-bidimensional parameters as in the H-minor-free case. A similar generalization of parameter-treewidth bounds beyond apex-minor-free graphs is not possible for all contraction-bidimensional parameters, e.g., dominating set [DFHT04a], but it would still be quite interesting to explore an analogous "theory of graph contractions" paralleling the Graph Minor Theory. Such a theory would be an interesting and powerful tool for handling problems that are closed under contractions but not minors, and therefore deserves more focus.

Another interesting direction is to obtain the best constant factors in terms of the fixed excluded minor H. These constants are particularly important in the context of the exponent in the running time of a fixed-parameter algorithm. At the heart of all such constant factors is the lead constant in Theorem 4. This factor must be $\Omega(\sqrt{|V(H)|} \lg |V(H)|)$, because otherwise such a bound would contradict the lower bound for general graphs. An upper bound near this lower bound (in particular, polynomial in $|V(H)|$) is not out of the question: the bound on the size of separators in [AST90] has a lead factor of $|V(H)|^{3/2}$. In fact, Alon, Seymour, and Thomas [AST90] suspect that the correct factor for separators is $\Theta(|V(H)|)$, which holds e.g. in bounded-genus graphs. We also suspect that the same bound holds for the factor in Theorem 4, which would imply the corresponding bound for separators.

A third interesting direction is to generalize the polynomial-time approximation schemes that come out of bidimensionality to more general algorithmic problems that do not correspond directly to bidimensional parameters. One general family of such problems arises when adding weights to vertices and/or edges, and the goal is e.g. to find the minimum-weight dominating set. It is difficult to define bidimensionality of the corresponding weighted parameter because its value is no longer well-defined on an $r \times r$ grid: the parameter value now depends on the weights of vertices in such a grid. Another family of such problems arises when placing constraints (e.g., on coverage or domination) only on subsets of vertices and/or edges. Examples of such problems include Steiner tree [AGK+98] and subset feedback vertex set [ENZ00]. Again it is difficult to define bidimensionality in such cases because the value of the parameter on a grid depends on which vertices and/or edges of the grid are in the subset.

Acknowledgments

We thank Fedor Fomin, Naomi Nishimura, Prabhakar Ragde, Paul Seymour, and Dimitrios Thilikos for their fruitful collaboration and discussions in this area.

References

[ABF+02] J. Alber, H. L. Bodlaender, H. Fernau, T. Kloks, and R. Niedermeier. Fixed parameter algorithms for dominating set and related problems on planar graphs. *Algorithmica*, 33(4):461–493, 2002.

[AFF+01] J. Alber, H. Fan, M. R. Fellows, H. Fernau, R. Niedermeier, F. A. Rosamond, and U. Stege. Refined search tree technique for DOMINATING SET on planar graphs. In *Proc. 26th Internat. Symp. Math. Found. Comput. Sci.*, LNCS 2136, pp. 111–122, 2001.

[AFN03] J. Alber, H. Fernau, and R. Niedermeier. Graph separators: a parameterized view. *J. Comput. System Sci.*, 67(4):808–832, Dec. 2003.

[AFN04] J. Alber, H. Fernau, and R. Niedermeier. Parameterized complexity: exponential speed-up for planar graph problems. *J. Algorithms*, 52(1):26–56, 2004.

[AGK+98] S. Arora, M. Grigni, D. Karger, P. Klein, and A. Woloszyn. A polynomial-time approximation scheme for weighted planar graph TSP. In *Proc. 9th ACM-SIAM Symp. Discr. Alg.*, pp. 33–41, 1998.

[AST90] N. Alon, P. Seymour, and R. Thomas. A separator theorem for nonplanar graphs. *J. Amer. Math. Soc.*, 3(4):801–808, 1990.

[Bak94] B. S. Baker. Approximation algorithms for NP-complete problems on planar graphs. *J. Assoc. Comput. Mach.*, 41(1):153–180, 1994.

[CGP02] Z.-Z. Chen, M. Grigni, and C. H. Papadimitriou. Map graphs. *J. ACM*, 49(2):127–138, 2002.

[CKL01] M.-S. Chang, T. Kloks, and C.-M. Lee. Maximum clique transversals. In *Proc. 27th Internat. Workshop Graph-Theor. Concepts in Comput. Sci.*, LNCS 2204, pp. 32–43, 2001.

[DFHT] E. D. Demaine, F. V. Fomin, M. Hajiaghayi, and D. M. Thilikos. Fixed-parameter algorithms for the (k, r)-center in planar graphs and map graphs. *ACM Trans. Alg.*. To appear. A preliminary version appears in *Proc. 30th Internat. Colloq. Automata Lang. Prog.*, LNCS 2719, 2003, pp. 829–844.

[DFHT04a] E. D. Demaine, F. V. Fomin, M. Hajiaghayi, and D. M. Thilikos. Bidimensional parameters and local treewidth. *SIAM J. Discrete Math.*, 2004. To appear. A preliminary version appears in *Proc. 6th Latin Amer. Symp. Theoret. Inf.*, LNCS 2976, 2004, pp. 109–118.

[DFHT04b] E. D. Demaine, F. V. Fomin, M. Hajiaghayi, and D. M. Thilikos. Subexponential parameterized algorithms on graphs of bounded genus and H-minor-free graphs. In *Proc. 15th ACM-SIAM Symp. Discr. Alg.*, pp. 823–832, 2004.

[DH04a] E. D. Demaine and M. Hajiaghayi. Diameter and treewidth in minor-closed graph families, revisited. *Algorithmica*, 40(3):211–215, Aug. 2004.

[DH04b] E. D. Demaine and M. Hajiaghayi. Equivalence of local treewidth and linear local treewidth and its algorithmic applications. In *Proc. 15th ACM-SIAM Symp. Discr. Alg.*, pp. 833–842, 2004.

[DH04c] E. D. Demaine and M. Hajiaghayi. Quickly deciding minor-closed parameters in general graphs. Manuscript, 2004.

[DH05a] E. D. Demaine and M. Hajiaghayi. Bidimensionality: New connections between FPT algorithms and PTASs. In *Proc. 16th ACM-SIAM Symp. Discr. Alg.*, 2005. To appear.

[DH05b] E. D. Demaine and M. Hajiaghayi. Graphs excluding a fixed minor have grids as large as treewidth, with combinatorial and algorithmic applications through bidimensionality. In *Proc. 16th ACM-SIAM Symp. Discr. Alg.*, 2005. To appear.

[DHN+04] E. D. Demaine, M. Hajiaghayi, N. Nishimura, P. Ragde, and D. M. Thilikos. Approximation algorithms for classes of graphs excluding single-crossing graphs as minors. *J. Comput. System Sci.*, 69(2):166–195, 2004.

[DHT] E. D. Demaine, M. Hajiaghayi, and D. M. Thilikos. Exponential speedup of fixed-parameter algorithms for classes of graphs excluding single-crossing graphs as minors. *Algorithmica*. To appear. A preliminary version appears in *Proc. 13th Internat. Symp. Alg. Comput.*, LNCS 2518, 2002, pp. 262–273.

[DHT04] E. D. Demaine, M. Hajiaghayi, and D. M. Thilikos. The bidimensional theory of bounded-genus graphs. In *Proc. 29th Internat. Symp. Math. Found. Comput. Sci.*, pp. 191–203, 2004.

[DJGT99] R. Diestel, T. R. Jensen, K. Y. Gorbunov, and C. Thomassen. Highly connected sets and the excluded grid theorem. *J. Combin. Theory Ser. B*, 75(1):61–73, 1999.

[ENZ00] G. Even, J. Naor, and L. Zosin. An 8-approximation algorithm for the subset feedback vertex set problem. *SIAM J. Comput.*, 30(4):1231–1252, 2000.

[Epp95] D. Eppstein. Subgraph isomorphism in planar graphs and related problems. In *Proc. 6th ACM-SIAM Symp. Discr. Alg.*, pp. 632–640, 1995.

[Epp00] D. Eppstein. Diameter and treewidth in minor-closed graph families. *Algorithmica*, 27(3-4):275–291, 2000.

[FG01] M. Frick and M. Grohe. Deciding first-order properties of locally tree-decomposable structures. *Journal of the ACM*, 48(6):1184–1206, 2001.

[FHL04] U. Feige, M. Hajiaghayi, and J. R. Lee. On improving approximate vertex separator and its algorithmic applications. Manuscript, 2004.

[FL88] M. R. Fellows and M. A. Langston. Nonconstructive tools for proving polynomial-time decidability. *Journal of the ACM*, 35(3):727–739, 1988.

[FRS87] H. Friedman, N. Robertson, and P. Seymour. The metamathematics of the graph minor theorem. In *Logic and combinatorics*, vol. 65 of *Contemp. Math.*, pp. 229–261. Amer. Math. Soc., Providence, RI, 1987.

[FT03] F. V. Fomin and D. M. Thilikos. Dominating sets in planar graphs: Branch-width and exponential speed-up. In *Proc. 14th ACM-SIAM Sympos. Discrete Algorithms*, pp. 168–177, 2003.

[GKL01] G. Gutin, T. Kloks, and C. M. Lee. Kernels in planar digraphs. In *Optimization Online*, 2001.

[Gro03] M. Grohe. Local tree-width, excluded minors, and approximation algorithms. *Combinatorica*, 23(4):613–632, 2003.

[HN02] M. Hajiaghayi and N. Nishimura. Subgraph isomorphism, log-bounded fragmentation and graphs of (locally) bounded treewidth. In *Proc. 27th Internat. Symp. Math. Found. Comput. Sci.*, LNCS 2420, pp. 305–318, 2002.

[KLL02] T. Kloks, C. M. Lee, and J. Liu. New algorithms for k-face cover, k-feedback vertex set, and k-disjoint set on plane and planar graphs. In *Proc. 28th Internat. Workshop Graph-Theor. Concepts Comput. Sci.*, LNCS 2573, pp. 282–295, 2002.

[KP02] I. Kanj and L. Perković. Improved parameterized algorithms for planar dominating set. In *Proc. 27th Internat. Symp. Math. Found. Comput. Sci.*, LNCS 2420, pp. 399–410, 2002.

[LT80] R. J. Lipton and R. E. Tarjan. Applications of a planar separator theorem. *SIAM J. Comput.*, 9(3):615–627, 1980.

[Ree97] B. A. Reed. Tree width and tangles: a new connectivity measure and some applications. In *Surveys in combinatorics*, vol. 241 of *London Math. Soc. Lecture Note Ser.*, pp. 87–162. Cambridge Univ. Press, 1997.

[RS] N. Robertson and P. D. Seymour. Graph minors. XX. Wagner's conjecture. *J. Combin. Theory Ser. B*. To appear.

[RS86a] N. Robertson and P. D. Seymour. Graph minors. II. Algorithmic aspects of tree-width. *J. Algorithms*, 7(3):309–322, 1986.

[RS86b] N. Robertson and P. D. Seymour. Graph minors. V. Excluding a planar graph. *J. Combin. Theory Ser. B*, 41(1):92–114, 1986.

[RS93] N. Robertson and P. Seymour. Excluding a graph with one crossing. In *Graph structure theory*, pp. 669–675. Amer. Math. Soc., 1993.

[RS95] N. Robertson and P. D. Seymour. Graph minors. XII. Distance on a surface. *J. Combin. Theory Ser. B*, 64(2):240–272, 1995.

[RS03] N. Robertson and P. D. Seymour. Graph minors. XVI. Excluding a non-planar graph. *J. Combin. Theory Ser. B*, 89(1):43–76, 2003.

[RST94] N. Robertson, P. D. Seymour, and R. Thomas. Quickly excluding a planar graph. *J. Combin. Theory Ser. B*, 62(2):323–348, 1994.

[Tho98] M. Thorup. Map graphs in polynomial time. In *Proc. 39th IEEE Sympos. Found. Comput. Sci.*, pp. 396–407, 1998.

[Wag37] K. Wagner. Über eine Eigenschaft der eben Komplexe. *Deutsche Math.*, 2:280–285, 1937.

Author Index

Lecture Notes in Computer Science

For information about Vols. 1–3305

please contact your bookseller or Springer

Vol. 3350: M. Hermenegildo, D. Cabeza (Eds.), Practical Aspects of Declarative Languages. VIII, 269 pages. 2005.

Vol. 3349: B.M. Chapman (Ed.), Shared Memory Parallel Programming with Open MP. X, 149 pages. 2005.

Vol. 3348: A. Canteaut, K. Viswanathan (Eds.), Progress in Cryptology - INDOCRYPT 2004. XIV, 431 pages. 2004.

Vol. 3347: R.K. Ghosh, H. Mohanty (Eds.), Distributed Computing and Internet Technology. XX, 472 pages. 2004.

Vol. 3346: R.H. Bordini, M. Dastani, J. Dix, A.E.F. Seghrouchni (Eds.), Programming Multi-Agent Systems. XIV, 249 pages. 2005. (Subseries LNAI).

Vol. 3345: Y. Cai (Ed.), Ambient Intelligence for Scientific Discovery. XII, 311 pages. 2005. (Subseries LNAI).

Vol. 3344: J. Malenfant, B.M. Østvold (Eds.), Object-Oriented Technology. ECOOP 2004 Workshop Reader. VIII, 215 pages. 2005.

Vol. 3342: E. Şahin, W.M. Spears (Eds.), Swarm Robotics. IX, 175 pages. 2005.

Vol. 3341: R. Fleischer, G. Trippen (Eds.), Algorithms and Computation. XVII, 935 pages. 2004.

Vol. 3340: C.S. Calude, E. Calude, M.J. Dinneen (Eds.), Developments in Language Theory. XI, 431 pages. 2004.

Vol. 3339: G.I. Webb, X. Yu (Eds.), AI 2004: Advances in Artificial Intelligence. XXII, 1272 pages. 2004. (Subseries LNAI).

Vol. 3338: S.Z. Li, J. Lai, T. Tan, G. Feng, Y. Wang (Eds.), Advances in Biometric Person Authentication. XVIII, 699 pages. 2004.

Vol. 3337: J.M. Barreiro, F. Martin-Sanchez, V. Maojo, F. Sanz (Eds.), Biological and Medical Data Analysis. XI, 508 pages. 2004.

Vol. 3336: D. Karagiannis, U. Reimer (Eds.), Practical Aspects of Knowledge Management. X, 523 pages. 2004. (Subseries LNAI).

Vol. 3335: M. Malek, M. Reitenspieß, J. Kaiser (Eds.), Service Availability. X, 213 pages. 2005.

Vol. 3334: Z. Chen, H. Chen, Q. Miao, Y. Fu, E. Fox, E.-p. Lim (Eds.), Digital Libraries: International Collaboration and Cross-Fertilization. XX, 690 pages. 2004.

Vol. 3333: K. Aizawa, Y. Nakamura, S. Satoh (Eds.), Advances in Multimedia Information Processing - PCM 2004, Part III. XXXV, 785 pages. 2004.

Vol. 3332: K. Aizawa, Y. Nakamura, S. Satoh (Eds.), Advances in Multimedia Information Processing - PCM 2004, Part II. XXXVI, 1051 pages. 2004.

Vol. 3331: K. Aizawa, Y. Nakamura, S. Satoh (Eds.), Advances in Multimedia Information Processing - PCM 2004, Part I. XXXVI, 667 pages. 2004.

Vol. 3330: J. Akiyama, E.T. Baskoro, M. Kano (Eds.), Combinatorial Geometry and Graph Theory. VIII, 227 pages. 2005.

Vol. 3329: P.J. Lee (Ed.), Advances in Cryptology - ASIACRYPT 2004. XVI, 546 pages. 2004.

Vol. 3328: K. Lodaya, M. Mahajan (Eds.), FSTTCS 2004: Foundations of Software Technology and Theoretical Computer Science. XVI, 532 pages. 2004.

Vol. 3327: Y. Shi, W. Xu, Z. Chen (Eds.), Data Mining and Knowledge Management. XIII, 263 pages. 2005. (Subseries LNAI).

Vol. 3326: A. Sen, N. Das, S.K. Das, B.P. Sinha (Eds.), Distributed Computing - IWDC 2004. XIX, 546 pages. 2004.

Vol. 3325: C.H. Lim, M. Yung (Eds.), Information Security Applications. XI, 472 pages. 2005.

Vol. 3323: G. Antoniou, H. Boley (Eds.), Rules and Rule Markup Languages for the Semantic Web. X, 215 pages. 2004.

Vol. 3322: R. Klette, J. Žunić (Eds.), Combinatorial Image Analysis. XII, 760 pages. 2004.

Vol. 3321: M.J. Maher (Ed.), Advances in Computer Science - ASIAN 2004. XII, 510 pages. 2004.

Vol. 3320: K.-M. Liew, H. Shen, S. See, W. Cai (Eds.), Parallel and Distributed Computing: Applications and Technologies. XXIV, 891 pages. 2004.

Vol. 3319: D. Amyot, A.W. Williams (Eds.), Telecommunications and beyond: Modeling and Analysis of Reactive, Distributed, and Real-Time Systems. XII, 301 pages. 2005.

Vol. 3318: E. Eskin, C. Workman (Eds.), Regulatory Genomics. VIII, 115 pages. 2005. (Subseries LNBI).

Vol. 3317: M. Domaratzki, A. Okhotin, K. Salomaa, S. Yu (Eds.), Implementation and Application of Automata. XII, 336 pages. 2005.

Vol. 3316: N.R. Pal, N.K. Kasabov, R.K. Mudi, S. Pal, S.K. Parui (Eds.), Neural Information Processing. XXX, 1368 pages. 2004.

Vol. 3315: C. Lemaître, C.A. Reyes, J.A. González (Eds.), Advances in Artificial Intelligence – IBERAMIA 2004. XX, 987 pages. 2004. (Subseries LNAI).

Vol. 3314: J. Zhang, J.-H. He, Y. Fu (Eds.), Computational and Information Science. XXIV, 1259 pages. 2004.

Vol. 3313: C. Castelluccia, H. Hartenstein, C. Paar, D. Westhoff (Eds.), Security in Ad-hoc and Sensor Networks. VIII, 231 pages. 2005.

Vol. 3312: A.J. Hu, A.K. Martin (Eds.), Formal Methods in Computer-Aided Design. XI, 445 pages. 2004.

Vol. 3311: V. Roca, F. Rousseau (Eds.), Interactive Multimedia and Next Generation Networks. XIII, 287 pages. 2004.

Vol. 3310: U.K. Wiil (Ed.), Computer Music Modeling and Retrieval. XI, 371 pages. 2005.

Vol. 3309: C.-H. Chi, K.-Y. Lam (Eds.), Content Computing. XII, 510 pages. 2004.

Vol. 3308: J. Davies, W. Schulte, M. Barnett (Eds.), Formal Methods and Software Engineering. XIII, 500 pages. 2004.

Vol. 3307: C. Bussler, S.-k. Hong, W. Jun, R. Kaschek, D.. Kinshuk, S. Krishnaswamy, S.W. Loke, D. Oberle, D. Richards, A. Sharma, Y. Sure, B. Thalheim (Eds.), Web Information Systems – WISE 2004 Workshops. XV, 277 pages. 2004.

Vol. 3306: X. Zhou, S. Su, M.P. Papazoglou, M.E. Orlowska, K.G. Jeffery (Eds.), Web Information Systems – WISE 2004. XVII, 745 pages. 2004.